In *The Quantum Theory of Fields*, Nobel Laureate Steven Weinberg combines his exceptional physical insight with his gift for clear exposition to provide a self-contained, comprehensive, and up-to-date introduction to quantum field theory. Volume II gives a current and self-contained account of the methods of quantum field theory, and how they have led to an understanding of the weak, strong, and electromagnetic interactions of the elementary particles. The presentation of modern mathematical methods is throughout interwoven with accounts of the problems of elemenatary particle physics and condensed matter physics to which they have been applied. Many topics are includded that are not usually found in books on quantum field theory. The book contains much original material, and is peppered with examples and insights from the author's experience as a leader of elementary particle physics. Problems are included at the end of each chapter.

From reviews of Volume I

'...an impressively lucid and thorough presentation of the subject... Weinberg manages to present difficult topics with richness of meaning and marvellous clarity. Full of valuable insights, his treatise is sure to become a classic, doing for quantum field theory what Dirac's *Quantum Mechanics* did for quantum mechanics. I eagerly await the publication of the second volume.'

S. S. Schweber, *Nature*

'For over twenty years there has been no good modern textbook on the subject. For all that time, Steven Weinberg has been promising to write one. That he has finally done it is cause for celebration among those who try to teach and try to learn the subject. Weinberg's book is for serious students of field theory... it is the first textbook to treat quantum field theory the way it is used by physicists today.'

Howard Georgi, *Science*

'Steven Weinberg, who contributed to the development of quantum chromodynamics and shared the Nobel Prize in Physics for his contributions to the electroweak theory, has written a definitive text on the physical foundations of quantum field theory. His book differs significantly from the long line of previous books on quantum field theory... To summarize, *Foundations* builds the structure of quantum field theory on the sure footing of physical insight. It is beautifully produced and meticulously edited... and it is a real bargain in price. If you want to learn quantum field theory, or have already learned it and want to have a definitive reference at hand, purchase this book.'

O. W. Greenberg, *Physics Today*

'In addition to a superb treatment of all the conventional topics there are numerous sections covering areas that are not normally emphasized, such as the subject of field redefinitions, higher-rank tensor fields and an unusually clear and thorough treatment of infrared effects... this latest book reinforces his high scholarly standards. It provides a unique exposition that will prove invaluable both to new research students as well as to experienced research workers. Together with Volume II, this will become a classic text on a subject of central importance to a wide area of theoretical physics.'

M. B. Green, *CERN Courier*

'I believe that what readers will find particularly helpful in this volume is the consistency of the whole approach, and the emphasis on quantities and properties that are directly useful to particle physicists. This is particularly true for those who are interested in the more phenomenological aspects. The reader only needs limited background knowledge, and a clear line is followed throughout the book, making it easy to follow. The author presents extremely thorough but elementary discusssions of important physical questions, some of which seem to be an original way of addressing the subject.'

J. Zinn-Justin, *Physics World*

'This is a well-written book by one of the masters of the subject...it is certainly destined to become a standard text book and should find its way to the shelves of every physics library.'

J. Madore, *Classical and Quantum Gravity*

'The book starts out with an excellent historical introduction, not found anywhere else, giving citations to many by now classic papers...a valuable reference work as well as a textbook for graduate students.'

G. Roepstorff, *Zentralblatt für Mathematik*

From reviews of Volume II

'It is a majestic exposition. The two volumes are structured in a logical way. Everything is explained with incisive clarity. Weinberg always goes to the heart of any argument, and includes many things that cannot be found elsewhere in the literature. Often I find myself thinking: "Ah! Now I understand that properly."...I find it hard to imagine a better treatment of quantum field theory than Weinberg's. All serious students and researchers will want to have these volumes on their shelves.'

John C. Taylor, *Nature*

'Weinberg's *Modern Applications* goes to the boundaries of our present understanding of field theory. It is unmatched by any other book on quantum field theory for its depth, generality and definitive character, and it will be an essential reference for serious students and researchers in elementary particle physics.'

O. W. Greenberg, *Physics Today*

'...Steven Weinberg is one of our most gifted makers of theoretical tools as well as a virtuoso in their use. His new book conveys both the satisfaction of understanding nature and the feel of the atelier, for the "modern applications" of its subtitle include both the derivation of physical consequences and the development of new tools for understanding and applying field theory itself...*Modern Applications* is a splendid book, with abundant useful references to the original literature. It is a very interesting read from cover to cover, for the wholeness Weinberg's personal perspective gives to quantum field theory and particle physics.'

Chris Quigg, *Science*

'Experienced researchers and beginning graduate students will delight in the gems of wisdom to be found in these pages. This book combines exposition of technical detail with physical insight in a unique manner that confirms the promise of Volume I and I have no doubt that these two volumes will rapidly constitute the classic treatment of this important subject.'

M. B. Green, *CERN Courier*

'...a valued reference and a mine of useful information for professional field theorists.'

Tom Kibble, *New Scientist*

'...a clear presentation of the subject, explaining the underlying concepts in much depth and in an accessible style. I expect that these volumes will become the first source we turn to when trying to answer the challenging questions asked by bright postgraduates when they first encounter quantum field theory...I have no doubt that *The Quantum Theory of Fields* will soon be found on the bookshelves of most particle theorists, and that it will be one of the main sources used in the preparation of lectures on the subject for postgraduate students.'

C. T. C. Sachrajda, *The Times Higher Education Supplement*

'...Weinberg has produced a masterpiece that will be a standard reference on the field for a long time to come.'

B. E. Y. Svensson, *Elementa*

From reviews of Volume III

'...has produced a treatise that many of us had long awaited, perhaps without fully realizing it...with the publication of *The Quantum Theory of Fields*, Vol. III, has performed an analogous service for supersymmetry...Although this volume is the third in a trilogy, it is quite different from its two predecessors, and it stands on its own...May a new generation of students imbibe its content and spirit.'

Physics Today

'The third volume of *The Quantum Theory of Fields* is a self-contained introduction to the world of supersymmetry and supergravity. It will be useful both for experienced researchers in the field and for students who want to take the first steps towards learning about supersymmetry. Unlike other books in this field, it covers the wide spectrum of possible applications of supersymmetry in physics.'

Hans Peter Nilles, *Nature*

'Weinberg is of course one of the creators of modern quantum field theory, as well as of its physical culmination, the standard model of all (nongravitational) interactions. It is...very timely that this latest part of his monograph, devoted to supersymmetry and supergravity, has just appeared. As a text, it has been pretested by Weinberg for a freestanding one-year graduate course; as a clear organizing reference to this extremely vast field, it will help the experts as well... Weinberg's style of presentation is as clear and meticulous as in his previous works.'

Stanley Deser, *Journal of General Relativity and Gravitation*

'Weinberg tries to be as elementary and clear as possible and steers clear of more sophisticated mathematical tools. Together with the previous volumes, this volume will serve as an invaluable reference to researchers and a textbook for graduate students.'

G. Roepstorff, *Zentralblatt MATH*

The Quantum Theory of Fields

Volume II
Modern Applications

Steven Weinberg
University of Texas at Austin

CAMBRIDGE
UNIVERSITY PRESS

CAMBRIDGE
UNIVERSITY PRESS

University Printing House, Cambridge CB2 8BS, United Kingdom

Published in the United States of America by Cambridge University Press, New York

Cambridge University Press is part of the University of Cambridge.

It furthers the University's mission by disseminating knowledge in the pursuit of education, learning and research at the highest international levels of excellence.

www.cambridge.org
Information on this title: www.cambridge.org/9780521550024

First published 1995
Eigth printing 2004
Paperback edition first published 2005
12th printing 2013

A catalogue record for this publication is available from the British Library

ISBN 978-0-521-55001-7 Volume 1 hardback
ISBN 978-0-521-67053-1 Volume 1 paperback
ISBN 978-0-521-55002-4 Volume 2 hardback
ISBN 978-0-521-67054-8 Volume 2 paperback
ISBN 978-0-521-66000-6 Volume 3 hardback
ISBN 978-0-521-67055-5 Volume 3 paperback
ISBN 978-0-521-67056-2 three-volume set

Contents

Sections marked with an asterisk are somewhat out of the book's main line of development and may be omitted in a first reading.

PREFACE TO VOLUME II xvii

NOTATION xx

15 NON-ABELIAN GAUGE THEORIES 1

15.1 Gauge Invariance 2

Gauge transformations □ Structure constants □ Jacobi identity □ Adjoint representation □ Yang–Mills theory □ Covariant derivatives □ Field strength tensor □ Finite gauge transformations □ Analogy with general relativity

15.2 Gauge Theory Lagrangians and Simple Lie Groups 7

Gauge field Lagrangian □ Metric □ Antisymmetric structure constants □ Simple, semisimple, and $U(1)$ Lie algebras □ Structure of gauge algebra □ Compact algebras □ Coupling constants

15.3 Field Equations and Conservation Laws 12

Conserved currents □ Covariantly conserved currents □ Inhomogeneous field equations □ Homogeneous field equations □ Analogy with energy-momentum tensor □ Symmetry generators

15.4 Quantization 14

Primary and secondary first-class constraints □ Axial gauge □ Gribov ambiguity □ Canonical variables □ Hamiltonian □ Reintroduction of A_α^0 □ Covariant action □ Gauge invariance of the measure

15.5 The De Witt–Faddeev–Popov Method 19

Generalization of axial gauge results □ Independence of gauge fixing functionals □ Generalized Feynman gauge □ Form of vertices

15.6 Ghosts 24

Determinant as path integral □ Ghost and antighost fields □ Feynman rules for ghosts □ Modified action □ Power counting and renormalizability

OUTLINE OF VOLUME I

Preface To Volume II

This volume describes the advances in the quantum theory of fields that have led to an understanding of the electroweak and strong interactions of the elementary particles. These interactions have all turned out to be governed by principles of gauge invariance, so we start here in Chapters 15–17 with gauge theories, generalizing the familiar gauge invariance of electrodynamics to non-Abelian Lie groups.

Some of the most dramatic aspects of gauge theories appear at high energy, and are best studied by the methods of the renormalization group. These methods are introduced in Chapter 18, and applied to quantum chromodynamics, the modern non-Abelian gauge theory of strong interactions, and also to critical phenomena in condensed matter physics. Chapter 19 deals with general spontaneously broken global symmetries, and their application to the broken approximate $SU(2) \times SU(2)$ and $SU(3) \times SU(3)$ symmetries of quantum chromodynamics. Both the renormalization group method and broken symmetries find some of their most interesting applications in the context of operator product expansions, discussed in Chapter 20.

The key to the understanding of the electroweak interactions is the spontaneous breaking of gauge symmetries, which are explored in Chapter 21 and applied to superconductivity as well as to the electroweak interactions. Quite apart from spontaneous symmetry breaking is the possibility of symmetry breaking by quantum-mechanical effects known as anomalies. Anomalies and various of their physical implications are presented in Chapter 22. This volume concludes with a discussion in Chapter 23 of extended field configurations, which can arise either as new ingredients in physical states, such as skyrmions, monopoles, or vortex lines, or as non-perturbative quantum corrections to path integrals, where anomalies play a crucial role.

It would not be possible to provide a coherent account of these developments if they were presented in a historical order. I have chosen instead to describe the material of this book in an order that seems to me to work best pedagogically — I introduce each topic at a point where the motivation as well as the mathematics can be understood with the least possible reference to material in subsequent chapters, even where logic might suggest a somewhat different order. For instance, instead of having one long chapter to introduce non-Abelian gauge theories, this material is split between Chapters 15 and 17, because Chapter 15 provides a motivation for the external field formalism introduced in Chapter 16, and this formalism is necessary for the work of Chapter 17.

In the course of this presentation, the reader will be introduced to various formal devices, including BRST invariance, the quantum effective action, and homotopy theory. The Batalin–Vilkovisky formalism is presented as an optional side track. It is introduced in Chapter 15 as a compact way of formulating gauge theories, whether based on open or closed symmetry algebras, and then used in Chapter 17 to study the cancellation of infinities in 'non-renormalizable' gauge theories, including general relativity, and in Chapter 22 to show that certain gauge theories are anomaly-free to all orders of perturbation theory. The effective field theory approach is extensively used in this volume, especially in applications to theories with broken symmetry, including the theory of superconductivity. I have struggled throughout for the greatest possible clarity of presentation, taking time to show detailed calculations where I thought it might help the reader, and dropping topics that could not be clearly explained in the space available.

The guiding aim of both Volumes I and II of this book is to explain to the reader why quantum field theory takes the form it does, and why in this form it does such a good job of describing the real world. Volume I outlined the foundations of the quantum theory of fields, emphasizing the reasons why nature is described at accessible energies by effective quantum field theories, and in particular by gauge theories. (A list of chapters of Volume I is given at the end of the table of contents of this volume.) The present volume takes quantum field theory and gauge invariance as its starting points, and concentrates on their implications.

This volume should be accessible to readers who have some familiarity with the fundamentals of quantum field theory. It is not assumed that the reader is familiar with Volume I (though it wouldn't hurt). Aspects of group theory and topology are explained where they are introduced.

Some of the formal methods described in this volume (such as BRST invariance and the renormalization group) have important applications in speculative theories that involve supersymmetry or superstrings. I am enthusiastic about the future prospects of these theories, but I have not included them in this book, because it seems to me that they require a whole book to themselves. (Perhaps supersymmetry and supergravity will be the subjects of a Volume III.) I have excluded some other interesting topics here, such as finite temperature field theory, lattice gauge calculations and the large N_c approximation, because they were not needed to provide either motivation or mathematical techniques for the rest of the book, and the book was long enough.

The great volume of the literature on quantum field theory and its applications makes it impossible for me to read or quote all relevant articles. I have tried to supply citations to the classic papers on each topic, as well as to papers that describe further developments of material covered here, and to references that present detailed calculations, data, or proofs referred to in the text. As before, the mere absence of a citation should not be interpreted as a claim that the material presented is original, but some of it is.

In my experience this volume provides enough material for a one-year course for graduate students on advanced topics in quantum field theory, or on elementary particle physics. Selected parts of Volumes I and II would be suitable as the basis of a compressed one-year course on both the foundations and the modern applications of quantum field theory. I have supplied problems for each chapter. Some of these problems aim simply at providing exercise in the use of techniques described in the chapter; others are intended to suggest extensions of the results of the chapter to a wider class of theories.

* * *

I must acknowledge my special intellectual debt to colleagues at the University of Texas, notably Luis Boya, Phil Candelas, Bryce and Cecile De Witt, Willy Fischler, Joaquim Gomis, and Vadim Kaplunovsky, and especially Jacques Distler. Also, Luis Alvarez-Gaumè, Sidney Coleman, John Dixon, Tony Duncan, Jürg Fröhlich, Arthur Jaffe, Marc Henneaux, Roman Jackiw, Joe Polchinski, Michael Tinkham, Cumrun Vafa, Don Weingarten, Edward Witten and Bruno Zumino gave valuable help with special topics. Jonathan Evans read through the manuscript of this volume, and made many valuable suggestions. For corrections to the first printing of this volume I am indebted to several students and colleagues, including Stephen Adler, Mark Byrd, Vincent Liu, Herbert Neuberger, Chun-yen Wang, and especially Michio Masujima. Thanks are due to Alyce Wilson, who prepared the illustrations and typed the LaTeX input files until I learned how to do it, to Terry Riley for finding countless books and articles, and to Jan Duffy for many helps. I am grateful to Maureen Storey and Alison Woollatt of Cambridge University Press for working to ready this book for publication, and especially to my editor, Rufus Neal, for his continued friendly good advice.

STEVEN WEINBERG

Austin, Texas
December, 1995

Notation

Latin indices i, j, k, and so on generally run over the three spatial coordinate labels, usually taken as 1, 2, 3. Where specifically indicated, they run over values 1, 2, 3, 4, with $x^4 = it$.

Greek indices μ, ν, etc. from the middle of the Greek alphabet generally run over the four spacetime coordinate labels 1, 2, 3, 0, with x^0 the time coordinate.

Greek indices α, β, etc. from the beginning of the Greek alphabet generally run over the generators of a symmetry algebra.

Repeated indices are generally summed, unless otherwise indicated.

The spacetime metric $\eta_{\mu\nu}$ is diagonal, with elements $\eta_{11} = \eta_{22} = \eta_{33} = 1$, $\eta_{00} = -1$.

The d'Alembertian is defined as $\Box \equiv \eta^{\mu\nu}\partial^2/\partial x^\mu \partial x^\nu = \nabla^2 - \partial^2/\partial t^2$, where ∇^2 is the Laplacian $\partial^2/\partial x^i \partial x^i$.

The 'Levi–Civita tensor' $\epsilon^{\mu\nu\rho\sigma}$ is defined as the totally antisymmetric quantity with $\epsilon^{0123} = +1$.

Spatial three-vectors are indicated by letters in boldface.

Three-vectors in isospin space are indicated by arrows.

A hat over any vector indicates the corresponding unit vector: Thus, $\hat{\mathbf{v}} \equiv \mathbf{v}/|\mathbf{v}|$.

A dot over any quantity denotes the time-derivative of that quantity.

Dirac matrices γ_μ are defined so that $\gamma_\mu\gamma_\nu + \gamma_\nu\gamma_\mu = 2\eta_{\mu\nu}$. Also, $\gamma_5 = i\gamma_0\gamma_1\gamma_2\gamma_3$, and $\beta = i\gamma^0 = \gamma_4$.

The step function $\theta(s)$ has the value +1 for $s > 0$ and 0 for $s < 0$.

The complex conjugate, transpose, and Hermitian adjoint of a matrix or vector A are denoted A^*, A^{T}, and $A^\dagger = A^{*\mathrm{T}}$, respectively. The Hermitian adjoint of an operator O is denoted $O^\dagger$, except where an asterisk is used to emphasize that a vector or matrix of operators is not transposed. +H.c. or +c.c. at the end of an expression indicates the addition of the Hermitian adjoint or complex conjugate of the foregoing terms. A bar on a Dirac spinor u is defined by $\bar{u} = u^\dagger \beta$. The antifield of a field χ in the Batalin–Vilkovisky formalism is denoted $\chi^\ddagger$ rather than χ^* to distinguish it from the ordinary complex conjugate or the antiparticle field.

Units are usually used with $\hbar$ and the speed of light taken to be unity. Throughout $-e$ is the rationalized charge of the electron, so that the fine structure constant is $\alpha = e^2/4\pi \simeq 1/137$.

Numbers in parenthesis at the end of quoted numerical data give the uncertainty in the last digits of the quoted figure. Where not otherwise indicated, experimental data are taken from 'Review of Particle Properties,' *Phys. Rev.* **D50**, 1173 (1994).

15
Non-Abelian Gauge Theories

The quantum field theories that have proved successful in describing the real world are all non-Abelian gauge theories, theories based on principles of gauge invariance more general than the simple $U(1)$ gauge invariance of quantum electrodynamics. These theories share with electrodynamics the attractive feature, outlined at the end of Section 8.1, that the existence and some of the properties of the gauge fields follow from a principle of invariance under local gauge transformations. In electrodynamics, fields $\psi_n(x)$ of charge e_n undergo the gauge transformation $\psi_n(x) \rightarrow \exp(ie_n\Lambda(x))\psi_n(x)$ with arbitrary $\Lambda(x)$. Since $\partial_\mu\psi_n(x)$ does not transform like $\psi_n(x)$, we must introduce a field $A_\mu(x)$ with the gauge transformation property $A_\mu(x) \rightarrow A_\mu(x)+\partial_\mu\Lambda(x)$, and use it to construct a gauge-covariant derivative $\partial_\mu\psi_n(x)-ie_nA_\mu(x)\psi_n(x)$, which transforms just like $\psi_n(x)$ and can therefore be used with $\psi_n(x)$ to construct a gauge-invariant Lagrangian. In a similar way, the existence and some of the properties of the gravitational field $g_{\mu\nu}(x)$ in general relativity follow from a symmetry principle, under general coordinate transformations.* Given these distinguished precedents, it was natural that local gauge invariance should be extended to invariance under local non-Abelian gauge transformations.

In the original 1954 work of Yang and Mills,[1] the non-Abelian gauge group was taken to be the $SU(2)$ group of isotopic spin rotations, and the vector fields analogous to the photon field were interpreted as the fields of strongly-interacting vector mesons of isotopic spin unity. This proposal immediately encountered the obstacle that these vector mesons would have to have zero mass, like photons, and it seemed that any such particles would already have been detected. Another problem was that, like all strong-interaction theories at that time, there was nothing that

* Of course, both local gauge invariance and general covariance can be realized in a trivial way, by taking $A_\mu(x)$ and $g_{\mu\nu}(x)$ to be non-dynamical c-number functions that simply characterize a choice of phase or coordinate system, respectively. These symmetries become physically significant when we treat $A_\mu(x)$ and $g_{\mu\nu}(x)$ as dynamical fields, over which we integrate in calculating S-matrix elements.

could be done with it; it seemed that the large coupling constant of the theory would preclude any use of perturbation theory.

Gauge theories were soon generalized to arbitrary non-Abelian gauge groups,[2] and their quantization continued to be studied mathematically, notably by Feynman,[3] Faddeev and Popov,[4] and De Witt,[5] in part as a warming up exercise for the harder problem of quantizing general relativity. They showed that the naive Feynman rules obtained by simply inspecting the Lagrangian need to be supplemented by additional 'ghost' loops. However, the physical relevance of these theories did not begin to be understood until the late 1960s. It eventually turned out that *all* of the observed interactions of elementary particles are generated by vector fields associated with local gauge symmetries; the corresponding spin 1 particles are either very heavy, as a result of a spontaneous breakdown of the gauge symmetry, or are 'trapped', as a result of the rise of the coupling constant at long distances. These matters will be the subjects respectively of Chapters 21 and 18. In this chapter we shall explore the formulation of the non-Abelian gauge theories, and the derivation of their Feynman rules.

15.1 Gauge Invariance

We assume that the Lagrangian of our theory is invariant under a set of infinitesimal transformations on the matter fields $\psi_\ell(x)$

$$\delta\psi_\ell(x) = i\,\epsilon^\alpha(x)(t_\alpha)_\ell{}^m\psi_m(x)\,, \tag{15.1.1}$$

with some set of independent constant matrices** t_α, and with real infinitesimal parameters $\epsilon^\alpha(x)$ which (as for gauge transformations in electrodynamics) are allowed to depend on position in spacetime. We assume that these symmetry transformations are the infinitesimal part of a Lie group; as shown in Section 2.2, this requires that the t_α obey commutation relations

$$[t_\alpha, t_\beta] = i\,C^\gamma{}_{\alpha\beta}t_\gamma\,, \tag{15.1.2}$$

where $C^\gamma{}_{\alpha\beta}$ are a set of real constants, known as the structure constants of the group. The antisymmetry of the commutator immediately tells us

** In this book we shall generally label symmetry generators with letters α, β, etc. from the beginning of the Greek alphabet, in order to keep these labels distinct from the indices μ, ν, etc. from the middle of the Greek alphabet that are used to label spacetime coordinates. Later, in dealing with broken symmetries, we will often use letters a, b, etc. from the beginning of the Latin alphabet to label generators of spontaneously broken symmetries, and letters i, j, etc. from the middle of the Latin alphabet to label generators of unbroken symmetries.

that the structure constants are similarly antisymmetric:

$$C^\gamma{}_{\alpha\beta} = -C^\gamma{}_{\beta\alpha} . \tag{15.1.3}$$

Also, from the Jacobi identity

$$0 = \Big[[t_\alpha, t_\beta], t_\gamma\Big] + \Big[[t_\gamma, t_\alpha], t_\beta\Big] + \Big[[t_\beta, t_\gamma], t_\alpha\Big] \tag{15.1.4}$$

we see that the Cs satisfy the further constraint

$$0 = C^\delta{}_{\alpha\beta} C^\epsilon{}_{\delta\gamma} + C^\delta{}_{\gamma\alpha} C^\epsilon{}_{\delta\beta} + C^\delta{}_{\beta\gamma} C^\epsilon{}_{\delta\alpha} . \tag{15.1.5}$$

Any set of constants $C^\gamma{}_{\alpha\beta}$ that satisfy Eqs. (15.1.3) and (15.1.5) define at least one set of matrices $t^{\rm A}{}_\alpha$:

$$(t^{\rm A}{}_\alpha)^\beta{}_\gamma \equiv -i C^\beta{}_{\gamma\alpha} , \tag{15.1.6}$$

that satisfy the commutation relations (15.1.2) with structure constants $C^\gamma{}_{\alpha\beta}$:

$$[t^{\rm A}{}_\alpha, t^{\rm A}{}_\beta] = i C^\gamma{}_{\alpha\beta} t^{\rm A}{}_\gamma . \tag{15.1.7}$$

This is known as the 'adjoint' (or 'regular') representation of the Lie algebra with structure constants $C^\alpha{}_{\beta\gamma}$.

For example, in the original Yang–Mills theory, the matter fields were the doublet consisting of proton and neutron fields ψ_p and ψ_n:

$$\psi = \begin{pmatrix} \psi_p \\ \psi_n \end{pmatrix}$$

and the t_α with $\alpha = 1, 2, 3$ were the isospin matrices

$$t_1 = \frac{1}{2}\begin{pmatrix} 0 & 1 \\ 1 & 0 \end{pmatrix} , \quad t_2 = \frac{1}{2}\begin{pmatrix} 0 & -i \\ i & 0 \end{pmatrix} , \quad t_3 = \frac{1}{2}\begin{pmatrix} 1 & 0 \\ 0 & -1 \end{pmatrix} .$$

These satisfy the commutation relations (15.1.2) with

$$C^\gamma{}_{\alpha\beta} = \epsilon_{\gamma\alpha\beta} ,$$

where as usual $\epsilon_{\gamma\alpha\beta}$ is $+1$ or -1 if γ, α, β is an even or an odd permutation of 1, 2, 3, respectively, and vanishes otherwise. We recognize this as the same as the Lie algebra (2.4.18) of the three-dimensional rotation group; the matrices t_α here furnish what we recognize as the spin 1/2 representation of this Lie algebra. The matrices (15.1.6) of the adjoint representation are here (in a basis with rows and columns labelled 1,2,3):

$$t^A_1 = \begin{bmatrix} 0 & 0 & 0 \\ 0 & 0 & -i \\ 0 & i & 0 \end{bmatrix} , \quad t^A_2 = \begin{bmatrix} 0 & 0 & i \\ 0 & 0 & 0 \\ -i & 0 & 0 \end{bmatrix} , \quad t^A_3 = \begin{bmatrix} 0 & -i & 0 \\ i & 0 & 0 \\ 0 & 0 & 0 \end{bmatrix} .$$

This is the spin 1 representation of the Lie algebra of the rotation group.

Now consider what is needed to make the Lagrangian invariant under the transformations (15.1.1). If there were no derivatives acting on the fields, the task would be easy — any function of the matter fields that was invariant under the transformation (15.1.1) with ϵ^α constant would also be invariant with ϵ^α arbitrary real functions of the spacetime coordinates. This is not the case if the Lagrangian involves derivatives of the fields (as it must), because with position–dependent functions $\epsilon^\alpha(x)$, the derivatives of the matter fields do not transform like the fields themselves. Differentiating Eq. (15.1.1) gives

$$\delta\Big(\partial_\mu\psi_\ell(x)\Big) = i\epsilon^\alpha(x)(t_\alpha)_\ell{}^m\Big(\partial_\mu\psi_m(x)\Big) + i\Big(\partial_\mu\epsilon^\alpha(x)\Big)(t_\alpha)_\ell{}^m\psi_m(x)\,. \quad (15.1.8)$$

To make the Lagrangian invariant, we need a field $A^\alpha{}_\mu$, whose transformation rule involves a term $\partial_\mu\epsilon^\alpha$, which can be used to cancel the second term in Eq. (15.1.8). Since this field carries an α-index, we would expect it also to undergo a matrix transformation like Eq. (15.1.1), but with t_α replaced with the adjoint representation matrices (15.1.6). Let us therefore tentatively take the transformation relation of these new 'gauge' fields as

$$\delta A^\beta{}_\mu = \partial_\mu\epsilon^\beta \ + \ i\epsilon^\alpha(t^{\rm A}{}_\alpha)^\beta{}_\gamma A^\gamma{}_\mu$$

or, using Eq. (15.1.6),

$$\delta A^\beta{}_\mu = \partial_\mu\,\epsilon^\beta \ + \ C^\beta{}_{\gamma\alpha}\epsilon^\alpha A^\gamma{}_\mu\,. \quad (15.1.9)$$

This allows us to construct a 'covariant derivative':†

$$(D_\mu\psi(x))_\ell = \partial_\mu\psi_\ell(x) - i\,A^\beta{}_\mu(x)(t_\beta)_\ell{}^m\psi_m(x)\,. \quad (15.1.10)$$

As planned, the term $\partial_\mu\epsilon^\beta$ in the transformation of $A^\beta{}_\mu$ in the second term of Eq. (15.1.10) cancels the term proportional to $\partial_\mu\epsilon^\beta$ in the transformation of the first term, leaving us with

$$\begin{aligned}\delta\,(D_\mu\,\psi)_\ell = i\,\epsilon^\alpha(t_\alpha)_\ell{}^m\partial_\mu\psi_m - i\,C^\beta{}_{\gamma\alpha}\,\epsilon^\alpha\,A^\gamma{}_\mu(t_\beta)_\ell{}^m\psi_m \\ +A^\gamma{}_\mu\,(t_\gamma)_\ell{}^m(t_\alpha)_m{}^n\psi_n\,\epsilon^\alpha\end{aligned}$$

or, using Eq. (15.1.2),

$$\delta(D_\mu\psi)_\ell \ = \ i\epsilon^\alpha\,(t_\alpha)_\ell{}^m\,(D_\mu\psi)_m\,, \quad (15.1.11)$$

so that $D_\mu\psi$ transforms just like ψ itself.

We also need to worry about derivatives of the gauge field. In order to eliminate the term $\partial_\nu\partial_\mu\epsilon^\beta$ in the transformation of $\partial_\nu A^\beta{}_\mu$, we antisymmetrize with respect to μ and ν, just as in electrodynamics. However, we still have terms in the transformation of $\partial_\nu A^\beta{}_\mu - \partial_\mu A^\beta{}_\nu$ proportional to first derivatives of $\epsilon(x)$, arising from the second term in Eq. (15.1.9). The easiest way to construct a 'covariant curl', $F^\gamma{}_{\nu\mu}$ in whose transformation

† As discussed in the next section, in writing Eq. (15.1.10) we are tacitly supposing that any coupling-constant factors like the electric charge are included in the t_β, and hence also in the structure constants.

rule all such derivatives of $\epsilon(x)$ cancel, is to consider the commutator of two covariant derivatives acting on a matter field ψ:

$$\Big([D_\nu, D_\mu]\psi\Big)_\ell = -i\,(t_\gamma)_\ell{}^m F^\gamma{}_{\nu\mu}\psi_m\,, \qquad (15.1.12)$$

where

$$F^\gamma{}_{\nu\mu} \equiv \partial_\nu A^\gamma{}_\mu - \partial_\mu A^\gamma{}_\nu + C^\gamma{}_{\alpha\beta}A^\alpha{}_\nu A^\beta{}_\mu\,. \qquad (15.1.13)$$

Eq. (15.1.12) makes it obvious that $F^\gamma{}_{\nu\mu}$ must transform just like a matter field that happens to belong to the adjoint representation:

$$\delta\, F^\beta{}_{\nu\mu} \equiv i\,\epsilon^\alpha (t^{\rm A}{}_\alpha)^\beta{}_\gamma F^\gamma{}_{\nu\mu} = \epsilon^\alpha C^\beta{}_{\gamma\alpha}F^\gamma{}_{\nu\mu}\,. \qquad (15.1.14)$$

The reader may check by direct calculation (using the relation (15.1.5)) that the quantity $F^\alpha{}_{\nu\mu}$ defined in (15.1.13) actually has the simple transformation rule (15.1.14).

For some purposes, it is useful to know that these infinitesimal gauge transformations can be upgraded to finite transformations. A group element can be parameterized by a set of real functions $\Lambda^\alpha(x)$ so that it acts on a general matter field $\psi_\ell(x)$ through the matrix transformation

$$\psi_\ell(x) \to \psi_{\ell\Lambda}(x) = \Big[\exp\Big(i\,t_\alpha\Lambda^\alpha(x)\Big)\Big]_\ell{}^m \psi_m(x)\,. \qquad (15.1.15)$$

We want the covariant derivative to transform in the same way:

$$(\partial_\mu - i\,t_\alpha A^\alpha{}_{\mu\Lambda})\psi_\Lambda = \exp(i\,t_\alpha\Lambda^\alpha)\,(\partial_\mu - i\,t_\alpha A^\alpha_\mu)\psi\,, \qquad (15.1.16)$$

so we must impose on $A^\alpha{}_\mu$ the transformation rule $A^\alpha_\mu \to A^\alpha_{\mu\Lambda}$, with

$$\partial_\mu \exp(i\,t_\beta\Lambda^\beta) - i\,t_\beta \exp(i\,t_\alpha\Lambda^\alpha)\,A^\beta_{\mu\Lambda} = -i\,\exp(i\,t_\alpha\Lambda^\alpha)t_\beta A^\beta$$

or in other words

$$t_\alpha\, A^\alpha{}_{\mu\Lambda} = \exp(i\,t_\beta\Lambda^\beta)\,t_\alpha A^\alpha{}_\mu\,\exp(-it_\beta\Lambda^\beta) - i\,\Big[\partial_\mu \exp(it_\beta\Lambda^\beta)\Big]\exp(-it_\beta\Lambda^\beta)\,. \qquad (15.1.17)$$

Eqs. (15.1.15) and (15.1.17) reduce to the previous transformation rules (15.1.1) and (15.1.9) in the limit where $\Lambda^\alpha(x)$ is an infinitesimal $\epsilon^\alpha(x)$.

From Eq. (15.1.17), we can see that by a suitable choice of $\Lambda^\beta(x)$, it is always possible to make $A^\alpha{}_{\mu\Lambda}(x)$ vanish at any *one* point, say $x = z$. (Simply take $\Lambda^\alpha(z)$ to vanish, and $\partial\Lambda^\alpha(x)/\partial x^\mu = -A^\alpha{}_\mu(x)$ at $x = z$.) Also, it is always possible to choose $\Lambda^\beta(x)$ so that any *one* spacetime component of $A^\alpha{}_{\mu\Lambda}(x)$ vanishes for all α everywhere in at least a finite domain around any given point. For instance, to make $A^\alpha{}_{3\Lambda}(x)$ vanish, we must solve the set of *ordinary* first-order differential equations for the parameters $\Lambda^\beta(x)$:

$$\partial_3 \exp(it_\beta\Lambda^\beta) = -i\,\exp(i\,t_\beta\Lambda^\beta)\,t_\alpha\,A^\alpha{}_3\,, \qquad (15.1.18)$$

which always have a solution in at least a finite domain around any ordinary point.

However, in general it is not possible to choose $\Lambda^\alpha(x)$ to make all four components $A^\alpha{}_{\mu\Lambda}(x)$ vanish in a finite region. For this purpose, we would have to be able to satisfy the *partial* differential equations

$$\partial_\mu \exp(it_\beta \Lambda^\beta) = -i \exp(i t_\beta \Lambda^\beta)\, t_\alpha\, A^\alpha{}_\mu \,, \qquad (15.1.19)$$

which cannot be solved unless certain integrability conditions are satisfied. In particular, if $A^\alpha{}_{\mu\Lambda}(x)$ vanishes everywhere then so does $F^\alpha{}_{\mu\nu\Lambda}(x)$, but since the field strength transforms homogeneously, $F^\alpha{}_{\mu\nu\Lambda}(x)$ can vanish only if $F^\alpha{}_{\mu\nu}(x)$ does. A gauge field $A^\alpha{}_\mu(x)$ is called a 'pure gauge' field if there exists a gauge transformation which makes it vanish everywhere. It is not difficult to show that the condition that $F^\alpha{}_{\mu\nu}$ should vanish everywhere is not only necessary but also sufficient for $A^\alpha{}_\mu(x)$ to be expressible in any simply connected region as a pure gauge field.[6]

* * *

There is a deep analogy between the construction here of objects that transform simply under gauge transformations and the construction in general relativity of objects that transform covariantly under general coordinate transformations. Just as we use the gauge field to construct covariant derivatives $D_\mu \psi_\ell$ of matter fields with the same gauge transformation properties as the matter fields themselves, so we use the affine connection $\Gamma^\mu{}_{\nu\lambda}(x)$ to construct covariant derivatives of tensors $T^{\rho\sigma\cdots}{}_{\kappa\lambda\cdots}$:

$$T^{\rho\cdots}{}_{\kappa\cdots;\nu} \equiv \partial_\nu T^{\rho\cdots}{}_{\kappa\cdots} + \Gamma^\rho{}_{\nu\lambda} T^{\lambda\cdots}{}_{\kappa\cdots} + \cdots - \Gamma^\mu{}_{\nu\kappa} T^{\rho\cdots}{}_{\mu\cdots} - \cdots \,,$$

which are themselves tensors. Also, from the derivatives of the gauge field we constructed a field strength $F^\alpha{}_{\mu\nu}$ with the gauge transformation property of a matter field belonging to the adjoint representation of the gauge group; correspondingly, from the derivatives of the affine connection we may construct a quantity:

$$R^\lambda{}_{\mu\nu\kappa} = \frac{\partial \Gamma^\lambda{}_{\mu\nu}}{\partial x^\kappa} - \frac{\partial \Gamma^\lambda{}_{\mu\kappa}}{\partial x^\nu} + \Gamma^\eta{}_{\mu\nu}\Gamma^\lambda{}_{\kappa\eta} - \Gamma^\eta{}_{\mu\kappa}\Gamma^\lambda{}_{\nu\eta} \,,$$

which transforms as a tensor, the Riemann–Christoffel curvature tensor. The commutator of two gauge-covariant derivatives D_μ and D_ν may be expressed in terms of the field-strength tensor $F^\alpha{}_{\mu\nu}$; similarly, the commutator of two covariant derivatives with respect to x^ν and x^κ may be expressed in terms of the curvature:

$$T^{\lambda\cdots}{}_{\mu\cdots;\nu;\kappa} - T^{\lambda\cdots}{}_{\mu\cdots;\kappa;\nu} = R^\lambda{}_{\sigma\nu\kappa}\, T^{\sigma\cdots}{}_{\mu\cdots} + \cdots - R^\sigma{}_{\mu\nu\kappa}\, T^{\lambda\cdots}{}_{\sigma\cdots} - \cdots \,.$$

The necessary and sufficient condition for the existence of a gauge in which the gauge field vanishes in a finite simply connected region is the vanishing of the field-strength tensor, and the necessary and sufficient condition for the existence of a coordinate system in which the affine connection vanishes in a finite simply connected region is the vanishing of the Riemann–Christoffel curvature tensor. The analogy breaks down in one important respect: in general relativity the affine connection is itself constructed from first derivatives of the metric tensor, while in gauge theories the gauge fields are not expressed in terms of any more fundamental fields.

15.2 Gauge Theory Lagrangians and Simple Lie Groups

The transformation rules of the gauge-field tensor $F^\alpha{}_{\mu\nu}$ and the matter fields ψ and their gauge-covariant derivatives do not involve the derivatives of the transformation parameters $\epsilon^\alpha(x)$, so if the Lagrangian is constructed solely from these ingredients, and if it is invariant under global transformations with ϵ^α constant, then it is invariant under gauge transformations with general position-dependent $\epsilon^\alpha(x)$. We therefore assume that the Lagrangian satisfies these conditions: that is,

$$\mathscr{L} = \mathscr{L}(\psi, D_\mu\psi, D_\nu D_\mu\psi, \cdots, F^\alpha{}_{\mu\nu}, D_\rho F^\alpha{}_{\mu\nu}\cdots) \tag{15.2.1}$$

with the invariance condition:

$$\begin{aligned}&\frac{\partial\mathscr{L}}{\partial\psi_\ell} i(t_\alpha)_\ell{}^m\psi_m + \frac{\partial\mathscr{L}}{\partial(D_\mu\psi_\ell)} i(t_\alpha)_\ell{}^m(D_\mu\psi)_m \\ &+ \frac{\partial\mathscr{L}}{\partial(D_\nu D_\mu\psi)_\ell} i(t_\alpha)_\ell{}^m(D_\nu D_\mu\psi)_m + \cdots + \frac{\partial\mathscr{L}}{\partial F^\beta{}_{\mu\nu}} C^\beta{}_{\gamma\alpha}F^\gamma{}_{\mu\nu} \\ &+ \frac{\partial\mathscr{L}}{\partial D_\rho F^\beta{}_{\mu\nu}} C^\beta{}_{\gamma\alpha}D_\rho F^\gamma{}_{\mu\nu} + \cdots = 0\,.\end{aligned} \tag{15.2.2}$$

On the other hand, the Lagrangian may not depend on the gauge field itself, except insofar as it appears in $F^\alpha{}_{\mu\nu}$ and in gauge-covariant derivatives D_μ. In particular, a mass term $-\frac{1}{2}m^2{}_{\alpha\beta}A_{\alpha\mu}A_\beta{}^\mu$ is ruled out.

We shall concentrate now on the terms in the Lagrangian that depend only on $F^\alpha{}_{\mu\nu}$. Just as in electrodynamics, for any massless particle of unit spin the Lagrangian must contain a free-particle term quadratic in $\partial_\mu A^\alpha{}_\nu - \partial_\nu A^\alpha{}_\mu$, and gauge invariance then dictates that this free-particle term should appear as part of a term quadratic in the field-strength tensor $F^\alpha{}_{\mu\nu}$. Lorentz invariance and parity conservation dictate its form as

$$\mathscr{L}_A = -\tfrac{1}{4}\, g_{\alpha\beta}F^\alpha{}_{\mu\nu}F^{\beta\mu\nu} \tag{15.2.3}$$

with a constant matrix $g_{\alpha\beta}$. If we do not assume parity (or CP or T) conservation, then we may also include in the Lagrangian a term

$$\mathscr{L}'_A = -\tfrac{1}{2}\,\theta_{\alpha\beta}\,\epsilon^{\mu\nu\rho\sigma}F^\alpha{}_{\mu\nu}F^\beta{}_{\rho\sigma}$$

with another constant matrix $\theta_{\alpha\beta}$. This term is actually a derivative, and therefore does not affect the field equations or the Feynman rules. Such a term would, however, have non-perturbative quantum mechanical effects, to be discussed in Section 23.6.

Before going on to consider the properties of the matrix $g_{\alpha\beta}$, it is worth drawing attention to the fact that it is not possible to introduce a kinematic term for the gauge field $A^{\alpha}{}_{\mu}(x)$ without also including interactions, the terms in Eq. (15.2.3) arising from the quadratic part of the field strength $F^{\alpha}{}_{\mu\nu}$ defined by Eq. (15.1.13). This is one more respect in which non-Abelian gauge theories resemble general relativity, where the kinematic part of the Lagrangian for the gravitational field is contained in the Einstein–Hilbert Lagrangian density $-\sqrt{g}R/8\pi G$, which also contains self-interactions of the field. The reasons in both cases are similar: the gravitational field interacts with itself because it interacts with anything that carries energy and momentum, and the gauge field interacts with itself because it interacts with anything that transforms according to a non-trivial representation (in this case the adjoint representation) of the gauge group. This is in contrast to the case of electrodynamics, where the photon does not carry electric charge, the quantum number with which it interacts, and it is consequently possible to introduce a kinematic term $-\frac{1}{4}F_{\mu\nu}F^{\mu\nu}$ for the electromagnetic field that does not entail interactions.

The numerical matrix $g_{\alpha\beta}$ may be taken symmetric, and must be taken real to give a real Lagrangian. In order for this term to satisfy the gauge-invariance requirement (15.2.2), we must have for all δ:

$$g_{\alpha\beta}F^{\alpha}{}_{\mu\nu}C^{\beta}{}_{\gamma\delta}F^{\gamma\mu\nu} = 0 \,.$$

In order for this to be true without having to impose any functional relations among the Fs, the matrix $g_{\alpha\beta}$ must satisfy the condition:

$$g_{\alpha\beta}C^{\beta}{}_{\gamma\delta} = -\, g_{\gamma\beta}C^{\beta}{}_{\alpha\delta} \,. \tag{15.2.4}$$

There is one more important condition on the matrix $g_{\gamma\beta}$. Just as in quantum electrodynamics, the rules of canonical quantization and the positivity properties of the quantum mechanical scalar product require that the matrix $g_{\alpha\beta}$ in the Lagrangian (15.2.3) must be positive-definite. (That is, $g_{\alpha\beta}u^{\alpha}u^{\beta}$ is positive for all real u, and vanishes for some real u only if $u^{\alpha} = 0$ for all α.) This is analogous to the requirement that in the kinematic Lagrangian $-\frac{1}{2}Z\partial_{\mu}\phi\partial^{\mu}\phi - \frac{1}{2}m^2\phi^2$ for a real scalar field ϕ, the constant Z must be positive-definite.

These requirements on the matrix $g_{\alpha\beta}$ have far reaching implications. They form one of a set of three equivalent conditions:

a: There exists a real symmetric positive-definite matrix $g_{\alpha\beta}$ that satisfies the invariance condition (15.2.4).

b: There is a basis for the Lie algebra (that is, a set of generators $\tilde{t}_{\alpha} = \mathscr{S}_{\alpha\beta}t_{\beta}$, with $\mathscr{S}$ a real non-singular matrix) for which the structure constants $\tilde{C}^{\alpha}{}_{\beta\gamma}$ are antisymmetric not only in the lower indices β and γ but in all three indices α, β and γ. (In this basis it is convenient to drop the distinction between upper and lower indices α, β, etc., and write $\tilde{C}_{\alpha\beta\gamma}$ in place of $\tilde{C}^{\alpha}{}_{\beta\gamma}$.)

c: The Lie algebra is the direct sum of commuting compact simple and $U(1)$ subalgebras.*

Appendix A of this chapter presents a proof of the equivalence of the conditions **a**, **b**, and **c**.[7]

Before going on to discuss the physical implications of this result, it will be useful to say a bit more about the condition of compactness. We will not use this here, but a compact Lie algebra consists of the generators of a compact Lie group: one for which the invariant volume of the group is finite. For instance, the rotation group is compact; the Lorentz group is not. As a simple example of a simple Lie algebra that is not compact, consider the commutation relations

$$[t_1, t_2] = -i\,t_3, \qquad [t_2, t_3] = i\,t_1, \qquad [t_3, t_1] = i\,t_2\,.$$

The structure constant here is real, but not completely antisymmetric; its non-vanishing components are

$$C^3{}_{12} = -C^3{}_{21} = -1, \quad C^1{}_{23} = -C^1{}_{32} = 1, \quad C^2{}_{31} = -C^2{}_{13} = 1\,.$$

The metric given by Eq. (15.A.10) is here diagonal, with elements:

$$g_{11} = g_{22} = -g_{33} = -2\,.$$

This is not a positive matrix, so the Lie algebra is not compact. It is in fact the Lie algebra of the non-compact group $O(2,1)$, the Lorentz group in two space and one time dimensions.

---

* Some definitions: A subalgebra $\mathscr{H}$ of a Lie algebra $\mathscr{G}$ is a linear space, spanned by certain real linear combinations $t_i = \mathscr{S}_{i\alpha} t_\alpha$ of the generators t_α of $\mathscr{G}$, such that $\mathscr{H}$ is itself a Lie algebra, in the sense that the commutators of the t_i with each other are of the form $[t_i, t_j] = i c^k{}_{ij} t_k$. A subalgebra $\mathscr{H}$ is called *invariant* if the commutator of any element of the whole algebra $\mathscr{G}$ with any element of the subalgebra $\mathscr{H}$ is in the subalgebra $\mathscr{H}$. A *simple* Lie algebra is one without invariant subalgebras. A $U(1)$ subalgebra of $\mathscr{G}$ is one with just a single generator that commutes with all generators of the whole algebra $\mathscr{G}$. A *semi-simple* Lie algebra is one that has no invariant Abelian subalgebras, i.e., invariant subalgebras whose generators all commute with each other. Semi-simple Lie algebras are direct sums of simple (but not $U(1)$) Lie algebras. A simple or semi-simple Lie algebra is said to be *compact* if the matrix $\mathrm{Tr}\{t^A{}_\alpha t^A{}_\beta\} = -C^\gamma{}_{\alpha\delta} C^\delta{}_{\beta\gamma}$ is positive-definite. The meaning and importance of the properties of simplicity and compactness will be discussed further below. In saying that a Lie algebra $\mathscr{G}$ is a direct sum of subgroups $\mathscr{H}_n$, it is meant that it is possible to find a basis for $\mathscr{G}$ with generators t_{na} for which the structure constants take the form

$$C^{\ell c}{}_{na\,mb} = \delta_{\ell m}\delta_{mn} C^{(n)c}{}_{ab}\,,$$

where $C^{(n)c}{}_{ab}$ is the structure constant of the subalgebra $\mathscr{H}_n$.

Two sets of generators that differ by a *real* non-singular linear transformation are considered to span the same Lie algebra, and generate the same group. This is not true for complex linear transformations of generators. In particular, any simple Lie algebra can be put into a compact form by a change of phase of the generators in a suitable basis. For instance, for the Lie algebra of the above example, it is only necessary to define new generators $t'_1 = it_1$, $t'_2 = it_2$, $t'_3 = t_3$, for which the commutation relations are

$$[t'_1, t'_2] = i\,t'_3, \quad [t'_2, t'_3] = i\,t'_1, \quad [t'_3, t'_1] = i\,t'_2\,.$$

The structure constant is now real and totally antisymmetric: $C^a{}_{bc} = \epsilon_{abc}$. Here $g_{ab} = 2\delta_{ab}$, and the algebra is compact. We recognize this, of course, as the familiar algebra of the compact group $O(3)$ of rotations in three dimensions. To see that this is always possible for any simple Lie algebra, note that the matrix g_{ab} defined by Eq. (15.A.10) is real, symmetric, and non-singular, so that by a real orthogonal transformation it may be put in a diagonal form with non-zero elements along the main diagonal. It is then only necessary to multiply all the generators that correspond in this basis to the negative diagonal elements of g_{ab} by factors i.

We note without proof that the finite-dimensional representations of compact Lie groups are all unitary, and the finite-dimensional representations of compact Lie algebras are correspondingly all Hermitian. Furthermore, it is easy to see that the only Lie algebras that can have any non-trivial representation by independent finite-dimensional Hermitian matrices t_α are direct sums of $U(1)$ and compact simple Lie algebras. To show this, we may simply define

$$g_{\alpha\beta} \equiv \mathrm{Tr}\{t_\alpha t_\beta\}\,.$$

This matrix is obviously positive-definite, because $g_{\alpha\beta}u^\alpha u^\beta = \mathrm{Tr}\{(u^\alpha t_\alpha)^2\}$ is positive for any real u^α and vanishes only if $u^\alpha t_\alpha = 0$, which is not possible unless all u^α vanish because the t^α are assumed independent. Furthermore this $g_{\alpha\beta}$ satisfies the invariance condition (15.2.4), as can be seen by multiplying the commutation relation (15.1.2) with t_δ and taking the trace; this gives

$$i\,C^\gamma{}_{\alpha\beta}\mathrm{Tr}\,\{t_\gamma t_\delta\} = \mathrm{Tr}\,\{[t_\alpha, t_\beta]\,t_\delta\} = \mathrm{Tr}\,\{t_\delta t_\alpha t_\beta - t_\beta t_\alpha t_\delta\}\,,$$

which is obviously antisymmetric in β and δ. Having verified **a**, we can rely on the above theorem to infer condition **c**, so that the Lie algebra must be a direct sum of compact simple and $U(1)$ subalgebras.

Let's now return to the physics of gauge theories. In this section we have inferred the existence of a positive symmetric real matrix $g_{\alpha\beta}$, that satisfies the invariance condition (15.2.4), from the necessity of constructing a suitable kinematic term in the Lagrangian for the gauge field, and in

Appendix A of this chapter we have shown that this result is equivalent to a condition on the Lie algebra, that it is a direct sum of compact simple and $U(1)$ subalgebras. For our purposes, the important thing about this result is that the simple Lie algebras are all of certain limited types and dimensionalities. For instance, it is easy to see that there is no simple Lie algebra with less than three generators, because in one or two dimensions there can be no non-zero totally antisymmetric structure constants with three indices. With three generators, an invariant subalgebra can be avoided by taking $C^3{}_{12}$, $C^2{}_{31}$, and $C^1{}_{23}$ all non-zero. In the basis in which the structure constant is real and totally antisymmetric, there is obviously only one possibility:

$$C_{\alpha\beta\gamma} = c\,\epsilon_{\alpha\beta\gamma} \;.$$

Here c is an arbitrary non-zero real constant, which can be eliminated by a change of scale of the generators, $t_\alpha \to t_\alpha/c$, so the Lie algebra is

$$[t_\alpha, t_\beta] = i\,\epsilon_{\alpha\beta\gamma} t_\gamma \;.$$

This may be recognized as the Lie algebra of the three-dimensional rotation group $O(3)$, and also of the group $SU(2)$ of unitary unimodular matrices in two dimensions, and was used as a basis for the original non-Abelian gauge theory of Yang and Mills. Continuing in the same way, it can be shown that there are no simple Lie algebras with 4, 5, 6, or 7 generators, one with 8 generators, and so on. Mathematicians (notably Killing and E. Cartan) have been able to catalog all simple Lie algebras. The compact forms of the simple Lie algebras form several infinite classes of algebras of the 'classical' Lie groups — the unitary unimodular, unitary orthogonal, and unitary symplectic groups — plus just five exceptional Lie algebras. This catalog is presented in Appendix B of this chapter.

It is also shown in Appendix A that under the equivalent conditions **a**, **b**, or **c**, the metric takes the form

$$g_{ma,nb} = g_m^{-2}\delta_{mn}\delta_{ab} \tag{15.2.5}$$

with real g_m, where m and n label the simple or $U(1)$ subalgebras, and a and b label the individual generators of these subalgebras. We can eliminate the constants g_m^{-2} by a rescaling of the gauge fields

$$A^\mu_{ma} \to \tilde{A}^\mu_{ma} \equiv g_m^{-1}\, A^\mu_{ma} \;, \tag{15.2.6}$$

but then in order to keep the same formulas (15.1.10) and (15.1.13) for $D_\mu\psi$ and $F_\alpha{}^{\mu\nu}$, we must also redefine the matrices t_α and the structure constants

$$t_{ma} \to \tilde{t}_{ma} = g_m t_{ma} \;, \tag{15.2.7}$$

$$C^{(m)}_{cab} \to \tilde{C}^{(m)}_{cab} = g_m C^{(m)}_{cab} \;. \tag{15.2.8}$$

That is, we can always *define* the scale of the gauge fields (now dropping the tildes) so that g_m in Eq. (15.2.5) is unity:

$$g_{\alpha\beta} = \delta_{\alpha\beta} \,, \tag{15.2.9}$$

but then the transformation matrices t_α and the structure constants $C_{\alpha\beta\gamma}$ contain an unknown multiplicative factor g_m for each simple or $U(1)$ subalgebra. These factors are the *coupling constants* of the gauge theory. Alternatively, it is sometimes more convenient to adopt some fixed though arbitrary normalization for the t_α and structure constants within each simple or $U(1)$ subalgebra, in which case the coupling constants appear in the gauge-field Lagrangian (15.2.3) as the factors g_m^{-2} in Eq. (15.2.5).

15.3 Field Equations and Conservation Laws

Using Eq. (15.2.9) for the matrix $g_{\alpha\beta}$ in Eq. (15.2.3), the full Lagrangian density is

$$\mathscr{L} = -\tfrac{1}{4} F_{\alpha\mu\nu} F_\alpha{}^{\mu\nu} + \mathscr{L}_M(\psi, D_\mu\psi) \,, \tag{15.3.1}$$

where in the absence of gauge fields $\mathscr{L}_M(\psi, \partial_\mu\psi)$ would be the 'matter' Lagrangian density. We could, in principle, include a dependence of $\mathscr{L}_M$ on $F_{\alpha\mu\nu}$ as well as higher covariant derivatives $D_\nu D_\mu\psi$, $D_\lambda F_{\alpha\mu\nu}$, etc., but we exclude these non-renormalizable terms here for the same reason as in electrodynamics: as discussed in Section 12.3, such terms would be highly suppressed at ordinary energies by negative powers of some very large mass. For this reason the standard model of the weak, electromagnetic and strong interactions has a Lagrangian of the general form (15.3.1).

The equations of motion of the gauge field are here

$$\partial_\mu \frac{\partial\mathscr{L}}{\partial(\partial_\mu A_{\alpha\nu})} = -\partial_\mu F_\alpha{}^{\mu\nu} = \frac{\partial\mathscr{L}}{\partial A_{\alpha\nu}}$$
$$= -F_\gamma{}^{\nu\mu} C_{\gamma\alpha\beta} A_{\beta\mu} - i\,\frac{\partial\mathscr{L}_M}{\partial D_\nu\psi}\,t_\alpha\psi$$

and so

$$\partial_\mu F_\alpha{}^{\mu\nu} = -\mathscr{J}_\alpha{}^\nu \,, \tag{15.3.2}$$

where $\mathscr{J}_\alpha{}^\nu$ is the current:

$$\mathscr{J}_\alpha{}^\nu \equiv -F_\gamma{}^{\nu\mu} C_{\gamma\alpha\beta} A_{\beta\mu} - i\,\frac{\partial\mathscr{L}_M}{\partial D_\nu\psi} t_\alpha\psi \,. \tag{15.3.3}$$

The current $\mathscr{J}_\alpha{}^\nu$ is conserved in the ordinary sense

$$\partial_\nu \mathscr{J}_\alpha{}^\nu = 0 \,, \tag{15.3.4}$$

as can be seen either from the Euler–Lagrange equations for ψ and the invariance condition (15.2.2) or, more easily, directly from the field equations (15.3.2).

The derivatives in Eqs. (15.3.2) and (15.3.4) are ordinary derivatives, not the gauge-covariant derivatives D_ν, so the gauge invariance of these equations is somewhat obscure. It can be made manifest by rewriting Eq. (15.3.2) in terms of the gauge-covariant derivative of the field strength

$$\begin{aligned} D_\lambda F_\alpha{}^{\mu\nu} &\equiv \partial_\lambda F_\alpha{}^{\mu\nu} - i(t^A_\beta)_{\alpha\gamma} A_{\beta\lambda} F_\gamma{}^{\mu\nu} \\ &= \partial_\lambda F_\alpha{}^{\mu\nu} - C_{\alpha\gamma\beta} A_{\beta\lambda} F_\gamma{}^{\mu\nu} \,. \end{aligned} \tag{15.3.5}$$

Then Eq. (15.3.2) reads

$$D_\mu F_\alpha{}^{\mu\nu} = -J_\alpha{}^\nu \,, \tag{15.3.6}$$

where $J_\alpha{}^\nu$ is the current of the matter fields alone

$$J_\alpha{}^\nu \equiv -i \frac{\partial \mathscr{L}_M}{\partial D_\nu \psi} t_\alpha \psi \,. \tag{15.3.7}$$

This is gauge-covariant, if $\mathscr{L}_M$ is gauge-invariant. Also, by operating on Eq. (15.3.6) with D_ν, using the commutation relation

$$[D_\nu, D_\mu] F^{\rho\sigma}_\alpha = -i(t^A{}_\gamma)_{\alpha\beta} F_{\gamma\nu\mu} F_\beta{}^{\rho\sigma} = -C_{\gamma\alpha\beta} F_{\gamma\nu\mu} F_\beta{}^{\rho\sigma} \,,$$

we see that $J_\alpha{}^\nu$ satisfies a gauge-covariant conservation law

$$D_\nu J_\alpha{}^\nu = 0 \,, \tag{15.3.8}$$

rather than the ordinary conservation law (15.3.4) obeyed by the full current $\mathscr{J}_\alpha{}^\nu$. Also, it is straightforward (using Eq. (15.1.5)) to derive the identities:

$$D_\mu \, F_{\alpha\nu\lambda} + D_\nu \, F_{\alpha\lambda\mu} + D_\lambda \, F_{\alpha\mu\nu} = 0 \,, \tag{15.3.9}$$

which hold whether or not the gauge fields satisfy the field equations.

These results serve to underscore the profound analogy mentioned in Section 15.1 between non-Abelian gauge theories and general relativity. In general relativity there is a matter energy-momentum tensor $T^\nu{}_\mu$, analogous to J^μ, which satisfies a generally covariant conservation law $T^\nu{}_{\mu;\nu} = 0$, and stands on the right-hand side of the Einstein field equations in their generally covariant form, $R^\nu{}_\mu - \frac{1}{2}\delta^\nu{}_\mu R = -8\pi G T^\nu{}_\mu$. However, $T^\nu{}_\mu$ is not conserved in the ordinary sense: $\partial_\nu T^\nu{}_\mu$ does not vanish. On the other hand, moving the non-linear terms on the left-hand side of the Einstein equation to the right-hand side gives a field equation[8]

$$\left(R^\nu{}_\mu - \frac{1}{2}\,\delta^\nu{}_\mu R\right)_{\text{LINEAR}} = -8\pi G\, \tau^\nu{}_\mu \,,$$

where $\tau^\nu{}_\mu$ is the non-tensor

$$\tau^\nu{}_\mu \equiv T^\nu{}_\mu + \frac{1}{8\pi G}\left(R^\nu{}_\mu - \frac{1}{2}\,\delta^\nu{}_\mu R\right)_{\text{NONLINEAR}} ,$$

analogous to $\mathscr{J}_\alpha{}^\nu$. Like $\mathscr{J}_\alpha{}^\nu$, $\tau^\nu{}_\mu$ is conserved in the ordinary sense

$$\partial_\nu \tau^\nu{}_\mu = 0$$

and may be regarded as the current of energy and momentum:

$$P_\mu = \int \tau^0{}_\mu\, d^3x \,.$$

It contains a purely gravitational term, because gravitational fields carry energy and momentum; without this term, $\tau^\nu{}_\mu$ could not be conserved. Similarly, $\mathscr{J}^\nu_\alpha$ contains a gauge-field term (the first term on the right in Eq. (15.3.3)) because for non-Abelian groups (those with $C^\gamma_{\alpha\beta} \neq 0$) the gauge fields carry the quantum numbers with which they interact. Because $\mathscr{J}_\alpha{}^\nu$ is conserved in the ordinary sense, it can be regarded as the current of these quantum numbers, with the symmetry generators given by the time-independent quantities

$$T_\alpha = \int \mathscr{J}_\alpha{}^0 d^3x \,. \tag{15.3.10}$$

(Also, the homogeneous equations (15.3.9) involve covariant derivatives, just as do the Bianchi identities of general relativity.) In contrast, none of these complications arises in quantum electrodynamics, because photons do not carry the quantum number, electric charge, with which they interact.

15.4 Quantization

We now proceed to quantize the gauge theories described in the previous two sections. The Lagrangian density is taken in the form (15.3.1):

$$\mathscr{L} = -\tfrac{1}{4}\, F_{\alpha\mu\nu}F_\alpha{}^{\mu\nu} + \mathscr{L}_M(\psi, D_\mu\psi) \,, \tag{15.4.1}$$

with

$$F_{\alpha\mu\nu} \equiv \partial_\mu A_{\alpha\nu} - \partial_\nu A_{\alpha\mu} + C_{\alpha\beta\gamma}A_{\beta\mu}A_{\gamma\nu} \,,$$
$$D_\mu\psi \equiv \partial_\mu\psi - it_\alpha A_{\alpha\mu}\psi \,.$$

We cannot immediately quantize this theory by setting commutators equal to i times the corresponding Poisson brackets. The problem is one of constraints. In the terminology of Dirac, described in Section 7.6, there is a primary constraint that

$$\Pi_{\alpha 0} \equiv \frac{\partial \mathscr{L}}{\partial(\partial_0 A^0_\alpha)} = 0 \tag{15.4.2}$$

and a secondary constraint provided by the field equation for A^0_α:

$$\begin{aligned} -\partial_\mu \frac{\partial \mathscr{L}}{\partial(\partial_\mu A_{\alpha 0})} + \frac{\partial \mathscr{L}}{\partial A_{\alpha 0}} &= \partial_\mu F_\alpha{}^{\mu 0} + F_\gamma{}^{\mu 0} C_{\gamma\alpha\beta} A_{\beta\mu} + J_\alpha{}^0 \\ &= \partial_k \Pi_\alpha{}^k + \Pi_\gamma{}^k C_{\gamma\alpha\beta} A_{\beta\, k} + J_\alpha{}^0 = 0\,, \end{aligned} \tag{15.4.3}$$

where $\Pi^k_\alpha \equiv \partial\mathscr{L}/\partial(\partial_0 A_{\alpha k}) = F^{k0}_\alpha$ is the 'momentum' conjugate to $A_{\alpha\, k}$, with k running over the values 1, 2, 3. The Poisson brackets of $\Pi_{\alpha 0}$ and $\partial_k \Pi_\alpha{}^k + \Pi_\gamma{}^k C_{\gamma\alpha\beta} A_{\beta\, k} + J_\alpha{}^0$ vanish (because the latter quantity is independent of $A_\alpha{}^0$), so these are first class constraints, which cannot be dealt with by replacing Poisson brackets with Dirac brackets.

As in the case of electrodynamics, we deal with these constraints by choosing a gauge. The Coulomb gauge adopted for electrodynamics would lead to painful complications here,* so instead we will work in what is known as *axial gauge*, based on the condition

$$A_{\alpha 3} = 0\,. \tag{15.4.4}$$

The canonical variables of the gauge field are then $A_{\alpha i}$, with i now running over the values 1 and 2, together with their canonical conjugates

$$\Pi_{\alpha i} \equiv \frac{\partial \mathscr{L}}{\partial(\partial_0 A_{\alpha i})} = -F_\alpha{}^{0i} = \partial_0 A_{\alpha i} - \partial_i A_{\alpha 0} + C_{\alpha\beta\gamma} A_{\beta 0} A_{\gamma i}\,. \tag{15.4.5}$$

The field $A_{\alpha 0}$ is not an independent canonical variable, but rather is defined in terms of the other variables by the constraint (15.4.3). To see this, note that the 'electric' field strengths $F_\alpha{}^{\mu 0}$ are

$$F_\alpha{}^{i0} = \Pi_{\alpha i}\,, \quad F_\alpha{}^{30} = \partial_3 A_\alpha{}^0\,, \tag{15.4.6}$$

so the constraint (15.4.3) reads

$$-(\partial_3)^2 A^0_\alpha = \partial_i \Pi_{\alpha i} + \Pi_{\gamma i} C_{\gamma\alpha\beta} A_{\beta i} + J_\alpha{}^0\,, \tag{15.4.7}$$

which can easily be solved (with reasonable boundary conditions) to give $A_\alpha{}^0$ as a functional of $\Pi_{\gamma i}$, $A_{\beta i}$, and $J_\alpha{}^0$. (We are using a summation

* In addition to purely algebraic complications, Coulomb gauge (like many other gauges) has a problem known as the Gribov ambiguity:[9] even with the condition that $\mathbf{A}_\alpha$ vanishes at spatial infinity, for each solution of the Coulomb gauge condition $\nabla \cdot \mathbf{A}_\alpha = 0$ there are other solutions that differ by finite gauge transformations. The Gribov ambiguity will not bother us here, because we quantize in axial gauge where it is absent, and we shall use other gauges like Lorentz gauge only to generate a perturbation series.

convention, with indices i, j, etc. summed over the values 1 and 2.) It should be noted that the canonical conjugate to the matter field ψ_ℓ is

$$\pi_\ell = \frac{\partial \mathscr{L}}{\partial(\partial_0 \psi_\ell)} = \frac{\partial \mathscr{L}_M}{\partial(D_0 \psi_\ell)}\,, \tag{15.4.8}$$

so the time component of the matter current can be expressed in terms of the canonical variables of the matter fields *alone*

$$J_\alpha{}^0 = -i\,\frac{\partial \mathscr{L}_m}{\partial(D_0\psi)_\ell}\,(t_\alpha)_\ell{}^m \psi_m = -i\,\pi_\ell (t_\alpha)_\ell{}^m \psi_m\,. \tag{15.4.9}$$

Hence Eq. (15.4.7) defines $A_\alpha{}^0$ at a given time as a functional of the canonical variables $\Pi_{\gamma i}$, $A_{\beta i}$, π_ℓ, and ψ_m at the same time.

Now that we have identified the canonical variables in this gauge, we can proceed to the construction of a Hamiltonian. The Hamiltonian density is

$$\begin{aligned}\mathscr{H} &= \Pi_{\alpha i}\partial_0 A_{\alpha i} + \pi_\ell \partial_0 \psi_\ell - \mathscr{L} \\ &= \Pi_{\alpha i}\left(F_{\alpha 0 i} + \partial_i A_{\alpha 0} - C_{\alpha\beta\gamma} A_{\beta 0} A_{\gamma i}\right) + \pi_\ell \partial_0 \psi_\ell \\ &\quad - \tfrac{1}{2}\, F_{\alpha 0 i} F_{\alpha 0 i} + \tfrac{1}{2}\, F_{\alpha ij} F_{\alpha ij} + \tfrac{1}{2}\, F_{\alpha i3} F_{\alpha i3} \\ &\quad - \tfrac{1}{2}\, F_{\alpha 03} F_{\alpha 03} - \mathscr{L}_M\,. \end{aligned} \tag{15.4.10}$$

Using Eqs. (15.4.4) and (15.4.6), this is

$$\begin{aligned}\mathscr{H} &= \mathscr{H}_M + \Pi_{\alpha i}(\partial_i A_{\alpha 0} - C_{\alpha\beta\gamma} A_{\beta 0} A_{\gamma i}) + \tfrac{1}{2}\,\Pi_{\alpha i}\Pi_{\alpha i} \\ &\quad + \tfrac{1}{4}\, F_{\alpha ij} F_{\alpha ij} + \tfrac{1}{2}\,\partial_3 A_{\alpha i}\partial_3 A_{\alpha i} - \tfrac{1}{2}\partial_3 A_{\alpha 0}\partial_3 A_{\alpha 0}\,, \end{aligned} \tag{15.4.11}$$

where $\mathscr{H}_M$ is the matter Hamiltonian density:

$$\mathscr{H}_M \equiv \pi_\ell\, \partial_0 \psi_\ell - \mathscr{L}_M\,. \tag{15.4.12}$$

Following the general rules derived in Section 9.2, we can now use this Hamiltonian density to calculate matrix elements as path integrals over $A_{\alpha i}$, $\Pi_{\alpha i}$, ψ_ℓ, and π_ℓ, with weighting factor $\exp(iI)$, where

$$I = \int d^4x \Big[\Pi_{\alpha i}\partial_0 A_{\alpha i} + \pi_\ell \partial_0 \psi_\ell - \mathscr{H} + \epsilon \text{ terms}\Big]\,, \tag{15.4.13}$$

in which the 'ϵ terms' serve only to supply the correct imaginary infinitesimal terms in propagator denominators. (See Section 9.2.) We note that Eqs. (15.4.7) and (15.4.9) give A_α^0 as a functional of the canonical variables, *linear* in $\Pi_{\alpha i}$ and π_ℓ. Inspection of Eq. (15.4.11) shows then (assuming $\mathscr{L}_M$ to be no more than quadratic in $D_\mu\psi$) that the integrand of the complete action (15.4.13) is no more than quadratic in $\Pi_{\alpha i}$ and π_ℓ. We could therefore carry out the path integral over these canonical 'momenta' by the usual rules of Gaussian integration. The trouble with this procedure is that the coefficients of the terms in Eq. (15.4.13) of second order in $\Pi_{\alpha i}$ are functions of the $A_{\alpha i}$, so the Gaussian integral would yield an awkward field-dependent determinant factor. Also, the whole formalism at this point looks hopelessly non-Lorentz-invariant.

Instead of proceeding in this way, we will apply a trick like that used in the path integral formulation of electrodynamics in Section 9.6. Note that if for a moment we think of $A_{\alpha 0}$ as an independent variable, then the action (15.4.13) is evidently quadratic in $A_{\alpha 0}$, with the coefficient of the second-order term $A_{\alpha 0}(x)A_{\beta 0}(y)$ equal to the field-independent kernel $(\partial_3)^2\delta^4(x-y)$. As we saw in the appendix to Chapter 9, the integral of such a Gaussian over $A_{\alpha 0}(x)$ is, up to a constant factor, equal to the value of the integrand at the stationary 'point' of the argument of the exponential. But the variational derivative of the action here is

$$\frac{\delta I}{\delta A_{\alpha 0}} = -\frac{\partial \mathscr{H}}{\partial A_{\alpha 0}} = J_\alpha{}^0 + \partial_i \Pi_{\alpha i} + C_{\beta\alpha\gamma}\Pi_{\beta i}A_{\gamma i} - \partial_3^2 A_{\alpha 0}\,,$$

so the stationary 'point' of the action is the solution of the constraint equation (15.4.7). Hence, instead of using for $A_{\alpha 0}$ the solution of Eq. (15.4.7), we can just as well treat it as an independent variable of integration.

With $A_{\alpha 0}$ now regarded as an independent variable, the Hamiltonian $\int d^3x \mathscr{H}$ is evidently quadratic in $\Pi_{\alpha i}$, with the coefficient of the second-order term $\Pi_{\alpha i}(x)\Pi_{\beta j}(y)$ given by the field-independent kernel $\frac{1}{2}\delta^4(x-y)\delta_{ij}$. Assuming that the same is true for the matter variable π_ℓ, we can evaluate path integrals over π_ℓ and $\Pi_{\alpha i}$ up to a constant factor by simply setting π_ℓ and $\Pi_{\alpha i}$ at the stationary 'points' of the action corresponding to Eq. (15.4.1):

$$0 = \frac{\delta I}{\delta \pi_\ell} = \partial_0 \psi_\ell - \frac{\partial \mathscr{H}_M}{\partial \pi_\ell}\,,$$

$$0 = \frac{\delta I}{\delta \Pi_{\alpha i}} = \partial_0 A_{\alpha i} - \Pi_{\alpha i} - \partial_i A_{\alpha 0} + C_{\alpha\beta\gamma}A_{\beta 0}A_{\gamma i} = F_{\alpha 0 i} - \Pi_{\alpha i}\,.$$

Inserting these back into Eq. (15.4.13) gives

$$\begin{aligned} I = \int d^4x \Big[&\mathscr{L}_M + \tfrac{1}{2}\, F_{\alpha 0i}\, F_{\alpha 0i} \\ &- \tfrac{1}{4}\, F_{\alpha ij}F_{\alpha ij} - \tfrac{1}{2}\partial_3 A_{\alpha i}\partial_3 A_{\alpha i} + \tfrac{1}{2}(\partial_3 A_{\alpha 0})^2\Big] \\ = \int d^4x\, \mathscr{L}\,, \end{aligned} \tag{15.4.14}$$

where $\mathscr{L}$ is the Lagrangian (15.3.1) with which we started! In other words, *we are to do path integrals over $\psi_\ell(x)$ and all four components of $A_{\alpha\mu}(x)$, with a manifestly covariant weighting factor* $\exp(iI)$ *given by Eqs. (15.4.14) and (15.3.1)*, but with the axial-gauge condition enforced by inserting a factor

$$\prod_{x,\alpha} \delta\Big(A_{\alpha 3}(x)\Big) . \tag{15.4.15}$$

As long as $\mathcal{O}_A, \mathcal{O}_B \cdots$ are gauge-invariant, we have

$$\begin{aligned}\langle T\{\mathcal{O}_A\mathcal{O}_B\cdots\}\rangle_{\text{VACUUM}} \propto & \int \left[\prod_{\ell,x} d\psi_\ell(x)\right] \left[\prod_{\alpha,\mu,x} dA_{\alpha\mu}(x)\right] \\ & \times \mathcal{O}_A\mathcal{O}_B\cdots \exp\{iI + \epsilon \text{ terms}\} \prod_{x,\alpha} \delta\left(A_{\alpha 3}(x)\right) ,\end{aligned} \tag{15.4.16}$$

with Lorentz- and gauge-invariant action I given by Eq. (15.4.14).

* * *

For future reference, we note that the volume element $\prod_{\alpha,\mu,x} dA_{\alpha\mu}(x)$ for the integration over gauge fields in (15.4.16) is gauge-invariant, in the sense that

$$\prod_{\alpha,\mu,x} dA_{\Lambda\,\alpha\mu}(x) = \prod_{\alpha,\mu,x} dA_{\alpha\mu}(x) , \tag{15.4.17}$$

where $A_{\Lambda\,\alpha\mu}(x)$ is the result of acting on $A_{\alpha\mu}(x)$ with a gauge transformation having transformation parameters $\Lambda_\alpha(x)$. It will be enough to show that this is true for transformations near the identity, say with infinitesimal transformation parameters $\lambda_\alpha(x)$. In this case,

$$A^\mu_{\lambda\,\alpha} = A^\mu_\alpha + \partial^\mu \lambda_\alpha + C_{\alpha\beta\gamma}\, A^\mu_\beta \lambda_\gamma ,$$

so the volume elements are related by

$$\prod_{\alpha,\mu,x} dA_{\lambda\,\alpha\mu}(x) = \text{Det}(\mathcal{N}) \prod_{\alpha,\mu,x} dA_{\alpha\mu}(x) ,$$

where $\mathcal{N}$ is the 'matrix':

$$\mathcal{N}_{\alpha\mu x,\beta\nu y} = \frac{\delta A_{\lambda\,\alpha\mu}(x)}{\delta A_{\beta\nu}(y)} = \delta^4(x-y)\,\delta^\nu_\mu \left[\delta_{\alpha\beta} + C_{\alpha\beta\gamma}\lambda_\gamma(x)\right] .$$

The determinant of $\mathcal{N}$ is unity to first order in λ_γ because the trace $C_{\alpha\alpha\gamma}$ vanishes.

In this chapter we shall assume that the volume element $\prod_{n,x} d\psi_n(x)$ for the integration over matter fields is also gauge-invariant. There are important subtleties here, to which we shall return in Chapter 22, but as shown there this assumption turns out to be valid in our present non-Abelian gauge theories of strong and electroweak interactions.

15.5 The De Witt–Faddeev–Popov Method

Our formula (15.4.16) for the path integral was derived in a gauge that is convenient for canonical quantization, but the Feynman rules that would be derived from this formula would hide the underlying rotational and Lorentz invariance of the theory. In order to derive manifestly Lorentz-invariant Feynman rules, we need to change the gauge.

We first note that Eq. (15.4.16) is (up to an unimportant constant factor) a special case of a general class of functional integrals, of the form:

$$\mathscr{I} = \int \left[\prod_{n,x} d\phi_n(x)\right] \mathscr{G}[\phi]\, B\Big[f[\phi]\Big]\, \mathrm{Det}\,\mathscr{F}[\phi]\,, \tag{15.5.1}$$

where $\phi_n(x)$ are a set of gauge and matter fields; $\prod_{n,x} d\phi_n(x)$ is a volume element; and $\mathscr{G}[\phi]$ is a functional of the $\phi_n(x)$, satisfying the gauge-invariance condition:

$$\mathscr{G}[\phi_\lambda] \prod_{n,x} d\phi_{\lambda n}(x) = \mathscr{G}[\phi] \prod_{n,x} d\phi_n(x)\,, \tag{15.5.2}$$

where $\phi_{\lambda n}(x)$ is the result of operating on ϕ with a gauge transformation having parameters $\lambda_\alpha(x)$. (Usually when this is satisfied both the functional $\mathscr{G}$ and the volume element are separately invariant, but Eq. (15.5.2) is all we need here.) Also, $f_\alpha[\phi; x]$ is a non-gauge-invariant 'gauge-fixing functional' of these fields that also depends on x and α; $B[f]$ is some numerical functional defined for general functions $f_\alpha(x)$ of x and α; and $\mathscr{F}$ is the 'matrix':

$$\mathscr{F}_{\alpha x, \beta y}[\phi] \equiv \left.\frac{\delta f_\alpha[\phi_\lambda; x]}{\delta \lambda_\beta(y)}\right|_{\lambda=0}\,. \tag{15.5.3}$$

(In accordance with our usual notation for functionals of functions or of functionals, $B\Big[f[\phi]\Big]$ is understood to depend on the values taken by $f_\alpha[\phi; x]$ for all values of the undisplayed variables α and x, with the displayed variable, the function $\phi_n(x)$, held fixed.) Eq. (15.5.1) does not represent the widest possible generalization of Eq. (15.4.16); we will see in Section 15.7 that there is a further generalization that is needed for some purposes. We start here with Eq. (15.5.1) because it will help to motivate the formalism of Section 15.7, and it is adequate for dealing with non-Abelian gauge theories in the most convenient gauges.

We now must check that the path integral (15.4.16) is in fact a special case of Eq. (15.5.1). In Eq. (15.4.16) the fields $\phi_n(x)$ consist of both $A_{\alpha\mu}(x)$ and matter fields $\psi_\ell(x)$, and

$$f_\alpha[A, \psi; x] = A_{\alpha 3}(x)\,, \tag{15.5.4}$$

$$B[f] = \prod_{x,\alpha} \delta\Big(f_\alpha(x)\Big)\,, \tag{15.5.5}$$

$$\mathscr{G}[A,\psi] = \exp\{iI + \epsilon \text{ terms}\}\, \mathscr{O}_A \mathscr{O}_B \cdots \,, \tag{15.5.6}$$

$$\prod_{n,x} d\phi_n(x) = \left[\prod_{\ell,x} d\psi_\ell(x)\right] \left[\prod_{\alpha,\mu,x} dA^\mu_\alpha(x)\right] . \tag{15.5.7}$$

(We are now dropping the distinction between upper and lower indices $\alpha, \beta, \cdots$.) Comparison of Eq. (15.4.16) with Eqs. (15.5.1)–(15.5.3) shows that these path integrals are indeed the same, aside from the factor Det $\mathscr{F}[\phi]$. For the particular gauge-fixing functional (15.5.4), this factor is field-independent: if $A^3_\alpha(x) = 0$, then the change in $A^3_\alpha(x)$ under a gauge transformation with parameters $\lambda_\alpha(x)$ is

$$A^3_{\lambda\alpha}(x) = \partial_3\, \lambda_\alpha(x) = \int d^4y\; \lambda_\alpha(y)\, \partial_3 \delta^4(x-y)\,,$$

so that here Eq. (15.5.3) is the field-independent 'matrix'

$$\mathscr{F}_{\alpha x, \beta y}[\phi] = \delta_{\alpha\beta}\, \partial_3 \delta^4(x-y)\,.$$

The determinant in Eq. (15.5.1) is therefore also field-independent in this gauge. As discussed in Chapter 9, field-independent factors in the functional integral affect only the vacuum-fluctuation part of expectation values and S-matrix elements, and so are irrelevant to the calculation of the connected parts of the S-matrix.

The point of recognizing the functional integral (15.4.16) for non-Abelian gauge theories as a special case of the general path integral (15.5.1) is that in this form we may freely change the gauge. Specifically, we have a theorem, that *the integral (15.5.1) is actually independent (within broad limits) of the gauge-fixing functional $f_\alpha[\phi; x]$, and depends on the choice of the functional $B[f]$ only through an irrelevant constant factor.*

Proof: Replace the integration variable ϕ everywhere in Eq. (15.5.1) with a new integration variable ϕ_Λ, with $\Lambda^\alpha(x)$ any arbitrary (but fixed) set of gauge transformation parameters:

$$\mathscr{I} = \int \left[\prod_{n,x} d\phi_{\Lambda n}(x)\right] \mathscr{G}[\phi_\Lambda] B\Big[f[\phi_\Lambda]\Big] \text{Det}\, \mathscr{F}[\phi_\Lambda]\,. \tag{15.5.8}$$

(This step is a mathematical triviality, like changing an integral $\int_{-\infty}^{\infty} f(x)dx$ to read $\int_{-\infty}^{\infty} f(y)dy$, and does not yet make use of our assumptions regarding gauge invariance.) Now use the assumed gauge invariance (15.5.2) of the measure $\Pi d\phi$ times the functional $\mathscr{G}[\phi]$ to rewrite this as

$$\mathscr{I} = \int \left[\prod_{n,x} d\phi_n(x)\right] \mathscr{G}[\phi] B\Big[f[\phi_\Lambda]\Big]\, \text{Det}\, \mathscr{F}[\phi_\Lambda]\,. \tag{15.5.9}$$

Since $\Lambda^\alpha(x)$ was arbitrary, the left-hand side here cannot depend on it. Integrating over $\Lambda^\alpha(x)$ with some suitable weight-functional $\rho[\Lambda]$ (to be chosen below) thus gives

$$\mathscr{I}\int\left[\prod_{\alpha,x}d\Lambda^\alpha(x)\right]\rho[\Lambda]=\int\left[\prod_{n,x}d\phi_n(x)\right]\mathscr{G}[\phi]C[\phi]\,, \tag{15.5.10}$$

where

$$C[\phi]\equiv\int\left[\prod_{\alpha,x}d\Lambda^\alpha(x)\right]\rho[\Lambda]\,B\Big[f[\phi_\Lambda]\Big]\text{Det}\,\mathscr{F}[\phi_\Lambda]\,. \tag{15.5.11}$$

Now, Eq. (15.5.3) gives

$$\mathscr{F}_{\alpha x,\beta y}[\phi_\Lambda]=\left.\frac{\delta f_\alpha[(\phi_\Lambda)_\lambda;x]}{\delta\lambda^\beta(y)}\right|_{\lambda=0}\,. \tag{15.5.12}$$

We are assuming that these transformations form a group; that is, we may write the result of performing the gauge transformation with parameters $\Lambda^\alpha(x)$ followed by the gauge transformation with parameters $\lambda^\alpha(x)$ as the action of a single 'product' gauge transformation with parameters $\tilde{\Lambda}^\alpha(x;\Lambda,\lambda)$,

$$(\phi_\Lambda)_\lambda=\phi_{\tilde{\Lambda}(\Lambda,\lambda)}\,. \tag{15.5.13}$$

Using the chain rule of partial (functional) differentiation, we have then

$$\mathscr{F}_{\alpha x,\beta y}[\phi_\Lambda]=\int\mathscr{J}_{\alpha x,\gamma z}[\phi,\Lambda]\mathscr{R}^{\gamma z}{}_{\beta y}[\Lambda]d^4z\,, \tag{15.5.14}$$

where

$$\mathscr{J}_{\alpha x,\gamma z}[\phi,\Lambda]\equiv\left.\frac{\delta f_\alpha[\phi_{\tilde{\Lambda}};x]}{\delta\tilde{\Lambda}^\gamma(z)}\right|_{\tilde{\Lambda}=\Lambda}=\frac{\delta f_\alpha[\phi_\Lambda;x]}{\delta\Lambda^\gamma(z)} \tag{15.5.15}$$

and

$$\mathscr{R}^{\gamma z}{}_{\beta y}[\Lambda]=\left.\frac{\delta\tilde{\Lambda}^\gamma(z;\Lambda,\lambda)}{\delta\lambda^\beta(y)}\right|_{\lambda=0}\,. \tag{15.5.16}$$

It follows that

$$\text{Det}\,\mathscr{F}[\phi_\Lambda]=\text{Det}\,\mathscr{J}[\phi,\Lambda]\;\;\text{Det}\,\mathscr{R}[\Lambda]\,. \tag{15.5.17}$$

We note that Det $\mathscr{J}[\phi,\Lambda]$ is nothing but the Jacobian of the transformation of integration variables from the $\Lambda^\alpha(x)$ to (for a fixed ϕ) the $f_\alpha[\phi_\Lambda;x]$. Hence, if we choose the weight-function $\rho(\Lambda)$ as

$$\rho(\Lambda)=1\Big/\text{Det}\,\mathscr{R}[\Lambda] \tag{15.5.18}$$

then

$$C[\phi]=\int\left[\prod_{\alpha,x}d\Lambda^\alpha(x)\right]\text{Det}\,\mathscr{J}[\phi,\Lambda]\,B\Big[f[\phi_\Lambda]\Big]$$

$$=\int\left[\prod_{\alpha,x}df_\alpha(x)\right]\,B\,[f]\equiv C\,, \tag{15.5.19}$$

which is clearly independent of ϕ. (Eq. (15.5.18) may be recognized by the reader as giving the invariant (Haar) measure on the space of group parameters.) We have then at last

$$\mathscr{I} = \frac{C\int\left[\prod_{n,x} d\phi_n(x)\right]\mathscr{G}[\phi]}{\int\left[\prod_{\alpha,x} d\Lambda^\alpha(x)\right]\rho[\Lambda]} . \tag{15.5.20}$$

This is clearly independent of our choice of $f_\alpha[\phi;x]$, which has been reduced to a mere variable of integration, and it depends on $B[f]$ only through the constant C, as was to be proved.

Before proceeding with the applications of this theorem, we should pause to note a tricky point in the derivation. The integrals in the numerator and denominator of Eq. (15.5.20) are both ill-defined for the same reason. Since $\mathscr{G}[\phi]$ is assumed to be gauge-invariant, its integral over ϕ cannot possibly converge; the integrand is constant along all 'orbits,' obtained by sending ϕ into ϕ_λ with all possible $\lambda^\alpha(x)$. Likewise, the integrand in the denominator is divergent, because $\rho(\Lambda)\Pi d\Lambda$ is nothing but the usual invariant volume element for integrating over the group, and this is also constant along 'orbits' $\Lambda \to \tilde{\Lambda}(\Lambda,\lambda)$. This divergence can be eliminated in both the numerator and denominator of Eq. (15.5.20) by formulating the theory on a finite spacetime lattice, in which case the volume of the gauge group is just the volume of the global Lie group itself times the number of lattice sites. Because the gauge-fixing factor $B[f]$ eliminates this divergence in the original definition (15.5.1) of the left-hand side of Eq. (15.5.20), we may presume that, as the number of lattice sites goes to infinity, it cancels between the numerator and denominator of the right-hand side of Eq. (15.5.20).

Now to the point. We have seen that the vacuum expectation value (15.4.16) in axial gauge is given by a functional integral of the general form (15.5.1). Armed with the above theorem, we conclude then that

$$\langle T\{\mathscr{O}_A\mathscr{O}_B\cdots\}\rangle_V \propto \int\left[\prod_{\ell,x} d\psi_\ell(x)\right]\left[\prod_{\alpha,\mu,x} dA^\mu{}_\alpha(x)\right]$$
$$\times\ \mathscr{O}_A\mathscr{O}_B\cdots\exp\{iI+\epsilon\text{ terms}\}\ B\Big[f[A,\psi]\Big]\text{Det}\,\mathscr{F}[A,\psi] \tag{15.5.21}$$

for (almost) any choice of $f_\alpha[A,\psi;x]$ and $B[f]$. We are now therefore free to use Eq. (15.5.21) to derive the Feynman rules in a more convenient gauge.

The path integrals that we understand how to calculate are of Gaussians times polynomials, so we will generally take

$$B[f] = \exp\left(-\frac{i}{2\xi}\int d^4x\, f_\alpha(x)f_\alpha(x)\right) \tag{15.5.22}$$

with arbitrary real parameter ξ. With this choice, the effect of the factor B in Eq. (15.5.21) is just to add a term to the effective Lagrangian

$$\mathscr{L}_{\rm EFF} = \mathscr{L} - \frac{1}{2\xi} f_\alpha f_\alpha \,. \tag{15.5.23}$$

The simplest Lorentz-invariant choice of the gauge-fixing function f_α is the same as in electrodynamics:

$$f_\alpha = \partial_\mu A^\mu_\alpha \,. \tag{15.5.24}$$

The bare gauge-field propagator can then be calculated just as in quantum electrodynamics. The free-vector-boson part of the effective action can be written

$$\begin{aligned} I_{0A} &= -\int d^4x \Big[\frac{1}{4}(\partial_\mu A_{\alpha\nu} - \partial_\nu A_{\alpha\mu})(\partial^\mu A_\alpha{}^\nu - \partial^\nu A_\alpha{}^\mu) \\ &\qquad + \frac{1}{2\xi}(\partial_\mu A_\alpha{}^\mu)(\partial_\nu A_\alpha{}^\nu) \,+\, \epsilon \text{ terms}\Big] \\ &= -\frac{1}{2}\int d^4x \, \mathscr{D}_{\alpha\mu x,\beta\nu y}\, A_\alpha{}^\mu(x) A_\beta{}^\nu(y) \,, \end{aligned}$$

where

$$\begin{aligned} \mathscr{D}_{\alpha\mu x,\beta\nu y} &= \eta_{\mu\nu}\, \frac{\partial^2}{\partial x^\lambda \partial y_\lambda}\, \delta^4(x-y)\delta_{\alpha\beta} \\ &\quad - \left(1 - \frac{1}{\xi}\right) \frac{\partial^2}{\partial x^\mu \partial y^\nu}\, \delta^4(x-y)\delta_{\alpha\beta} \,+\, \epsilon \text{ terms} \\ &= (2\pi)^{-4}\delta_{\alpha\beta} \int d^4p \left[\eta_{\mu\nu}(p^2 - i\epsilon) - \left(1 - \frac{1}{\xi}\right) p_\mu p_\nu\right] e^{ip\cdot(x-y)} \,. \end{aligned}$$

Taking the reciprocal of the matrix in square brackets, we find the propagator:

$$\begin{aligned} \Delta_{\alpha\mu,\beta\nu}(x,y) &= (\mathscr{D}^{-1})_{\alpha\mu x,\beta\nu y} \\ &= (2\pi)^{-4} \int d^4p \left[\eta_{\mu\nu} + (\xi - 1)\frac{p_\mu p_\nu}{p^2}\right] \frac{e^{ip\cdot(x-y)}}{p^2 - i\epsilon} \,. \end{aligned} \tag{15.5.25}$$

This is a generalization of both Landau and Feynman gauges, which are recovered by taking $\xi = 0$ and $\xi = 1$, respectively. For $\xi \to 0$, the functional (15.5.22) oscillates very rapidly except near $f_\alpha = 0$, so this functional acts like a delta-function imposing the Landau gauge condition $\partial_\mu A^\mu = 0$, leading naturally to a propagator satisfying the corresponding condition $\partial^\mu \Delta_{\alpha\mu,\beta\nu} = 0$. For non-zero values of ξ the functional $B[f]$ does not pick out gauge fields satisfying any specific gauge condition on the field $A_{\alpha\mu}$, but it is common to refer to the propagator (15.5.25) as being in a 'generalized Feynman gauge' or 'generalized ξ-gauge'. It is often a good strategy to calculate physical amplitudes with ξ left arbitrary, and then at the end of the calculation check that the results are ξ-independent.

With one qualification, the Feynman rules are now obvious: the contributions of vertices are to be read off from the interaction terms in the original Lagrangian $\mathscr{L}$, with gauge-field propagators given by Eq. (15.5.25), and matter-field propagators calculated as before. To be specific, the trilinear interaction term in $\mathscr{L}$

$$-\tfrac{1}{2}\, C_{\alpha\beta\gamma}(\partial_\mu A_{\alpha\nu} - \partial_\nu A_{\alpha\mu})A_\beta{}^\mu A_\gamma{}^\nu$$

corresponds to a vertex to which are attached three vector boson lines. If these lines carry (incoming) momenta p, q, k and Lorentz and gauge-field indices $\mu\alpha, \nu\beta, \lambda\gamma$, then according to the momentum-space Feynman rules, the contribution of such a vertex to the integrand is

$$i(2\pi)^4\delta^4(p+q+k)\,[-i\,C_{\alpha\beta\gamma}]\,\Big[p_\nu\eta_{\mu\lambda} - p_\lambda\eta_{\mu\nu} + q_\lambda\eta_{\nu\mu} - q_\mu\eta_{\nu\lambda} + k_\mu\eta_{\lambda\nu} - k_\nu\eta_{\lambda\mu}\Big]\,. \tag{15.5.26}$$

Also, the A^4 interaction term in $\mathscr{L}$,

$$-\tfrac{1}{4}\, C_{\epsilon\alpha\beta}C_{\epsilon\gamma\delta}\, A_{\alpha\mu}A_{\beta\nu}A_\gamma{}^\mu A_\delta{}^\nu\,,$$

corresponds to a vertex to which are attached four vector boson lines. If these lines carry (incoming) momenta p, q, k, ℓ, and Lorentz and gauge indices $\mu\alpha$, $\nu\beta$, $\rho\gamma$, and $\sigma\delta$, then the contribution of such a vertex to the integrand is

$$i(2\pi)^4\delta^4(p+q+k+\ell) \times \Big[-C_{\epsilon\alpha\beta}\,C_{\epsilon\gamma\delta}(\eta_{\mu\rho}\eta_{\nu\sigma} - \eta_{\mu\sigma}\eta_{\nu\rho})$$
$$-C_{\epsilon\alpha\gamma}\,C_{\epsilon\delta\beta}(\eta_{\mu\sigma}\eta_{\rho\nu} - \eta_{\mu\nu}\eta_{\sigma\rho}) - C_{\epsilon\alpha\delta}\,C_{\epsilon\beta\gamma}(\eta_{\mu\nu}\eta_{\rho\sigma} - \eta_{\mu\rho}\eta_{\sigma\nu})\Big]\,. \tag{15.5.27}$$

(Recall that the structure constants $C_{\alpha\beta\gamma}$ contain coupling constant factors, so the factors (15.5.26) and (15.5.27) are respectively of first and second order in coupling constants.)

The one complication in the Feynman rules with which we have not yet dealt is the presence in Eq. (15.5.21) of the factor Det $\mathscr{F}$, which for general gauges is *not* a constant. We now turn to a consideration of this factor.

15.6 Ghosts

We now consider the effect of the factor Det $\mathscr{F}$ in Eq. (15.5.21) on the Feynman rules for a non-Abelian gauge theory. In order to be able to treat this effect as a modification of the Feynman rules, recall that as shown in

Section 9.5, the determinant of any matrix $\mathscr{F}_{\alpha x,\beta y}$ may be expressed as a path integral

$$\text{Det}\,\mathscr{F} \propto \int \left[\prod_{\alpha,x} d\omega^*_\alpha(x)\right] \left[\prod_{\alpha,x} d\omega_\alpha(x)\right] \exp(iI_{GH})\,, \tag{15.6.1}$$

where

$$I_{GH} \equiv \int d^4x\, d^4y\; \omega^*_\alpha(x)\,\omega_\beta(y)\,\mathscr{F}_{\alpha x,\beta y}\,. \tag{15.6.2}$$

Here ω^*_α and ω_α are a set of independent anticommuting classical variables, and the constant of proportionality is field-independent. (We have to choose the ω_α and ω^*_α field variables to be fermionic in order to reproduce the factor Det $\mathscr{F}$; had we chosen these field variables to be bosonic, the path integral (15.6.1) would have been proportional to (Det $\mathscr{F})^{-1}$.) The fields ω^*_α and ω_α are not necessarily related by complex conjugation; indeed, in Section 15.7 we shall see that for some purposes we need to assume that ω^*_α and ω_α are independent *real* variables. The whole effect of the factor Det $\mathscr{F}$ is the same as that of including $I_{GH}[\omega,\omega^*]$ in the full effective action, and integrating over 'fields' ω and ω^*. That is, for arbitrary gauge-fixing functionals $f_\alpha(x)$,

$$\langle T\{\mathscr{O}_A\cdots\}\rangle_V \propto \int \left[\prod_{n,x} d\psi_n(x)\right] \left[\prod_{\alpha,\mu,x} dA_{\alpha\mu}(x)\right]$$
$$\times \left[\prod_{\alpha,x} d\omega_\alpha(x)\, d\omega^*_\alpha(x)\right] \exp\Big(iI_{\text{MOD}}[\psi,A,\omega,\omega^*]\Big)\,\mathscr{O}_A\cdots\,, \tag{15.6.3}$$

where I_{MOD} is a modified action

$$I_{\text{MOD}} = \int d^4x\,\left[\mathscr{L} - \frac{1}{2\xi}\,f_\alpha f_\alpha\right] + I_{GH}\,. \tag{15.6.4}$$

The fields ω_α and ω^*_α are Lorentz scalars (at least in covariant gauges) but satisfy Fermi statistics. The connection between spin and statistics is not really violated here, because there are no particles described by these fields that can appear in initial or final states. For that reason, ω_α and ω^*_α are called the fields of 'ghost' and 'antighost' particles. Inspection of Eq. (15.6.2) shows that the action respects the conservation of a quantity known as 'ghost number,' equal to +1 for ω_α, −1 for ω^*_α, and zero for all other fields.

The Feynman rules for the ghosts are simplest in the case in which the 'matrix' $\mathscr{F}$ may be expressed as

$$\mathscr{F} = \mathscr{F}_0 + \mathscr{F}_1\,, \tag{15.6.5}$$

where $\mathscr{F}_0$ is field-independent and of zeroth order in coupling constants, while $\mathscr{F}_1$ is field-dependent and proportional to one or more coupling constant factors. In this case, the ghost propagator is just

$$\Delta_{\alpha\beta}(x,y) = -(\mathscr{F}_0^{-1})_{\alpha x,\beta y} \tag{15.6.6}$$

and the ghost vertices are to be read off from the interaction term

$$I'_{GH} = \int d^4x\, d^4y\; \omega^*_\alpha(x)\, \omega_\beta(y) (\mathscr{F}_1)_{\alpha x,\beta y}\,. \tag{15.6.7}$$

For instance, in the generalized ξ-gauge discussed in the previous section, we have

$$f_\alpha = \partial_\mu A^\mu_\alpha \tag{15.6.8}$$

and for infinitesimal gauge parameters λ_α, Eq. (15.1.9) gives:

$$A^\mu_{\alpha\lambda} = A^\mu_\alpha + \partial^\mu \lambda_\alpha + C_{\alpha\gamma\beta}\lambda_\beta A^\mu_\gamma$$

so that

$$\mathscr{F}_{\alpha x,\beta y} = \left.\frac{\delta \partial_\mu A^\mu_{\alpha\lambda}(x)}{\delta \lambda_\beta(y)}\right|_{\lambda=0}$$
$$= \Box\, \delta^4(x-y) + C_{\alpha\gamma\beta}\frac{\partial}{\partial x^\mu}\left[A^\mu_\gamma(x)\, \delta^4(x-y)\right]\,. \tag{15.6.9}$$

This is of the form (15.6.5), with

$$(\mathscr{F}_0)_{\alpha x,\beta y} = \Box \delta^4(x-y)\, \delta_{\alpha\beta}\,, \tag{15.6.10}$$

$$(\mathscr{F}_1)_{\alpha x,\beta y} = -C_{\alpha\beta\gamma}\frac{\partial}{\partial x^\mu}\left[A^\mu_\gamma(x)\delta^4(x-y)\right]\,. \tag{15.6.11}$$

From Eqs. (15.6.6) and (15.6.10), we see that the ghost propagator is

$$\Delta_{\alpha\beta}(x,y) = \delta_{\alpha\beta}(2\pi)^{-4}\int d^4p\, (p^2 - i\epsilon)^{-1}\, e^{ip\cdot(x-y)}\,, \tag{15.6.12}$$

so in this gauge the ghosts behave like spinless fermions of zero mass, transforming according to the adjoint representation of the gauge group. Using Eqs. (15.6.7) and (15.6.11) and integrating by parts, we find that the ghost interaction term in the action is now

$$I'_{\rm GH} = \int d^4x\, C_{\alpha\beta\gamma}\frac{\partial \omega^*_\alpha}{\partial x^\mu} A^\mu_\gamma\, \omega_\beta\,. \tag{15.6.13}$$

This interaction corresponds to vertices to which are attached one outgoing ghost line, one incoming ghost line, and one vector boson line. If these lines carry (incoming) momenta p, q, k respectively and gauge group indices α, β, γ respectively, and the gauge field carries a vector index μ, then the contribution of such a vertex to the integrand is given by the momentum-space Feynman rules as

$$i(2\pi)^4\delta^4(p+q+k)\, \times\, ip_\mu C_{\alpha\beta\gamma}\,. \tag{15.6.14}$$

The ghosts propagate around loops, with single vector boson lines attached at each vertex along the loops, and with an extra minus sign supplied for each loop as is usual for fermionic field variables.

The extra minus sign for ghost loops suggests that each ghost field ω_α together with the associated antighost field ω^*_α represents something like a negative degree of freedom. These negative degrees of freedom are necessary because in using covariant gauge field propagators we are really over-counting; the physical degrees of freedom are the components of $A^\mu_\alpha(x)$, *less* the parameters $\Lambda_\alpha(x)$ needed to describe a gauge transformation.

In summary, the modified action (15.6.4) may be written in generalized ξ-gauge as

$$I_{\rm MOD} = \int d^4x \; \mathscr{L}_{\rm MOD} \tag{15.6.15}$$

with a modified Lagrangian density:

$$\begin{aligned}\mathscr{L}_{\rm MOD} = \mathscr{L}_{\rm M} - \frac{1}{4}F^{\mu\nu}_\alpha F_{\alpha\mu\nu} - \frac{1}{2\xi}\,(\partial_\mu A^\mu_\alpha)(\partial_\nu A^\nu_\alpha) \\ -\partial_\mu\omega^*_\alpha\,\partial^\mu\omega_\alpha + C_{\alpha\beta\gamma}(\partial_\mu\omega^*_\alpha)\,A^\mu_\gamma\,\omega_\beta\;.\end{aligned} \tag{15.6.16}$$

It is important that this Lagrangian is *renormalizable* (if the matter Lagrangian $\mathscr{L}_{\rm M}$ is), in the elementary sense that its terms involve products of fields and their derivatives of total dimensionality (in powers of mass) four or less. (The kinematic term $-\partial_\mu\omega^*_\alpha\,\partial^\mu\omega_\alpha$ in Eq. (15.6.16) fixes the dimensionality of the fields ω and ω^* to be mass to the power unity, just like ordinary scalar and gauge fields.) However, there is more to renormalizability than power counting; it is necessary also that there be a counterterm to absorb every divergence. In the next section we shall consider a remarkable symmetry that will be used in Section 17.2 to show that non-Abelian gauge theories are indeed renormalizable in this sense, and that can even take the place of the Faddeev–Popov–De Witt approach that we have been following.

15.7 BRST Symmetry

Although the Faddeev–Popov–De Witt method described in the previous two sections makes the Lorentz invariance of the theory manifest, it still rests on a choice of gauge, and hence naturally it hides the underlying gauge invariance of the theory. This is a serious problem in trying to prove the renormalizability of the theory — gauge invariance restricts the form of the terms in the Lagrangian that are available as counterterms to absorb ultraviolet divergences, but once we choose a gauge, how do we know that gauge invariance still restricts the ways that the infinities can appear?

Remarkably, however, even after we choose a gauge, the path integral still does have a symmetry related to gauge invariance. This symmetry was discovered by Becchi, Rouet, and Stora,[10] (and independently by Tyutin,[11]) in 1975, several years after the work of Faddeev and Popov and De Witt, and is known in honor of its discoverers as BRST symmetry. This symmetry will be presented more-or-less as it was originally discovered, as a by-product of the method of Faddeev, Popov, and De Witt, but as we shall see it can also be regarded as a replacement for the Faddeev–Popov–De Witt approach.

We have seen in Eqs. (15.6.3) and (15.6.4) that the Feynman rules for a non-Abelian gauge theory may be obtained from a path integral over matter, gauge, and ghost fields, with a modified action, which we may write

$$I_{\rm MOD} = I_{\rm EFF} + I_{GH} = \int d^4x\, \mathscr{L}_{\rm MOD}\,, \tag{15.7.1}$$

$$\mathscr{L}_{\rm MOD} \equiv \mathscr{L} - \frac{1}{2\xi} f_\alpha f_\alpha + \omega^*_\alpha \Delta_\alpha\,, \tag{15.7.2}$$

where we have now introduced the quantity

$$\Delta_\alpha(x) \equiv \int d^4y\, \mathscr{F}_{\alpha x, \beta y}[A, \psi]\, \omega_\beta(y)\,. \tag{15.7.3}$$

This is for the choice

$$B[f] \propto \exp\left(-\frac{i}{2\xi}\int d^4x\, f_\alpha f_\alpha\right) \tag{15.7.4}$$

of the gauge-fixing functional in Eq. (15.5.21). For our present purposes, it will be helpful to rewrite $B[f]$ as a Fourier integral:

$$B[f] = \int \left[\prod_{\alpha,x} dh_\alpha(x)\right] \exp\left[\frac{i\xi}{2}\int d^4x\, h_\alpha h_\alpha\right] \exp\left[i\int d^4x\, f_\alpha h_\alpha\right]\,. \tag{15.7.5}$$

We must now do our path integrals over the field h_α (often known as a 'Nakanishi–Lautrup' field[11a]) as well as over matter, gauge, ghost and antighost fields, with a new modified action

$$I_{\rm NEW} = \int d^4x \left(\mathscr{L} + \omega^*_\alpha \Delta_\alpha + h_\alpha f_\alpha + \tfrac{1}{2}\xi h_\alpha h_\alpha\right)\,. \tag{15.7.6}$$

This modified action is not gauge-invariant — indeed, it had better not be, if we are to be able to use it in path integrals. However, it *is* invariant under a 'BRST' symmetry transformation, parameterized by an infinitesimal constant θ that *anticommutes* with ω_α, ω^*_α, and all fermionic matter fields. For a given θ, the BRST transformation is

$$\delta_\theta \psi = it_\alpha \theta \omega_\alpha \psi \,, \tag{15.7.7}$$

$$\delta_\theta A_{\alpha\mu} = \theta D_\mu \omega_\alpha = \theta[\partial_\mu \omega_\alpha + C_{\alpha\beta\gamma} A_{\beta\mu} \omega_\gamma] \,, \tag{15.7.8}$$

$$\delta_\theta \omega^*_\alpha = -\theta h_\alpha \,, \tag{15.7.9}$$

$$\delta_\theta \omega_\alpha = -\tfrac{1}{2}\theta \, C_{\alpha\beta\gamma} \omega_\beta \omega_\gamma \,, \tag{15.7.10}$$

$$\delta_\theta h_\alpha = 0 \,. \tag{15.7.11}$$

(Recall that in fermionic path integrals, there is no connection between ω_α and ω^*_α, so that Eq. (15.7.9) does not need to be the adjoint of Eq. (15.7.10).) Because h_α is BRST-invariant, we could if we like replace the Gaussian factor $\exp(\frac{1}{2}i\xi \int h_\alpha h_\alpha)$ in Eq. (15.7.5) with an arbitrary smooth functional of h_α, yielding an arbitrary functional $B[f]$, without affecting the BRST invariance of the action. However, for the purposes of diagrammatic calculation and renormalization it will help to leave $B[f]$ as a Gaussian.

In checking the invariance of the action (15.7.1), it will be very useful first to note that the transformation (15.7.7)–(15.7.11) is *nilpotent*; that is, if F is any functional of $\psi, A, \omega, \omega^*$, and h, and we define sF by

$$\delta_\theta F \equiv \theta s F \tag{15.7.12}$$

then*

$$\delta_\theta(sF) = 0 \tag{15.7.13}$$

or equivalently

$$s(sF) = 0 \,. \tag{15.7.14}$$

It is straightforward to verify this nilpotence when δ_θ acts on a single field. First, acting on a matter field,

$$\begin{aligned}\delta_\theta s \psi = it_\alpha \, \delta_\theta(\omega_\alpha \psi) &= -\tfrac{1}{2} i \, C_{\alpha\beta\gamma} t_\alpha \theta \omega_\beta \omega_\gamma \psi - t_\alpha t_\beta \omega_\alpha \theta \omega_\beta \psi \\ &= -\tfrac{1}{2} i \, C_{\alpha\beta\gamma} t_\alpha \theta \omega_\beta \omega_\gamma \psi + t_\alpha t_\beta \theta \omega_\alpha \omega_\beta \psi \,.\end{aligned}$$

The product $\omega_\alpha \omega_\beta$ in the second term on the right is antisymmetric in α and β, so we can replace $t_\alpha t_\beta$ in this term with $\frac{1}{2}[t_\alpha, t_\beta]$, and this term thus cancels the first term:

$$ss\psi = 0 \,. \tag{15.7.15}$$

* In the original work on BRST symmetry the functional $B[f]$ was left in the form (15.7.4), so that h_α was replaced in Eq. (15.7.9) with $-f_\alpha/\xi$, and the BRST transformation was only nilpotent when acting on functions of ω_α and the gauge and matter fields, but not of ω^*_α.

Next, acting on a gauge field, we have

$$\begin{aligned}
\delta_\theta s A_{\alpha\mu} &= \delta_\theta D_\mu \omega_\alpha \\
&= \partial_\mu \delta_\theta \omega_\alpha + C_{\alpha\beta\gamma}\delta_\theta A_{\beta\mu}\omega_\gamma + C_{\alpha\beta\gamma}A_{\beta\mu}\delta_\theta\omega_\gamma \\
&= \theta\Big(-\tfrac{1}{2}C_{\alpha\beta\gamma}\partial_\mu(\omega_\beta\omega_\gamma) + C_{\alpha\beta\gamma}(\partial_\mu\omega_\beta)\omega_\gamma \\
&\quad + C_{\alpha\beta\gamma}C_{\beta\delta\epsilon}A_{\delta\mu}\omega_\epsilon\omega_\gamma - \tfrac{1}{2}C_{\alpha\beta\gamma}C_{\gamma\delta\epsilon}A_{\beta\mu}\omega_\delta\omega_\epsilon\Big) \\
&= \theta\Big(\tfrac{1}{2}C_{\alpha\beta\gamma}(\partial_\mu\omega_\beta)\omega_\gamma + \tfrac{1}{2}C_{\alpha\beta\gamma}(\partial_\mu\omega_\gamma)\omega_\beta \\
&\quad - C_{\alpha\beta\gamma}C_{\gamma\delta\epsilon}A_{\delta\mu}\omega_\epsilon\omega_\beta - \tfrac{1}{2}C_{\alpha\beta\gamma}C_{\gamma\delta\epsilon}A_{\beta\mu}\omega_\delta\omega_\epsilon\Big)\,.
\end{aligned}$$

The first two terms of the final expression cancel because $C_{\alpha\beta\gamma}$ is antisymmetric in β and γ, and the third and fourth terms cancel because of the Jacobi identity (15.1.5), so

$$ssA_{\alpha\mu} = 0\,. \tag{15.7.16}$$

Eqs. (15.7.9) and (15.7.11) show immediately that

$$ss\omega^*_\alpha = 0 \tag{15.7.17}$$

and

$$ssh_\alpha = 0\,. \tag{15.7.18}$$

Finally,

$$\begin{aligned}
\delta_\theta s\omega_\alpha &= -\tfrac{1}{2}C_{\alpha\beta\gamma}\delta_\theta(\omega_\beta\omega_\gamma) \\
&= \tfrac{1}{4}\theta\left(C_{\alpha\beta\gamma}C_{\beta\delta\epsilon}\,\omega_\delta\omega_\epsilon\omega_\gamma - C_{\alpha\beta\gamma}C_{\gamma\delta\epsilon}\,\omega_\beta\omega_\delta\omega_\epsilon\right) \\
&= \tfrac{1}{2}\theta\, C_{\alpha\beta\gamma}C_{\gamma\delta\epsilon}\omega_\delta\omega_\epsilon\omega_\beta\,.
\end{aligned}$$

But the product $\omega_\beta\omega_\delta\omega_\epsilon$ is antisymmetric, so the Jacobi identity (15.1.5) shows that this too vanishes

$$ss\omega_\alpha = 0\,. \tag{15.7.19}$$

Now consider a product of two fields ϕ_1 and ϕ_2, either or both of which may be ψ, A, ω, ω^*, or h, not necessarily at the same point in spacetime. Then

$$\delta_\theta(\phi_1\phi_2) = \theta(s\phi_1)\phi_2 + \phi_1\theta(s\phi_2) = \theta\Big[(s\phi_1)\phi_2 \pm \phi_1 s\phi_2\Big]\,,$$

where the sign $\pm$ is plus if ϕ_1 is bosonic, minus if ϕ_1 is fermionic. That is,

$$s(\phi_1\phi_2) = (s\phi_1)\phi_2 \pm \phi_1 s\phi_2\,.$$

Since as we have seen $\delta_\theta(s\phi_1) = \delta_\theta(s\phi_2) = 0$, the effect of a BRST transformation on $s(\phi_1\phi_2)$ is

$$\delta_\theta s(\phi_1\phi_2) = (s\phi_1)\theta(s\phi_2) \pm \theta(s\phi_1)(s\phi_2)\,.$$

But $s\phi$ always has statistics opposite to ϕ, so moving θ to the left in the first term on the right-hand side introduces a sign factor $\mp$:

$$\delta_\theta s(\phi_1\phi_2) = \theta\Big[\mp(s\phi_1)(s\phi_2) \pm (s\phi_1)(s\phi_2)\Big] = 0\,.$$

Continuing in this way, we see that BRST transformations are nilpotent acting on any product of fields at arbitrary spacetime points:

$$\delta_\theta\, s(\phi_1\phi_2\phi_3\cdots) = 0\,.$$

Any functional $F[\phi]$ can be written as a sum of multiple integrals of such products with c-number coefficients, so likewise

$$\delta_\theta\, sF[\phi] = \theta ssF[\phi] = 0\,. \tag{15.7.20}$$

This completes the proof of the nilpotency of the BRST transformation.

Now let us return to the verification of the BRST invariance of the action (15.7.6). First note that for any functional of matter and gauge fields alone, the BRST transformation is just a gauge transformation with infinitesimal gauge parameter

$$\lambda_\alpha(x) = \theta\omega_\alpha(x)\,. \tag{15.7.21}$$

Therefore the first term in Eq. (15.7.6) is automatically BRST-invariant:

$$\delta_\theta \int d^4x\, \mathscr{L} = 0\,. \tag{15.7.22}$$

To calculate the effect of a BRST transformation on the rest of the action (15.7.6), note that its effect on the gauge-fixing function is just the gauge transformation (15.7.21), so

$$\begin{aligned}\delta_\theta f_\alpha[x;A,\psi] &= \int \frac{\delta f_\alpha[x;A_\lambda,\psi_\lambda]}{\delta\lambda^\beta(y)}\bigg|_{\lambda=0} \theta\,\omega_\beta(y)\,d^4y \\ &= \theta\int \mathscr{F}_{\alpha x,\beta y}[A,\psi]\,\omega_\beta(y)\,d^4y\end{aligned}$$

or in terms of the quantity (15.7.3)

$$\delta_\theta f_\alpha[x;A,\psi] = \theta\Delta_\alpha(x;A,\psi,\omega)\,. \tag{15.7.23}$$

(Note that $\mathscr{F}$ is a bosonic quantity, so there is no sign change in moving θ to the left here.) Also recall that $\delta_\theta\omega^*_\alpha = -\theta h_\alpha$ and $\delta_\theta h_\alpha = 0$. Therefore the terms in the integrand of the 'new' action (15.7.6) other than $\mathscr{L}$ may be written

$$\omega^*_\alpha\Delta_\alpha + h_\alpha f_\alpha + \tfrac{1}{2}\xi h_\alpha h_\alpha = -s(\omega^*_\alpha f_\alpha + \tfrac{1}{2}\xi\omega^*_\alpha h_\alpha) \tag{15.7.24}$$

or in other words

$$I_{\text{NEW}} = \int d^4x\, \mathscr{L} + s\Psi\,, \tag{15.7.25}$$

where

$$\Psi \equiv -\int d^4x\,(\omega^*_\alpha f_\alpha + \tfrac{1}{2}\xi\omega^*_\alpha h_\alpha)\ . \tag{15.7.26}$$

The nilpotence of the BRST transformation tells us immediately that the term $s\Psi$ as well as $\int d^4x\,\mathscr{L}$ is BRST-invariant.

In a sense the converse of this result also applies: we shall see in Section 17.2 that a renormalizable Lagrangian that obeys BRST invariance and the other symmetries of the Lagrangian (15.7.25) must take the form of Eq. (15.7.25), aside from changes in the values of various constant coefficients. But this is not enough to establish the renormalizability of these theories. BRST symmetry transformations act non-linearly on the fields, and in this case there is no simple connection between the symmetries of the Lagrangian and the symmetries of matrix elements and Greens functions. Using the external field methods developed in the next chapter, it will be shown in Section 17.2 that the ultraviolet divergent terms in Feynman amplitudes (though not the finite parts) do obey a sort of renormalized BRST invariance, which allows the proof of renormalizability to be completed.

Eq. (15.7.25) shows that the physical content of any gauge theory is contained in the *kernel* of the BRST operator (that is, in a general BRST-invariant term $\int d^4x\,\mathscr{L} + s\Psi$), modulo terms in the *image* of the BRST transformation (that is, terms of the form $s\Psi$). The kernel modulo the image of any nilpotent transformation is said to form the *cohomology* of the transformation. There is another sense in which the physical content of a gauge theory may be identified with the cohomology of the BRST operator.[12] It is a fundamental physical requirement that matrix elements between physical states should be independent of our choice of the gauge-fixing function f_α, or in other words, of the functional Ψ in Eq. (15.7.25). The change in any matrix element $\langle\alpha|\beta\rangle$ due to a change $\tilde\delta\Psi$ in Ψ is

$$\tilde\delta\langle\alpha|\beta\rangle = i\langle\alpha|\tilde\delta I_{\rm NEW}|\beta\rangle = i\langle\alpha|s\tilde\delta\Psi|\beta\rangle\ . \tag{15.7.27}$$

(We use a tilde here to distinguish this arbitrary change in the gauge-fixing function from a BRST transformation or a gauge transformation.) We can introduce a fermionic BRST 'charge' Q, defined so that for any field operator Φ,

$$\delta_\theta\Phi = i[\theta\,Q, \Phi] = i\theta\,[Q,\Phi]_\mp\ ,$$

or in other words,

$$[Q,\Phi]_\mp = is\Phi\ , \tag{15.7.28}$$

the sign being $-$ or $+$ according as Φ is bosonic or fermionic. The nilpotence of the BRST transformation then gives

$$0 = -ss\Phi = [Q,[Q,\Phi]_\mp]_\pm = [Q^2,\Phi]_-\ .$$

For this to be satisfied for all operators Φ, it is necessary for Q^2 either to vanish or be proportional to the unit operator. But Q^2 cannot be proportional to the unit operator, since it has a non-vanishing ghost quantum number**, so it must vanish:

$$Q^2 = 0\,. \tag{15.7.29}$$

From Eqs. (15.7.27) and (15.7.28), we have

$$\tilde{\delta}\langle\alpha|\beta\rangle = \langle\alpha|[Q, \tilde{\delta}\Psi]|\beta\rangle\,. \tag{15.7.30}$$

In order for this to vanish for all changes $\tilde{\delta}\Psi$ in Ψ, it is necessary that

$$\langle\alpha|Q = Q|\beta\rangle = 0\,. \tag{15.7.31}$$

Thus physical states are in the kernel of the nilpotent operator Q. Two physical states that differ only by a state vector in the image of Q, that is, of form $Q|\cdots\rangle$, evidently have the same matrix element with all other physical states, and are therefore physically equivalent. Hence *independent* physical states correspond to states in the kernel of Q, modulo the image of Q — that is, they correspond to the cohomology of Q.

To see how this works in practice, let us consider the simple example of pure electrodynamics.† Taking the gauge-fixing function as $f = \partial_\mu A^\mu$ and integrating over the auxiliary field h, the BRST transformation (15.7.8)–(15.7.10) is here

$$s A_\mu = \partial_\mu \omega\,, \qquad s\,\omega^* = \partial_\mu A^\mu/\xi\,, \qquad s\,\omega = 0\,. \tag{15.7.32}$$

We expand the fields in normal modes††

$$\begin{aligned} A^\mu(x) &= (2\pi)^{-3/2}\int \frac{d^3p}{\sqrt{2p^0}}\left[a^\mu(\mathbf{p})\,e^{ip\cdot x} + a^{\mu *}(\mathbf{p})\,e^{-ip\cdot x}\right]\,,\\ \omega(x) &= (2\pi)^{-3/2}\int \frac{d^3p}{\sqrt{2p^0}}\left[c(\mathbf{p})\,e^{ip\cdot x} + c^*(\mathbf{p})\,e^{-ip\cdot x}\right]\,,\\ \omega^*(x) &= (2\pi)^{-3/2}\int \frac{d^3p}{\sqrt{2p^0}}\left[b(\mathbf{p})\,e^{ip\cdot x} + b^*(\mathbf{p})\,e^{-ip\cdot x}\right]\,. \end{aligned} \tag{15.7.33}$$

** Recall that the ghost quantum number is defined as +1 for ω_α, −1 for ω^*_α, and 0 for all gauge and matter fields.

† Eqs. (15.6.11) and (15.6.7) show that because the structure constants vanish in electrodynamics, the ghosts here are not coupled to other fields. Nevertheless, electrodynamics provides a good example of the use of BRST symmetry in identifying physical states. Indeed, in analyzing the physicality conditions on 'in' and 'out' states we ignore interactions, so for this purpose a non-Abelian gauge theory is treated like several copies of quantum electrodynamics.

†† Just as $\omega^*(x)$ is not to be thought of as the Hermitian adjoint of $\omega(x)$, b^* and c^* are not the adjoints of c and b. But since $A^\mu(x)$ is Hermitian, $\omega(x)$ is Hermitian if Q is.

Matching coefficients of $e^{\pm ip\cdot x}$ on both sides of Eq. (15.7.28) yields

$$\begin{aligned} &[Q, a^\mu(\mathbf{p})]_- = -p^\mu c(\mathbf{p})\,, \qquad [Q, a^{\mu *}(\mathbf{p})]_- = p^\mu c^*(\mathbf{p})\,, \\ &[Q, b(\mathbf{p})]_+ = p^\mu a_\mu(\mathbf{p})/\xi\,, \qquad [Q, b^*(\mathbf{p})]_+ = p^\mu a^*_\mu(\mathbf{p})/\xi\,, \\ &[Q, c(\mathbf{p})]_+ = [Q, c^*(\mathbf{p})]_+ = 0\,. \end{aligned} \tag{15.7.34}$$

Consider any state $|\psi\rangle$ satisfying the physicality condition (15.7.31):

$$Q|\psi\rangle = 0\,. \tag{15.7.35}$$

The states $|e,\psi\rangle = e_\mu a^{\mu *}(\mathbf{p})|\psi\rangle$ with one additional photon then satisfy the physicality condition $Q|e,\psi\rangle = 0$ if $e_\mu p^\mu = 0$. Also, the state $|\psi\rangle' \equiv b^*(\mathbf{p})|\psi\rangle$ satisfies

$$Q|\psi\rangle' = p^\mu a^*_\mu(\mathbf{p})|\psi\rangle/\xi\,, \tag{15.7.36}$$

so $|e+\alpha p,\psi\rangle = |e,\psi\rangle + \xi\alpha Q|\psi\rangle'$, and is therefore physically equivalent to $|e,\psi\rangle$. From this we conclude that e^μ is physically equivalent to $e^\mu + \alpha p^\mu$, which is the usual 'gauge-invariance' condition on photon polarization vectors. On the other hand,

$$Qb^*(\mathbf{p})|\psi\rangle = p^\mu a^*_\mu(\mathbf{p})|\psi\rangle \neq 0\,,$$

so $b^*|\psi\rangle$ does not satisfy the physicality condition (15.7.31). Also, for any e_μ with $e\cdot p \neq 0$,

$$c^*(\mathbf{p})|\psi\rangle = Qe_\mu a^{*\mu}(\mathbf{p})|\psi\rangle/e\cdot p$$

so $c^*|\psi\rangle$ is BRST-exact, and hence equivalent to zero. *Thus the physical Hilbert space is free of ghosts and antighosts.*

To maintain Lorentz invariance, we must interpret all four components of $a^\mu(\mathbf{p})$ as annihilation operators, in the sense that

$$0 = a_\mu(\mathbf{p})|0\rangle\,, \tag{15.7.37}$$

where $|0\rangle$ is the BRST-invariant vacuum state. But the canonical commutation relations derived from the BRST-invariant action (say, with $\xi = 1$) give

$$[a_\mu(\mathbf{p}), a^*_\nu(\mathbf{p}')]_- = \eta_{\mu\nu}\delta^3(\mathbf{p}-\mathbf{p}')\,, \tag{15.7.38}$$

corresponding to the propagator in Feynman gauge. This violates the usual positivity rules of quantum mechanics, because Eqs. (15.7.37) and (15.7.38) yield[13]

$$\langle 0|a_0(\mathbf{p})\,a^*_0(\mathbf{p}')|0\rangle = -\langle 0|0\rangle\,. \tag{15.7.39}$$

Nevertheless we can rest assured that all amplitudes among physical states satisfy the usual positivity conditions, because these states satisfy Eq. (15.7.31), and for such states the transition amplitudes are the same as they would be in a more physical gauge like Coulomb or axial gauge, where there is no problem of positivity or unitarity.

The Faddeev–Popov–De Witt formalism described so far necessarily yields an action that is bilinear in the ghost fields ω^*_α and ω_α. This is adequate for renormalizable Yang–Mills theories with the gauge-fixing function $f_\alpha = \partial_\mu A^\mu_\alpha$, but not in more general cases. For instance, as we shall see in Section 17.2, in other gauges renormalizable Yang–Mills theories need $\omega^*\omega^*\omega\omega$ terms in the Lagrangian density to serve as counterterms for the ultraviolet divergences in loop graphs with four external ghost lines.

Fortunately the Faddeev–Popov–De Witt formalism represents only one way of generating a class of equivalent Lagrangians that yield the same unitary S-matrix. The BRST formalism provides a more general approach, that dispenses altogether with the Faddeev–Popov–De Witt formalism. In this approach, one takes the action to be the most general local functional of matter, gauge, ω^A, ω^{*A} and h^A fields with ghost number zero that is invariant under the BRST transformation (15.7.7)–(15.7.11) and under any other global symmetries of the theory. (For renormalizable theories one would also limit the Lagrangian density to operators of dimensionality four or less, but this restriction plays no role in the following discussion.) In the next section we shall prove, in a context more general than Yang–Mills theories, that the most general action of this sort is the sum of a functional of the matter and gauge fields (collectively called ϕ) alone, plus a term given by the action of the BRST operator s on an arbitrary functional Ψ of ghost number -1:

$$I_{\rm NEW}[\phi, \omega, \omega^*, h] = I_0[\phi] + s\,\Psi[\phi, \omega, \omega^*, h]\,, \tag{15.7.40}$$

as for instance in the Faddeev–Popov–De Witt action (15.7.25), but with $s\Psi$ now not necessarily bilinear in ghost and antighost fields.

By the same argument as before, the S-matrix elements for states that are annihilated by the BRST generator Q are independent of the choice of Ψ in Eq. (15.7.40), so if there is any choice of Ψ for which the ghosts decouple, then the ghosts decouple in general. In Yang–Mills theories, such a Ψ is provided by quantization of the theory in axial gauge, so in such theories ghosts decouple for arbitrary choices of the functional $\Psi[\phi, \omega, \omega^*, h]$, not just those choices like (15.7.26) that are generated by the Faddeev–Popov–De Witt formalism.

We can go further, and free ourselves of all dependence on canonical quantization in Lorentz-non-invariant gauges like axial gauge. Again, take the action to be the most general functional of gauge, matter, ω^A, ω^{*A} and h^A fields with ghost number zero, that is invariant under the BRST transformation (15.7.7)–(15.7.11) and under any other global symmetries of the theory, including Lorentz invariance. From the BRST invariance of the action we can infer the existence of a conserved nilpotent BRST generator Q. With the ghost and antighost fields treated as Hermitian, Q is also Hermitian. The space of physical states is defined as above as consisting of states annihilated by Q, with two states treated as equivalent if their difference is Q acting on another state. It has been shown that for Yang–Mills theories this space is free of ghosts and antighosts and has a positive-definite norm, and that the S-matrix in this space is unitary.[13a]

This procedure is known as *BRST quantization*. It has been extended to theories with other local symmetries, such as general relativity and string theories. Unfortunately, it seems so far to be necessary to give separate proofs in each case that the BRST-cohomology is ghost-free and that the S-matrix acting in this space is unitary. The key point in these proofs is that, for each negative-norm degree of freedom, such as the time components of the gauge fields in Yang–Mills theories, there is one local symmetry that allows this degree of freedom to be transformed away.

* * *

Although we shall not use it here, there is a beautiful geometric interpretation[14] of the ghosts and the BRST symmetry that should be mentioned. The gauge fields A^μ_α may be written as one-forms $A_\alpha \equiv A_{\alpha\mu} dx^\mu$, where dx^μ are a set of anticommuting c-numbers. (See Section 5.8.) This can be combined with the ghost to compose a one-form $\mathscr{A}_\alpha \equiv A_\alpha + \omega_\alpha$ in an extended space. Also, the ordinary exterior derivative $d \equiv dx^\mu \, \partial/\partial x^\mu$ may be combined with the BRST operator s to form an exterior derivative $\mathscr{D} \equiv d + s$ in this space, which is nilpotent because $s^2 = d^2 = sd + ds = 0$.

The next chapter will introduce external field methods, which will be used along with the BRST symmetry in Chapter 17 to complete the proof of the renormalizability of non-Abelian gauge theories.

15.8 Generalizations of BRST Symmetry*

The BRST symmetry described in the previous section has a useful generalization to the quantization of a wide class of theories, including general relativity and string theories. In all these cases, we deal with an action $I[\phi]$ and measure $[d\phi] \equiv \prod_r d\phi^r$ that are invariant under a set of infinitesimal transformations

$$\phi^r \to \phi^r + \epsilon^A \delta_A \phi^r \,. \tag{15.8.1}$$

* This section lies somewhat out of the book's main line of development, and may be omitted in a first reading.

This is an abbreviated 'De Witt' notation, with r and A including spacetime coordinates as well as discrete labels, and sums including integrals over these coordinates. For instance, for the gauge transformation (15.1.9), the index A consists of a group index α and a spacetime coordinate x, with $\epsilon^{\alpha x} \equiv \epsilon^{\alpha}(x)$, while the index r consists of a vector index μ as well as a group index α and a spacetime coordinate x, with $\phi^{\mu\alpha x} \equiv A^{\alpha}{}_{\mu}(x)$; in the notation of Eq. (15.8.1), the variation $\delta_A\phi^r$ in the transformation (15.1.9) reads

$$\delta_{\beta\, y}\phi^{\mu\alpha x} = \delta^{\beta}_{\alpha}\frac{\partial}{\partial x^{\mu}}\delta^4(x-y) + C^{\beta}{}_{\gamma\alpha}\phi^{\mu\gamma x}\delta^4(x-y)\,.$$

As in the special case of Yang–Mills theories discussed in the previous section, BRST invariance can be used as a substitute for the Faddeev–Popov–De Witt formulation of these theories, one that is applicable even where the Faddeev–Popov–De Witt approach fails. Nevertheless, in order to motivate the introduction of BRST invariance, we shall begin here with the Faddeev–Popov–De Witt formulation of theories with general local symmetries, and then go on to consider further generalizations.

By following the same arguments used to derive Eq. (15.5.21), we obtain the general Faddeev–Popov–De Witt theorem:

$$\frac{C}{\Omega}\int [d\phi]\, e^{iI[\phi]}\, V[\phi] = \int [d\phi]\, e^{iI[\phi]}\, B[f[\phi]]\, \mathrm{Det}(\delta_A f_B[\phi])\, V[\phi]\,, \qquad (15.8.2)$$

where $V[\phi]$ is an arbitrary functional of ϕ^r that is invariant under the gauge transformations (15.8.1); $f_A[\phi]$ are a set of gauge-fixing functionals** of the ϕ^r, chosen so that the 'matrix' $\delta_A f_B[\phi]$ has a non-vanishing determinant, and $B[f]$ is a more-or-less arbitrary functional of the f_A (as for instance $\prod_A \delta(f_A)$). The constant Ω is the volume of the gauge group, and the constant C is defined (as in Eq. (15.5.19)) by

$$C \equiv \int [df]\, B[f]\,. \qquad (15.8.3)$$

As we have seen, in gauge theories the importance of Eq. (15.8.2) is that it tells us that the integral on the right-hand side is independent of the choice of the gauge-fixing functionals f_A, and depends on $B[f]$ only through the constant C. Where some meaning can be given to the usually infinite group volume Ω, as in gauge theories on a finite spacetime

** We use the same letters A, B, etc. to label the f_A as the gauge variations δ_A, in order to emphasize that there must be as many gauge-fixing functionals as there are independent gauge transformations. However, in some cases like string theory it is natural to use gauge-fixing functionals f^a for which, although the index a runs over 'as many' values as the index A on the gauge variations δ_A, the values taken by these indices are quite different. No change is needed in the present formalism as long as we can define $f_A = c_{Aa}f^a$, with c_{Aa} field-independent and non-singular.

lattice, Eq. (15.8.2) can also have value as a formula for the integral on the left-hand side.

To define a nilpotent BRST transformation, we must first express the functional $B[f]$ as a Fourier transform

$$B[f] = \int [dh] \; \exp(ih^A f_A)\,\mathscr{B}[h] \,, \tag{15.8.4}$$

where $[dh] \equiv \prod_A dh^A$. Also, the determinant can be expressed as an integral over fermionic c-number fields† ω^{*A} and ω^A:

$$\text{Det}(\delta_A f_B[\phi]) \propto \int [d\omega^*]\,[d\omega] \; \exp\left(i\omega^{*B}\omega^A\delta_A f_B\right) , \tag{15.8.5}$$

where $[d\omega^*] \equiv \prod_A d\omega^{*A}$ and $[d\omega] \equiv \prod_A d\omega^A$, and as usual '$\propto$' means proportional with field-independent factors. Inserting these in Eq. (15.8.2) gives a general formula for the gauge-fixed path integral

$$\int [d\phi] \; \exp\left(iI[\phi]\right) B[f[\phi]] \; \text{Det}(\delta_A f_B[\phi])\, V[\phi]$$

$$\propto \int [d\phi]\,[dh]\,[d\omega^*]\,[d\omega] \; \exp\left(iI_{\text{NEW}}[\phi, h, \omega, \omega^*]\right) \mathscr{B}[h] \; V[\phi] \,, \tag{15.8.6}$$

where I_{NEW} is the new total action:

$$I_{\text{NEW}}[\phi, h, \omega, \omega^*] = I[\phi] + h^A f_A[\phi] + \omega^{*B}\omega^A\delta_A f_B[\phi] \,. \tag{15.8.7}$$

As mentioned in Section 15.6, we can think of the ghost fields as compensation for the fact that we are integrating over all ϕ^r, including those ϕ^r that differ only by gauge transformations (15.8.1). Because ghosts are fermions, loops of ghost lines carry extra minus signs that allow these loops to compensate for the integration over gauge-equivalent ϕs. But for this to work, there must be just as many ghost fields ω^A as there are independent gauge transformations. That is, since the ω^A are independent, the gauge transformations (15.8.1) must all be independent. This is the case for gauge transformations in Yang–Mills theory and coordinate transformations in general relativity, but not always. The classic example of a theory with non-independent gauge transformations is the theory of p-form gauge fields, described in Section 8.8. A p-form A (an antisymmetric tensor of rank p) undergoes a gauge transformation $A \to A + d\phi$, where ϕ is a $(p-1)$-form, and $d\phi$ is its exterior derivative (the antisymmetrized derivative). Because d is nilpotent, for $p \geq 2$ we can shift ϕ by an amount $d\psi$ without changing the gauge transformation, so there is a sort of invariance under gauge transformations of gauge

† It is common in string theories and elsewhere to find the ghost fields ω^{*A} and ω^A written b^A (or b_A) and c^A, respectively.

transformations, in which the transformation parameters are the $(p-2)$-forms ψ. In such cases we must compensate for introducing too many ghosts by also introducing 'ghosts of ghosts.'[15] For $p \geq 3$ we need to compensate further by introducing 'ghosts of ghosts of ghosts,' and so on. In what follows we shall assume that the gauge transformations (15.8.1) are all independent, so that the ghost fields ω^A (and the antighosts ω^{*A}) are all we need.

Although the original symmetry (15.8.1) has been eliminated by the insertion of the non-gauge-invariant functional $B[f]$, the new total action has an exact symmetry under the infinitesimal BRST transformations

$$\chi \to \chi + \theta\, s\chi\,, \tag{15.8.8}$$

where χ is any of the ϕ^r, ω^A, ω^{A*}, or h^A; θ is an infinitesimal anticommuting c-number; and s is the *Slavnov operator*

$$s = \omega^A\, \delta_A \phi^r \frac{\delta_L}{\delta\phi^r} - \frac{1}{2}\omega^B\omega^C f^A{}_{BC}\frac{\delta_L}{\delta\omega^A} - h^A\frac{\delta_L}{\delta\omega^{*A}}\,. \tag{15.8.9}$$

In Eq. (15.8.9) the subscript L denotes left differentiation, defined so that if $\delta F = \delta\chi\, G$, then $\delta_L F/\delta\chi = G$, and $f^A{}_{BC}$ is the structure constant†† appearing in the commutation relation

$$[\delta_B, \delta_C] = f^A{}_{BC}\delta_A\,. \tag{15.8.10}$$

The $f^A{}_{BC}$ are field-independent in non-Abelian gauge theories and in string theories, though not always, but the BRST formalism is not limited to this case. A straightforward calculation gives

$$\begin{aligned} s^2 = \frac{1}{2}\omega^A\omega^B &\left[\delta_A\phi^s\frac{\delta_L(\delta_B\phi^r)}{\delta\phi^s} - \delta_B\phi^s\frac{\delta_L(\delta_A\phi^r)}{\delta\phi^s} - f^C{}_{AB}\delta_C\phi^r\right]\frac{\delta_L}{\delta\phi^r} \\ -\frac{1}{2}\omega^B\omega^C\omega^D &\left[f^E{}_{BC}f^A{}_{DE} + \delta_D\phi^r\frac{\delta_L f^A{}_{BC}}{\delta\phi^r}\right]\frac{\delta_L}{\delta\omega^A}\,. \end{aligned} \tag{15.8.11}$$

Hence the condition that the BRST transformation be nilpotent is equivalent to the commutation relation (15.8.10), together with a consistency condition

$$f^E{}_{[BC}f^A{}_{D]E} + \delta_{[D}\phi^r\left(\delta_L f^A{}_{BC]}/\delta\phi^r\right) = 0\,, \tag{15.8.12}$$

where the brackets in subscripts indicate antisymmetrization with respect to the enclosed indices B, C, and D. Eq. (15.8.12) may be derived from the commutation relation (15.8.10) in the same way as the usual Jacobi

†† For instance, for a gauge transformation acting on a matter field $\psi(x)$ we have $\delta_{\beta y}\psi(x) = it_\beta\psi(x)\delta^4(x-y)$, and so $\delta_{\beta y}\delta_{\gamma z}\psi(x) = -t_\gamma t_\beta\psi(x)\delta^4(x-y)\delta^4(x-z)$. Hence in this case we have $f^{\alpha x}{}_{\beta y\gamma z} = C^\alpha{}_{\beta\gamma}\delta^4(x-y)\delta^4(x-z)$.

identity, and takes the place of the Jacobi identity for symmetries with field-dependent structure constants.

To show that the transformation (15.8.8) is a symmetry of $I_{\rm NEW}$, we note (recalling that θ anticommutes with ω^{*A}) that Eq. (15.8.7) may be rewritten

$$I_{\rm NEW}[\phi,h,\omega,\omega^*] = I[\phi] - s(\omega^{*A} f_A)\,. \tag{15.8.13}$$

The term $I[\phi]$ is BRST-invariant, because on the fields ϕ^r a BRST transformation is just a gauge transformation (15.8.1) with ϵ^A replaced with $\theta\omega^A$, which commutes with all ϕ^r. The term $s(\omega^{*A} f_A)$ is BRST-invariant because BRST transformations are nilpotent.

For several reasons we may need to consider a wider class of actions than those that can be constructed by the Faddeev–Popov–De Witt approach, by simply requiring that the action is invariant under the BRST transformation (15.8.8). As a step toward showing that such an action yields physically sensible results, we shall now prove the general result (already used in the previous section) that the most general BRST-invariant functional of ghost number zero is the sum of a functional of the ϕ alone, plus a term given by the action of the BRST operator s on an arbitrary functional Ψ of ghost number -1:

$$I_{\rm NEW}[\phi,\omega,\omega^*,h] = I_0[\phi] + s\,\Psi[\phi,\omega,\omega^*,h] \tag{15.8.14}$$

as for instance in the Faddeev–Popov–De Witt action (15.8.13). In brief, the BRST cohomology consists of gauge-invariant functionals $I[\phi]$ of the fields ϕ^r alone.

To prove Eq. (15.8.14), we note that the BRST transformation (15.8.8)–(15.8.9) does not change the total number of h^A and ω^{*A} fields, so if we expand I in a series of terms I_N that contain definite total numbers N of h^A and ω^{*A} fields, then there can be no cancellations in sI between terms with different N, so each term must be separately BRST-invariant:

$$sI_N = 0\,. \tag{15.8.15}$$

We next introduce what is called a Hodge operator:

$$t \equiv \omega^{*A}\frac{\delta}{\delta h^A}\,. \tag{15.8.16}$$

It is straightforward to check the anticommutation relation

$$\{s,t\} = -\omega^{*A}\frac{\delta_L}{\delta\omega^{*A}} - h^A\frac{\delta}{\delta h^A}\,. \tag{15.8.17}$$

Applying the operator $\{s,t\}$ to I_N and using Eq. (15.8.15) then gives

$$stI_N = -NI_N\,, \tag{15.8.18}$$

so each I_N except for I_0 is BRST-exact, in the sense that it may be written as the operator s acting on some other functional. The complete functional I may therefore be written in the form $I_0 + s\Psi$, with

$$\Psi = -\sum_{N=1}^{\infty} \frac{tI_N}{N} . \tag{15.8.19}$$

The term I_0 is by definition independent of ω^{*A} and h^A, and since we assume it has zero ghost number it must also be independent of ω^A, as was to be proved.

To show the invariance of physical matrix elements under changes in the definition of the gauge-fixing functional Ψ, we define a fermionic 'charge' Q, such that the change under a BRST transformation of any operator Φ is

$$\delta_\theta \Phi = i[\theta\, Q, \Phi] = i\theta\, [Q, \Phi]_\mp \tag{15.8.20}$$

with the top or bottom sign in $[x, y]_\mp \equiv xy \mp yx$ according as Φ is bosonic or fermionic. Just as in the previous section, the nilpotence of the BRST transformation then tells us that $Q^2 = 0$. Matrix elements of gauge-invariant operators between physical states will be independent of the choice of Ψ if and only if the physical states $|\alpha\rangle$ and $\langle\beta|$ satisfy

$$Q|\alpha\rangle = \langle\beta|Q = 0\,, \tag{15.8.21}$$

so that physically distinguishable physical states are again in one-to-one correspondence with elements of the cohomology of Q. The general BRST-invariant action (15.8.14) will therefore yield physically sensible results for any gauge-fixing functional Ψ if we can find *some* Ψ, like axial gauge in Yang–Mills theories, in which the ghosts do not interact with other fields. If this ghost-free choice of Ψ is inconvenient for actual calculation, as for instance axial gauge is inconvenient because it violates Lorentz invariance, we can adopt any gauge-fixing Ψ we like, and still be confident that there is a unitary S-matrix with no ghosts in initial or final states.

This approach works well in string theories, where so-called light-cone quantization takes the place of axial gauge. But in other theories like general relativity there is no way of choosing a coordinate system in which the ghosts decouple. Such theories may be dealt with by the BRST-quantization method described at the end of the previous section, using BRST invariance to prove that the S-matrix in a physical ghost-free Hilbert space is unitary.

The discovery[17] of invariance under an 'anti-BRST' symmetry[18] showed that, despite appearances, there is a similarity between the roles of ω^A and ω^{*A}, which remains somewhat mysterious.

15.9 The Batalin–Vilkovisky Formalism*

This section will describe a powerful formalism, widely known as the Batalin–Vilkovisky[19] method. It is developed in the Lagrangian framework, but has its roots in the earlier Batalin–Fradkin–Vilkovisky formalism,[20] which had been derived in the Hamiltonian framework. (The two schemes have been proved perturbatively equivalent.[21]) As we shall see in Section 17.1, the same formal machinery had been developed even earlier by Zinn-Justin[22] in order to deal with the renormalization of gauge theories. There are at least three areas where this formalism has proved invaluable:

(i) Up to this point, we have considered only irreducible symmetries with an algebra that closes in the sense of Eq. (15.8.10). In some theories, such as supergravity (without auxiliary fields)[23] the algebra is open: it closes only when the field equations are satisfied, so that terms appear in Eq. (15.8.10) proportional to $\delta I/\delta\chi^n$. Similar terms will also then appear in the consistency conditions (15.8.12). Eq. (15.8.11) then shows that in such theories s^2 will not vanish, but rather will equal a linear combination of the derivatives $\delta I/\delta\chi^n$. As we will see in this section, the Batalin–Vilkovisky method allows us to deal with very general gauge theories, including those with open or reducible gauge symmetry algebras.

(ii) As mentioned above, essential aspects of the Batalin–Vilkovisky formalism were originally developed by Zinn-Justin in order to prove the renormalizability of gauge theories. The crucial point, to be explained in Section 17.1, is that although the sum of all one-particle-irreducible diagrams in a background field does not obey the BRST symmetries of the original action, it does share one of the key properties of the action, known as the master equation.

(iii) The Batalin–Vilkovisky method provides a convenient way of analyzing the possible violations of symmetries of the action by quantum effects. It is used for this purpose in Section 22.6.

The starting point of the Batalin–Vilkovisky formalism is the introduction of what are called 'antifields,' one for each field in the theory. We let χ^n run over all the fields ϕ^r, ω^A, ω^{*A}, and h^A, and for each χ^n we introduce an external antifield** $\chi^\ddagger_n$, with the same Bose or Fermi statistics

* This section lies somewhat out of the book's main line of development, and may be omitted in a first reading.

** The symbol ‡ is used here in place of the more usual * in order to emphasize that it has nothing whatever to do with complex conjugation or charge conjugation. In particular, the antighost field ω^{*A} is *not* the same as the antifield $\omega^\ddagger_A$ of ω^A.

and opposite ghost number as that of the BRST-transformed field $s\chi^n$. That is, $\chi_n^\ddagger$ has the opposite statistics to χ^n, and a ghost number equal to $-\text{gh}(\chi^n) - 1$, where $\text{gh}(\chi^n)$ is the ghost number of χ^n. In the simplest cases, including Yang–Mills theories and quantum gravity, the original gauge-invariant action $I[\phi]$ is supplemented with a term coupling the antifields $\chi_n^\ddagger$ to the $s\chi^n$, giving an action

$$S[\chi, \chi^\ddagger] \equiv I[\phi] + (s\chi^n)\,\chi_n^\ddagger\,. \tag{15.9.1}$$

This satisfies what is called the *master equation*

$$0 = \frac{\delta_R S}{\delta \chi_n^\ddagger}\frac{\delta_L S}{\delta \chi^n}\,, \tag{15.9.2}$$

with 'R' and 'L' here denoting right- and left-differentiation. To check this, note that the terms in Eq. (15.9.2) of zeroth order in antifields just yield the condition of gauge invariance

$$0 = \left(s\phi^r\right)\frac{\delta_L I}{\delta \phi^r} = \omega^A\,\delta_A I[\phi]\,, \tag{15.9.3}$$

while the terms linear in the antifields $\phi_n^\ddagger$ provide the condition of nilpotence:

$$0 = \left(s\chi^m\right)\frac{\delta_L (s\chi^n)}{\delta \chi^m} = s^2\chi^n\,. \tag{15.9.4}$$

The $\chi_n^\ddagger$ are external fields, and must be given suitable values before we use $S[\chi, \chi^\ddagger]$ to calculate the S-matrix. For this purpose, we introduce an arbitrary fermionic functional $\Psi[\chi]$ with ghost number -1, and set†

$$\chi_n^\ddagger = \frac{\delta \Psi[\chi]}{\delta \chi^n}\,. \tag{15.9.5}$$

Then Eq. (15.9.1) becomes

$$S[\phi, \delta\Psi/\delta\chi] = I[\phi] + (s\chi^n)\,\delta\Psi[\chi]/\delta\chi^n = I[\phi] + s\Psi[\chi]\,. \tag{15.9.6}$$

Comparison with Eq. (15.8.14) shows that this is the same as the gauge-fixed action $I_{\text{NEW}}[\chi]$. Thus, using the same arguments as in the previous section, physical matrix elements are not affected by small changes in Ψ. The action (15.8.7) constructed by the Faddeev–Popov–De Witt method corresponds to the choice $\Psi = -\omega^{*A} f_A$, for which $\phi_r^\ddagger = \omega^{*A}\,\delta f_A/\delta\phi^r$, $\omega_C^\ddagger = 0$, and $\omega_A^{*\ddagger} = -f_A$.

So far, nothing new has been accomplished. The first new point is that the master equation (15.9.2) can be used for more general theories

† It is not necessary to distinguish between left- and right-differentiation here, because either χ^n or $\delta\Psi/\delta\chi^n$ must be bosonic.

by letting $S[\chi, \chi^\ddagger]$ be a non-linear functional of the antifields $\chi^\ddagger_n$. (For reducible theories we must also include ghosts for ghosts among the χ^n, as discussed in the previous section, along with their antifields.) As above, we take the statistics of $\chi^\ddagger_n$ to be opposite to that of χ^n, and its ghost number equal to $-\text{gh}(\chi^n) - 1$, and we require $S[\chi, \chi^\ddagger]$ to be a bosonic operator of ghost number zero. Because ω^{*A} and h_A have *linear* BRST transformations they are unaffected by the complications affecting the other χ^n (in this connection, see Section 16.4), so they and their antifields enter in the action $S[\chi, \chi^\ddagger]$ in the same way as in Eq. (15.9.1). That is,

$$S = S_{\text{min}}[\phi, \omega, \phi^\ddagger, \omega^\ddagger] - h^A \, \omega^{*\ddagger}_A \,, \qquad (15.9.7)$$

where $\phi^\ddagger_r$, $\omega^\ddagger_A$, and $\omega^{*\ddagger}_A$ are the antifields of ϕ^r, ω^A, and ω^{*A}, with ghost numbers -1, -2, and 0, respectively, and $S_{\text{min}}[\phi, \omega, \phi^\ddagger, \omega^\ddagger]$ is some bosonic functional of ghost number zero. The last term in (15.9.7) has no effect on the master equation, so S_{min} satisfies the master equation by itself.††

Because S_{min} has ghost number zero, its expansion in powers of antifields must take the form

$$\begin{aligned} S_{\text{min}} = {} & I[\phi] + \omega^A f^r_A[\phi] \, \phi^\ddagger_r + \tfrac{1}{2}\omega^A \omega^B f^C{}_{AB}[\phi] \, \omega^\ddagger_C \\ & + \tfrac{1}{2}\omega^A \omega^B f^{rs}{}_{AB}[\phi] \, \phi^\ddagger_r \, \phi^\ddagger_s + \omega^A \omega^B \omega^C f^{rD}{}_{ABC}[\phi] \, \phi^\ddagger_r \omega^\ddagger_D \\ & + \tfrac{1}{2}\omega^A \omega^B \omega^C \omega^D f^{EF}{}_{ABCD}[\phi] \, \omega^\ddagger_E \omega^\ddagger_F + \dots \end{aligned} \qquad (15.9.8)$$

The term in the master equation (15.9.2) of zeroth order in antifields (and hence first order in ω^A) yields

$$0 = f^r_A[\phi] \, \frac{\delta I[\phi]}{\delta \phi^r} \,, \qquad (15.9.9)$$

which is just the statement that $I[\phi]$ is invariant under the transformation

$$\phi^r \to \phi^r + \epsilon^A f^r_A[\phi] \qquad (15.9.10)$$

with ϵ^A arbitrary infinitesimals. The term in the master equation proportional to $\phi^\ddagger_s$ on the right and $\omega^A \omega^B$ on the left yields

$$\begin{aligned} 0 = {} & f^r_A[\phi] \, \frac{\delta f^s_B[\phi]}{\delta \phi^r} - f^r_B[\phi] \, \frac{\delta f^s_A[\phi]}{\delta \phi^r} + f^C{}_{AB}[\phi] \, f^s_C[\phi] \\ & + \frac{\delta I[\phi]}{\delta \phi^r} \, f^{rs}_{AB}[\phi] \,, \end{aligned} \qquad (15.9.11)$$

which, when the field equations $\delta I / \delta \phi = 0$ are satisfied, becomes the commutation relation (with structure constant $f^C{}_{AB}[\phi]$) for the transformation (15.9.10). The other term in the master equation linear in antifields

†† The fields ϕ^r, ω^A, $\phi^\ddagger_r$, $\omega^\ddagger_A$ are sometimes called *minimal variables*, while fields like ω^{*A} and h_A which together with their antifields enter bilinearly as in Eq. (15.9.7), are called *trivial pairs*.

is proportional to $\omega^A\omega^B\omega^C$ on the left and $\omega_D^\ddagger$ on the right; it yields

$$0 = f^r_{[A}[\phi]\,\frac{\delta f^D{}_{BC]}[\phi]}{\delta\phi^r} - f^E{}_{[AB}[\phi]\,f^D{}_{C]E}[\phi] + f^{rD}_{ABC}[\phi]\,\frac{\delta I[\phi]}{\delta\phi^r}\,, \qquad (15.9.12)$$

where square brackets in the subscripts indicate antisymmetrization with respect to the enclosed indices A, B, and C. When the field equations are satisfied this becomes the generalized Jacobi identity (15.8.12). Eq. (15.9.11) is necessary for the consistency of the symmetry condition (15.9.9) (assuming that the f^r_A furnish a complete set of gauge symmetries), while Eq. (15.9.12) is necessary for the consistency of the commutation relations (15.9.11). Note that the terms in Eqs. (15.9.11) and (15.9.12) that arise from the terms in $S_{\rm min}$ quadratic in antifields are proportional to $\delta I[\phi]/\delta\chi$, so they vanish when the field equations are satisfied, and in this sense are characteristic of open symmetry algebras. Terms in the master equation of second or higher order in antifields involve terms in $S_{\rm min}$ that are of third and/or higher order in antifields. These provide consistency conditions for Eqs. (15.9.11) and (15.9.12), consistency conditions for these consistency conditions, and so on. It is a virtue of the Batalin–Vilkovisky formalism that all these consistency conditions are incorporated in the one master equation.

The master equation may be reinterpreted as a statement of invariance of S under a generalized BRST transformation. In order to see this, and for future purposes, it is useful to introduce a formal device known as the *antibracket*. Returning now to our previous notation, the antibracket of two general functionals $F[\chi,\chi^\ddagger]$ and $G[\chi,\chi^\ddagger]$ is defined by

$$(F,G) \equiv \frac{\delta_R F}{\delta\chi^n}\,\frac{\delta_L G}{\delta\chi^\ddagger_n} - \frac{\delta_R F}{\delta\chi^\ddagger_n}\,\frac{\delta_L G}{\delta\chi^n}\,. \qquad (15.9.13)$$

Note that right- and left-functional derivatives of a bosonic functional like S with respect to a bosonic or fermionic field variable are equal to each other or negatives of each other, respectively. Since just one of either $\chi^\ddagger_n$ or χ^n is always fermionic and the other bosonic, it follows that for the antibracket (S,S) the second term on the right-hand side in Eq. (15.9.13) changes sign if we reverse left- and right-differentiation

$$\frac{\delta_R S}{\delta\chi^\ddagger_n}\,\frac{\delta_L S}{\delta\chi^n} = -\frac{\delta_L S}{\delta\chi^\ddagger_n}\,\frac{\delta_R S}{\delta\chi^n} = -\frac{\delta_R S}{\delta\chi^n}\,\frac{\delta_L S}{\delta\chi^\ddagger_n}\,.$$

(The last step is allowed because, since one of the factors here is bosonic, their order is immaterial.) We see that the second term on the right in Eq. (15.9.13) for (S,S) is the negative of the first. The master equation (15.9.2) may therefore be written as the requirement that the antibracket of S with itself vanishes:

$$(S,S) = 0\,. \qquad (15.9.14)$$

This is a non-trivial requirement, because the antibracket has the general symmetry property

$$(F,G) = \pm(G,F)\,, \tag{15.9.15}$$

the sign being +1 where F and G are both bosonic, and −1 otherwise. In particular, (F,F) automatically vanishes if F is fermionic, but not if F is bosonic.

The generalized BRST transformation is defined by

$$\hat{\delta}_\theta \chi^n = \theta \frac{\delta_R S}{\delta \chi^\ddagger_n} = -\theta(S,\chi^n)\,, \tag{15.9.16}$$

$$\hat{\delta}_\theta \chi^\ddagger_n = -\theta \frac{\delta_R S}{\delta \chi^n} = -\theta(S,\chi^\ddagger_n)\,, \tag{15.9.17}$$

where θ is a fermionic infinitesimal constant. (When S is of the form (15.9.1), the transformation of χ is the same as the original BRST transformation $\hat{\delta}_\theta \chi^n = \theta s \chi^n$.) To evaluate the effect of this transformation on general functionals, we note that the antibracket acts as a derivative, in the sense that

$$(F,GH) = (F,G)\,H \pm G\,(F,H)\,, \tag{15.9.18}$$

where the sign is −1 if G is fermionic and F is bosonic, and +1 otherwise. Hence if G and H are arbitrary functionals of χ and $\chi^\ddagger$ with $\hat{\delta}_\theta G = -\theta(S,G)$ and $\hat{\delta}_\theta H = -\theta(S,H)$, then

$$\hat{\delta}_\theta(GH) = -\theta(S,G)H - G\theta(S,H) = -\theta[(S,G)H \pm G(S,H)]\,,$$

where the sign is + or − if G is bosonic or fermionic. Taking F in Eq. (15.9.18) equal to the bosonic functional S, we see then that

$$\hat{\delta}_\theta(GH) = -\theta(S,GH)\,.$$

Together with Eqs. (15.9.16) and (15.9.17), this shows that for any functional F formed as a sum of products of fields and antifields

$$\hat{\delta}_\theta F = -\theta(S,F)\,. \tag{15.9.19}$$

The master equation (15.9.14) may be interpreted as the statement that these generalized BRST transformations leave S invariant

$$\hat{\delta}_\theta S = -\theta(S,S) = 0\,. \tag{15.9.20}$$

Like the original BRST transformation, this symmetry transformation is nilpotent. To see this, we use the Jacobi identity for the antibracket:

$$\pm(F,(G,H)) + \text{cyclic permutations} = 0\,, \tag{15.9.21}$$

where the sign in the first term is − if F and H are bosonic and + otherwise, with corresponding signs for the other two cyclic permutations of F, G, and H. Taking $F = G = S$, Eq. (15.9.21) becomes

$$0 = \mp(S,(S,H)) \mp (H,(S,S)) + (S,(H,S) = \mp 2(S,(S,H)) \mp (H,(S,S))\,,$$

where the sign is $-$ or $+$ if H is bosonic or fermionic. The master equation (15.9.14) then yields the nilpotency condition

$$(S,(S,H)) = 0\,. \tag{15.9.22}$$

Because of this symmetry, the solution of the master equation is not unique. For instance, Eq. (15.9.22) shows that for any given solution S we can find another solution given by the infinitesimal transformation

$$S' = S + (\delta F, S)\,, \tag{15.9.23}$$

with δF an infinitesimal functional of χ and $\chi^\ddagger$ that is arbitrary, except that it must be fermionic and have ghost number -1 in order that S' should be bosonic and have ghost number zero. In particular, taking δF to be a fermionic functional $\epsilon\Psi$ of χ^n alone gives

$$S'[\chi,\chi^\ddagger] = S[\chi,\chi^\ddagger] + \epsilon \frac{\delta\Psi[\chi]}{\delta\chi^n}\frac{\delta_R S}{\delta\chi_n^\ddagger} = S\left[\chi, \chi^\ddagger + \epsilon\frac{\delta\Psi}{\delta\chi}\right]\,. \tag{15.9.24}$$

These infinitesimal transformations may be trivially integrated, and show that the master equation is still satisfied if we shift the antifields to new variables $\chi_n^{\ddagger\prime} \equiv \chi_n^\ddagger - \delta\Psi/\delta\chi^n$.

The transformation (15.9.23) is a special case of what are usually called canonical transformations, and will here be called 'anticanonical transformations' to distinguish them from the canonical transformations of Chapter 7. An anticanonical transformation is any transformation of fields and antifields, finite or infinitesimal, that leaves unchanged the fundamental antibracket relations:

$$(\chi^n, \chi_m^\ddagger) = \delta^n_m\,, \qquad (\chi^n,\chi^m) = (\chi_n^\ddagger, \chi_m^\ddagger) = 0\,. \tag{15.9.25}$$

For instance, consider the infinitesimal anticanonical transformation generated by an infinitesimal fermion generator δF, under which any bosonic or fermionic functional G is transformed into

$$G \to G' = G + (\delta F, G)\,. \tag{15.9.26}$$

It is easy to show that this does not change the fundamental antibrackets (15.9.25). For this purpose, note that the antibracket (G,H) of two functionals G and H is transformed into (G',H'), which to first order in infinitesimals is

$$(G',H') = (G,H) + ((\delta F),G),H) + (G,(\delta F,H))\,.$$

Using the Jacobi identity (15.9.21), this is

$$(G',H') = (G,H) \pm (\delta F,(G,H))\,, \tag{15.9.27}$$

with the sign + if G and H are both bosonic, and − otherwise. In particular, if (G, H) is a c-number then it is unchanged by an anticanonical transformation. (This is another way of seeing that the transformation (15.9.23) leaves the master equation unchanged.) The fields and antifields have c-number antibrackets (15.9.25), so the same must be true for the transformed fields and antifields.

To calculate the S-matrix we must give some definite value to the antifields. As in the simple case of closed gauge algebras where S is of the linear form (15.9.5), we can do this by taking the antifields in the form (15.9.5); that is, we calculate the S-matrix using the 'gauge-fixed' action

$$I_\Psi[\chi] = S\left[\chi, \frac{\delta\Psi[\chi]}{\delta\chi}\right] , \tag{15.9.28}$$

where $\Psi[\chi]$ is a fermionic functional of ghost number −1. According to the remarks following Eq. (15.9.24), this is the same as taking the canonically transformed antifields $\chi^{\ddagger'}$ equal to zero.

The gauge-fixed action is invariant under a BRST transformation that acts on fields χ^n alone:

$$\delta_\theta \chi^n = \theta s \chi^n , \qquad \text{where} \qquad s\chi^n = \left(\frac{\delta_R S[\chi, \chi^\ddagger]}{\delta \chi^\ddagger_n}\right)_{\chi^\ddagger = \delta\Psi/\delta\chi} . \tag{15.9.29}$$

To check this, note that

$$s I_\Psi[\chi] = \left(\frac{\delta_R S[\chi, \chi^\ddagger]}{\delta \chi^\ddagger_n} \frac{\delta_L S[\chi, \chi^\ddagger]}{\delta \chi^n}\right)_{\chi^\ddagger = \delta\Psi/\delta\chi} + \left(\frac{\delta_R S[\chi, \chi^\ddagger]}{\delta \chi^\ddagger_m} \frac{\delta_L^2 \Psi[\chi]}{\delta \chi^m\, \delta\chi^n} \frac{\delta_L S[\chi, \chi^\ddagger]}{\delta \chi^\ddagger_n}\right)_{\chi^\ddagger = \delta\Psi/\delta\chi} .$$

The first term on the right-hand side vanishes as a result of the master equation, and the second because the summand is antisymmetric‡ in m and n.

For closed algebras with S of the form (15.9.1), the transformation (15.9.29) is the original BRST transformation $\delta_\theta \chi^n = \theta s \chi^n$. But for general open algebras, the transformation (15.9.29) is not the same as the original BRST transformation, and in general is not even nilpotent unless the field equations are satisfied. Instead, from the terms in the master equation of

‡ For χ^n and χ^m both bosonic this is because $\delta S/\delta\chi^n$ and $\delta S/\delta\chi^m$ anticommute. For one of χ^n and χ^m fermionic and the other bosonic, this is because the right- and left-derivatives of S with respect to whichever χ is fermionic have opposite signs. The terms with χ^n and χ^m both fermionic are antisymmetric because for these terms $\delta_L^2\Psi/\delta\chi^m\delta\chi^n$ is antisymmetric.

first order in the shifted antifields $\chi_n^{\ddagger'} = \chi_n^{\ddagger} - \delta\Psi[\chi]/\delta\chi^n$, we find

$$s^2\chi^m = \mp\left(\frac{\delta_L}{\delta\chi_m^{\ddagger}}\frac{\delta_R}{\delta\chi_n^{\ddagger}}I[\chi,\chi^{\ddagger}]\right)_{\chi^{\ddagger}=\delta\Psi/\delta\chi}\frac{\delta I_\Psi[\chi]}{\delta\chi^n}\,, \tag{15.9.30}$$

with the sign $-$ or $+$ if χ^m is bosonic or fermionic, respectively. Again, we see that the dependence on the field equations characteristic of open gauge algebras is associated with terms in S quadratic in the antifields.

Until now we have been considering only the formulation of a classical field theory based on an open or closed gauge algebra. Now we must consider how quantum mechanical calculations are done in such theories. Physical matrix elements may be calculated by functional integrals weighted with $\exp(iI_\Psi[\chi])$, where as explained above, $I_\Psi[\chi]$ is obtained from $S[\chi,\chi^{\ddagger}]$ by setting $\chi_n^{\ddagger} = \delta\Psi[\chi]/\delta\chi^n$, or in other words $\chi_n^{\ddagger'} = 0$. We wish to evaluate the effect of a change in $\Psi[\chi]$ on these matrix elements. First consider the vacuum–vacuum amplitude

$$Z_\Psi = \int\left[\prod d\chi\right]\,\exp(iI_\Psi[\chi])\,. \tag{15.9.31}$$

Under a shift $\delta\Psi[\chi]$ in $\Psi[\chi]$, this changes by an amount

$$\delta Z = i\int\left[\prod d\chi\right]\,\exp(iI_\Psi[\chi])\left(\frac{\delta_R S[\chi,\chi^{\ddagger}]}{\delta\chi_n^{\ddagger}}\right)_{\chi^{\ddagger}=\delta\Psi/\delta\chi}\left(\frac{\delta\,(\delta\Psi[\chi])}{\delta\chi^n}\right)\,. \tag{15.9.32}$$

Integrating by parts in field space, this becomes

$$\delta Z = \int\left[\prod d\chi\right]\,\exp(iI_\Psi[\chi]) \times\left\{\frac{\delta_R S[\chi,\chi^{\ddagger}]}{\delta\chi_n^{\ddagger}}\frac{\delta_L I_\Psi[\chi]}{\delta\chi^n} - i\Delta S[\chi,\chi^{\ddagger}]\right\}_{\chi^{\ddagger}=\delta\Psi/\delta\chi}\delta\Psi[\chi]\,, \tag{15.9.33}$$

where

$$\Delta \equiv \frac{\delta_R}{\delta\chi_n^{\ddagger}}\frac{\delta_L}{\delta\chi^n}\,. \tag{15.9.34}$$

We see that the condition for the Ψ independence of the vacuum–vacuum amplitude is in general *not* the master equation (15.9.2), but rather what is called the *quantum master equation*

$$(S,S) - 2i\Delta S = 0 \quad \text{at} \quad \chi^{\ddagger} = \delta\Psi/\delta\chi\,. \tag{15.9.35}$$

In cgs units a factor $1/\hbar$ would accompany each factor of $S[\chi,\chi^{\ddagger}]$, so the second term in Eq. (15.9.35) would have a coefficient $-2i\hbar$ in place of $-2i$. Thus whenever the quantum master equation (15.9.35) is satisfied, the term in S of zeroth order in $\hbar$ satisfies the original master equation (15.9.2). Usually it is easy to construct an action that satisfies the classical master equation, so that the first term in Eq. (15.9.35) vanishes, and the question is then whether the second term also vanishes. The case where the quantum master equation is not satisfied by a local action is considered in Chapter 22, on anomalies.

Assuming the quantum master equation (15.9.35) to be satisfied, the change in the vacuum expectation value of an operator $\mathcal{O}[\chi]$ due to a change $\delta\Psi$ in Ψ is

$$\delta\langle\mathcal{O}\rangle = \frac{-i}{Z_\Psi}\int\left[\prod d\chi\right]\,\exp\Big(iI_\Psi[\chi]\Big)\,\frac{\delta_R\mathcal{O}[\chi]}{\delta\chi^n}\left(\frac{\delta_R S(\chi,\chi^\ddagger)}{\delta\chi^\ddagger_n}\right)_{\chi^\ddagger=\delta\Psi/\delta\chi}\delta\Psi[\chi]\,. \tag{15.9.36}$$

The coefficient of the exponential in the integrand in Eq. (15.9.36) is just $s\mathcal{O}[\chi]$. We see that expectation values of operators that are invariant‡‡ under the generalized BRST transformation (15.9.16)–(15.9.17) are unaffected by a change in the gauge-fixing fermion Ψ. Corresponding results hold for vacuum expectation values of two or more operators.

Appendix A A Theorem Regarding Lie Algebras

In this appendix we consider a general Lie algebra $\mathcal{G}$, with generators t_α and structure constants $C^\alpha{}_{\beta\gamma}$, and prove the equivalence of three conditions:

a: There exists a real symmetric positive-definite matrix $g_{\alpha\beta}$ that satisfies the invariance condition

$$g_{\alpha\beta}C^\beta{}_{\gamma\delta} = -\,g_{\gamma\beta}C^\beta{}_{\alpha\delta}\,. \tag{15.A.1}$$

(This is the condition (15.2.4) shown in Section 15.2 to be necessary on physical grounds.)

b: There is a basis for the Lie algebra (that is, a set of generators $\tilde{t}_\alpha = \mathcal{S}_{\alpha\beta}t_\beta$, with $\mathcal{S}$ a real non-singular matrix) for which the structure constants $\tilde{C}^\alpha{}_{\beta\gamma}$ are antisymmetric not only in the lower indices β and γ but in all three indices α, β and γ.

c: The Lie algebra $\mathcal{G}$ is the direct sum of commuting compact simple and $U(1)$ subalgebras $\mathcal{H}_m$.

‡‡ For open gauge theories there are generally no operators other than constants that are invariant under the transformations (15.9.16)–(15.9.17). Instead one should consider operators $O(\chi,\chi^\ddagger)$ that are invariant under the nilpotent 'quantum BRST operator' σ, defined by $\sigma O = (O,S) - i\Delta O$. Where O depends only on χ, this condition reduces to Eq. (15.9.36). If $\sigma O = 0$ then the expectation value of $O(\Psi,\delta\Psi/\delta\chi)$ is unaffected by a small change in the gauge-fixing fermion Ψ.[24]

We shall prove the equivalence of statements **a**, **b**, and **c** by showing that **a** implies **b**, **b** implies **c**, and **c** implies **a**. As a by-product, we shall also show that if these conditions are satisfied, then it is possible to choose the generators of $\mathscr{G}$ as t_{ma}, with m labelling the simple or $U(1)$ subalgebra $\mathscr{H}_m$ to which t_{ma} belongs and a labelling the individual generators in this subalgebra, and with the matrix $g_{ma,nb}$ that satisfies Eq. (15.A.1) taking the form

$$g_{ma,nb} = g_m^{-2}\delta_{mn}\delta_{ab} , \tag{15.A.2}$$

where the g_m^{-2} are arbitrary real positive constants.

First, let us assume **a**, the existence of a real symmetric positive-definite $g_{\alpha\beta}$ satisfying the invariance condition (15.A.1). We may then define new generators

$$\tilde{t}_\alpha \equiv (g^{-1/2})_{\alpha\beta} t_\beta , \tag{15.A.3}$$

in which the existence of a real inverse square-root matrix $g^{-1/2}$ is guaranteed by the positive-definiteness of $g_{\alpha\beta}$. These satisfy a Lie algebra

$$\left[\tilde{t}_\alpha, \tilde{t}_\beta\right] = i\, \tilde{C}_{\alpha\beta\gamma}\tilde{t}_\gamma , \tag{15.A.4}$$

where

$$\tilde{C}_{\gamma\alpha\beta} \equiv (g^{-1/2})_{\alpha\alpha'}(g^{-1/2})_{\beta\beta'}(g^{+1/2})_{\gamma\gamma'}C^{\gamma'}{}_{\alpha'\beta'} . \tag{15.A.5}$$

(In this basis it is convenient to drop the distinction between upper and lower indices α, β, etc., and write $\tilde{C}_{\gamma\alpha\beta}$ in place of $\tilde{C}^{\gamma}{}_{\alpha\beta}$.) Then Eq. (15.A.1) tells us that $\tilde{C}_{\alpha\gamma\delta}$ is antisymmetric in α and γ as well as in γ and δ, and hence is totally antisymmetric, verifying **b**.

Next let us assume **b**, the existence of a basis for the Lie algebra for which the structure constants are totally antisymmetric. In this basis, the matrices $(\tilde{t}^{A}{}_\alpha)_{\beta\gamma} \equiv -i\tilde{C}_{\beta\gamma\alpha}$ of the adjoint representation are imaginary and antisymmetric, and hence Hermitian. According to a general theorem about Hermitian matrices,* the $\tilde{t}^{A}{}_\alpha$ are then either irreducible or totally reducible.

* If a set of matrices H_α is not irreducible, then by definition there must be a set of vectors u_n that span a subspace (other than the whole space) that is left invariant by the H_α: that is, for all α and n, $H_\alpha u_n = \sum_m (C_\alpha)_{mn} u_m$. In this case we can adopt a basis consisting of the vectors u_n together with a set v_k that span the space orthogonal to all the u_n. If the H_α are Hermitian then $(u_n, H_\alpha v_k) = \sum_m (C_\alpha)^*_{mn}(u_m, v_k) = 0$, so the space spanned by the v_k is also invariant under the H_α: $H_\alpha v_k = \sum_\ell (D_\alpha)_{\ell k} v_\ell$. In this basis the matrices H_α are simultaneously reduced to a block-diagonal form:

$$H_\alpha = \begin{pmatrix} C_\alpha & 0 \\ 0 & D_\alpha \end{pmatrix} .$$

Continuing in this way, we can completely reduce the matrices H_α to a block-diagonal form with irreducible matrices in the blocks.

By an irreducible set of $N \times N$ matrices $\tilde{t}^A{}_\alpha$ is meant a set for which there exists no subspace of dimensionality $< N$ that is left invariant by all the $\tilde{t}^A{}_\alpha$ — that is, no set of less than N non-zero vectors $(u_r)_\beta$ for which $(\tilde{t}^A{}_\alpha)_{\beta\gamma}(u_r)_\gamma$ is for each α and r a linear combination of vectors $(u_s)_\beta$. Since the matrices $(\tilde{t}^A{}_\alpha)_{\beta\gamma}$ are proportional to the structure constants in this basis, this is equivalent to the statement that there is no set of linear combinations $\mathscr{T}_r \equiv (u_r)_\gamma \tilde{t}^A{}_\gamma$ that is closed under commutation with all the $\tilde{t}^A{}_\alpha$, that is, for which $[\tilde{t}^A{}_\alpha, \mathscr{T}_r]$ is for each α and r a linear combination of the $\mathscr{T}_s$. Such a set of matrices $\mathscr{T}_r$ would furnish the generators of an invariant subalgebra of the full Lie algebra; the absence of such a set means that the Lie algebra is simple.

By a totally reducible set of matrices $\tilde{t}^A{}_\alpha$, is meant one that by a suitable choice of basis may be written as block-diagonal supermatrices

$$(\tilde{t}^A{}_\alpha)_{ma,nb} = [t^{A(m)}{}_\alpha]_{ab}\,\delta_{mn}\,, \tag{15.A.6}$$

where the submatrices $t^{A(m)}{}_\alpha$ are either irreducible or vanish.** Adopting this basis also for the Lie algebra itself, the structure constants are then

$$\tilde{C}_{\ell c,ma,nb} = i(\tilde{t}^A{}_{\ell c})_{ma,nb} = i\,(t^{A(m)}{}_{\ell c})_{ab}\,\delta_{mn}\,. \tag{15.A.7}$$

But since this is totally antisymmetric, and proportional to δ_{mn}, it must also be proportional to $\delta_{\ell n}$ and $\delta_{\ell m}$:

$$\tilde{C}_{\ell c,ma,nb} = \delta_{\ell n}\,\delta_{\ell m}\,C^{(\ell)}_{cab}\,. \tag{15.A.8}$$

In other words, for *any* representation $t^{(m)}{}_a \equiv t_{ma}$ of the Lie algebra in this basis, we have

$$[t^{(m)}{}_a, t^{(n)}{}_b] = i\,\delta_{mn} C^{(m)}_{cab} t^{(m)}{}_c\,, \tag{15.A.9}$$

with $C^{(m)}_{cab}$ real and totally antisymmetric in the indices a, b, c. The fact that it is possible to construct a basis in which the generators fall into sets $t^{(m)}$, with the commutators of the generators in one set with each other given by a linear combination of generators in that set, and with all members of one set commuting with all members of any other set, is what we mean when we say that the Lie algebra is a *direct sum* of the subalgebras $t^{(m)}$. For each m, the set of matrices of the adjoint representation $t^{(m)A}$ is either irreducible or zero, corresponding to a subalgebra that is either simple or consists of so-called $U(1)$ generators, which commute with all generators of the whole algebra.

We have thus shown that the most general Lie algebra with totally antisymmetric real structure constants is a direct sum of one or more

** Here m and n label the blocks along the main diagonal, and a and b label rows and columns within these blocks. Also, m is not summed, and the range of the indices a, b, etc. in general depends on m.

simple and/or $U(1)$ Lie algebras. Furthermore, the simple subalgebras are compact, in the sense that each matrix $-C^{(m)}_{acd}C^{(m)}_{bdc}$ is positive-definite, because for any real vector u_a, $-C^{(m)}_{acd}C^{(m)}_{bdc}u_a u_b = \sum_{cd}[\sum_a u_a C^{(m)}_{acd}]^2$ is a sum of positive quantities, which cannot vanish unless $u_a = 0$, since for $u_a \neq 0$ the condition that $\sum_a u_a C^{(m)}_{acd} = 0$ would imply that $\sum_a u_a t^{(m)}{}_a$ is itself an invariant Abelian subalgebra, in contradiction with the fact that the $t^{(m)}{}_a$ form a simple Lie algebra. This completes the proof of **c**.

Finally, let us assume **c**, that we have a Lie algebra that is the direct sum of a set of simple or $U(1)$ Lie algebras; that is, in some basis $t^{(m)}_a = \mathscr{S}_{ma,\alpha}t_\alpha$ with real non-singular $\mathscr{S}$, we have

$$[t^{(m)}_a, t^{(n)}_b] = i\delta_{nm}C^{(m)c}{}_{ab}t^{(m)}_c \ ,$$

where each subalgebra $t^{(m)}$ is either simple or commutes with everything. We furthermore assume that the simple subalgebras are compact, in the sense that the matrices

$$g^{(m)}_{ab} \equiv -C^{(m)c}{}_{ad}C^{(m)d}{}_{bc} \tag{15.A.10}$$

are positive-definite. To construct a real positive-definite matrix $g_{\alpha\beta}$ satisfying Eq. (15.A.1), in the basis of generators $t_{ma} = t^{(m)}_a$, we take

$$g_{ma,nb} \equiv g^{(m)}_{ab}\delta_{mn} \ , \tag{15.A.11}$$

where $g^{(m)}_{ab}$ is taken as the matrix (15.A.10) when $t^{(m)}$ is a simple subalgebra, and is taken as an arbitrary real symmetric positive-definite matrix when $t^{(m)}$ is a direct sum of one or more $U(1)$ subalgebras. The matrix (15.A.11) is obviously real, symmetric, and positive-definite because each $g^{(m)}_{ab}$ is. To check the requirement (15.A.1), recall the Jacobi identity for the structure constants:

$$C^{(m)c}{}_{ad}\,C^{(m)d}{}_{be} + C^{(m)c}{}_{bd}\,C^{(m)d}{}_{ea} + C^{(m)c}{}_{ed}\,C^{(m)d}{}_{ab} = 0 \tag{15.A.12}$$

and contract with $C^{(m)e}{}_{fc}$. After renaming the summation indices in the third term so that $c \to d \to e \to c$ and using Eq. (15.A.10), the result for the simple subalgebras may be written:

$$g^{(m)}_{df}C^{(m)d}{}_{ab} = C^{(m)c}{}_{ad}C^{(m)d}{}_{be}C^{(m)e}{}_{cf} - C^{(m)c}{}_{fd}C^{(m)d}{}_{be}C^{(m)e}{}_{ca} \, .$$

The important point is that this shows that the left-hand side is antisymmetric in a and f:

$$g^{(m)}_{df}C^{(m)d}{}_{ab} = -g^{(m)}_{da}C^{(m)d}{}_{fb} \, . \tag{15.A.13}$$

The same result holds trivially for the $U(1)$ subalgebras, where the structure constants vanish. The symmetry condition (15.A.1) follows immediately from Eq. (15.A.13), thus completing the proof of **a**. This concludes our proof of the equivalence of statements **a**, **b**, and **c**.

Now we come back to the matrix $g_{\alpha\beta}$. With totally antisymmetric structure constants, the invariance condition (15.2.4) may be expressed as the statement that this matrix commutes with all the matrices in the adjoint representation of the Lie algebra

$$[g, t^{A}{}_{\gamma}] = 0 . \qquad (15.A.14)$$

We have seen that all $(t^{A}{}_{\gamma})_{\alpha\beta}$ can be put in block-diagonal form, with irreducible (or zero) submatrices along the main diagonal. A well-known theorem[25] tells us that then $g_{\alpha\beta}$ must also be block-diagonal, with blocks of the same size and position as in the $t^{A}{}_{\gamma}$, and with the submatrix in each block proportional to the unit matrix. (Where two of the submatrices in the $t^{A}{}_{\gamma}$ are equivalent, in the sense that they can be related by a similarity transformation, it may be necessary to make a suitable change of basis in order to bring the submatrices of $g_{\alpha\beta}$ into a form proportional to unit matrices.) In the notation of Eq. (15.A.11), the metric is then given by Eq. (15.A.2).

Appendix B The Cartan Catalog

We present here without proof the complete catalog of simple Lie algebras, worked out in its final form by E. Cartan.[26] These will be presented here in their 'compact' form — that is, with generators that can be faithfully represented by finite-dimensional Hermitian matrices. The Lie algebras will be labelled with a subscript $n \geq 1$ indicating their 'rank' — the number of independent commuting linear combinations of generators.

A_n: This is the algebra of the special unitary group $SU(n+1)$, the group of all unitary ($U^\dagger = U^{-1}$) unimodular ($\text{Det}\, U = 1$) matrices in $n+1$ dimensions. Any such matrix that is infinitesimally close to the identity may be expressed as

$$U = 1 + iH ,$$

with infinitesimal H satisfying the conditions

$$H^\dagger = H , \quad \text{Tr}\, H = 0 ,$$

so A_n is the Lie algebra of all Hermitian traceless matrices in $n+1$ dimensions. Any set of commuting Hermitian matrices may be simultaneously diagonalized, and the maximum number of independent diagonal *traceless* matrices in $n+1$ dimensions is n, so this is the rank of A_n. Any Hermitian matrix in $n+1$ dimensions has $(n+1)^2$ independent real parameters ($n+1$ real numbers on the main diagonal, and $n(n+1)/2$ complex numbers above the main diagonal, equal to the complex conjugates of those below it), and one of these is eliminated by the tracelessness condition, so the dimensionality of A_n is

$$d(A_n) = (n+1)^2 - 1 = n(n+2)\,.$$

All of the A_n are simple.

$\boldsymbol{B_n}$: This is the algebra of the unitary orthogonal group $O(2n+1)$, consisting of all unitary ($\mathcal{O}^\dagger = \mathcal{O}^{-1}$) and orthogonal ($\mathcal{O}^{\mathrm{T}} = \mathcal{O}^{-1}$) and hence real matrices in $2n+1$ dimensions. (The restriction to unitary matrices is sometimes indicated by calling the group $UO(2n+1)$.) Any such matrix $\mathcal{O}$ that is infinitesimally close to the identity may be expressed as

$$\mathcal{O} = 1 + iA\,,$$

where A is an infinitesimal matrix satisfying the conditions

$$A^* = -A = A^{\mathrm{T}}\,.$$

(It would make no difference here if we restricted ourselves to the subgroup $SO(2n+1)$, for which $\mathcal{O}$ is subject to the further condition $\mathrm{Det}\,\mathcal{O} = 1$, since any orthogonal matrix $\mathcal{O}$ that is close to the identity would have $\mathrm{Det}\,\mathcal{O} = 1$ anyway.) Any set of commuting imaginary antisymmetric $(2n+1)$-dimensional matrices may (by a common orthogonal transformation) be put in the form of a supermatrix

$$\begin{bmatrix} a_1\sigma_2 & & & & 0 \\ & a_2\sigma_2 & & & \\ & & \ddots & & \\ & & & a_n\sigma_2 & \\ 0 & & & & 0 \end{bmatrix}$$

where σ_2 is the usual 2×2 matrix

$$\sigma_2 = \begin{bmatrix} 0 & -i \\ i & 0 \end{bmatrix}$$

and $a_1, \cdots, a_n$ are real. The rank of B_n is thus evidently n. An imaginary antisymmetric matrix is completely specified by the imaginary numbers above the real diagonal, so its dimensionality is

$$d(B_n) = \frac{(2n+1)(2n)}{2} = n(2n+1)\,.$$

All of the B_n are simple.

There is an alternative definition of $O(N)$ that will help in understanding the motivation for the next large set of simple Lie algebras. Instead of defining $O(N)$ as the group of N-dimensional real matrices that satisfy the orthogonality condition $\mathcal{O}^{\mathrm{T}}\mathcal{O} = 1$, it can just as well be defined as the group of N-dimensional real matrices $\mathscr{M}$ that satisfy the condition

$$\mathscr{M}^{\mathrm{T}}\mathscr{P}\mathscr{M} = \mathscr{P} ,$$

where $\mathscr{P}$ is an arbitrary positive-definite real symmetric matrix. This is because any such $\mathscr{P}$ may always be expressed as $\mathscr{P} = \mathscr{R}^{\mathrm{T}}\mathscr{R}$ with $\mathscr{R}$ some real non-singular matrix, so there is a similarity transformation that takes any matrix $\mathscr{M}$ satisfying the above condition into a real orthogonal matrix, $\mathscr{R}\mathscr{M}\mathscr{R}^{-1}$. Thus we may let $\mathscr{P}$ be various different real symmetric positive-definite matrices without changing the group.

$\boldsymbol{C_n}$: This is the algebra of the unitary symplectic group $USp(2n)$, the group of unitary matrices $\mathscr{M}$ that leave an *antisymmetric* non-singular matrix $\mathscr{A}$ invariant:

$$\mathscr{M}^{\mathrm{T}}\mathscr{A}\mathscr{M} = \mathscr{A} ,$$

$$\mathscr{A}^{\mathrm{T}} = -\mathscr{A} \quad , \quad \mathrm{Det}\,\mathscr{A} \neq 0 .$$

(Note that in d dimensions, $\mathrm{Det}\,\mathscr{A} = \mathrm{Det}\,\mathscr{A}^{\mathrm{T}} = (-)^d\mathrm{Det}\,\mathscr{A}$, so if d is odd then $\mathrm{Det}\,\mathscr{A}$ must vanish. Thus there is no $USp(d)$ unless d is even.) Any such antisymmetric non-singular (perhaps complex) matrix $\mathscr{A}$ may be written in the standard form

$$\mathscr{A} = \mathscr{R}^{\mathrm{T}}\mathscr{A}_0\mathscr{R} ,$$

where $\mathscr{R}$ is unitary and $\mathscr{A}_0$ is the supermatrix:

$$\mathscr{A}_0 = \begin{bmatrix} 0 & 1 \\ -1 & 0 \end{bmatrix} .$$

(Alternatively, we could take $\mathscr{A}_0$ as a block-diagonal supermatrix, with σ_2 s along the main diagonal.) Hence $USp(2n)$ may be described as the group of unitary matrices $\mathscr{S}$ satisfying

$$\mathscr{S}^{\mathrm{T}}\mathscr{A}_0\mathscr{S} = \mathscr{A}_0$$

since by a unitary transformation any such $\mathscr{S}$ can be transformed into a unitary matrix $\mathscr{M} = \mathscr{R}^{-1}\mathscr{S}\mathscr{R}$ that satisfies the previous condition $\mathscr{M}^{\mathrm{T}}\mathscr{A}\mathscr{M} = \mathscr{A}$. Any such matrix $\mathscr{S}$ that is infinitesimally far from the identity may be written

$$\mathscr{S} = 1 + i\mathscr{H} ,$$

where $\mathscr{H}$ is an infinitesimal matrix satisfying the conditions

$$\mathscr{H}^{\dagger} = \mathscr{H}, \quad \mathscr{H}^{\mathrm{T}}\mathscr{A}_0 + \mathscr{A}_0\mathscr{H} = 0 .$$

The most general $2n$-dimensional matrix $\mathscr{H}$ satisfying these conditions may be written as a supermatrix

$$\mathscr{H} = \begin{bmatrix} \mathscr{A} & \mathscr{B} \\ \mathscr{B}^* & -\mathscr{A}^* \end{bmatrix} ,$$

where the n-dimensional complex submatrices $\mathscr{A}, \mathscr{B}$ satisfy

$$\mathscr{A}^\dagger = \mathscr{A} , \qquad \mathscr{B}^\mathrm{T} = \mathscr{B} .$$

A maximal set of commuting generators have $\mathscr{A}$ diagonal and $\mathscr{B}$ zero, and so are of the form

$$\mathscr{H} = \begin{bmatrix} a_1 & & & & & 0 \\ & \ddots & & & & \\ & & a_n & & & \\ & & & -a_1 & & \\ & & & & \ddots & \\ 0 & & & & & -a_n \end{bmatrix}$$

with $a_1, \cdots, a_n$ real. The rank of C_n is thus evidently n. The dimensionality of C_n is the number n^2 of independent real parameters in the Hermitian matrix $\mathscr{A}$, plus the number $2n(n+1)/2$ of independent real parameters in the complex symmetric matrix $\mathscr{B}$

$$d(C_n) = n^2 + 2 \times n(n+1)/2 = n(2n+1) .$$

All of the C_n algebras are simple.

$\boldsymbol{D_n}$: This is the algebra of the unitary orthogonal group $O(2n)$, consisting of all unitary orthogonal matrices in $2n$ dimensions. The discussion of B_n can be carried over to D_n, except that here any set of commuting generators can be put in the form

$$\begin{bmatrix} a_1\sigma_2 & & & 0 \\ & a_2\sigma_2 & & \\ & & \ddots & \\ 0 & & & a_n\sigma_2 \end{bmatrix}$$

so the rank is still n. Also, the dimensionality here is

$$d(D_n) = \frac{(2n)(2n-1)}{2} = n(2n-1) .$$

All of the D_n are simple except for D_1, which is the Abelian algebra with just one generator, and D_2, which is the direct sum $B_1 + B_1$.

Exceptional Lie Algebras: In addition to the above classical Lie algebras, there are just five special cases, the algebras $G_2(d = 14)$; $F_4(d = 52)$; $E_6(d = 78)$; $E_7(d = 133)$; $E_8(d = 248)$.

Not all of the classical Lie algebras are really different. There are just four isomorphisms

$$A_1 = B_1 = C_1 , \qquad C_2 = B_2 , \qquad A_3 = D_3 .$$

These correspond to isomorphisms among Lie groups of which these are the algebras. However, isomorphisms like $B_1 = A_1$, $B_2 = C_2$, and $D_3 = A_3$ do not mean that $SO(3)$ is isomorphic to $SU(2)$, or that $SO(5)$ is isomorphic to $USp(4)$, or that $SO(6)$ is isomorphic to $SU(4)$. Instead, $SU(2), USp(4)$, and $SU(4)$ are the simply connected *covering groups* of $SO(3)$, $SO(5)$, and $SO(6)$. (Covering groups are discussed in Chapter 2.) Nevertheless, the isomorphisms of the Lie algebras make it especially easy to construct the double-valued fundamental spinor representations of $SO(3)$, $SO(5)$, and $SO(6)$; they are just the defining representations of $SU(2)$, $USp(4)$, and $SU(4)$, respectively. Also $SO(4)$ is isomorphic to $SO(3) \times SO(3)$, so its double-valued spinor representation is just the defining representation of $SU(2) \times SU(2)$. For $d \geq 7$ the double-valued spinor representations of $SO(d)$ must be constructed by other means. The simplest technique uses Clifford algebras, discussed in Section 5.4.

Problems

1. Derive the Bianchi identity

$$D_\mu F_{\alpha\nu\lambda} + D_\nu F_{\alpha\lambda\mu} + D_\lambda F_{\alpha\mu\nu} = 0 .$$

2. Suppose we use generalized Coulomb gauge in a non-Abelian gauge theory, taking the gauge-fixing function as $f_\alpha = \nabla \cdot \mathbf{A}_\alpha$. Derive the ghost Lagrangian. What is the ghost propagator? (Take $B[f] = \exp(-i\int d^4x\, f_\alpha f_\alpha/2\xi)$.)

3. Suppose that in electrodynamics we use a gauge-fixing function $f = \partial_\mu A^\mu + cA_\mu A^\mu$, with c an arbitrary constant. Derive the ghost Lagrangian. (Take $B[f] = \exp(-i\int d^4x\, f^2/2\xi)$.) What is the ghost propagator?

4. Show that there is no simple Lie algebra with just four generators.

5. Show that if a field $\psi_\ell(x)$ belonging to a representation of a gauge group with generator matrices t_α varies along a path $x^\mu = x^\mu(\tau)$ according to the differential equation

$$\frac{d\psi(\tau)}{d\tau} = it_\alpha \psi(\tau) A_{\alpha\mu}\left(x(\tau)\right) \frac{dx^\mu(\tau)}{d\tau} ,$$

(with both $\psi_\ell(x)$ and $A_{\alpha\mu}(x)$ classical c-number fields) then the change in ψ around any *small* closed path $\mathscr{P}$ around a point X^μ is proportional to

$$t_\alpha \psi(X)\, F_{\alpha\mu\nu}(X) \oint_{\mathscr{P}} x^\mu(\tau)\, \frac{dx^\nu(\tau)}{d\tau}\, d\tau\,.$$

Find the constant of proportionality. Use this result to show that if $F_{\alpha\mu\nu}(x)$ vanishes everywhere then it is possible by a gauge transformation to make $A_{\alpha\mu}(x)$ vanish throughout at least a finite region. (Hint: Follow the analogous argument in electrodynamics, or in general relativity, as in Ref. 6.)

6. Applying path-integral methods to a general non-Abelian gauge theory, calculate the propagator of the gauge field $A_{\alpha\mu}(x)$ if we choose a gauge-fixing functional $f_\alpha = n_\mu A^\mu_\alpha$, with n_μ arbitrary constants. (Take $B[f] = \exp(-i \int d^4x\, f_\alpha f_\alpha / 2\xi)$.) What is the ghost Lagrangian? What is the ghost propagator? What is the ghost interaction vertex?

7. Suppose we adopt BRST invariance instead of gauge invariance as a fundamental physical principle. Derive the most general Lagrangian density constructed from sums of products of fields and field derivatives of dimensionality (mass)^d with $d \leq 4$, constructed from $A_{\alpha\mu}$, ω_α, ω^*_α, h_α, and/or their derivatives, that is invariant (up to total derivatives) under Lorentz transformations, ghost number phase transformations ($\omega_\alpha \to e^{i\theta}\omega_\alpha$, $\omega^*_\alpha \to e^{-i\theta}\omega^*_\alpha$), global gauge transformations (ϵ_α constant), and BRST transformations.

8. Show that the antibracket satisfies the symmetry condition (15.9.15) and the Jacobi identity (15.9.21).

9. Show that if a functional O satisfies the condition $(O, S) = i\Delta S$ and the action S satisfies the quantum master equation then the quantum average $\langle O \rangle$ is independent of the gauge-fixing functional Ψ.

References

1. C. N. Yang and R. L. Mills, *Phys. Rev.* **96**, 191 (1954). O. Klein had come close to a formulation of the $SU(2)$ Yang–Mills theory in a talk at a 1938 conference at Warsaw, reported in *New Theories in Physics* (International Institute of Intellectual Cooperation, Paris, 1939). For a critical discussion of Klein's theory, see D. J. Gross, 'Oscar Klein and Gauge Theory,' talk at the 1994 Oscar Klein symposium at Stockholm, Princeton preprint PUPT–1508.

2. R. Utiyama, *Phys. Rev.* **101**, 1597 (1956).

3. R. P. Feynman, *Acta Phys. Polonica* **24**, 697 (1963).

4. L. D. Faddeev and V. N. Popov, *Phys. Lett.* **25B**, 29 (1967).

5. B. S. De Witt, *Phys. Rev.* **162**, 1195, 1239 (1967).

6. The proof is easily adapted from the standard proof of the corresponding result in general relativity that, for the existence of a coordinate transformation to a flat metric in a finite simply connected region, it is necessary and sufficient that the Riemann–Christoffel curvature tensor should vanish. See, for example, S. Weinberg, *Gravitation and Cosmology* (Wiley, New York, 1972): Sections 6.3 and 6.4.

7. The proof that **a** (with $g_{\alpha\beta} = \delta_{\alpha\beta}$) implies **b** and that **b** implies **c** was given by M. Gell-Mann and S. L. Glashow, *Ann. Phys. (N.Y.)* **15**, 437 (1961).

8. See, for example, S. Weinberg, *ibid.*, Section 7.6.

9. V. Gribov, *Nucl. Phys.* **B139**, 1 (1978). Also see R. Jackiw, I. Muzinich, and C. Rebbi, *Phys. Rev.* **D17**, 1576 (1978); R. Jackiw, in *New Frontiers in High Energy Physics*, eds. B. Kursunoglu, A. Perlmutter, and L. Scott (Plenum, New York, 1978); N. Christ and T. D. Lee, *Phys. Rev.* **D22**, 939 (1980); R. Jackiw, in *Current Algebra and Anomalies* (World Scientific, Singapore, 1985): footnote 50.

10. C. Becchi, A. Rouet, and R. Stora, *Comm. Math. Phys.* **42**, 127 (1975); in *Renormalization Theory*, eds. G. Velo and A. S. Wightman (Reidel, Dordrecht, 1976); *Ann. Phys.* **98**, 287 (1976).

11. I. V. Tyutin, Lebedev Institute preprint N39 (1975).

11*a*. N. Nakanishi, *Prog. Theor. Phys.* **35**, 1111 (1966); B. Lautrup, *Mat. Fys. Medd. Kon. Dan. Vid.-Sel. Medd.* **35**, 29 (1967).

12. The original source of this argument is unknown to me. I learned it from J. Polchinski.

13. S. N. Gupta, *Proc. Phys. Soc.* **63**, 681 (1950); **64**, 850 (1951); K. Bleuler, *Helv. Phys. Acta* **3**, 567 (1950); K. Bleuler and W. Heitler, *Progr. Theor. Phys.* **5**, 600 (1950).

13*a*. G. Curci and R. Ferrari, *Nuovo Cimento* **35**, 273 (1976); T. Kugo and I. Ojima, *Prog. Theor. Phys.* **60**, 1869 (1978); *Prog. Theor. Phys. Suppl.* **66**, 1 (1979). Also see Ref. 11*a* for the case of electrodynamics.

14. J. Thierry-Mieg, *J. Math. Phys.* **21**, 2834 (1980); R. Stora, in *Progress in Gauge Theories: Proceedings of a Symposium at Cargèse*, 1983, eds. G. 't Hooft, A. Jaffe, H. Lehmann, P. K. Mitter, I. M. Singer, and R. Stora (Plenum, New York, 1984): p. 543; L. Bonora and P. Cotta-Ramusino, *Commun. Math. Phys.* **87**, 589 (1983); J. Mañes, R. Stora, and B. Zumino, *Commun. Math. Phys.* **102**, 157 (1985); A. S. Schwarz, *Commun. Math. Phys.* **155**, 249 (1993); P. M. Lavrov, P. Yu Moshhin and A. A. Reshetnyak, Tomsk preprint hep-th/9507104; P. M. Lavrov, Tomsk preprint hep-th/9507105.

15. W. Siegel, *Phys. Lett.* **93B**, 170 (1980); T. Kimura, *Prog. Theor. Phys.* **64**, 357 (1980); **65**, 338 (1981).

16. G. Barnich, F. Brandt, and M. Henneaux, *Commun. Math. Phys.* **174**, 57 (1995); Section 14.

17. G. Curci and R. Ferrari, *Nuovo Cimento* **32A**, 151 (1976); I. Ojima, *Prog. Theor. Phys. Supp.* **64**, 625 (1980); L. Baulieu and J. Thierry-Mieg, *Nucl. Phys.* **B197**, 477 (1982).

18. The anti-BRST transformation was extended to general local symmetries by L. Alvarez-Gaumè and L. Baulieu, *Nucl. Phys.* **B212**, 255 (1983). It is induced by the operator:

$$\bar{s} = \omega^{*A}\,\delta_A\phi^r\,\frac{\delta_L}{\delta\phi^r} - \frac{1}{2}\omega^{*B}\,\omega^{*C}\,f^A{}_{BC}\,\frac{\delta_L}{\delta\omega^{*A}} - \omega^{*B}\,h^C\,f^A{}_{BC}\,\frac{\delta}{\delta h^A}$$
$$+\Big[-f^A{}_{BC}\omega^{*B}\omega^C + h^A\Big]\frac{\delta_L}{\delta\omega^A}\,.$$

It is straightforward to show that this transformation is nilpotent:

$$\bar{s}^2 = 0\,.$$

Also, the BRST and anti-BRST transformations anticommute. Furthermore, there is a cohomology theorem (unpublished) for the anti-BRST transformations, like that for the BRST transformations: the most general functional I of ϕ^r, ω^A, ω^{*A}, and h^A that satisfies the anti-BRST invariance condition $\bar{s}I = 0$ and has ghost number zero is of the form

$$I[\phi,\omega,\omega^*,h] = I_0[\phi] + \bar{s}\,\bar{\Psi}[\phi,\omega,\omega^*,h]\,.$$

(Just use the Hodge operator $\bar{t} \equiv \omega^A\,\delta/\delta h^A$ in place of t.) Anti-BRST symmetry can be an aid in enumerating possible terms in Lagrangians and Greens functions, but I do not know of any case where it is indispensable.

19. I. A. Batalin and G. A. Vilkovisky, *Phys. Lett.* **B102**, 27 (1981); *Nucl. Phys.* **B234**, 106 (1984); *J. Math. Phys.* **26**, 172 (1985). Also see B.

L. Voronov and I. V. Tyutin, *Theor. Math. Phys.* **50**, 218, 628 (1982). For a lucid review, see J. Gomis, J. París and S. Samuel, *Phys. Rep.* **259**, 1 (1995).

20. E. S. Fradkin and G. A. Vilkovisky, *Phys. Lett.* **B55**, 224 (1975); CERN report TH2332 (1977); I. A. Batalin and G. A. Vilkovisky, *Phys. Lett.* **B69**, 309 (1977); I. S. Fradkin and T. E. Fradkina, *Phys. Lett.* **B72**, 334 (1977). Also see M. Henneaux and C. Teitelboim, *Quantization of Gauge Systems* (Princeton University Press, Princeton, 1992).

21. I. A. Batalin and I. V. Tyutin, *Phys. Lett.* **B356**, 373 (1995).

22. J. Zinn-Justin, in *Trends in Elementary Particle Theory – International Summer Institute on Theoretical Physics in Bonn 1974* (Springer-Verlag, Berlin, 1975): p. 2.

23. D. Z. Freedman, P. van Nieuwenhuizen, and S. Ferrara, *Phys. Rev.*, **D13**, 3214 (1976); R. E. Kallosh, *Nucl. Phys.* **B141**, 141 (1978); B. de Wit and J. W. van Holten, *Phys. Lett.*, **B79**, 389 (1979); P. van Nieuwenhuizen *Phys. Rep.*, **68**, 189 (1981).

24. M. Henneaux and C. Teitelboim, Ref. 20: Section 18.1.4.

25. E. P. Wigner, *Group Theory* (Academic Press, New York, 1959): pp. 76–8.

26. The original reference is E. Cartan, *Sur la Structure des Groupes de Transformations Finis et Continue*, (Paris, 1894; 2nd edn 1933). For a textbook treatment, see, for example, R. Gilmore, *Lie Groups, Lie Algebras, and Some of Their Applications* (Wiley, New York, 1974): Chapter 9.

16
External Field Methods

It is often useful to consider quantum field theories in the presence of a classical external field. One reason is that in many physical situations, there really is an external field present, such as a classical electromagnetic or gravitational field, or a scalar field with a non-vanishing vacuum expectation value. (As we shall see in Chapter 19, such scalar fields can play an important role in the spontaneous breakdown of symmetries of the Lagrangian.) But even where there is no actual external field present in a problem, some calculations are greatly facilitated by considering physical amplitudes in the presence of a fictitious external field. This chapter will show that it is possible to take all multiloop effects into account by summing 'tree' graphs whose vertices and propagators are taken from a *quantum effective action*, which is nothing but the one-particle-irreducible connected vacuum–vacuum amplitude in the presence of an external field. It will turn out in the next chapter that this provides an especially handy way both of completing the proof of the renormalizabilty of non-Abelian gauge theories begun in Chapter 15, and of calculating the charge renormalization factors that we need in order to establish the crucial property of asymptotic freedom in quantum chromodynamics.

16.1 The Quantum Effective Action

Consider a quantum field theory with action $I[\phi]$, and suppose we 'turn on' a set of classical currents $J_r(x)$ coupled to the fields $\phi^r(x)$ of the theory. The complete vacuum–vacuum amplitude in the presence of these currents is then

$$Z[J] \equiv \langle \text{VAC, out}|\text{VAC, in}\rangle_J$$
$$= \int \left[\prod_{s,y} d\phi^s(y)\right] \exp\left(iI[\phi] + i\int d^4x\, \phi^r(x)\, J_r(x) + \epsilon \text{ terms}\right) . \tag{16.1.1}$$

(The fields $\phi^r(x)$ need not be scalars. They might even be fermionic, though until Section 16.4 we shall not bother to keep track of the signs that would appear in that case.) The Feynman rules for calculating $Z[J]$ are just the same as for calculating the vacuum–vacuum amplitude $Z[0]$ in the absence of the external current, except that the Feynman diagrams now contain vertices of a new kind, to which a *single* ϕ^r-line is attached. Such a vertex labelled with a coordinate x contributes a 'coupling' factor $iJ_r(x)$ to the integrand of the position-space Feynman amplitude. Equivalently, we could say that in the expansion of $Z[J]$ in powers of J, the coefficient of the term proportional to $iJ_r(x)\,iJ_s(y)\cdots$ is just the sum of diagrams with external lines (including propagators) corresponding to the fields $\phi^r(x)$, $\phi^s(y)$, etc. In particular, the first derivative gives the vacuum matrix element of the quantum mechanical operator $\Phi^r(x)$ corresponding to $\phi^r(x)$:

$$\left[\frac{\delta}{\delta J_r(y)} Z[J]\right]_{J=0} = i\int\left[\prod_{r,x} d\phi^r(x)\right]\,\phi^r(y)\exp\{iI[\phi]+\epsilon \text{ terms}\}$$
$$= i\,\langle \text{VAC, out}|\Phi^r(y)|\text{VAC, in}\rangle_{J=0}\,. \qquad (16.1.2)$$

We sometimes work with functionals $Z[J]$ defined by (16.1.1) where the $\phi^r(x)$ are not elementary fields (that is, fields appearing in the action) but products of such fields. Where $\phi^r(x)$ is a product of N elementary fields, the new vertices in the Feynman rules for $Z[J]$ have N lines attached. Some of the results of this chapter (including Eq. (16.1.2)) apply in this case, but where Feynman diagrams are involved, it will be tacitly assumed that the $\phi^r(x)$ are elementary.

Now, $Z[J]$ is given by the sum of *all* vacuum–vacuum amplitudes in the presence of the current J, including disconnected as well as connected diagrams, but not counting as different those diagrams that differ only by a permutation of vertices in the same or different connected subdiagrams. A general diagram that consists of N connected components will contribute to $Z[J]$ a term equal to the product of the contributions of these components, divided by the number $N!$ of permutations of vertices that merely permute all the vertices in one connected component with all the vertices in another.* Hence, the sum of all graphs is

$$Z[J] = \sum_{N=0}^{\infty}\frac{1}{N!}(iW[J])^N = \exp(iW[J])\,, \qquad (16.1.3)$$

* The contribution of a Feynman diagram with N connected components containing $n_1, n_2, \cdots n_N$ vertices will be proportional to a factor $1/(n_1+\cdots n_N)!$ from the Dyson expansion, and a factor $(n_1+\cdots n_N)!/N!$ equal to the number of permutations of these vertices, counting as identical those permutations that merely permute all the vertices in one component with all the vertices in another.

where $iW[J]$ is the sum of all *connected* vacuum–vacuum amplitudes, again not counting as different those diagrams that differ only by a permutation of vertices.

For many purposes, it will be useful to go one step further, and in place of $W(J)$ work with the sum of all connected *one-particle-irreducible* graphs. (A one-particle-irreducible graph is one that cannot be disconnected by cutting through any one internal line.) We can give a formal expression for this sum as follows. First, define $\phi^r_J(x)$ as the vacuum expectation value of the operator $\Phi^r(x)$ in the presence of the current J:

$$\phi^r_J(x) \equiv \frac{\langle \text{VAC, out}|\Phi^r(x)|\text{VAC, in}\rangle_J}{\langle \text{VAC, out}|\text{VAC, in}\rangle_J} = -\frac{i}{Z[J]}\frac{\delta}{\delta J_r(x)}Z[J] \tag{16.1.4}$$

or in terms of the sum of connected graphs

$$\phi^r_J(x) = \frac{\delta}{\delta J_r(x)} W[J] . \tag{16.1.5}$$

This formula can be inverted. Define $J_{\phi r}(x)$ as the current for which (16.1.4) has a prescribed value $\phi^r(x)$:

$$\phi^r_J(x) = \phi^r(x) \;\text{ if }\; J_r(x) = J_{\phi r}(x) .$$

The *quantum effective action*[1] $\Gamma[\phi]$ is defined (as a functional of ϕ, not J) by the Legendre transformation

$$\Gamma[\phi] \equiv -\int d^4x\, \phi^r(x) J_{\phi r}(x) + W[J_\phi] . \tag{16.1.6}$$

We will soon show that $\Gamma[\phi]$ is the sum of all connected one-particle-irreducible graphs in the presence of the current J_ϕ. However, let us first take a look at another aspect of its physical significance.

Note that the variational derivative of $\Gamma[\phi]$ is

$$\frac{\delta \Gamma[\phi]}{\delta \phi^s(y)} = -\int d^4x\, \phi^r(x)\,\frac{\delta J_{\phi r}(x)}{\delta \phi^s(y)} - J_{\phi s}(y) + \int d^4x \left[\frac{\delta W[J]}{\delta J_r(x)}\right]_{J=J_\phi} \frac{\delta J_{\phi r}(x)}{\delta \phi^s(y)}$$

or, using Eq. (16.1.5),

$$\frac{\delta \Gamma[\phi]}{\delta \phi^s(y)} = -J_{\phi s}(y) . \tag{16.1.7}$$

Thus $\Gamma[\phi]$ is the 'effective action', in the sense that the possible values for the external fields $\phi^s(y)$ in the *absence* of a current J are given by the stationary 'points' of Γ:

$$\frac{\delta \Gamma[\phi]}{\delta \phi^s(y)} = 0 \;\text{ for }\; J = 0 . \tag{16.1.8}$$

This may be compared with the classical field equations, which just require that the actual action $I[\phi]$ be stationary. Hence Eq. (16.1.8) may be regarded as the equation of motion for the external field ϕ, taking quantum corrections into account.

Not only does $\Gamma[\phi]$ provide the quantum-corrected field equations; it is an effective action in the sense that $iW[J]$ may be calculated as a sum of connected *tree* graphs for the vacuum–vacuum amplitude, with vertices calculated as if the action were $\Gamma[\phi]$ instead of $I[\phi]$. By a tree graph is meant one that becomes disconnected if we cut any internal line. All the effects of loop diagrams are taken into account by using $\Gamma[\phi]$ in place of $I[\phi]$.

To see this,[2] let us consider the quantity $W_\Gamma[J,g]$ that we would get instead of $W[J]$, if we used an action $g^{-1}\Gamma[\phi]$ in place of $I[\phi]$:

$$\exp\{iW_\Gamma[J,g]\} \equiv \int \prod_{r,x} d\phi^r(x)\, \exp\left\{ig^{-1}\left[\Gamma[\phi] + \int d^4x\, \phi^r(x) J_r(x)\right] + \epsilon \text{ terms}\right\}, \tag{16.1.9}$$

with arbitrary constant g. The propagator here is the inverse of the coefficient of the term in $g^{-1}\Gamma(\phi)$ quadratic in ϕ, and is hence proportional to g, while all vertices make a contribution proportional to $1/g$, so a graph with V vertices (including those produced by the current J) and I internal lines (including those attached to the J vertices) is proportional to g^{I-V}. For any connected graph, the number of loops is $L = I - V + 1$, so the L-loop term in $W_\Gamma[J,g]$ has the g dependence

$$\Big(W_\Gamma[J,g]\Big)_{L\text{ loops}} \propto g^{L-1}\,. \tag{16.1.10}$$

Equivalently, we may write (at least formally)

$$W_\Gamma[J,g] = \sum_{L=0}^{\infty} g^{L-1} W_\Gamma^{(L)}[J]\,, \tag{16.1.11}$$

where (as can be seen by setting $g = 1$), the quantity $W_\Gamma^{(L)}[J]$ is the L-loop contribution to the connected vacuum amplitude $W_\Gamma[J,1]$ that we would obtain if we used $\Gamma[\phi]$ (*without* a factor g) in place of the action $I[\phi]$.

Now, we are specially interested here in the sum of *tree* graphs, those without loops, calculated with vertices and propagators calculated as if the action were $\Gamma[\phi]$ instead of $I[\phi]$. In our present notation, this is $W_\Gamma^{(0)}[J]$. In order to isolate the $L = 0$ term in Eq. (16.1.11), consider the limit $g \to 0$. In this limit, the path integral (16.1.9) is dominated by the point of stationary phase,

$$\exp\Big\{i\, W_\Gamma[J,g]\Big\} \propto \exp\left\{ig^{-1}\left[\Gamma[\phi_J] + \int d^4x\, \phi_J^r(x) J_r(x)\right]\right\}, \tag{16.1.12}$$

where, because of its definition as the field produced by the current J, the field ϕ_J is the stationary point of the exponent, in the sense that

$$\left.\frac{\delta\Gamma[\phi]}{\delta\phi^r(x)}\right|_{\phi=\phi_J} = -J_r(x)\,. \tag{16.1.13}$$

The proportionality factor in Eq. (16.1.12) is in general a functional of J, but it is a power series in g starting with terms of order g^0. Hence taking the logarithm of both sides and isolating the terms of order g^{-1} gives

$$W_\Gamma^{(0)}[J] = \Gamma[\phi_J] + \int d^4x\,\phi_J^r(x)\,J_r(x)\,. \tag{16.1.14}$$

By setting $\phi = \phi_J$ in Eq. (16.1.6), we see that the right-hand side of Eq. (16.1.14) is just $W[J]$:

$$W_\Gamma^{(0)}[J] = W[J]\,. \tag{16.1.15}$$

To recapitulate, this says that $W[J]$ may be calculated by using $\Gamma[\phi]$ in place of $I[\phi]$ (the subscript Γ) and keeping only tree (0-loop) graphs:

$$iW[J] = \int_{\substack{\text{CONNECTED}\\ \text{TREE}}} \left[\prod_{r,x} d\phi^r(x)\right] \exp\left\{i\Gamma[\phi] + i\int \phi^r(x)\,J_r(x)\,d^4x\right\}\,. \tag{16.1.16}$$

Now, any connected graph for $iW[J]$ can be regarded as a tree, whose vertices consist of one-particle-irreducible subgraphs. Thus in order for Eq. (16.1.16) to be correct, $i\Gamma[\phi]$ must be the sum of all one-particle-irreducible connected graphs with arbitrary numbers of external lines, each external line corresponding to a factor ϕ rather than a propagator or wave function. For this reason, the coefficients in an expansion of $\Gamma[\phi]$ in powers of fields and their derivatives around some fixed field ϕ_0 may be regarded as *renormalized* coupling constants, with the renormalization 'point' specified by ϕ_0 rather than by some set of momenta.

Equivalently, $i\Gamma[\phi_0]$ for some fixed field $\phi_0^r(x)$ may be expressed as the sum of one-particle-irreducible graphs for the vacuum–vacuum amplitude, calculated with a shifted action $I[\phi + \phi_0]$:

$$i\Gamma[\phi_0] = \int_{\substack{\text{1PI}\\ \text{CONNECTED}}} \left[\prod_{r,x} d\phi^r(x)\right] \exp\{iI[\phi + \phi_0]\}\,. \tag{16.1.17}$$

This is because any place where ϕ_0 appears in any of the vertices or propagators within the one-particle-irreducible graphs in Eq. (16.1.17) is also a place where an external ϕ-line could be attached. (The restriction to one-particle-irreducible graphs plays an essential role in Eq. (16.1.17); without this restriction we could shift the variable of integration, yielding an integral that would be manifestly independent of ϕ_0.) In place of Eq. (16.1.17), it is often convenient to write

$$\exp\Big[i\Gamma[\phi_0]\Big] = \int_{\text{1PI}} \Big[\prod_{rx} d\phi^r(x)\Big] \exp\Big\{iI[\phi+\phi_0]\Big\}, \qquad (16.1.18)$$

in which we evaluate the path integral including all graphs, connected or not, in which each connected component is one-particle-irreducible.

* * *

This formalism provides a simple method of summing tree graphs. As one example, consider the relation between the complete two-point function $\Delta^{rx,sy}$ and its one-particle-irreducible part $\Pi_{rx,sy}$. From Eqs. (16.1.5) and (16.1.7), we find

$$\Delta^{rx,sy} \equiv \frac{\delta^2 W[J]}{\delta J_r(x)\,\delta J_s(y)} = \frac{\delta \phi^r_J(x)}{\delta J_s(y)}, \qquad (16.1.19)$$

$$\Pi_{rx,sy} \equiv \frac{\delta^2 \Gamma[\phi]}{\delta \phi^r(x)\,\delta \phi^s(y)} = -\frac{\delta J_{\phi\, r}(x)}{\delta \phi^s(y)}. \qquad (16.1.20)$$

It follows immediately that the 'matrices' Δ and Π are related by

$$\Delta = -\Pi^{-1}. \qquad (16.1.21)$$

This is the counterpart of the familiar relation (10.3.15) between propagators and self-energy parts, with the extra term q^2+m^2 in the denominator in Eq. (10.3.15) representing the zeroth-order term in the one-particle-irreducible two-point function.

16.2 Calculation of the Effective Potential

To see how the formalism of the previous section works in practice, consider a simple example, the renormalizable theory of a single real scalar field $\phi(x)$, with action

$$I[\phi] = -\int d^4x \left[\lambda + \tfrac{1}{2}\partial_\rho\phi\,\partial^\rho\phi + \tfrac{1}{2}m^2\phi^2 + \tfrac{1}{24}\,g\phi^4\right]. \qquad (16.2.1)$$

(We are here including a 'cosmological constant' $-\lambda$ in the Lagrangian density, for reasons that will become apparent). Suppose for simplicity that we wish to calculate $\Gamma[\phi_0]$ for a position-independent field $\phi_0(x) = \phi_0$. Then every term in $\Gamma[\phi_0]$ will contain a factor of the volume of

Figure 16.1. Feynman diagrams with zero, one, or two loops for the quantum effective action of the theory of a neutral scalar field ϕ with interaction ϕ^4.

spacetime

$$\mathscr{V}_4 = \int d^4x = \delta^4(p-p)(2\pi)^4 \tag{16.2.2}$$

arising from the momentum conservation delta-function. We will therefore write, for ϕ_0 constant

$$\Gamma[\phi_0] = -\mathscr{V}_4 V(\phi_0)\,, \tag{16.2.3}$$

where $V(\phi_0)$ is an ordinary function, known as the *effective potential.* In this section we will calculate the effective potential to one-loop order. This was originally done by Coleman and E. Weinberg[3] in a study of spontaneous symmetry breaking, the subject of Chapters 19 and 21. The results were also used by them in one of the early applications of the renormalization group, to be described in Section 18.2.

Shifting ϕ by ϕ_0, the action in Eq. (16.1.18) is:*

$$I[\phi+\phi_0] = -\mathscr{V}_4\left[\lambda + \tfrac{1}{2}m^2\phi_0^2 + \tfrac{1}{24}g\phi_0^4\right] - [m^2\phi_0 + \tfrac{1}{6}g\,\phi_0^3]\int d^4x\,\phi$$
$$-\int d^4x\left[\tfrac{1}{2}\partial_\rho\phi\partial^\rho\phi + \tfrac{1}{2}\mu^2(\phi_0)\phi^2\right] - \int d^4x\left[\tfrac{1}{6}g\,\phi_0\phi^3 + \tfrac{1}{24}g\,\phi^4\right], \tag{16.2.4}$$

where μ^2 is the field-dependent mass

$$\mu^2(\phi_0) = m^2 + \tfrac{1}{2}g\phi_0^2\,. \tag{16.2.5}$$

Note that there now appear new interactions proportional to ϕ (which have no effect on one-particle-irreducible graphs) and also ϕ^3, as well as terms with the same structure as those in the original action.

The Feynman diagrams for $\Gamma[\phi_0]$ with up to two loops are shown in Figure 16.1. The zero-loop term in the vacuum–vacuum amplitude is just given by the constant term in $I[\phi+\phi_0]$

$$i\Gamma^{(0\text{ loop})}[\phi_0] = -i\mathscr{V}_4\left(\lambda + \frac{1}{2}m^2\phi_0^2 + \frac{g}{24}\phi_0^4\right). \tag{16.2.6}$$

* There are limitations on the applicability of perturbation theory for $m^2 < 0$, discussed in the following section.

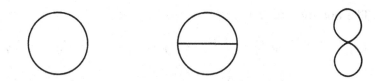

The one-loop term is given by

$$\exp\left(i\Gamma^{(1\text{ loop})}[\phi_0]\right) = \int \prod_x d\phi(x)\,\exp\Big\{-\tfrac{1}{2}i\int d^4x\Big[\partial_\rho\phi\partial^\rho\phi + \mu^2(\phi_0)\phi^2\Big] + \epsilon\text{ terms}\Big\}\,. \tag{16.2.7}$$

We learned how to calculate such integrals in Chapter 9; the result is given by Eq. (9.A.18):

$$i\Gamma^{(1\text{ loop})}[\phi_0] = \ln\text{Det}\left(\frac{iK}{\pi}\right)^{-1/2} = -\frac{1}{2}\,\text{Tr}\,\ln\left(\frac{iK}{\pi}\right)\,, \tag{16.2.8}$$

where here

$$K_{x,y} = \left[\frac{\partial^2}{\partial x^\lambda \partial y_\lambda} + \mu^2(\phi_0) - i\epsilon\right]\delta^4(x-y)\,. \tag{16.2.9}$$

As usual, to calculate such traces it is helpful to diagonalize the 'matrix' K by passing to momentum space:

$$K_{p,q} = \int \frac{d^4x}{(2\pi)^2}\,e^{-ip\cdot x}\,\frac{d^4y}{(2\pi)^2}\,e^{iq\cdot y}K_{x,y} = \left(p^2 + \mu^2(\phi_0) - i\epsilon\right)\delta^4(p-q)\,. \tag{16.2.10}$$

The logarithm of this diagonal matrix is just the diagonal matrix with logarithms along the main diagonal:

$$\left[\ln\left(\frac{iK}{\pi}\right)\right]_{p,q} = \ln\left[\frac{i}{\pi}\,(p^2 + \mu^2(\phi_0) - i\epsilon)\right]\,\delta^4(p-q) \tag{16.2.11}$$

and its trace is then

$$i\Gamma^{(1\text{ loop})}[\phi_0] = -\frac{1}{2}\int d^4p\left[\ln\left(\frac{iK}{\pi}\right)\right]_{p,p} = -\frac{\mathscr{V}_4}{2(2\pi)^4}\int d^4p\,\ln\left(\frac{i}{\pi}\left(p^2 + \mu^2(\phi_0) - i\epsilon\right)\right)\,. \tag{16.2.12}$$

Putting together Eqs. (16.2.6) and (16.2.12), we have the effective potential to one-loop order

$$V(\phi_0) = \lambda + \frac{1}{2}m^2\phi_0^2 + \frac{g}{24}\phi_0^4 + \mathscr{I}\left(\mu^2(\phi_0)\right)\,, \tag{16.2.13}$$

where

$$\mathscr{I}(\mu^2) \equiv \frac{-i}{2(2\pi)^4}\int d^4p\,\ln\left(\frac{i}{\pi}[p^2 + \mu^2 - i\epsilon]\right)\,. \tag{16.2.14}$$

It is painfully obvious that this formula for the effective potential contains ultraviolet divergences. Fortunately, these are naturally absorbed into a renormalization of the parameters of the theory. Although the integral (16.2.14) is divergent, simple power-counting shows that it is made convergent by differentiating three times with respect to μ^2:

$$\mathscr{I}'''(\mu^2) = -\frac{i}{(2\pi)^4}\int d^4p\,(p^2+\mu^2-i\epsilon)^{-3}\,.$$

Once again, the $-i\epsilon$ term tells us that we must rotate the p^0 contour counterclockwise, so that $p^0 = ip_4$, with p_4 running from $-\infty$ to $+\infty$:

$$\mathscr{I}'''(\mu^2) = \frac{1}{(2\pi)^4}\int_0^\infty \frac{2\pi^2k^3dk}{(k^2+\mu^2)^3} = \frac{1}{32\pi^2\mu^2}\,.$$

Integrating thrice, we have then

$$\mathscr{I}(\mu^2) = \frac{\mu^4\ln\mu^2}{64\pi^2} + A + B\,\mu^2 + C\mu^4\,.$$

The constants A, B, C are not determined by this method of calculation, which is hardly a serious problem since they are obviously infinite anyway. We eliminate these constants by defining 'renormalized' values for λ, m^2, and g:

$$\lambda_R \equiv \lambda + A + Bm^2 + Cm^4\,,$$
$$m_R^2 \equiv m^2 + gB + 2gm^2C\,,$$
$$g_R \equiv g + 6g^2C\,.$$

Our final result for the potential to one-loop order is then

$$V(\phi_0) = \lambda_R + \frac{1}{2}m_R^2\phi_0^2 + \frac{g_R}{24}\phi_0^4 + \frac{\mu^4(\phi_0)\ln\mu^2(\phi_0)}{64\pi^2}\,, \tag{16.2.15}$$

where $\mu(\phi_0)$ is the field-dependent mass defined by Eq. (16.2.5), which to this order can be calculated using m_R and g_R in place of m and g:

$$\mu^2(\phi) = m_R^2 + \tfrac{1}{2}g_R\phi^2\,.$$

Similar results hold if the theory contains a complex spin $\frac{1}{2}$ fermion field $\psi(x)$ that interacts with the scalar ϕ. For instance, if this interaction Hamiltonian density has the simple form $G\phi\bar{\psi}\psi$, then the mass $M(\phi_0)$ of the fermion in the presence of a constant scalar background field ϕ_0 is of the form:

$$M(\phi_0) = M(0) + G\phi_0\,.$$

It is easy to see that in this case the potential (16.2.15) then receives an additional term:

$$V(\phi_0) = \lambda_R + \frac{1}{2}m_R^2\phi_0^2 + \frac{g_R}{24}\phi_0^4 + \frac{\mu^4(\phi_0)\ln\mu^2(\phi_0)}{64\pi^2} - \frac{M^4(\phi_0)\ln M^2(\phi_0)}{16\pi^2}\,. \tag{16.2.16}$$

The numerical coefficient in the new term is four times that of the $\mu^4 \ln \mu^2$ term in Eq. (16.2.15), because the fermion and antifermion described by ψ and $\bar{\psi}$ are assumed here to be different, and because each has two spin states, while the sign is opposite because (as shown in Chapter 9) fermionic path integrals of Gaussians give a result proportional to the determinant of the matrix coefficient in the exponent, in contrast to bosonic path integrals, which give a result proportional to the inverse of this determinant.

16.3 Energy Interpretation

The effective action $\Gamma[\phi]$ and potential $V[\phi]$ have an important interpretation in terms of the energy and energy density respectively.[4] To see this, suppose that we turn on a current $J^n(\mathbf{x}, t)$ that rises smoothly from zero at $t = -\infty$ to a finite value $\mathscr{J}^n(\mathbf{x})$, and remains at that value for a long time T, after which it drops smoothly again to zero at $t = +\infty$. The effect of this perturbation is to convert the vacuum to a state with a definite energy $E[\mathscr{J}]$ (a functional of $\mathscr{J}^n(\mathbf{x})$), in which it remains for a time T, after which it returns to the vacuum. However, although the 'out' vacuum is the same physical state as the 'in' vacuum, the state vectors differ by the phase $\exp(-iE[\mathscr{J}]T)$ accumulated during the time T:

$$\langle \text{VAC, out}|\text{VAC, in}\rangle_J = \exp\Big(-iE[\mathscr{J}]T\Big) . \tag{16.3.1}$$

Comparing with Eqs. (16.1.1) and (16.1.3), this gives

$$W[J] = -E[\mathscr{J}]T . \tag{16.3.2}$$

To see the connection between this energy and the effective action, suppose we seek the state Ω_ϕ that minimizes the energy expectation value

$$\langle H \rangle_\Omega = \frac{(\Omega, H\Omega)}{(\Omega, \Omega)} , \tag{16.3.3}$$

subject to the condition that the quantum fields $\Phi_n(\mathbf{x}, t)$ have the time-independent expectation value $\phi_n(\mathbf{x})$

$$\frac{(\Omega, \Phi_n(\mathbf{x}, t)\Omega)}{(\Omega, \Omega)} = \phi_n(\mathbf{x}) . \tag{16.3.4}$$

It is convenient also to impose the condition that Ω is normalized

$$(\Omega, \Omega) = 1 . \tag{16.3.5}$$

To minimize the expectation value (16.3.3) subject to the constraints (16.3.4) and (16.3.5), we use the method of Lagrange multipliers, and instead minimize the quantity

$$(\Omega, H\Omega) - \alpha(\Omega, \Omega) - \int d^3x\, \beta^n(\mathbf{x}) \Big(\Omega, \Phi_n(\mathbf{x})\Omega\Big) \tag{16.3.6}$$

with no constraints on Ω. This gives

$$H\Omega = \alpha\Omega + \int d^3x\, \beta^n(\mathbf{x})\Phi_n(\mathbf{x})\Omega\,. \tag{16.3.7}$$

Both α and $\beta^n(\mathbf{x})$ are to be chosen in such a way as to satisfy the constraints (16.3.4) and (16.3.5) and therefore depend functionally on the prescribed expectation value $\phi_n(\mathbf{x})$.

Now, we have said that in the presence of a current $\mathscr{J}^n(\mathbf{x})$, the Hamiltonian $H - \int d^3x\, \mathscr{J}^n(\mathbf{x})\,\Phi_n(\mathbf{x})$ has an eigenvalue $E[\mathscr{J}]$:

$$\left[H - \int d^3x\, \mathscr{J}^n(\mathbf{x})\,\Phi_n(\mathbf{x})\right] \Psi_{\mathscr{J}} = E[\mathscr{J}]\Psi_{\mathscr{J}} \tag{16.3.8}$$

with a normalized eigenvector $\Psi_{\mathscr{J}}$. Furthermore, since slowly turning on this current converts the vacuum into this energy eigenstate, we can presume that $E[\mathscr{J}]$ is the lowest energy eigenstate in the presence of this current. Therefore Eqs. (16.3.4), (16.3.5), and (16.3.7) are satisfied by

$$\Omega = \Psi_{\mathscr{J}_\phi}, \tag{16.3.9}$$

$$\alpha = E[\mathscr{J}_\phi]\,, \tag{16.3.10}$$

$$\beta^n(\mathbf{x}) = \mathscr{J}^n_\phi(\mathbf{x})\,, \tag{16.3.11}$$

where $\mathscr{J}_\phi(\mathbf{x})$ is the current for which $\Phi(\mathbf{x})$ has an expectation value $\phi(\mathbf{x})$ in the state $\Psi_{\mathscr{J}}$.

Setting $\mathscr{J} = \mathscr{J}_\phi$ in Eq. (16.3.8) and taking the scalar product with $\Psi_{\mathscr{J}_\phi}$, the minimum energy of states in which the fields Φ_n are constrained to have the expectation values ϕ_n is seen to be

$$\langle H\rangle_\Omega = E[\mathscr{J}_\phi] + \int d^3x\, \mathscr{J}^n_\phi(\mathbf{x})\phi_n(\mathbf{x})\,. \tag{16.3.12}$$

Recalling Eq. (16.3.2) and the assumed form of $J(x)$, this is

$$\langle H\rangle_\Omega = \frac{1}{T}\Big[-W[J_\phi] + \int d^4x\, J^n_\phi(\mathbf{x})\phi_n(\mathbf{x})\Big] = -\frac{1}{T}\,\Gamma[\phi]\,. \tag{16.3.13}$$

As noted in the previous section, if the field $\phi(x)$ has a constant value ϕ over a large spacetime volume $\mathscr{V}_4 = \mathscr{V}_3 T$, then we may write the effective action in terms of an effective potential $V(\phi)$:

$$\Gamma[\phi] = -\mathscr{V}_3 T\, V(\phi)\,. \tag{16.3.14}$$

In this case Eq. (16.3.13) tells us that the energy density is

$$\langle H\rangle_\Omega/\mathscr{V}_3 = V(\phi)\,. \tag{16.3.15}$$

This is the main result: *$V(\phi)$ is the minimum of the expectation value of the energy density for all states constrained by the condition that the scalar fields Φ_n have expectation values ϕ_n.* One consequence is that in the absence of external currents the vacuum state will relax to a state in which the potential $V(\phi)$ is not only *stationary*, which is required by the field equations (16.1.8), but also a *minimum.*

This result helps to resolve a problem in the interpretation of the quantum effective potential. The Euclidean version of the path-integral formalism (described in Appendix A of Chapter 23) makes it manifest that the two-point function Δ is positive (in the matrix sense), so according to Eq. (16.1.2) the same is true of $-\Pi = \Delta^{-1}$. Together with Eq. (16.2.3), this implies that for a single scalar field the effective potential $V(\phi)$ must have a positive (or zero) second derivative with respect to ϕ. More generally, the effective potential must be *convex*:[5]

$$V\Big(\lambda\phi_1 + (1-\lambda)\phi_2\Big) \leq \lambda V(\phi_1) + (1-\lambda)V(\phi_2) \text{ for } 0 \leq \lambda \leq 1\,.$$

But inspection of Eq. (16.2.13) shows that for $m^2 < 0$ and $g > 0$ the zero-loop approximation to the effective potential for the scalar field theory with action (16.2.1) has a *negative-definite* second derivative when ϕ is between the two minima of the effective potential at $\pm\sqrt{6|m^2|/g}$. This contradiction arises because the derivation of perturbation theory implicitly relies on the existence of a stable vacuum, but when $V''(\phi) < 0$, the field ϕ is at a value of $\tilde{\phi}$ where $V(\tilde{\phi}) - J_\phi\tilde{\phi}$ is a *maximum* rather than a minimum, which means that the vacuum state in the presence of the current J_ϕ is unstable.

So what is the true effective potential for this scalar field theory when $m^2 < 0$ and ϕ lies between the two minima of the potential? The result of this section is that we must find the state of minimum energy in which the expectation value of the operator Φ equals ϕ. As long as ϕ is between the two minima of the potential, we can give Φ an expectation value ϕ by taking the state as a suitable linear combination of the two states where ϕ is at the minima $\phi \simeq \pm\sqrt{6|m^2|/g}$. The energy in this state is equal to the energy at the minima, so this state clearly minimizes the energy. (Interference terms here vanish in the limit of infinite volume, for reasons explained in Section 19.1.) Thus the effective potential between the two minima of the potential is a *constant*, satisfying the requirement that it have a nonpositive second derivative. The same argument shows that in more general theories where the potential has two local minima of unequal energy, the potential between these minima is linear.[6]

16.4 Symmetries of the Effective Action

In some but not all cases, the symmetries of the action $I[\phi]$ are automatically also symmetries of the effective action $\Gamma[\phi]$. For instance, in the example of Section 16.2, the action (16.2.1) has a symmetry under the discrete transformation $\phi \to -\phi$. Thus it follows from their definition that $Z[J]$ and $W[J]$ are even under the corresponding reflection $J \to -J$. Eq. (16.1.5) then shows that $\phi_{-J} = -\phi_J$, and hence $J_{-\phi} = -J_\phi$, so Eq. (16.1.6) shows that $\Gamma[\phi]$ is even under $\phi \to -\phi$. This is borne out by the one-loop result (16.2.15). The fermion-loop contribution in (16.2.16) also exhibits the symmetry under $\phi \to -\phi$ in the special case where $M(0) = 0$, because in this case the action is invariant under the combined transformation $\phi \to -\phi$, $\psi \to \gamma_5\psi$.

We encounter problems in establishing the renormalizability of a theory, unless we can show that the symmetries that we impose on the action also apply to the effective action. For instance, in the example above, if $I[\phi]$ were assumed to be even in ϕ but $\Gamma[\phi]$ turned out not to be, then the coefficients of the terms in Γ proportional to $\int d^4x\,\phi$ and $\int d^4x\,\phi^3$ would be divergent, but the symmetry of the action would not allow us to introduce counterterms to absorb these infinities.

With this motivation, let's now turn to the important class of symmetries generated by infinitesimal transformations

$$\chi^n(x) \to \chi^n(x) + \epsilon F^n[x;\chi]\,, \tag{16.4.1}$$

where F^n is a function of x^μ that depends functionally on χ^n. (For instance, $F^n[x;\chi]$ may be an ordinary function of the χ^n and their derivatives at the point x.) We are now using the symbol χ^n rather than ϕ^r to denote the different types of fields, to emphasize that these include not only ordinary gauge and matter fields (which in the next chapter will be denoted $\phi^r(x)$), but all other fields appearing in the gauge-fixed action, including ghost fields. We repeat that these $\chi^n(x)$ may be of any type, not necessarily scalars.

We assume that both the action and the measure are invariant under the symmetry transformation (16.4.1):

$$I[\chi + \epsilon F] = I[\chi]\,, \tag{16.4.2}$$

$$\prod_{n,x} d\Big(\chi^n(x) + \epsilon F^n[x;\chi]\Big) = \prod_{n,x} d\chi^n(x)\,. \tag{16.4.3}$$

(It is actually sufficient for only the product $(\prod_{n,x} d\chi^n(x))\exp(iI)$ to be invariant, but where this is true usually Eqs. (16.4.2) and (16.4.3) both apply.) Replacing the integration variables in Eq. (16.1.1) with $\chi^n(x) +$

$\epsilon F^n[x;\chi]$, we have then

$$
\begin{aligned}
Z[J] &= \int \left[\prod_{n,x} d\Big(\chi^n(x) + \epsilon F^n[x;\chi]\Big)\right] \\
&\quad \times \exp\left\{iI[\chi + \epsilon F] + i\int d^4x\, \Big(\chi^n(x) + \epsilon F^n[x;\chi]\Big)J_n(x)\right\} \\
&= \int \left[\prod_{n,x} d\chi^n(x)\right] \exp\left\{iI[\chi] + i\int d^4x\, \Big(\chi^n(x) + \epsilon F^n[x;\chi]\Big)J_n(x)\right\} \\
&= Z[J] + i\epsilon \int \left(\prod_{n,x} d\chi^n(x)\right) \int F^n[y;\chi]J_n(y)\, d^4y \\
&\quad \times \exp\left\{iI[\chi] + i\int d^4x\, \chi^n(x)J_n(x)\right\}
\end{aligned}
$$

and hence

$$
\int d^4y\, \langle F^n(y)\rangle_J J_n(y) = 0\,, \tag{16.4.4}
$$

where $\langle\ \ \rangle_J$ denotes the quantum average in the presence of the current $J_n(x)$,

$$
\begin{aligned}
Z[J]\langle F^n(y)\rangle_J \equiv \int \left(\prod_{n,x} d\chi^n(x)\right) F^n[y;\chi] \\
\times \exp\left\{iI[\chi] + i\int d^4x\, \chi^n(x)J_n(x)\right\}\,,
\end{aligned} \tag{16.4.5}
$$

normalized so that $\langle 1\rangle_J = 1$. But recall that $J_n(y)$ is given in terms of the effective action $\Gamma[\chi]$ by Eq. (16.1.7)

$$
J_{n,\chi}(y) = -\frac{\delta\Gamma[\chi]}{\delta\chi^n(y)}\,.
$$

Therefore Eq. (16.4.4) may be written as

$$
0 = \int d^4y\, \langle F^n(y)\rangle_{J_\chi} \frac{\delta\Gamma[\chi]}{\delta\chi^n(y)}\,. \tag{16.4.6}
$$

In other words, $\Gamma[\chi]$ is invariant under the infinitesimal transformation

$$
\chi^n(y) \to \chi^n(y) + \epsilon\langle F^n(y)\rangle_{J_\chi}\,. \tag{16.4.7}
$$

Such symmetry conditions are known as *Slavnov–Taylor identities.*[7]

Is this the symmetry transformation with which we started? It is for one very important class of infinitesimal symmetry transformations: those that are *linear*. For such symmetries F is

$$
F^n[x;\chi] = s^n(x) + \int t^n{}_m(x,y)\chi^m(y)d^4y\,. \tag{16.4.8}
$$

(In the most common case $s^n(x)$ vanishes and $t^n{}_m(x,y)$ is a constant matrix times $\delta^4(x-y)$.) For any linear F, we have

$$\langle F^n(x)\rangle_J = s^n(x) + \int t^n{}_m(x,y)\langle \chi^m(y)\rangle_J\, d^4y\,.$$

But for any fixed χ, J_χ is defined as the value of the current J that makes $\langle \chi^m(y)\rangle_J$ equal to $\chi^m(y)$, so

$$\langle F^n(x)\rangle_{J_\chi} = s^n(x) + \int t^n{}_m(x,y)\chi^m(y)\, d^4y = F^n[x;\chi]\,. \qquad (16.4.9)$$

Hence Eq. (16.4.6) requires that $\Gamma[\chi]$ be invariant under all the functional linear transformations $\chi^n \to \chi^n + \epsilon F^n$ that leave $I[\chi]$ and the measure invariant.

We occasionally have to deal with symmetry transformations that are not linear. One important example is provided by the BRST transformation discussed in Section 15.7. For non-linear transformations, the symmetry transformation (16.4.7) under which the effective action is invariant is *not* generally the same as the assumed symmetry transformation (16.4.1) that leaves the original action invariant, because the average of a non-linear functional of fields is not generally the same as the functional of the average fields. Indeed, the form of $\langle F\rangle_{J_\chi}$ as a functional of χ depends in general on the dynamics of the system, and is usually non-local. This complication will be dealt with in the next chapter by the method of antibrackets.

* * *

Up to now, we have tacitly been assuming that the fields χ^n and the corresponding transformation functions F^n and currents J_n are all bosonic. We will need to take note of the sign factors that appear when some of these are fermionic, as in particular for supersymmetry or BRST transformations, where ϵ is fermionic and χ^n and F^n have opposite statistics. With currents inserted to the right of fields, as in Eqs. (16.1.1) and (16.4.5), Eqs. (16.1.5) and (16.1.7) hold in the form

$$\frac{\delta_R W[J]}{\delta J_m(y)} = \chi_J^m(y)\,, \qquad (16.4.10)$$

$$\frac{\delta_L \Gamma[\chi]}{\delta \chi^m(y)} = -J_{\chi,m}(y)\,, \qquad (16.4.11)$$

with the subscripts R and L indicating that the derivative is to act from the right or left. In consequence, the Slavnov–Taylor identity (16.4.6) should be written

$$0 = \int d^4y\, \langle F^n(y)\rangle_{J_\chi} \frac{\delta_L \Gamma[\chi]}{\delta \chi^n(y)}\,. \qquad (16.4.12)$$

Problems

1. Consider a theory of real pseudoscalars $\phi(x)$ and complex Dirac fields $\psi(x)$, with masses M and m respectively, and interaction $g\bar{\psi}\gamma_5\psi\phi$. Evaluate the effective potential for ϕ = constant, $\psi = 0$, to one-loop order.

2. Derive general formulas for $\delta^3 W[J]/\delta J_n(x)\,\delta J_m(y)\,\delta J_\ell(z)$ and $\delta^4 W[J]/\delta J_n(x)\,\delta J_m(y)\,\delta J_\ell(z)\,\delta J_k(w)$ in terms of the variational derivatives of $\Gamma[\phi]$ with respect to ϕ. Show which Feynman diagrams correspond to each term in these formulas.

3. Calculate the effective potential to one-loop order for the theory of a neutral scalar field ϕ with interaction Lagrangian density $g\phi^3/6$ in six spacetime dimensions.

4. Suppose that the action $I[\phi]$ is invariant under a *finite* matrix transformation $\phi_n(x) \to \sum_m M_{nm}\phi_m(x)$. Under which transformation of the currents is $W[J]$ then invariant? Use this result to derive a symmetry property of $\Gamma[\phi]$.

References

1. The effective action $\Gamma[\phi]$ was introduced by J. Goldstone, A. Salam, and S. Weinberg, *Phys. Rev.* **127**, 965 (1962), who defined it perturbatively, as a sum over one-particle-irreducible connected diagrams. The non-perturbative definition (16.1.6) was first given independently by B. De Witt, in *Relativity, Groups, and Topology – Lectures Delivered at Les Houches during the 1963 Session of the Summer School of Theoretical Physics*, C. De Witt and B. De Witt, eds. (Gordon and Breach, New York, 1964); G. Jona-Lasinio, *Nuovo Cimento* **34**, 1790 (1964).

2. S. Coleman, *Aspects of Symmetry* (Cambridge University Press, Cambridge, 1985): pp. 135–6.

3. S. Coleman and E. Weinberg, *Phys. Rev.* **D7**, 1888 (1973).

4. K. Symanzik, *Comm. Math. Phys.* **16**, 48 (1970); S. Coleman, *Aspects of Symmetry* (Cambridge University Press, Cambridge, 1985): pp 139–42.

5. K. Symanzik, Ref. 4; J. Iliopoulos, C. Itzykson, and A. Martin, *Rev. Mod. Phys.* **47**, 165 (1975).

6. Y. Fujimoto, L. O'Raifeartaigh, and G. Parravicini, *Nucl. Phys.* **B212**, 268 (1983); R. W. Haymaker and J. Perez-Mercader, *Phys. Rev.* **D27**, 1948 (1983); C. M. Bender and F. Cooper, *Nucl. Phys.* **B224**, 403 (1983); M. Hindmarsh and D. Johnston, *J. Math. Phys.* **A19**, 141 (1986); V. Branchina, P. Castorina, and D. Zappalà, *Phys. Rev.* **D41**, 1948 (1990); K. Cahill, *Phys. Rev.* **D52**, 4704 (1995).

7. A. A. Slavnov, *Theor. Math. Phys.* **10**, 152 (1972) [English translation: *Theor. and Math. Phys.* **10**, 99 (1972)]; J. C. Taylor, *Nucl. Phys.* **B33**, 436 (1971).

17

Renormalization of Gauge Theories

We now return to gauge theories, and use the external field formalism described in the previous chapter to study the renormalizability of these theories and to carry out an important calculation.

17.1 The Zinn-Justin Equation

In this section the BRST symmetry described in Section 15.7 will be used to demonstrate a fundamental property of the quantum effective action, first derived by Zinn-Justin.[1] According to the general rules outlined in Section 16.4, the BRST invariance of the action $I[\chi]$ imposes on the effective action $\Gamma[\chi]$ the condition

$$\int d^4x \, \langle \Delta^n(x) \rangle_{J_\chi} \frac{\delta_L \Gamma[\chi]}{\delta \chi^n(x)} = 0 \,, \tag{17.1.1}$$

where the change in $\chi^n(x)$ under a BRST transformation with infinitesimal fermionic parameter θ is

$$\delta_\theta \chi^n(x) = \theta \Delta^n(x) \tag{17.1.2}$$

and $\langle \cdots \rangle$ here denotes a vacuum expectation value taken in the presence of a current J_χ that makes the vacuum expectation value of the operator fields $X^n(x)$ equal to the c-number functions $\chi^n(x)$. The implied sum over n runs over all of the fields in the BRST formalism; that is, over ω_α, ω^*_α, and h_α as well as the gauge and matter fields that in Section 15.8 we have collectively called ϕ^r. Because $\Delta^n(x)$ is quadratic in the fields when χ^n is a gauge or matter field or ω_α, Eq. (17.1.1) does *not* in general tell us that the effective action is invariant under the same BRST transformations as the action itself.

To handle this complication, we employ a trick that proves useful in dealing with any sort of nilpotent symmetry transformation. First, we introduce a set of c-number external fields $K_n(x)$, and define a new effective action by

$$\Gamma[\chi, K] \equiv W[J_{\chi,K}, K] - \int d^4x\, \chi^n(x)\, J_{\chi,K\, n}(x)\,, \tag{17.1.3}$$

where the connected vacuum persistence amplitude W is here calculated with the gauge-fixed action* $I + \int d^4x\, \Delta^n K_n$:

$$e^{iW[J,K]} \equiv \int \left[\prod_{n,x} d\chi^n(x)\right] \exp\left(iI + i\int d^4x\, \Delta^n K_n + i\int d^4x\, \chi^n J_n\right) \tag{17.1.4}$$

and $J_{\chi,K}$ is the current required to give the fields the expectation values χ in the presence of the external fields K:

$$\left.\frac{\delta_R W[J,K]}{\delta J_n(x)}\right|_{J=J_{\chi,K}} \equiv \chi^n(x)\,. \tag{17.1.5}$$

(The K_n must have the same fermionic or bosonic statistics as Δ^n, which is opposite to that of χ^n.) Since the BRST transformation is nilpotent, the quantities $\Delta^n(x)$ are BRST-invariant, so in the same way as in Section 16.4 we can show that the new effective action $\Gamma[\chi, K]$ satisfies a BRST-invariance condition:

$$\int d^4x\, \langle \Delta^n(x)\rangle_{J_{\chi,K},K} \frac{\delta_L \Gamma[\chi, K]}{\delta \chi^n(x)} = 0\,, \tag{17.1.6}$$

where $\langle \cdots \rangle_{J,K}$ denotes a vacuum expectation value calculated in the presence of the current J and the external fields K:

$$\langle \mathscr{O}[\chi]\rangle_{J,K} = \frac{\int \left[\prod_{n,x} d\chi^n(x)\right] \mathscr{O}[\chi] \exp\left(iI + i\int d^4x\, \Delta^n K_n + i\int d^4x\, \chi^n J_n\right)}{\int \left[\prod_{n,x} d\chi^n(x)\right] \exp\left(iI + i\int d^4x\, \Delta^n K_n + i\int d^4x\, \chi^n J_n\right)}\,. \tag{17.1.7}$$

It is convenient to express the expectation value of Δ^n as a variational derivative of the effective action. Taking the right variational derivative of Eq. (17.1.3) with respect to K gives

$$\frac{\delta_R \Gamma[\chi, K]}{\delta K_n(x)} = \left.\frac{\delta_R W[J,K]}{\delta K_n(x)}\right|_{J=J_{\chi,K}} + \int d^4y\, \left.\frac{\delta_R W[J,K]}{\delta J_m(y)}\right|_{J=J_{\chi,K}} \frac{\delta_R J_{\chi,K\, m}(y)}{\delta K_n(x)}$$
$$- \int d^4y\, \chi^m(y) \frac{\delta_R J_{\chi,K\, m}(y)}{\delta K_n(x)}\,.$$

Using Eq. (17.1.5) we see that the last two terms cancel, and using the

---

* Here I is the action $I_{\rm NEW}$ modified as described in Section 15.7 to depend on the ghost and antighost fields ω_α and ω^*_α and on the auxiliary field h_α, with the subscript 'NEW' dropped from now on.

definitions (17.1.4) and (17.1.7) gives us then our desired relation

$$\frac{\delta_R \Gamma[\chi, K]}{\delta K_n(x)} = \frac{\delta_R W[J,K]}{\delta K_n(x)}\bigg|_{J=J_{\chi,K}} = \langle \Delta^n(x) \rangle_{J_{\chi,K},K} \,. \tag{17.1.8}$$

The BRST symmetry condition (17.1.6) may now be written as a simple condition, the *Zinn-Justin equation*, involving the effective action alone:

$$\int d^4x \, \frac{\delta_R \Gamma[\chi, K]}{\delta K_n(x)} \frac{\delta_L \Gamma[\chi, K]}{\delta \chi^n(x)} = 0 \,. \tag{17.1.9}$$

As remarked after Eq. (15.9.3), the interchange of fields and antifields (or in this case χ^n and K_n) results simply in a change of sign of the left-hand side of Eq. (17.1.9), so this can be written as

$$(\Gamma, \Gamma) = 0 \,, \tag{17.1.10}$$

where the antibracket is calculated here with K_n in place of the antifield of χ^n:

$$(F, G) \equiv \int d^4x \, \frac{\delta_R F[\chi, K]}{\delta \chi^n(x)} \frac{\delta_L G[\chi, K]}{\delta K_n(x)} - \int d^4x \, \frac{\delta_R F[\chi, K]}{\delta K_n(x)} \frac{\delta_L G[\chi, K]}{\delta \chi^n(x)} \,. \tag{17.1.11}$$

This is formally the same as the Batalin–Vilkovisky 'master equation' discussed in Section 15.9, but appears here as a constraint on the quantum effective action $\Gamma[\chi, K]$ rather than on the fundamental action $S[\chi, \chi^\ddagger]$. The Zinn-Justin equation (17.1.10) will be used in the next two sections to show how to renormalize gauge theories, and in Section 22.6 to study anomalies in these theories.

17.2 Renormalization: Direct Analysis

The simplest non-Abelian gauge theories are renormalizable in the 'Dyson' sense that the operators in the Lagrangian density all have dimensionality (in powers of mass) four or less. As we saw in Chapter 12, this guarantees that the infinities in the quantum effective action only appear in terms that could be cancelled by the counterterms in interactions of dimensionality four or less. But there is more to renormalizability than this. The Lagrangian density is constrained by gauge invariance and other symmetries. For a theory to be renormalizable, it is necessary that the infinities in the quantum effective action satisfy the same constraints, up to possible renormalizations of the fields.[2]

The effective action $\Gamma[\chi, K]$ is a complicated functional of both χ and K, about which the symmetry condition (17.1.9) says complicated things, but fortunately matters are much simpler for the infinite terms in Γ. We will write the action $S[\chi, K] \equiv I[\chi] + \int d^4x \, \Delta^n K_n$ as the sum of a term $S_R[\chi, K]$

in which masses and coupling constants are set equal to their renormalized values, plus a correction $S_\infty[\chi, K]$, which contains the counterterms that we intend to cancel the infinities from loop graphs. Both S_R and S_∞ must be taken to have the symmetries of the original action $S[\chi, K]$, so the question is whether the infinite parts of the higher-order contributions to Γ share the same symmetries, so that they can be cancelled by the counterterms in S_∞.

We may expand Γ in a series of terms Γ_N that arise both from diagrams with just N loops and also from graphs with $N - M$ loops (where $1 \leq M \leq N$) involving various counterterms in $S_\infty[\chi, K]$ that will be used to cancel infinities in graphs with a total of M loops:

$$\Gamma[\chi, K] = \sum_{N=0}^{\infty} \Gamma_N[\chi, K] . \tag{17.2.1}$$

The symmetry condition (17.1.10) then reads,* for each N,

$$\sum_{N'=0}^{N} (\Gamma_{N'}, \Gamma_{N-N'}) = 0 . \tag{17.2.2}$$

In the sum (17.2.1) the leading term is just $\Gamma_0[\chi, K] = S_R[\chi, K]$, which of course is finite. Suppose that, for all $M \leq N - 1$, all infinities arising from M-loop graphs have been cancelled by counterterms in S_∞ . Then infinities can appear in Eq. (17.2.2) only in the $N' = 0$ and $N' = N$ terms, which are equal, and the infinite part of this condition tells us that the infinite part $\Gamma_{N,\infty}$ of Γ_N is subject to the condition that

$$(S_R, \Gamma_{N,\infty}) = 0 . \tag{17.2.3}$$

This is a symmetry principle generated by S_R, just like that described by Eqs. (15.9.16) and (15.9.17). Note in particular that the transformation $X \mapsto (S_R, X)$ acts on the external fields K_n as well as the fields χ^n.

Up to this point, we have used none of the special properties of a renormalizable Yang–Mills theory. Now note that, according to the general power-counting rules of renormalization theory, with all infinities cancelled in subgraphs of Γ_N, the infinite part $\Gamma_{N,\infty}[\chi, K]$ of $\Gamma_N[\chi, K]$ can only be a sum of products of fields (including K) and their derivatives of

* Such order-by-order relations may be derived formally by repeating the reasoning of Section 16.1: When the action S_R is replaced with $g^{-1}S_R$, the contribution of a connected L-loop graph with I internal lines and V vertices is multiplied by a factor $g^{V-I} = g^{L-1}$. If the counterterms in S_∞ associated with N-loop diagrams are also provided with factors g^N, then Γ_L is the value for $g = 1$ of the term in Γ of order g^{L-1}. Eq. (17.2.2) then follows by requiring Eq. (17.1.10) to hold in each order in g. In cgs units the action has the same dimensions as $\hbar$, and so appears in the path integral multiplied with a factor $1/\hbar$, so we can also use $\hbar$ as a loop-counting parameter in place of g.

dimensionality (in powers of mass) four or less. Finally, the arguments of Section 16.4 show that $\Gamma[\chi, K]$ and hence also $\Gamma_{N,\infty}[\chi, K]$ are invariant under all of the *linearly* realized symmetry transformations under which the action is invariant. (As described below, these are: Lorentz transformations, *global* gauge transformations, antighost translations, and the ghost phase transformations associated with ghost number conservation. Of course, the auxiliary fields K_n must be assigned suitable transformation properties under these symmetry transformations.) These conditions together with Eq. (17.2.3) will suffice to tell us all we need to know about the structure of $\Gamma_{N,\infty}[\chi, K]$.

To implement these conditions, we need to know the dimensionalities of the external fields K_n. If a field χ^n has dimensionality d_n (that is, d_n powers of mass), then Δ^n correspondingly has dimensionality $d_n + 1$ (as can be seen by inspection of the BRST transformation rules (15.7.7)–(15.7.11)), so in order for $\int d^4x\, K_n \Delta^n$ to be dimensionless, K_n must have dimensionality $3 - d_n$. The fields $A^{\alpha\mu}$, ω^α, and $\omega^{\alpha *}$ all have dimensionalities $d_n = +1$, so the corresponding K_n all have dimensionalities +2. (We do not introduce any external field corresponding to h^α, because this field is BRST-invariant.) Any spin 1/2 matter fields ψ_ℓ have dimensionalities 3/2, and the corresponding K_n thus also have dimensionalities 3/2. Thus a dimension four quantity like $\Gamma_{N,\infty}[\chi, K]$ is at most quadratic in any of the K_n. Furthermore terms that are of second order in the K_n cannot involve any other fields, except that a term of second order in the K_n for spin 1/2 matter fields may involve at most one additional field of dimensionality unity.

We can now use ghost number conservation to show that in fact $\Gamma_{N,\infty}[\chi, K]$ does not contain any terms of second order in the K_n. For this purpose, we also need the ghost quantum numbers of the K_n. If χ^n has ghost quantum number γ_n, then Δ^n has ghost quantum number $\gamma_n + 1$, so K_n must be assigned ghost quantum number $-\gamma_n - 1$. The ghost quantum numbers of the fields $A^{\alpha,\mu}$, ψ^ℓ, ω^α, and $\omega^{\alpha *}$ are respectively 0, 0, +1, and −1, so the corresponding external fields K_n have ghost quantum numbers −1, −1, −2, and 0, respectively. This rules out any terms in $\Gamma_{N,\infty}[\chi, K]$ that are of second order in the K_n, with the one possible exception of a term of second order in the external fields K^*_α associated with $\omega^{\alpha *}$ (and involving no other fields). However, these last terms are also forbidden, for a different reason. The BRST transformation of $\omega^{\alpha *}$ is linear in the fields, with

$$\Delta^{\alpha *} = -h^\alpha \,, \tag{17.2.4}$$

so here

$$\frac{\delta_L \Gamma_{N,\infty}[\chi, K]}{\delta K^*_\alpha} = \langle \Delta^{\alpha *} \rangle_{J_\chi, K} = -h^\alpha$$

is independent of K^*_α. It follows that $\Gamma_{N,\infty}[\chi, K]$ is linear in $K^{\alpha,*}$, and depends on $K^{\alpha,*}$ only through a term $-\int d^4x\, K^*_\alpha h^\alpha$. (Both K^*_α and h^α are bosonic, so their order is immaterial.) In particular, for $N > 0$, $\Gamma_{N,\infty}[\chi, K]$ is independent of $K^{\alpha *}$.

We have seen that $\Gamma_{N,\infty}[\chi, K]$ is at most linear in all of the K_n. We will write it as

$$\Gamma_{N,\infty}[\chi, K] = \Gamma_{N,\infty}[\chi, 0] + \int d^4x\, \mathscr{D}^n_N[\chi; x]\, K_n(x)\,. \tag{17.2.5}$$

We also recall that S_R has the K dependence:

$$S_R[\chi, K] = S_R[\chi] + \int d^4x\, \Delta^n[\chi; x] K_n(x)\,.$$

The terms in Eq. (17.2.3) of zeroth and first order in K therefore give**

$$\int d^4x \left[\Delta^n[\chi; x]\, \frac{\delta_L \Gamma_{N,\infty}[\chi, 0]}{\delta\chi^n(x)} + \mathscr{D}^n_N[\chi; x]\, \frac{\delta_L S_R[\chi]}{\delta\chi^n(x)}\right] = 0 \tag{17.2.6}$$

and

$$\int d^4x \left[\Delta^n(\chi; x)\, \frac{\delta_L \mathscr{D}^m_N(\chi; y)}{\delta\chi^n(x)} + \mathscr{D}^n_N(\chi; x) \frac{\delta_L \Delta^m(\chi; y)}{\delta\chi^n(x)}\right] = 0\,, \tag{17.2.7}$$

respectively. These relations may be made more perspicuous by introducing the quantities

$$\Gamma^{(\epsilon)}_N[\chi] \equiv S_R[\chi] + \epsilon\Gamma_{N,\infty}[\chi, 0]\,, \tag{17.2.8}$$

and

$$\Delta^{(\epsilon)n}_N(x) \equiv \Delta^n(x) + \epsilon\mathscr{D}^n_N(x)\,, \tag{17.2.9}$$

with ϵ infinitesimal. Then (17.2.6) (together with the BRST invariance of S_R) just says that $\Gamma^{(\epsilon)}_N[\chi]$ is invariant under the transformation

$$\chi^n(x) \to \chi^n(x) + \theta\Delta^{(\epsilon)n}_N(x)\,, \tag{17.2.10}$$

while Eq. (17.2.7) (together with the nilpotence of the original BRST transformation) tells us that this transformation is *nilpotent*.

We must now consider what form this nilpotent transformation may take. As already mentioned, $\Gamma_{N,\infty}$ consists only of terms of dimensionality four or less, so $\mathscr{D}^n_N$ and hence $\Delta^{(\epsilon)n}_N(x)$ have at most the dimensionality

** The second terms in Eqs. (17.2.6) and (17.2.7) have been put into the form shown here by recalling that, because χ^n and K_n have opposite statistics, for any bosonic functionals A and B,

$$\frac{\delta_R A}{\delta\chi^n}\frac{\delta_L B}{\delta K_n} = -\frac{\delta_L A}{\delta\chi^n}\frac{\delta_R B}{\delta K_n} = -\frac{\delta_R B}{\delta K_n}\frac{\delta_L A}{\delta\chi^n}\,.$$

of the original BRST transformation function $\Delta^n(x)$. Also, $\mathscr{D}^n_N$ and hence also $\Delta^{(\epsilon)n}_N(x)$ must have the same Lorentz transformation properties and ghost quantum numbers as $\Delta^n(x)$. Hence the most general form of the transformation (17.2.10) is

$$\begin{aligned}\psi &\to \psi + i\theta\omega^\alpha T_\alpha \psi \,, \\ A_{\alpha\mu} &\to A_{\alpha\mu} + \theta\left[B_{\alpha\beta}\partial_\mu\omega_\beta + D_{\alpha\beta\gamma}A_{\beta\mu}\omega_\gamma\right] \,, \\ \omega_\alpha &\to \omega_\alpha - \tfrac{1}{2}\theta E_{\alpha\beta\gamma}\omega_\beta\omega_\gamma \,,\end{aligned}$$

where T_α is some matrix acting on the spinor fields, and $B_{\alpha\beta}$, $D_{\alpha\beta\gamma}$, and $E_{\alpha\beta\gamma}$ are constants, with $E_{\alpha\beta\gamma}$ antisymmetric in β and γ. Also, the transformations of ω^*_α and h_α are linear, and are therefore unchanged:

$$\omega^*_\alpha \to \omega^*_\alpha - \theta h_\alpha \,, \qquad h_\alpha \to h_\alpha \,.$$

Next we impose the condition of nilpotence. The most important requirement is that $E_{\alpha\beta\gamma}\omega_\beta\omega_\gamma$ should be invariant. This yields the requirement that $E_{\alpha\beta\gamma}E_{\beta\delta\epsilon}\omega_\delta\omega_\epsilon\omega_\gamma$ should vanish, so that the part of $E_{\alpha\beta\gamma}E_{\beta\delta\epsilon}$ that is totally antisymmetric in δ, ϵ, γ vanishes:

$$E_{\alpha\beta\gamma}E_{\beta\delta\epsilon} + E_{\alpha\beta\epsilon}E_{\beta\gamma\delta} + E_{\alpha\beta\delta}E_{\beta\epsilon\gamma} = 0 \,.$$

But this just tells us that $E_{\alpha\beta\gamma}$ is the structure constant of some Lie algebra $\mathscr{E}$. Because $E_{\alpha\beta\gamma}$ goes to the structure constant $C_{\alpha\beta\gamma}$ of the original gauge Lie algebra $\mathscr{A}$ for $\epsilon \to 0$, $\mathscr{E}$ must be the same as $\mathscr{A}$, and the structure constants $E_{\alpha\beta\gamma}$ can differ from the original $C_{\alpha\beta\gamma}$ only by a multiplicative factor:†

$$E_{\alpha\beta\gamma} = \mathscr{Z}C_{\alpha\beta\gamma} \,.$$

(This is for simple gauge groups; in the general case we would have a separate factor $\mathscr{Z}$ for each simple subgroup.)

Next we turn to the condition that the transformation (17.2.10) be nilpotent when acting on the gauge fields. The requirement that $B_{\alpha\beta}\partial_\mu\omega_\beta + D_{\alpha\beta\gamma}A_{\beta\mu}\omega_\gamma$ be invariant tells us that

$$D_{\alpha\beta\gamma}D_{\beta\delta\epsilon} - D_{\alpha\beta\epsilon}D_{\beta\delta\gamma} = E_{\beta\epsilon\gamma}D_{\alpha\delta\beta} = \mathscr{Z}C_{\beta\epsilon\gamma}D_{\alpha\delta\beta}$$

and

$$B_{\alpha\beta}E_{\beta\gamma\delta} = D_{\alpha\beta\delta}B_{\beta\gamma} \,.$$

† The requirement of global gauge invariance rules out any non-trivial similarity transformation in the relation between $E_{\alpha\beta\gamma}$ and $C_{\alpha\beta\gamma}$.

The first condition has the unique solution††

$$D_{\alpha\beta\gamma} = \mathscr{Z} C_{\alpha\beta\gamma} \,.$$

The second condition tells us that the matrix $B_{\alpha\beta}$ commutes with the adjoint representation of the gauge group, and hence (since we have chosen the structure constants totally antisymmetric) must be proportional to a Kronecker delta, with a coefficient we shall call $\mathscr{Z}\mathscr{N}$:

$$B_{\alpha\beta} = \mathscr{Z}\mathscr{N}\delta_{\alpha\beta} \,.$$

Finally, the condition that the transformation (17.2.10) be nilpotent when acting on the fermion fields (if any) requires that $\omega^\alpha T_\alpha \psi$ is invariant. This tells us that

$$[T_\beta, T_\gamma] = iE_{\alpha\beta\gamma} T_\alpha \,,$$

so T_α differs from the generator t_α in the original Lagrangian only by a factor $\mathscr{Z}$:

$$T_\alpha = \mathscr{Z} t_\alpha \,.$$

We have thus seen that, apart from the appearance of the new constants $\mathscr{Z}$ and $\mathscr{N}$, the transformation (17.2.10) is just the BRST transformation with which we started:

$$\psi \to \psi + i\mathscr{Z}\theta\omega^\alpha t_\alpha \psi \,, \tag{17.2.11}$$

$$A_{\alpha\mu} \to A_{\alpha\mu} + \mathscr{Z}\theta \left[\mathscr{N}\partial_\mu\omega_\alpha + C_{\alpha\beta\gamma}A_{\beta\mu}\omega_\gamma\right] \,, \tag{17.2.12}$$

$$\omega_\alpha \to \omega_\alpha - \tfrac{1}{2}\mathscr{Z}\,\theta\, C_{\alpha\beta\gamma}\omega_\beta\omega_\gamma \,, \tag{17.2.13}$$

$$\omega^*_\alpha \to \omega^*_\alpha - \theta h_\alpha \,, \tag{17.2.14}$$

$$h_\alpha \to h_\alpha \,. \tag{17.2.15}$$

Now we must use this symmetry to constrain the structure of the corrected action (17.2.8). Since this contains only the original renormalized action plus the infinite part of the N-loop contribution, it must be the integral of a Lagrangian density

$$\Gamma_N^{(\epsilon)} = \int d^4x\, \mathscr{L}_N^{(\epsilon)} \tag{17.2.16}$$

with $\mathscr{L}_N^{(\epsilon)}$ a local function of fields and field derivatives of dimensionality (in powers of mass) no greater than 4. Furthermore, as we found in Section 16.4, $\mathscr{N}_N^{(\epsilon)}$ must be invariant under all the symmetries of the

†† The matrix $(D_\gamma)_{\alpha\beta} \equiv D_{\gamma\alpha\beta}/\mathscr{Z}$ satisfies the commutation relations of the gauge Lie algebra, $[D_\gamma, D_\epsilon] = C_{\beta\epsilon\gamma}D_\beta$. But the only representation of a simple Lie algebra with the same dimensionality and $\mathscr{A}$ transformation properties as the adjoint representation of $\mathscr{A}$ is the adjoint representation itself.

original Lagrangian that act *linearly* on the fields. To identify these symmetries, recall that in generalized ξ-gauge, the 'new' Lagrangian density in Eq. (15.7.6) takes a form given by replacing the term $-(\partial_\mu A^\mu_\alpha)(\partial_\nu A^\nu_\alpha)/2\xi$ in Eq. (15.6.16) with the terms $h_\alpha f_\alpha + \frac{1}{2}\xi h_\alpha h_\alpha$ in Eq. (15.7.6):

$$\mathscr{L}_{\rm NEW} = \mathscr{L}_{\rm M} - \tfrac{1}{4}F^{\mu\nu}_\alpha F_{\alpha\mu\nu} - \partial_\mu\omega^*_\alpha\,\partial^\mu\omega_\alpha + C_{\alpha\beta\gamma}(\partial_\mu\omega^*_\alpha)\,A^\mu_\gamma\,\omega_\beta + h_\alpha\partial_\mu A^\mu_\alpha + \tfrac{1}{2}\xi h_\alpha h_\alpha\,. \tag{17.2.17}$$

Inspection of this formula reveals the following linear symmetries:
(1) **Lorentz invariance.**
(2) **Global gauge invariance** – that is, invariance under the transformations

$$\delta\psi_\ell(x) = i\,\epsilon^\alpha(t_\alpha)_\ell{}^m\psi_m(x)\,, \tag{17.2.18}$$

$$\delta A^\beta{}_\mu(x) = C_{\beta\gamma\alpha}\epsilon^\alpha A^\gamma{}_\mu(x)\,, \tag{17.2.19}$$

$$\delta\omega_\beta(x) = C_{\beta\gamma\alpha}\epsilon^\alpha\omega_\gamma(x)\,, \tag{17.2.20}$$

$$\delta\omega^*_\beta(x) = C_{\beta\gamma\alpha}\epsilon^\alpha\omega^*_\gamma(x)\,, \tag{17.2.21}$$

$$\delta h_\beta(x) = C_{\beta\gamma\alpha}\epsilon^\alpha h_\gamma(x)\,, \tag{17.2.22}$$

with constant infinitesimal parameters ϵ^α.
(3) **Antighost translation invariance** — that is, invariance under the transformation

$$\omega^*_\alpha(x) \to \omega^*_\alpha(x) + c_\alpha\,, \tag{17.2.23}$$

with arbitrary constant parameters c_α.
(4) **Ghost number conservation** — that is, the conservation of a ghost number equal to +1 for ω_α, −1 for ω^*_α, and 0 for all other fields.

We shall now proceed to work out the structure of the most general Lagrangian density that is renormalizable, in the sense that it consists only of terms of dimensionality +4 or less, that has these linearly acting symmetries, and that is invariant under the modified BRST transformations (17.2.11)–(17.2.15).

From Eq. (17.2.17) we may conclude that the fields A^μ_α, ω_α, ω^*_α, and h_α have the dimensionalities +1, +1, +1, and +2, respectively (in powers of mass). Note also that ghost number conservation requires that ω and ω^* come in pairs, while antighost translation invariance dictates that ω^* always appears as a derivative. Each pair of ω and $\partial_\mu\omega^*$ fields adds +3 to the dimensionality, so renormalizability rules out any term with more than one such pair. With one such pair we can have at most one more derivative or one additional gauge field, and Lorentz invariance dictates that we must have one or the other. The only renormalizable allowed interactions involving ghost fields are then linear combinations of terms of the form $\partial_\mu\omega^*_\alpha\,\partial^\mu\omega_\beta$ or $\partial_\mu\omega^*_\alpha\,A^\mu_\gamma\,\omega_\beta$.

Next, let us consider the terms that involve the field h_α and possibly other fields but not ω or ω^*. This field has dimensionality +2, so renormalizability and Lorentz invariance allows this field to appear only[‡] multiplied with another h_β or $\partial_\mu A^\mu_\beta$ or $A^\mu_\beta A_{\mu\gamma}$.

Finally, the Lagrangian will contain renormalizable terms involving only the matter and gauge fields. We will call the sum of these terms $\mathscr{L}_{\psi A}$. Putting this all together and using global gauge invariance, the most general renormalizable interaction allowed by the assumed symmetries (aside from BRST invariance) takes the form:

$$\mathscr{L}_N^{(\epsilon)} = \mathscr{L}_{\psi A} + \tfrac{1}{2}\xi' h_\alpha h_\alpha + c h_\alpha \partial_\mu A^\mu_\alpha - e_{\alpha\beta\gamma} h_\alpha A^\mu_\beta A_{\gamma\mu}$$
$$- Z_\omega (\partial_\mu \omega^*_\alpha)(\partial_\mu \omega_\alpha) - d_{\alpha\beta\gamma}(\partial_\mu \omega^*_\alpha)\omega_\beta A^\mu_\gamma \,, \qquad (17.2.24)$$

where ξ', Z_ω, c, $d_{\alpha\beta\gamma}$, and $e_{\alpha\beta\gamma}$ are unknown constants, with no constraints except obvious symmetry properties such as global gauge invariance and $e_{\alpha\beta\gamma} = e_{\alpha\gamma\beta}$. (As mentioned earlier, we are assuming for simplicity that the gauge group is simple, but the extension to a direct sum of simple and $U(1)$ gauge groups would be trivial; for instance, instead of one term proportional to $h_\alpha h_\alpha$ we would have a sum of such terms, one for each simple subgroup of the gauge group.)

Now we impose BRST invariance. The cancellation of terms in $\delta \mathscr{L}_N^{(\epsilon)}$ proportional to $\theta \partial_\mu h_\alpha \partial^\mu \omega_\alpha$ tells us that

$$c = Z_\omega / \mathscr{Z}\mathscr{N} \,. \qquad (17.2.25)$$

The cancellation of terms in $\delta \mathscr{L}_N^{(\epsilon)}$ proportional to $\theta\, \partial_\mu h_\alpha\, \omega_\beta A^\mu_\gamma$ (or $\theta \partial_\mu \omega^*_\alpha \omega_\beta \partial^\mu \omega_\gamma$) requires that

$$d_{\alpha\beta\gamma} = -(Z_\omega/\mathscr{N})\, C_{\alpha\beta\gamma} \,. \qquad (17.2.26)$$

The terms in $\delta \mathscr{L}_N^{(\epsilon)}$ proportional to $\theta \partial_\mu \omega^*_\alpha\, \omega_\beta\, \omega_\gamma\, A^\mu_\delta$ automatically then cancel by virtue of the Jacobi identity for the structure constants. The cancellation of terms in $\delta \mathscr{L}_N^{(\epsilon)}$ proportional to $\theta h_\alpha \partial_\mu \omega_\beta A^\mu_\gamma$ (or $\theta h_\alpha A^\mu_\beta A_{\gamma\mu} \omega_\delta$) yields

$$e_{\alpha\beta\gamma} = 0 \,. \qquad (17.2.27)$$

Finally, the effect of the infinitesimal transformation (17.2.10) on matter and gauge fields is the same as a local gauge transformation with gauge parameters $\epsilon_\alpha = \mathscr{Z}\mathscr{N}\theta\omega_\alpha$ and gauge couplings renormalized by a factor $1/\mathscr{N}$ (that is, with t_α and $C_{\alpha\beta\gamma}$ replaced with $\tilde{t}_\alpha \equiv t_\alpha/\mathscr{N}$ and $\tilde{C}_{\alpha\beta\gamma} \equiv C_{\alpha\beta\gamma}/\mathscr{N}$), so the cancellation of terms in $\delta \mathscr{L}_N^{(\epsilon)}$ with only one factor of ω_α and no factors h_α or ω^*_α simply tells us that the Lagrangian $\mathscr{L}_{\psi,A}$ for these fields is *gauge-invariant*, with a renormalized gauge coupling. We

[‡] In a theory with scalar fields, we could also have renormalizable terms with h_α multiplied with one or two scalar fields. Such terms cause no trouble, but for brevity they will not be considered here.

conclude then that the most general renormalizable Lagrangian density allowed by our assumed symmetry principles is

$$\mathscr{L}_N^{(\epsilon)} = -\tfrac{1}{4}Z_A \tilde{F}_\alpha^{\mu\nu}\tilde{F}_{\alpha\mu\nu} - Z_\psi \bar{\psi}\gamma^\mu[\partial_\mu - i\tilde{t}_\alpha A_{\alpha\mu}]\psi - \bar{\psi}m_\psi\psi + \tfrac{1}{2}\xi' h_\alpha h_\alpha$$
$$+(Z_\omega/\mathcal{N}\mathscr{Z})h_\alpha\partial_\mu A_\alpha^\mu - Z_\omega(\partial_\mu\omega_\alpha^*)(\partial_\mu\omega_\alpha) + Z_\omega\tilde{C}_{\alpha\beta\gamma}(\partial_\mu\omega_\alpha^*)\omega_\beta A_\gamma^\mu\,, \tag{17.2.28}$$

where the tilde on $\tilde{F}_\alpha^{\mu\nu}$ indicates that the field strength is to be calculated using the renormalized structure constant $\tilde{C}_{\alpha\beta\gamma} \equiv C_{\alpha\beta\gamma}/\mathcal{N}$. But apart from the appearance of a number of new constant coefficients, this is the same Lagrangian with which we started. The new constants in this Lagrangian (including the gauge coupling constant) may be freely shifted by adjusting the Nth-order terms in the corresponding constants in the original unrenormalized Lagrangian. In particular, we can adjust these terms to make $\Gamma_N^{(\epsilon)} = S_R$, in which case $\Gamma_{N,\infty} = 0$, completing the proof.

* * *

In the above proof we made important use of the accidental invariance of the gauge-fixed Lagrangian (17.2.17) under the antighost translation transformation (17.2.23). This symmetry would not be present for gauge-fixing functions other than $f_\alpha = \partial_\mu A_\alpha^\mu$, which are not spacetime derivatives. One frequently cited example which preserves Lorentz invariance and global gauge invariance is $f_\alpha = \partial_\mu A_\alpha^\mu + a_{\alpha\beta\gamma}A_\beta^\mu A_{\gamma\mu}$, where $a_{\alpha\beta\gamma}$ is a constant matrix, symmetric in β and γ, which transforms as a tensor under global gauge transformations. (Such constant tensors exist for all $SU(N)$ groups with $N \geq 3$.) Another more important example is the background gauge-fixing functional to be introduced in Section 17.4.

The absence of antighost translation invariance does not affect our argument that $\mathscr{L}_N^{(\epsilon)}$ is a Lorentz- and global gauge-invariant local function of fields and field derivatives of dimensionality no greater than 4, that is invariant under the renormalized BRST transformation (17.2.11)–(17.2.15). But without antighost translation invariance there are new terms in $\mathscr{L}_N^{(\epsilon)}$ that satisfy these conditions. Since the transformation (17.2.11)–(17.2.15) is nilpotent, we can construct such terms as $s'F$, where the transformation (17.2.11)–(17.2.15) is written as $\chi^n \to \chi^n + \theta s'\chi^n$, and F is an arbitrary Lorentz- and global gauge-invariant function of ghost number -1. One such term is

$$a_{\alpha\beta\gamma}s'\left(\omega_\alpha^* A_{\beta\mu}A_\gamma^\mu\right) = -a_{\alpha\beta\gamma}\left[h_\alpha A_{\beta\mu}A_\gamma^\mu + 2\mathscr{Z}\omega_\alpha^*(\mathcal{N}\partial_\mu\omega_\beta + C_{\beta\delta\epsilon}A_{\delta\mu}\omega_\delta)A_\gamma^\mu\right].$$

This causes no trouble: it is just a renormalized version of the usual ghost and gauge-fixing terms arising from terms $a_{\alpha\beta\gamma}A_\beta^\mu A_{\gamma\mu}$ in the gauge-fixing function f_α. But there is also another possible term of the form

$$b_{\alpha\beta\gamma}s'\left(\omega_\alpha^*\omega_\beta^*\omega_\gamma\right) = -b_{\alpha\beta\gamma}\left[2h_\alpha\omega_\beta^*\omega_\gamma + \tfrac{1}{2}\mathscr{Z}C_{\gamma\delta\epsilon}\omega_\alpha^*\omega_\beta^*\omega_\delta\omega_\epsilon\right],$$

where $b_{\alpha\beta\gamma}$ is a constant, antisymmetric in α and β, which transforms as a tensor under global gauge transformations. (Such tensors exist for any Lie group; for instance, we could take $b_{\alpha\beta\gamma}$ to be proportional to $C_{\alpha\beta\gamma}$.) But the Faddeev–Popov–De Witt method cannot yield four-ghost interactions in the Lagrangian, so there are no counterterms available to absorb ultraviolet divergences in this term. This is not just a technical obstacle to proving renormalizability; for gauge-fixing functions like $f_\alpha = \partial_\mu A^\mu_\alpha + a_{\alpha\beta\gamma} A^\mu_\beta A_{\gamma\mu}$, one-loop graphs actually do yield divergences in four-ghost amplitudes that cannot be cancelled by counterterms in the Faddeev–Popov–De Witt Lagrangian.

Aside from avoiding gauge-fixing functions other than $f_\alpha = \partial_\mu A^\mu_\alpha$, the only solution to this problem seems to be the one mentioned in Section 15.7. We must give up the Faddeev–Popov–De Witt approach, and instead take the action from the beginning as the most general renormalizable function of the gauge, matter, ghost, and auxiliary fields that is invariant under BRST and the other symmetries of the theory. According to the arguments of Section 15.8, the action can be written in the form $I_0 + s\Psi$ with I_0 ghost-free, and the S-matrix is independent of Ψ, so we can justify this procedure by quantizing the gauge theory in axial gauge, where ghosts decouple, and then taking Ψ to be anything we like. In particular, we can include $s(\omega^* \omega^* \omega)$ terms in the action that can serve as counterterms to the divergences in four-ghost vertices.

17.3 Renormalization: General Gauge Theories*

The proof of the renormalizability of non-Abelian gauge theories in the previous section relied on a 'brute force' analysis of the possible terms in the action of dimensionality four or less. But as we saw in Chapter 12, this limitation on the dimensionality of terms in the action can be at best a good approximation. The successful renormalizable quantum field theories that are used to describe the strong, weak, and electromagnetic interactions are almost certainly effective field theories, accompanied with terms of dimensionality $d > 4$; these terms are normally not observed because they are suppressed by $4 - d$ powers of some very large mass, perhaps of order 10^{16}– 10^{18} GeV. Gravitation too can be described by an effective field theory, in which the Lagrangian density contains not only the Einstein–Hilbert term $-\sqrt{g}R/16\pi G$, but also all scalars constructed from four or more derivatives of the gravitational field. We need to

* This section lies somewhat out of the book's main line of development, and may be omitted in a first reading.

show that gauge theories of this sort, which are not renormalizable in the power-counting sense, are nonetheless renormalizable in the modern sense that the ultraviolet divergences are governed by the gauge symmetries in such a way that there is a counterterm available to cancel every infinity.[3]

For this purpose, let us return to the action $S[\chi, \chi^\ddagger]$ introduced in Section 15.9, taken as a function of independent fields χ^n (including gauge and matter fields ϕ^r and ghost fields ω^A as well as the non-minimal fields ω^{A*} and h^A), together with their antifields $\chi_n^\ddagger$. In theories like quantum gravity or Yang–Mills theories, that are based on a closed gauge algebra with structure constants $f^C{}_{AB}$, this action is constrained to be of the form

$$S = I[\phi] + \omega^A f^r_A[\phi]\, \phi_r^\ddagger + \tfrac{1}{2}\omega^A \omega^B f^C{}_{AB}[\phi]\, \omega_C^\ddagger - h^A\, \omega_A^{*\ddagger} \,, \tag{17.3.1}$$

where $I[\phi]$ is invariant under the infinitesimal gauge transformations $\phi^r \to \phi^r + \epsilon^A f^r_A[\phi]$. (As in Section 15.9, the indices r, A, etc. include a spacetime coordinate, over which we integrate in sums over these indices.)

We shall not limit ourselves here to actions of this form, but we shall suppose that the local symmetries of the theory are imposed by requiring that the action must obey some 'structural constraints' on its antifield dependence, of which Eq. (17.3.1) provides just one example. The structural constraints are assumed to be linear, in the sense that if they are satisfied for $S + \mathscr{S}_1$ and for $S + \mathscr{S}_2$, then for any constants α_1, α_2 they are satisfied for $S + \alpha_1\mathscr{S}_1 + \alpha_2\mathscr{S}_2$. We also impose the quantum master equation (15.9.35)

$$(S, S) - 2i\hbar\Delta S = 0 \,, \tag{17.3.2}$$

with ΔS defined by Eq. (15.9.34), and the factor $\hbar$ now made explicit as a 'loop-counting' parameter, in the sense described in the footnote of the previous section.

The action is taken as a power series in $\hbar$

$$S = S_R + \hbar S_1 + \hbar^2 S_2 + \cdots \,, \tag{17.3.3}$$

where S_R is an action of the same general form as S, but with all coupling parameters replaced with finite renormalized values, and the S_N are a set of infinite counterterms. The action S is supposed to satisfy the quantum master equation (17.3.2) for all $\hbar$, so S_R satisfies the classical master equation

$$(S_R, S_R) = 0 \,, \tag{17.3.4}$$

while the counterterms satisfy

$$\Big(S_R, S_N\Big) = -\tfrac{1}{2} \sum_{M=1}^{N-1} \Big(S_M, S_{N-M}\Big) + i\Delta S_{N-1} \,. \tag{17.3.5}$$

The counterterms S_N are not by themselves sufficient to cancel the ultraviolet divergences in loop graphs. As a generalization of the conventional renormalization of fields, we also have to introduce a set of renormalized fields and antifields, defined in terms of the original fields and antifields by an arbitrary anticanonical transformation. An infinitesimal anticanonical transformation may be defined in terms of an infinitesimal generating functional δF by Eq. (15.9.26), so a sequence of anticanonical transformations $G(t) \to G(t+\delta t) = G(t) + (F(t)\delta t, G(t))$ (where $G(t)$ is any functional of fields and antifields, and $F(t)$ is the generating functional) leads to a finite canonical transformation $G \to \tilde{G} \equiv G(1)$, with

$$\frac{d}{dt}G(t) = (F(t), G(t))\,, \qquad G(0) = G\,. \tag{17.3.6}$$

If $F(t)$ is given by a power series

$$F(t) = \hbar F_1 + \tfrac{1}{2}\hbar^2 t F_2 + \cdots\,, \tag{17.3.7}$$

then Eqs. (17.3.3), (17.3.6), and (17.3.7) yield a transformed action

$$\tilde{S} = S_R + \hbar\Big[S_1 + (F_1, S_R)\Big] + \hbar^2\Big[S_2 + (F_1, S_1) + (F_2, S_R) + \tfrac{1}{2}(F_1, (F_1, S_R))\Big] + \cdots\,. \tag{17.3.8}$$

The question is whether we can use what freedom we have to choose the F_N and S_N so as to cancel all infinities arising from loop graphs.

As already noted, the first term S_R in Eq. (17.3.3) is automatically finite. Suppose that by cancellation of infinities with S_M and F_M for $M < N$ it has been possible to eliminate all infinities in the terms Γ_M of order $\hbar^M$ in the quantum effective action with $M < N$. As we saw in the previous section, the Zinn-Justin equation (derived here by setting $\chi^\ddagger_n = K_n + \delta\Psi/\delta\chi^n$) tells us in this case that the infinite part $\Gamma_{N,\infty}$ of the term in the quantum effective action of order $\hbar^N$ satisfies the condition

$$\Big(S_R, \Gamma_{N,\infty}\Big) = 0\,. \tag{17.3.9}$$

The field and antifield variables χ^n and $\chi^\ddagger_n$ are related to the variables χ^n and K_n by an anticanonical transformation, which preserves all antibrackets, so the antibracket in Eq. (17.3.9) may be calculated in terms of χ^n and $\chi^\ddagger_n$ instead of χ^n and K_n.

The condition (17.3.5) satisfied by the counterterm S_N is not the same as the condition (17.3.9) satisfied by $\Gamma_{N,\infty}$. However, given any S^0_N that satisfies Eq. (17.3.5), we can find a class of other solutions

$$S_N = S^0_N + S'_N\,, \tag{17.3.10}$$

where S'_N is arbitrary, except for the condition that $S_R + S'_N$ like $S_R + S^0_N$ satisfies the same symmetry conditions as S_R, and that

$$\Big(S_R, S'_N\Big) = 0 \tag{17.3.11}$$

so as not to invalidate Eq. (17.3.5). The infinite part of the Nth-order term in the quantum effective action may therefore be written

$$\Gamma_{N,\infty} = S'_{N,\infty} + (F_{N,\infty}, S_R) + X_{N,\infty}\,, \tag{17.3.12}$$

where X_N consists of terms from loop graphs, as well as from the term S_N^0 and various terms in Γ that involve S_M and F_M for $M < N$. For instance, for $N = 2$ Eq. (17.3.8) gives

$$X_2 = S_2^0 + 2(F_1, S_1) + (F_1, (F_1, S_R)) + \text{two-loop terms involving only } S_R$$
$$+ \text{ one-loop terms involving } S_R,\ S_1 \text{ and } F_1\,.$$

For our purposes the only thing we need to know about X_N is that it does not involve S'_N or F_N, and that it is invariant under any linearly realized global symmetries of S_R.

Now, because $(S_R, S_R) = 0$, the operation $F \mapsto (S_R, F)$ is nilpotent; for all F,

$$\Big(S_R, \big(S_R, F\big)\Big) = 0\,. \tag{17.3.13}$$

Hence it follows from Eqs. (17.3.9) and (17.3.11)–(17.3.13) that

$$\Big(S_R, X_{N,\infty}\Big) = 0\,. \tag{17.3.14}$$

Any term in $X_{N,\infty}$ of the form (S_R, Y) may be cancelled in Eq. (17.3.12) by choosing $F_{N,\infty}$ equal to Y. Thus the space of possible remaining infinite terms in $X_{N,\infty}$ that need to be cancelled by the counterterm $S'_{N,\infty}$ consists of those functionals X that satisfy $(S_R, X) = 0$, counting as equivalent functionals that differ only by terms of the form (S_R, Y). In other words, the infinities in Γ_N that need to be cancelled by the counterterms S'_N belong to the *cohomology* of the mapping $X \mapsto (S_R, X)$.

The possible form of the counterterm S'_N is limited by the requirement that $S_R + S'_N$ must satisfy whatever structural constraints are imposed on the action S. Thus we can complete the proof of renormalizability if we can show that the cohomology of the mapping $X \mapsto (S_R, X)$ consists only of functionals that satisfy this structural constraint.

In the case of quantum gravity coupled to the Yang–Mills fields of a semisimple gauge symmetry, the symmetries of the theory are implemented by the structural constraint (17.3.1). In this case there is a theorem[4] that states that the cohomology of the mapping $X \mapsto (S_R, X)$ (on the space of local functionals, rather than spacetime-dependent functions, of ghost number zero) consists** of functionals $A[\phi]$ that are invariant under the

** Strictly speaking, this is valid if one requires that the constant coefficients in S do not take special values for which S would be invariant under a larger group of local symmetries. For instance, this excludes the case $S = 0$.

gauge transformation $\phi^r \to \phi^r + \epsilon^A f^r_A[\phi]$ with structure constants $f^C{}_{AB}$. Any Nth-order infinity of this sort may be cancelled with a counterterm S'_N of the same form, so although these theories are not renormalizable in the conventional power-counting sense, they are renormalizable in the sense that all infinities can be eliminated by a choice of parameters in the original bare action $I[\phi]$ and by a suitable renormalization of fields and antifields.

In other theories the cohomology of the map $X \mapsto (S_R, X)$ contains additional terms. This does not necessarily require a weakening of the structural constraints, because the additional terms in the cohomology may not correspond to actual ultraviolet divergences. For instance, in gauge theories with $U(1)$ factors the cohomology contains terms[4] corresponding to a redefinition of the action of the $U(1)$ gauge symmetry on the various fields of the theory, which if infinite would require us to weaken the structural constraints by leaving the normalization of the transformation functions $f^r_A[\phi]$ in Eq. (17.3.1) arbitrary. But this infinity is forbidden by the same soft-photon theorems that tell us that ratios of $U(1)$ couplings to various fields (like the ratios of the various lepton charges in quantum electrodynamics) are unaffected by radiative corrections. (See Section 10.4.) Where the extra terms in the cohomology do contain ultraviolet divergences, it is necessary to weaken the structural constraint imposed on S in order that there should be a counterterm for every possible ultraviolet divergence. It is not known whether this will always be possible; if not, some theories may have to be rejected because of their unremovable ultraviolet divergences.

17.4 Background Field Gauge

We next turn to a method of calculation that explicitly preserves a sort of gauge invariance, and that therefore proves extremely convenient, especially in one-loop calculations. We consider the effective action $\Gamma[A, \psi, \omega, \omega^*]$ as a functional* of classical external gauge, matter, ghost and antighost fields: $A_{\alpha\mu}(x)$, $\psi_\ell(x)$, $\omega_\alpha(x)$, $\omega^*_\alpha(x)$. Even though ghosts and antighosts never appear in initial or final states, we are considering background ghost and antighost fields as well as gauge and matter fields in order to deal with parts of diagrams that have external ghost or antighost lines.

* We are now returning to the specific choice of the gauge-fixing functional $B[f]$ as the Gaussian (15.7.4), and we are integrating out the auxiliary field h_α, so that the gauge-fixing term in the modified Lagrangian is just $-f_\alpha f_\alpha/2\xi$.

As described in Section 16.1, $\Gamma[A, \psi, \omega, \omega^*]$ is the sum of connected one-particle-irreducible graphs for the vacuum–vacuum amplitude, calculated in a theory in which the quantum fields $A', \psi', \omega', \omega'^*$ over which we integrate are replaced in the action with shifted fields $A + A', \psi + \psi'$, $\omega + \omega'$, $\omega^* + \omega'^*$, the path integral being taken over primed fields with the unprimed fields held fixed. We are free to choose the gauge-fixing function $f_\alpha(x)$ pretty much any way we like; instead of our previous choice $f_\alpha = \partial_\mu A'^\mu_\alpha$ (or $\partial_\mu[A'^\mu_\alpha + A^\mu_\alpha]$) we shall now take[5]

$$f_\alpha = \partial_\mu A'^\mu_\alpha + C_{\alpha\beta\gamma} A_{\beta\mu} A'^\mu_\gamma \,. \tag{17.4.1}$$

The reason for this choice is that it makes the gauge-fixing term $f_\alpha f_\alpha$ invariant under a formal transformation, in which the background field A^μ_α transforms as a gauge field, while the quantum field A'^μ_α transforms homogeneously, like an ordinary matter field that happens to belong to the adjoint representation of the gauge group

$$\delta A^\mu_\alpha = \partial^\mu \epsilon_\alpha - C_{\alpha\beta\gamma}\, \epsilon_\beta\, A^\mu_\gamma \,, \tag{17.4.2}$$

$$\delta A'^\mu_\alpha = -C_{\alpha\beta\gamma}\, \epsilon_\beta\, A'^\mu_\gamma \,. \tag{17.4.3}$$

The transformation properties of f_α can be seen most easily by writing it as a new sort of covariant derivative

$$f_\alpha \equiv \bar{D}_\mu A'^\mu_\alpha \,, \tag{17.4.4}$$

where for any field ϕ_α in the adjoint representation

$$\bar{D}_\mu \phi_\alpha \equiv \partial_\mu \phi_\alpha + C_{\alpha\beta\gamma} A_{\beta\mu} \phi_\gamma \,. \tag{17.4.5}$$

We see that under the transformation (17.4.2), (17.4.3), the function (17.4.1) transforms just like A'_α:

$$\delta f_\alpha = -C_{\alpha\beta\gamma}\, \epsilon_\beta f_\gamma \,, \tag{17.4.6}$$

so the term $f_\alpha f_\alpha$ in the modified Lagrangian is invariant

$$\delta(f_\alpha f_\alpha) = -2C_{\alpha\beta\gamma}\, f_\alpha\, \epsilon_\beta\, f_\gamma = 0 \,. \tag{17.4.7}$$

Also, the original Lagrangian density $\mathscr{L}$ depends on A and A' only through the sum $A + A'$, which under the combined transformation (17.4.2), (17.4.3) undergoes an ordinary gauge transformation

$$\delta(A^\mu_\alpha + A'^\mu_\alpha) = \partial^\mu \epsilon_\alpha - C_{\alpha\beta\gamma}\, \epsilon_\beta (A^\mu_\gamma + A'^\mu_\gamma) \,. \tag{17.4.8}$$

If we transform the background and quantum matter fields by

$$\delta\psi = i\, t_\alpha \epsilon_\alpha \psi \,, \tag{17.4.9}$$

$$\delta\psi' = i\, t_\alpha \epsilon_\alpha \psi' \,, \tag{17.4.10}$$

then also

$$\delta(\psi + \psi') = i\, t_\alpha \epsilon_\alpha (\psi + \psi') \,. \tag{17.4.11}$$

The original Lagrangian $\mathscr{L}$ is invariant under the original gauge transformations (17.4.8), (17.4.11), and only depends on $A + A'$ and $\psi + \psi'$, so it is also invariant under the new formal transformations (17.4.2), (17.4.3), (17.4.9), (17.4.10).

It will be useful to make this invariance property more explicit, by writing $\mathscr{L}$ in terms of the background-covariant derivative $\bar{D}_\mu$. In general, we have

$$\begin{aligned}\mathscr{L} = &-\tfrac{1}{4}\Big(\partial_\mu[A_{\alpha\nu} + A'_{\alpha\nu}] - \partial_\nu[A_{\alpha\mu} + A'_{\alpha\mu}] + C_{\alpha\beta\gamma}[A_{\beta\mu} + A'_{\beta\mu}][A_{\gamma\nu} + A'_{\gamma\nu}]\Big)^2 \\ &+ \mathscr{L}_M\Big(\psi + \psi',\ \partial_\mu(\psi + \psi') - i\,t_\alpha(A_{\alpha\mu} + A'_{\alpha\mu})(\psi + \psi')\Big) \\ = &-\tfrac{1}{4}\Big(F_{\alpha\mu\nu} + \bar{D}_\mu A'_{\alpha\nu} - \bar{D}_\nu A'_{\alpha\mu} + C_{\alpha\beta\gamma}A'_{\beta\mu}A'_{\gamma\nu}\Big)^2 \\ &+ \mathscr{L}_M\Big(\psi + \psi',\ \bar{D}_\mu(\psi + \psi') - i\,t_\alpha A'_{\alpha\mu}(\psi + \psi')\Big)\,, \end{aligned} \qquad (17.4.12)$$

where, as in Eq. (17.4.5),

$$\bar{D}_\mu A'_{\alpha\nu} \equiv \partial_\mu A'_{\alpha\nu} + C_{\alpha\beta\gamma}A_{\beta\mu}A'_{\gamma\nu}\,, \qquad (17.4.13)$$

$$\bar{D}_\mu \psi \equiv \partial_\mu \psi - i\,t_\alpha A_{\alpha\mu}\psi\,, \qquad (17.4.14)$$

and $F_{\alpha\mu\nu}$ is the background field strength

$$F_{\alpha\mu\nu} \equiv \partial_\mu A_{\alpha\nu} - \partial_\nu A_{\alpha\mu} + C_{\alpha\beta\gamma}A_{\beta\mu}A_{\gamma\nu}\,. \qquad (17.4.15)$$

(The square in the first term of $\mathscr{L}$ is intended to imply obvious index contractions.) Clearly $\mathscr{L}$ is invariant under the new transformations (17.4.2), (17.4.3), (17.4.9), (17.4.10), because it involves $A_{\alpha\mu}$ only in the field strength $F_{\alpha\mu\nu}$ and in background covariant derivatives $\bar{D}_\mu$ of 'matter' fields $A'_{\alpha\mu}$, ψ' and ψ.

This new transformation should be carefully distinguished from a true gauge transformation. Such a transformation can have no effect on A or ψ, which are just prescribed classical background fields, and induces an ordinary gauge transformation on $A + A'$ and $\psi + \psi'$, so

$$\delta_{\rm TRUE} A^\mu_\alpha = 0\,, \qquad (17.4.16)$$

$$\begin{aligned}\delta_{\rm TRUE} A'^\mu_\alpha &= \partial^\mu \epsilon_\alpha - C_{\alpha\beta\gamma}\epsilon_\beta(A^\mu_\gamma + A'^\mu_\gamma) \\ &= \bar{D}_\mu \epsilon_\alpha - C_{\alpha\beta\gamma}\epsilon_\beta A'^\mu_\gamma \end{aligned} \qquad (17.4.17)$$

and

$$\delta_{\rm TRUE}\,\psi = 0\,, \qquad (17.4.18)$$

$$\delta_{\rm TRUE}\,\psi' = i\,t_\alpha \epsilon_\alpha(\psi + \psi')\,. \qquad (17.4.19)$$

Of course, this is the same as the formal transformations (17.4.2), (17.4.3), (17.4.9), (17.4.10) in its effect on $A + A'$ and $\psi + \psi'$, and therefore also leaves the original Lagrangian $\mathscr{L}$ invariant. However, for our new choice

(17.4.1) of f_α, the term $f_\alpha f_\alpha$ does *not* depend only on $A + A'$, and is *not* invariant under (17.4.16) and (17.4.17). Instead,

$$\delta_{\rm TRUE} f_\alpha = \bar{D}_\mu(\bar{D}^\mu \epsilon_\alpha - C_{\alpha\beta\gamma}\epsilon_\beta A'^\mu_\gamma) \tag{17.4.20}$$

with $\bar{D}_\mu$ given by (17.4.5).

Finally, let us consider the ghost Lagrangian in this new gauge. The quantity (15.7.3) in the ghost action is given in general by just replacing ϵ_α with the ghost field $\omega_\alpha + \omega'_\alpha$ in $\delta_{\rm TRUE} f_\alpha$:

$$\Delta_\alpha = \bar{D}_\mu\left[\bar{D}^\mu(\omega_\alpha + \omega'_\alpha) - C_{\alpha\beta\gamma}(\omega_\beta + \omega'_\beta)A'^\mu_\gamma\right] . \tag{17.4.21}$$

The ghost Lagrangian in Eq. (15.6.2) is therefore

$$\mathscr{L}_{\rm GH} = (\omega^*_\alpha + \omega'^*_\alpha)\,\bar{D}_\mu\left[\bar{D}^\mu(\omega_\alpha + \omega'_\alpha) - C_{\alpha\beta\gamma}(\omega_\beta + \omega'_\beta)\,A'^\mu_\gamma\right] \tag{17.4.22}$$

or, integrating by parts,

$$\mathscr{L}_{\rm GH} = -\left(\bar{D}_\mu(\omega^*_\alpha + \omega'^*_\alpha)\right)\left(\bar{D}^\mu(\omega_\alpha + \omega'_\alpha) - C_{\alpha\beta\gamma}(\omega_\beta + \omega'_\beta)A'^\mu_\gamma\right) . \tag{17.4.23}$$

This is manifestly invariant under the joint transformations (17.4.2), (17.4.3), supplemented now with transformations on ω and ω':

$$\delta\omega_\alpha = -C_{\alpha\beta\gamma}\epsilon_\beta\omega_\gamma , \tag{17.4.24}$$

$$\delta\omega'_\alpha = -C_{\alpha\beta\gamma}\epsilon_\beta\omega'_\gamma , \tag{17.4.25}$$

and likewise

$$\delta\omega^*_\alpha = -C_{\alpha\beta\gamma}\epsilon_\beta\omega^*_\gamma , \tag{17.4.26}$$

$$\delta\omega'^*_\alpha = -C_{\alpha\beta\gamma}\epsilon_\beta\omega'^*_\gamma . \tag{17.4.27}$$

We see that the formal combined transformation (17.4.2), (17.4.3), (17.4.9), (17.4.10), and (17.4.24)–(17.4.27) leave invariant the complete Lagrangian in the modified action (15.6.4):

$$\mathscr{L}_{\rm MOD} = \mathscr{L} - \frac{1}{2\xi} f_\alpha f_\alpha + \mathscr{L}_{GH} . \tag{17.4.28}$$

We are integrating over A', ψ', ω', and ω'^* with a measure that is presumed to be invariant under the simple matrix transformations (17.4.3), (17.4.10), (17.4.25), and (17.4.27), so the effective action $\Gamma[A, \psi, \omega, \omega^*]$ is invariant under the remaining transformations (17.4.2), (17.4.9), (17.4.24), and (17.4.26). In other words, it is gauge-invariant in the same sense as the original action $I[A, \psi, \omega, \omega^*]$.

This formal gauge invariance sets powerful constraints on the infinities that can occur in the effective action. The ultraviolet divergences in Γ appear in the coefficients of terms whose dimensionalities are [mass]^d with $d \leq 4$, but here these terms are invariant under the background gauge transformations (17.4.2), (17.4.9), (17.4.24), and (17.4.26). For instance, in a gauge theory based on a simple gauge group, with spin 1/2 fermions belonging to an irreducible representation of this group, the only such terms are of the form

$$\Gamma_\infty = \int d^4x\, \mathscr{L}_\infty \,, \tag{17.4.29}$$

$$\mathscr{L}_\infty = -\tfrac{1}{4} L_A\, F_{\alpha\mu\nu} F_\alpha^{\mu\nu} - L_\psi \bar{\psi} \gamma^\mu \bar{D}_\mu \psi$$
$$-m\, L_m \bar{\psi}\psi - L_\omega (\bar{D}_\mu \omega_\alpha^*)(\bar{D}^\mu \omega_\alpha) \,, \tag{17.4.30}$$

where here $F_{\alpha\mu\nu}$, $\bar{D}_\mu\psi$, $\bar{D}_\mu\omega_\alpha$, and $\bar{D}_\mu\omega_\alpha^*$ are constructed entirely from background fields:**

$$F_{\alpha\mu\nu} \equiv \partial_\mu A_{\alpha\nu} - \partial_\nu A_{\alpha\mu} + C_{\alpha\beta\gamma} A_{\beta\mu} A_{\gamma\nu} \,, \tag{17.4.31}$$

$$\bar{D}_\mu \psi \equiv \partial_\mu \psi - i\, t_\alpha A_{\alpha\mu} \psi \,, \tag{17.4.32}$$

$$\bar{D}_\mu \omega_\alpha \equiv \partial_\mu \omega_\alpha + C_{\alpha\beta\gamma} A_{\beta\mu} \omega_\gamma \,, \tag{17.4.33}$$

$$\bar{D}_\mu \omega_\alpha^* \equiv \partial_\mu \omega_\alpha^* + C_{\alpha\beta\gamma} A_{\beta\mu} \omega_\gamma^* \,. \tag{17.4.34}$$

Dimensional analysis leads us to expect that the constants L_A, L_ψ, L_m, and L_ω are logarithmically divergent.

To deal with these infinities, we note that the Lagrangian (17.4.12) contains a purely classical piece

$$\mathscr{L}_{\rm CLASS} = -\tfrac{1}{4} F_{\alpha\mu\nu} F_\alpha^{\mu\nu} - \bar{\psi}(\gamma^\mu \bar{D}_\mu + m)\psi - (\bar{D}_\mu \omega_\alpha^*)(\bar{D}^\mu \omega_\alpha) \tag{17.4.35}$$

obtained by dropping all terms in $\mathscr{L}_{\rm MOD}$ that involve the quantum fields $A', \psi', \omega', \omega'^*$. We define renormalized fields

$$A^R_{\alpha\mu} \equiv \sqrt{1 + L_A}\, A_{\alpha\mu} \,, \tag{17.4.36}$$

$$\psi^R_\ell \equiv \sqrt{1 + L_\psi}\, \psi_\ell \,, \tag{17.4.37}$$

$$\omega^R_\alpha \equiv \sqrt{1 + L_\omega}\, \omega_\alpha \,, \tag{17.4.38}$$

$$\omega^{R*}_\alpha \equiv \sqrt{1 + L_\omega}\, \omega^*_\alpha \,, \tag{17.4.39}$$

so that the sum of the terms (17.4.30) and (17.4.35) takes the form

$$\mathscr{L}_{\rm CLASS} + \mathscr{L}_\infty = -\tfrac{1}{4} F^R_{\alpha\mu\nu} F^{R\mu\nu}_\alpha - \bar{\psi}^R \gamma^\mu D^R_\mu \psi^R$$
$$-m^R \bar{\psi}^R \psi^R - (D^R_\mu \omega^{*R}_\alpha)(D^{\mu R} \omega^R_\alpha) \,, \tag{17.4.40}$$

where m^R is the renormalized mass

$$m^R \equiv m(1 + L_m)\big/(1 + L_\psi) \,, \tag{17.4.41}$$

** The condition of a simple gauge group insures that there is just a single kinetic term for A and for ω, proportional to $F_{\alpha\mu\nu}F_\alpha^{\mu\nu}$ and $\bar{D}_\mu\omega_\alpha^*\bar{D}^\mu\omega_\alpha$, respectively, while the condition that ψ transforms irreducibly insures that there is just a single kinetic term and a single mass term for ψ. It would be easy to treat more general possibilities, at the cost of a slight complication in notation. We are also implicitly using the conservation of ghost number.

and

$$F^R_{\alpha\mu\nu} \equiv \partial_\mu A^R_{\alpha\nu} - \partial_\nu A^R_{\alpha\mu} + C^R_{\alpha\beta\gamma} A^R_{\beta\mu} A^R_{\gamma\nu} \,, \tag{17.4.42}$$

$$D^R_\mu \psi^R \equiv \partial_\mu \psi^R - i\, t^R_\alpha A^R_{\alpha\mu} \psi^R \,, \tag{17.4.43}$$

$$D^R_\mu \omega^R_\alpha \equiv \partial_\mu \omega^R_\alpha + C^R_{\alpha\beta\gamma} A^R_{\beta\mu} \omega^R_\gamma \,, \tag{17.4.44}$$

$$D^R_\mu \omega^{*R}_\alpha \equiv \partial_\mu \omega^{*R}_\alpha + C^R_{\alpha\beta\gamma} A^R_{\beta\mu} \omega^{*R}_\gamma \,. \tag{17.4.45}$$

The renormalized structure constants and group generators here are just

$$C^R_{\alpha\beta\gamma} \equiv (1 + L_A)^{-1/2} C_{\alpha\beta\gamma} \,, \tag{17.4.46}$$

$$t^R_\alpha \equiv (1 + L_A)^{-1/2}\, t_\alpha \,. \tag{17.4.47}$$

Because we assumed that the Lie algebra here is simple, the structure constant $C_{\alpha\beta\gamma}$ and group generator t_α are fixed by the group structure except for a single common factor, the unrenormalized gauge coupling constant g. Eqs. (17.4.46) and (17.4.47) thus simply tell us that the gauge coupling constant g^R factor in C^R and t^R is renormalized by

$$g^R = g(1 + L_A)^{-1/2} \,. \tag{17.4.48}$$

This result exhibits the particular virtue of the background field gauge. In a general gauge we would encounter independent renormalization factors for the gauge field and the gauge coupling constant, and we would have to calculate two separate amplitudes (say, the vacuum polarization and three-gauge-field vertex function) in order to be able to sort these out. In background field gauge, background field gauge invariance ties these two renormalizations together by requiring that the infinite terms in the effective Lagrangian involve the field strength in its original form (17.4.31), and so we can calculate the charge renormalization factor by studying just one gauge field amplitude.

17.5 A One-Loop Calculation in Background Field Gauge

As an exercise, we are now going to calculate the one-loop renormalization factor for the gauge coupling constant in a general non-Abelian gauge theory. As we will see in the next chapter, this provides an essential input in so-called 'renormalization group' calculations of physical processes at high energy; the results we obtain here will be used there to demonstrate the asymptotic freedom of non-Abelian gauge theories.

The method to be employed here is somewhat novel. Usually, one considers the effective action in a spacetime-dependent background gauge field, and calculates the terms quadratic in this field, extracting a factor $(q_\mu A_{\alpha\nu} - q_\nu A_{\alpha\mu})^2$ (where q is the gauge field four-momentum), and only then isolating the logarithmic divergence by setting $q = 0$ in the coefficient of this factor. Instead we shall follow the much simpler course of taking the gauge field to be spacetime-independent from the beginning. In this case the terms in the effective action that are quadratic or cubic in the gauge field of course vanish, but there is a non-vanishing quartic term that is ultraviolet divergent, and which can be used to calculate the coupling constant renormalization factor $(1 + L_A)^{-1/2}$. In this way our one-loop calculation becomes a matter of simple matrix algebra. Note that this procedure can only work in background field gauge; otherwise there would be independent logarithmic divergences in the parts of the effective action that are quartic and quadratic in the background gauge field.

With this motivation, we turn to the calculation of the one-loop effective action in a background field for which $A_{\alpha\mu}$ is constant and $\psi = \omega = \omega^* = 0$. For such a background field, the full modified Lagrangian is*

$$\mathscr{L}_{\rm MOD} = \mathscr{L} + \mathscr{L}_f + \mathscr{L}_{GH}\,, \tag{17.5.1}$$

$$\mathscr{L} = -\tfrac{1}{4}\Big(F_{\alpha\mu\nu} + \bar{D}_\mu A'_{\alpha\nu} - \bar{D}_\nu A'_{\alpha\mu} + C_{\alpha\beta\gamma}A'_{\beta\mu}A'_{\gamma\nu}\Big)^2$$
$$- \bar{\psi}'(\bar{\not{D}} - i\,t_\alpha \not{A}'_\alpha + m)\psi'\,, \tag{17.5.2}$$

$$\mathscr{L}_f \equiv -\frac{1}{2\xi}f_\alpha f_\alpha = -\frac{1}{2\xi}(\bar{D}_\mu A'^\mu_\alpha)^2\,, \tag{17.5.3}$$

$$\mathscr{L}_{GH} = -(\bar{D}_\mu \omega'^*_\alpha)(\bar{D}^\mu \omega'_\alpha - C_{\alpha\beta\gamma}\omega'_\beta A'^\mu_\gamma)\,. \tag{17.5.4}$$

One-loop graphs for vacuum–vacuum amplitudes are calculated from the part of the action that is quadratic in the quantum fields $A', \psi', \omega', \omega'^*$ over which one integrates. Keeping only such quadratic terms, we have

$$\mathscr{L}_{\rm QUAD} = -\frac{1}{4}(\bar{D}_\mu A'_{\alpha\nu} - \bar{D}_\nu A'_{\alpha\mu})^2 - \frac{1}{2}F^{\mu\nu}_\alpha C_{\alpha\beta\gamma}A'_{\beta\mu}A'_{\gamma\nu}$$
$$-\bar{\psi}'(\bar{\not{D}} + m)\psi' - \frac{1}{2\xi}(\bar{D}_\mu A'^\mu_\alpha)^2 - (\bar{D}_\mu\omega'^*_\alpha)(\bar{D}^\mu\omega'_\alpha)\,. \tag{17.5.5}$$

The corresponding action may be put in the general quadratic form:

$$I_{\rm QUAD} \equiv \int d^4x\,\mathscr{L}_{\rm QUAD}$$
$$= -\tfrac{1}{2}\int d^4x\,d^4y\,A'^\mu_\alpha(x)A'^\nu_\beta(y)\,\mathscr{D}^A_{x\alpha\mu,y\beta\nu} - \int d^4x\,d^4y\,\bar{\psi}'_k(x)\psi'_\ell(y)\,\mathscr{D}^\psi_{xk,y\ell}$$
$$- \int d^4x\,d^4y\,\omega'^*_\alpha(x)\omega'_\beta(y)\,\mathscr{D}^\omega_{x\alpha,y\beta}\,, \tag{17.5.6}$$

* See Eqs. (17.4.12), (17.4.4), and (17.4.23). We are here specializing to the case of matter fields forming a multiplet of spin $\frac{1}{2}$ fermions. The squares in $\mathscr{L}$ and $\mathscr{L}_f$ include obvious index contractions.

with

$$\mathscr{D}^A_{x\alpha\mu,y\beta\nu} \equiv \eta_{\mu\nu}\Big(-\delta_{\gamma\alpha}\frac{\partial}{\partial x_\lambda} + C_{\gamma\delta\alpha}A^\lambda_\delta(x)\Big)\Big(-\delta_{\gamma\beta}\frac{\partial}{\partial y^\lambda} + C_{\gamma\epsilon\beta}A_{\epsilon\lambda}(y)\Big)\delta^4(x-y)$$
$$-\Big(-\delta_{\gamma\alpha}\frac{\partial}{\partial x^\nu} + C_{\gamma\delta\alpha}A_{\delta\nu}(x)\Big)\Big(-\delta_{\gamma\beta}\frac{\partial}{\partial y^\mu} + C_{\gamma\epsilon\beta}A_{\epsilon\mu}(y)\Big)\delta^4(x-y)$$
$$+F_{\gamma\mu\nu}(x)C_{\gamma\alpha\beta}\delta^4(x-y)$$
$$+\frac{1}{\xi}\Big(-\delta_{\gamma\alpha}\frac{\partial}{\partial x^\mu} + C_{\gamma\delta\alpha}A_{\delta\mu}(x)\Big)\Big(-\delta_{\gamma\beta}\frac{\partial}{\partial y^\nu} + C_{\gamma\epsilon\beta}A_{\epsilon\nu}(y)\Big)\delta^4(x-y) \tag{17.5.7}$$

$$\mathscr{D}^\psi_{xk,y\ell} \equiv \left(-\gamma^\mu\frac{\partial}{\partial y^\mu} - it_\alpha A_\alpha(y) + m\right)_{k\ell}\delta^4(x-y)\,, \tag{17.5.8}$$

$$\mathscr{D}^\omega_{x\alpha,y\beta} = \left(-\delta_{\gamma\alpha}\frac{\partial}{\partial x^\lambda} + C_{\gamma\delta\alpha}A_{\delta\lambda}(x)\right)\left(-\delta_{\gamma\beta}\frac{\partial}{\partial y_\lambda} + C_{\gamma\epsilon\beta}A^\lambda_\epsilon(y)\right)\delta^4(x-y)\,. \tag{17.5.9}$$

(The minus signs in front of $\partial/\partial x$ and $\partial/\partial y$ drop out when we integrate by parts.)

The one-loop contribution to the effective action is given (as in Section 16.2) by

$$\exp\left(i\Gamma^{1\text{ loop}}[A]\right) \propto \int_{1PI}(\prod dA')(\prod d\psi')(\prod d\bar\psi')(\prod d\omega')(\prod d\omega^{*\prime})$$
$$\times\ \exp\left(i\,I_{\text{QUAD}}[A',\psi',\bar\psi',\omega',\omega'^*;A]\right)$$
$$\propto (\text{Det}\mathscr{D}^A)^{-1/2}(\text{Det}\mathscr{D}^\psi)^{+1}(\text{Det}\mathscr{D}^\omega)^{+1}\,. \tag{17.5.10}$$

(The exponents $-1/2$ and $+1$ appear because A' is a real boson field, while ψ', $\bar\psi'$, ω', and ω'^* are distinct fermionic fields.) The calculation of such determinants is generally not an easy task. However, it becomes much simpler in the case of constant external fields, where the $\mathscr{D}$s can be diagonalized by passing to momentum space.

Let's therefore now consider the case of constant background field $A_{\alpha\mu}$. For non-Abelian gauge theories such a constant field *cannot* be removed by a gauge transformation, as shown by the non-zero values of various gauge-covariant fields

$$F_{\alpha\mu\nu} = C_{\alpha\beta\gamma}A_{\beta\mu}A_{\gamma\nu}\,, \tag{17.5.11}$$
$$D_\lambda F_{\delta\mu\nu} = C_{\delta\epsilon\alpha}C_{\alpha\beta\gamma}A_{\epsilon\lambda}A_{\beta\mu}A_{\gamma\nu}\,, \tag{17.5.12}$$

and so on. Lorentz and gauge invariance tell us how to express the part of $\Gamma[A]$ of a given dimensionality as the integral of a finite number of local functions of $F_{\alpha\mu\nu}$, $D_\lambda F_{\alpha\mu\nu}$, etc.; the coefficients of the terms in this expression can be inferred by comparing the contribution that these terms make to $\Gamma[A]$ for constant background field $A_{\alpha\mu}$ with the results of a perturbative expansion.

We transform each of the 'matrices' $\mathscr{D}^A$, $\mathscr{D}^\Psi$, and $\mathscr{D}^\omega$ to a momentum basis by the usual normalized Fourier transform

$$\mathscr{D}_{q\cdots,p\cdots} = \int \frac{d^4x}{(2\pi)^2}\, e^{-iq\cdot x} \int \frac{d^4y}{(2\pi)^2}\, e^{ip\cdot y}\, \mathscr{D}_{x\cdots,y\cdots}\,. \tag{17.5.13}$$

With A constant, this gives

$$\mathscr{D}_{q\cdots,p\cdots} = \delta^4(p-q)\,\mathscr{M}_{\cdots,\cdots}(q)\,, \tag{17.5.14}$$

where $\cdots$ denotes discrete indices, and the $\mathscr{M}$ are finite q-dependent matrices

$$\begin{aligned}\mathscr{M}^A_{\alpha\mu,\beta\nu}(q) = {}& \eta_{\mu\nu}(-iq_\lambda\delta_{\gamma\alpha} + A_{\delta\lambda}C_{\gamma\delta\alpha})(iq^\lambda\delta_{\gamma\beta} + A_\epsilon{}^\lambda C_{\gamma\epsilon\beta})\\ & -(-iq_\nu\delta_{\gamma\alpha} + A_{\delta\nu}C_{\gamma\delta\alpha})\,(iq_\mu\delta_{\gamma\beta} + A_{\epsilon\mu}C_{\gamma\epsilon\beta})\\ & +F_{\gamma\mu\nu}C_{\gamma\alpha\beta}\\ & +(-iq_\mu\delta_{\gamma\alpha} + A_{\delta\mu}C_{\gamma\delta\alpha})\,(iq_\nu\delta_{\gamma\beta} + A_{\epsilon\nu}C_{\gamma\epsilon\beta})/\xi\\ & +\epsilon \text{ terms}\,, \end{aligned}\tag{17.5.15}$$

$$\mathscr{M}^\Psi_{k\ell}(q) = (i\,\not{q} - it_\alpha \not{A}_\alpha + m)_{k\ell} + \epsilon \text{ terms}\,, \tag{17.5.16}$$

$$\begin{aligned}\mathscr{M}^\omega_{\alpha\beta}(q) = {}& (-iq_\lambda\delta_{\gamma\alpha} + A_{\delta\lambda}C_{\gamma\delta\alpha})\,(iq^\lambda\delta_{\gamma\beta} + A_\epsilon{}^\lambda C_{\gamma\epsilon\beta})\\ & +\epsilon \text{ terms}\,,\end{aligned}\tag{17.5.17}$$

with $F_{\gamma\mu\nu}$ given by Eq. (17.5.11). From Eq. (17.5.10) we have then

$$\begin{aligned} i\Gamma^{(1\text{ loop})}[A] &= -\tfrac{1}{2}\ln \mathrm{Det}\mathscr{D}^A + \ln \mathrm{Det}\mathscr{D}^\Psi + \ln \mathrm{Det}\mathscr{D}^\omega\\ &= -\tfrac{1}{2}\mathrm{Tr}\ln \mathscr{D}^A + \mathrm{Tr}\ln \mathscr{D}^\Psi + \mathrm{Tr}\ln \mathscr{D}^\omega\\ &= \delta^4(p-p)\int d^4q\Big[-\tfrac{1}{2}\mathrm{tr}\,\ln \mathscr{M}^A(q) + \mathrm{tr}\,\ln \mathscr{M}^\Psi(q)\\ &\qquad + \mathrm{tr}\ln \mathscr{M}^\omega(q)\Big]\,. \end{aligned}\tag{17.5.18}$$

We denote traces by 'tr' instead of 'Tr' in the last line of Eq. (17.5.18) (and from now on in this section) to indicate that these are the usual traces of finite matrices rather than of integral operators.

Since we are aiming here at a calculation of the infinite factor L_A multiplying FF terms in the effective action, let's isolate the term in (17.5.18) that is of fourth order in the background field A. For this purpose, it is convenient to divide each $\mathscr{M}$ into terms $\mathscr{M}_n$ containing $n = 0, 1$, or 2 factors of A:

$$\mathscr{M} = \mathscr{M}_0 + \mathscr{M}_1 + \mathscr{M}_2\,. \tag{17.5.19}$$

It is then elementary algebra to show that the term in (17.5.18) of fourth

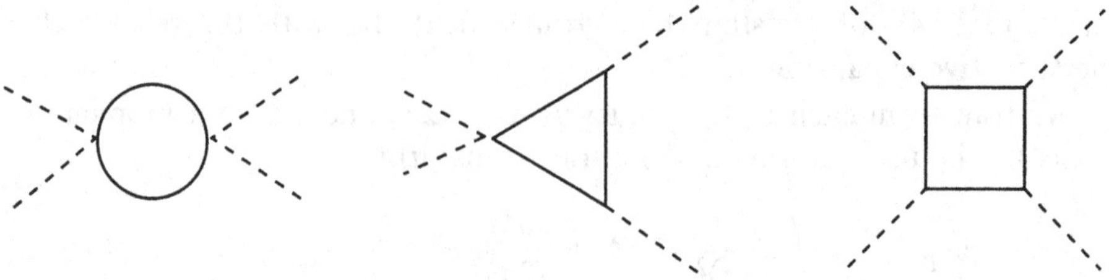

Figure 17.1. One-loop Feynman diagrams for the term in the quantum effective action that is quartic in a constant background gauge field A^μ_α. Here solid lines represent internal gauge, ghost, or matter lines; dashed lines indicate factors of A^μ_α. These three diagrams correspond to the three terms in Eq. (17.5.20).

order in $A_{\alpha\mu}$ is

$$[\text{tr}\ln \mathscr{M}]_{A^4} = \text{tr}\Big\{ -\tfrac{1}{2}[\mathscr{M}_0^{-1}\mathscr{M}_2]^2 + [\mathscr{M}_0^{-1}\mathscr{M}_1]^2\mathscr{M}_0^{-1}\mathscr{M}_2 - \tfrac{1}{4}[\mathscr{M}_0^{-1}\mathscr{M}_1]^4\Big\}\,. \tag{17.5.20}$$

(To see this, insert factors ϵ and ϵ^2 multiplying $\mathscr{M}_1$ and $\mathscr{M}_2$ in Eq. (17.5.19), differentiate tr ln $\mathscr{M}$ four times with respect to ϵ, divide by 4! and set $\epsilon = 0$.) The $\mathscr{M}_0^{-1}$ factors here are just the usual propagators; for $\xi = 1$, these are

$$[\mathscr{M}_0^A(q)]^{-1}_{\alpha\mu,\beta\nu} = \delta_{\alpha\beta}\eta_{\mu\nu}(q^2 - i\epsilon)^{-1}\,, \tag{17.5.21}$$

$$[\mathscr{M}_0^\psi(q)]^{-1}_{k\ell} = [i\,\not{q} + m]^{-1}_{k\ell}\,, \tag{17.5.22}$$

$$[\mathscr{M}_0^\omega(q)]^{-1}_{\alpha,\beta} = \delta_{\alpha\beta}(q^2 - i\epsilon)^{-1}\,. \tag{17.5.23}$$

Indeed, the three terms in Eq. (17.5.20) just correspond to the three Feynman diagrams shown in Figure 17.1; the present method of calculation saves us from having to think about signs and combinatoric factors.

For the A loop, Eq. (17.5.15) gives for $\xi = 1$:

$$[\mathscr{M}_1^A(q)]_{\alpha\mu,\beta\nu} = -2\eta_{\mu\nu}q^\lambda[\mathscr{A}_\lambda]_{\alpha\beta}\,,$$

$$[\mathscr{M}_2^A(q)]_{\alpha\mu,\beta\nu} = +\Big[\eta_{\mu\nu}\mathscr{A}_\lambda\mathscr{A}^\lambda - \mathscr{A}_\nu\mathscr{A}_\mu - \mathscr{A}_\mu\mathscr{A}_\nu\Big]_{\alpha\beta} + F_{\gamma\mu\nu}C_{\gamma\alpha\beta}\,,$$

where $\mathscr{A}_\lambda$ is the matrix

$$[\mathscr{A}_\lambda]_{\alpha\beta} \equiv -i\,C_{\alpha\beta\gamma}A_{\gamma\lambda}$$

for which

$$\begin{aligned}[\mathscr{A}_\lambda, \mathscr{A}_\rho]_{\alpha\beta} &= -C_{\alpha\delta\gamma}C_{\delta\beta\epsilon}(A_{\gamma\lambda}A_{\epsilon\rho} - A_{\gamma\rho}A_{\epsilon\lambda})\\ &= -(C_{\alpha\delta\gamma}C_{\delta\beta\epsilon} + C_{\alpha\delta\epsilon}C_{\delta\beta\gamma})\,A_{\gamma\lambda}A_{\epsilon\rho}\\ &= +C_{\alpha\delta\beta}C_{\delta\epsilon\gamma}A_{\gamma\lambda}A_{\epsilon\rho} = C_{\alpha\delta\beta}F_{\delta\rho\lambda}\,.\end{aligned}$$

The integrals have the structure

$$\int d^4q\, q^\mu q^\nu f(q^2) = \tfrac{1}{4}\eta^{\mu\nu}\int d^4q\, q^2 f(q^2)\,,$$

$$\int d^4q\, q^\mu q^\nu q^\rho q^\sigma f(q^2) = \tfrac{1}{24}[\eta^{\mu\nu}\eta^{\rho\sigma} + \eta^{\mu\rho}\eta^{\nu\sigma} + \eta^{\mu\sigma}\eta^{\nu\rho}]\int d^4q\,(q^2)^2 f(q^2)\,.$$

We then find, for $\xi = 1$:

$$\int d^4q\,\mathrm{tr}\left\{\left[\mathscr{M}_0^A(q)^{-1}\mathscr{M}_2^A(q)\right]^2\right\} = 4\mathscr{I}\,\mathrm{tr}[\mathscr{A}^\lambda\mathscr{A}_\lambda\mathscr{A}^\eta\mathscr{A}_\eta] + 4\mathscr{I}\,C_{\gamma\alpha\beta}C_{\delta\alpha\beta}F_{\gamma\mu\nu}F_\delta^{\mu\nu}\,,$$

$$\int d^4q\,\mathrm{tr}\left\{\left[\mathscr{M}_0^A(q)^{-1}\mathscr{M}_1^A(q)\right]^2\mathscr{M}_0^A(q)^{-1}\mathscr{M}_2^A(q)\right\} = 4\mathscr{I}\,\mathrm{tr}[\mathscr{A}^\lambda\mathscr{A}_\lambda\mathscr{A}^\eta\mathscr{A}_\eta]\,,$$

$$\int d^4q\,\mathrm{tr}\left\{\left[\mathscr{M}_0^A(q)^{-1}\mathscr{M}_1^A(q)\right]^4\right\} = \tfrac{8}{3}\mathscr{I}\,\mathrm{tr}\left[2\mathscr{A}^\lambda\mathscr{A}_\lambda\mathscr{A}^\eta\mathscr{A}_\eta + \mathscr{A}^\lambda\mathscr{A}^\eta\mathscr{A}_\lambda\mathscr{A}_\eta\right],$$

where $\mathscr{I}$ is the divergent integral

$$\mathscr{I} \equiv \int d^4q\,[q^2 - i\epsilon]^{-2}\,, \tag{17.5.24}$$

whose significance is discussed below. Putting this together in Eq. (17.5.20), we have

$$\int d^4q\left[\mathrm{tr}\ln\mathscr{M}^A(q)\right]_{A^4} = \tfrac{2}{3}\mathscr{I}\,\mathrm{tr}\left[\mathscr{A}^\lambda\mathscr{A}_\lambda\mathscr{A}^\eta\mathscr{A}_\eta - \mathscr{A}^\lambda\mathscr{A}^\eta\mathscr{A}_\lambda\mathscr{A}_\eta\right] - 2\mathscr{I}\,C_{\gamma\alpha\beta}C_{\delta\alpha\beta}\,F_{\gamma\mu\nu}\,F_\delta^{\mu\nu}\,.$$

Both terms are actually of the same form, and when combined yield

$$\int d^4q\left[\mathrm{tr}\ln\mathscr{M}^A(q)\right]_{A^4} = -\tfrac{5}{3}\mathscr{I}\,C_{\gamma\alpha\beta}C_{\delta\alpha\beta}F_{\gamma\mu\nu}F_\delta^{\mu\nu}\,. \tag{17.5.25}$$

Now jumping ahead to the ghost loop, we see from Eq. (17.5.17) that

$$[\mathscr{M}_1^\omega(q)]_{\alpha\beta} = -2[\mathscr{A}^\lambda]_{\alpha\beta}\,q_\lambda\,, \tag{17.5.26}$$

$$[\mathscr{M}_2^\omega(q)]_{\alpha\beta} = [\mathscr{A}^\lambda\mathscr{A}_\lambda]_{\alpha\beta}\,. \tag{17.5.27}$$

We then find

$$\int d^4q\,\mathrm{tr}\left\{\left[\mathscr{M}_0^\omega(q)^{-1}\mathscr{M}_2^\omega(q)\right]^2\right\} = \mathscr{I}\,\mathrm{tr}[\mathscr{A}^\lambda\mathscr{A}_\lambda\mathscr{A}^\eta\mathscr{A}_\eta]\,,$$

$$\int d^4q\,\mathrm{tr}\left\{\left[\mathscr{M}_0^\omega(q)^{-1}\mathscr{M}_1^\omega(q)\right]^2\mathscr{M}_0^\omega(q)^{-1}\mathscr{M}_2^\omega(q)\right\} = \mathscr{I}\,\mathrm{tr}[\mathscr{A}^\lambda\mathscr{A}_\lambda\mathscr{A}^\eta\mathscr{A}_\eta]\,,$$

$$\int d^4q\,\mathrm{tr}\left\{\left[\mathscr{M}_0^\omega(q)^{-1}\mathscr{M}_1^\omega(q)\right]^4\right\} = \tfrac{2}{3}\mathscr{I}\,\mathrm{tr}[2\mathscr{A}^\lambda\mathscr{A}_\lambda\mathscr{A}^\eta\mathscr{A}_\eta + \mathscr{A}^\lambda\mathscr{A}^\eta\mathscr{A}_\lambda\mathscr{A}_\eta]\,.$$

Thus for the ghost loop the integral of the quantity (17.5.20) is

$$\int d^4q\,[\mathrm{tr}\ln\mathscr{M}^{\omega}(q)]_{A^4} = \tfrac{1}{6}\mathscr{I}\,\mathrm{tr}[\mathscr{A}^{\lambda}\mathscr{A}_{\lambda}\mathscr{A}^{\eta}\mathscr{A}_{\eta} - \mathscr{A}^{\lambda}\mathscr{A}^{\eta}\mathscr{A}_{\lambda}\mathscr{A}_{\eta}]$$
$$= \tfrac{1}{12}\mathscr{I}C_{\gamma\alpha\beta}C_{\delta\alpha\beta}F_{\gamma\mu\nu}F_{\delta}^{\mu\nu}\,. \qquad (17.5.28)$$

Finally, the vertices in the matter loop are

$$[\mathscr{M}_1^{\psi}(q)]_{k\ell} = -i\,(t_\alpha A\!\!\!/_\alpha)_{k\ell}\,, \qquad [\mathscr{M}_2^{\psi}(q)]_{k\ell} = 0\,,$$

so here there is only one term in Eq. (17.5.20):

$$\int d^4q\,[\mathrm{tr}\ln\mathscr{M}^{\psi}(q)]_{A^4} = -\tfrac{1}{4}\int d^4q\,\mathrm{tr}\left\{\left[(i\,q\!\!\!/ + m)^{-1}t_\alpha A\!\!\!/_\alpha\right]^4\right\}.$$

We are interested in the ultraviolet-divergent part of this integral, so we may drop the mass (which is negligible for large q^ν) and write

$$\int d^4q\,[\mathrm{tr}\ln\mathscr{M}^{\psi}(q)]_{A^4} = -\tfrac{1}{4}\,\mathrm{tr}\{t_\alpha t_\beta t_\gamma t_\delta\}\,A_{\alpha\mu}A_{\beta\nu}A_{\gamma\rho}A_{\delta\sigma}$$
$$\times\int d^4q\,\frac{\mathrm{tr}\{\,q\!\!\!/\gamma^\mu\,q\!\!\!/\gamma^\nu\,q\!\!\!/\gamma^\rho\,q\!\!\!/\gamma^\sigma\}}{(q^2 - i\epsilon)^4}$$
$$= -\frac{\mathscr{I}}{96}\,\mathrm{tr}\{t_\alpha t_\beta t_\gamma t_\delta\}A_{\alpha\mu}A_{\beta\nu}A_{\gamma\rho}A_{\delta\sigma}$$
$$\times\,\mathrm{tr}\left\{2\gamma_\lambda\gamma^\mu\gamma^\lambda\gamma^\nu\gamma_\eta\gamma^\rho\gamma^\eta\gamma^\sigma + \gamma_\lambda\gamma^\mu\gamma_\eta\gamma^\nu\gamma^\lambda\gamma^\rho\gamma^\eta\gamma^\sigma\right\}, \qquad (17.5.29)$$

where $\mathscr{I}$ is the same divergent integral as in Eq. (17.5.24). To calculate the traces of Dirac matrices, we use the anticommutation relations of these matrices to write

$$\mathrm{tr}\left\{2\gamma_\lambda\gamma^\mu\gamma^\lambda\gamma^\nu\gamma_\eta\gamma^\rho\gamma^\eta\gamma^\sigma + \gamma_\lambda\gamma^\mu\gamma_\eta\gamma^\nu\gamma^\lambda\gamma^\rho\gamma^\eta\gamma^\sigma\right\}$$
$$= 8\,\mathrm{tr}\{\gamma^\mu\gamma^\nu\gamma^\rho\gamma^\sigma\} - 4\,\mathrm{tr}\{\gamma^\nu\gamma^\mu\gamma^\rho\gamma^\sigma\} - 4\,\mathrm{tr}\{\gamma^\mu\gamma^\rho\gamma^\nu\gamma^\sigma\}$$
$$= -64\eta^{\mu\rho}\eta^{\nu\sigma} + 32\eta^{\mu\nu}\eta^{\rho\sigma} + 32\eta^{\mu\sigma}\eta^{\nu\rho}\,.$$

Eq. (17.5.29) then gives

$$\int d^4q\left[\mathrm{tr}\ln\mathscr{M}^{\psi}(q)\right]_{A^4} = \tfrac{1}{3}\mathscr{I}\,\mathrm{tr}\left\{[t_\alpha, t_\beta][t_\gamma, t_\delta]\right\}A_{\alpha\mu}A_{\beta\nu}A_\gamma^\mu A_\delta^\nu$$
$$= -\tfrac{1}{3}\mathscr{I}F_{\gamma\mu\nu}F_\delta^{\mu\nu}\,\mathrm{tr}\{t_\gamma t_\delta\}\,. \qquad (17.5.30)$$

Using Eqs. (17.5.25), (17.5.28), and (17.5.30) in Eq. (17.5.18) gives at last

$$\Gamma_{A^4}^{(1\text{ loop})} = \frac{-i\mathscr{I}}{(2\pi)^4}\int d^4x\,F_{\gamma\mu\nu}F_\delta^{\mu\nu}\left[\left(\frac{5}{6} + \frac{1}{12}\right)C_{\gamma\alpha\beta}C_{\delta\alpha\beta} - \frac{1}{3}\,\mathrm{tr}\{t_\gamma t_\delta\}\right], \qquad (17.5.31)$$

where we have expressed the momentum-space delta-function in (17.5.14) as

$$\delta^4(p-p) = (2\pi)^{-4}\int d^4x\, 1\,. \tag{17.5.32}$$

It is important that the result turns out to depend on $A_{\alpha\mu}$ only through the field strength (17.5.11), as required by background gauge invariance.

Let us now (for the first time in this section) use the assumed simplicity of the gauge group and irreducibility of the matter field multiplet. In this case

$$C_{\gamma\alpha\beta}C_{\delta\alpha\beta} = g^2 C_1\,\delta_{\gamma\delta}\,, \tag{17.5.33}$$

$$\mathrm{tr}\{t_\gamma t_\delta\} = g^2 C_2 \delta_{\gamma\delta}\,, \tag{17.5.34}$$

where g is the common gauge coupling constant appearing as a factor in $C_{\gamma\alpha\beta}$ and t_γ, and C_1 and C_2 are numerical constants that characterize the gauge group and the representation of this group provided by the matter multiplet. For instance, in the original Yang–Mills theory the gauge group is $SU(2)$ (or equivalently $SO(3)$) and the structure constants are

$$C_{\gamma\alpha\beta} = g\epsilon_{\alpha\beta\gamma}$$

with α, β, and γ running over the values 1, 2, 3. Comparing with Eq. (17.5.33), we see that here

$$C_1 = 2\,.$$

Also in this theory the matter field forms a doublet with t_α given by $g/2$ times the usual Pauli matrix σ_α, so

$$C_2 = 1/2\,.$$

Somewhat more generally, for the group $SU(N)$ with n_f fermions in the defining representation, with a conventional normalization of generators we have**

$$C_1 = N\,, \qquad C_2 = n_f/2\,. \tag{17.5.35}$$

Returning now to the general case, Eqs. (17.5.33) and (17.5.34) give

$$\Gamma^{(1\text{ loop})}_{A^4} = \frac{-ig^2\mathscr{I}}{(2\pi)^4}\int d^4x F_{\gamma\mu\nu}F^{\mu\nu}_\gamma\left[\frac{11}{12}C_1 - \frac{1}{3}C_2\right]. \tag{17.5.36}$$

That is, the infinite constant L_A in Eq. (17.4.30) is

$$L_A = \frac{4ig^2\mathscr{I}}{(2\pi)^4}\left(\frac{11}{12}C_1 - \frac{1}{3}C_2\right). \tag{17.5.37}$$

** For $SU(3)$ the t_α are taken as $g/2$ times the Gell-Mann matrices λ_α used in Section 19.7, so that $C_{\alpha\beta\gamma} = (g/2)f_{\alpha\beta\gamma}$.

It remains to say a word about the interpretation of the divergent integral $\mathscr{I}$. First, before we try to integrate over the three-momentum $\mathbf{q}$, we can rotate the contour of integration of q^0 in Eq. (17.5.24) to the imaginary axis; as usual, the $-i\epsilon$ in the denominator forces us to rotate counterclockwise, so that $q^0 = iq^4$, with q^4 running from $-\infty$ to $+\infty$. The integral is then

$$\mathscr{I} = i\int_0^\infty \frac{2\pi^2 q^3 dq}{q^4}\,, \tag{17.5.38}$$

where q is now the magnitude of the Euclidean four-vector (q^1, q^2, q^3, q^4). To go further, we evidently need some method of regulating the integral. The simplest way of dealing with the ultraviolet divergence is just to cut off the integral for q above a scale Λ. However, we also need a lower cut-off to deal with the infrared divergence. This is provided by the physics of the situation. If the momenta of the four vector particles is not zero, then a momentum flows through the internal lines of the diagrams, providing an infrared cut-off at the scale μ of these momenta. Similarly, if we evaluate the fourth variational derivative of $\Gamma[A]$ with respect to A not at $A = 0$ but at a finite A, then the propagators of the internal lines do not blow up at zero momenta, and we have an infrared cut-off at a scale $\mu \sim gA$. Either way, $\mathscr{I}$ takes the form

$$\mathscr{I} = 2\pi^2 i\int_\mu^\Lambda \frac{dq}{q} = 2\pi^2 i\ \ln\left(\frac{\Lambda}{\mu}\right) \tag{17.5.39}$$

and hence

$$L_A = -\frac{g^2}{2\pi^2}\left(\frac{11}{12}C_1 - \frac{1}{3}C_2\right)\ln\left(\frac{\Lambda}{\mu}\right) + O(g^4)\,. \tag{17.5.40}$$

Eq. (17.4.48) then gives the renormalized coupling as

$$g_R = g\left[1 + \frac{g^2}{4\pi^2}\ln\left(\frac{\Lambda}{\mu}\right)\left(\frac{11}{12}C_1 - \frac{1}{3}C_2\right) + O(g^4)\right]\,. \tag{17.5.41}$$

We note that while in quantum electrodynamics the radiative corrections discussed in Section 11.2 decrease the physical coupling g_R relative to the bare coupling g, in non-Abelian gauge theories they *increase* the physical coupling over the bare coupling, provided that the fermion multiplet is small enough so that $C_2 < 11C_1/4$. The importance of this point will be explored in Chapter 18.

Alternatively, we can deal with the ultraviolet divergence by the methods of dimensional regularization, discussed in Section 11.2. Here, in place of Eq. (17.5.39), we write

$$\mathscr{I} = i\int_0^\infty \frac{2\pi^2 q^{d-1} dq}{(q^2+\mu^2)^2}\,, \tag{17.5.42}$$

where d is a complex dimensionality, allowed to approach 4 at the end of the calculation, and μ is an infrared cut-off, again taken of the order of the external momenta (or of the background fields times g). As long as d is complex with Re $d < 0$ and $\mu^2 > 0$, this has the finite value

$$\mathscr{I} = -i\pi^2 \left(\frac{d}{2} - 1\right) \mu^{d-4}\pi \Big/ \sin\left[\left(\frac{d}{2} - 2\right)\pi\right] .$$

Analytically continuing to $d \to 4$, this is

$$\mathscr{I} \to -2i\pi^2 \left[\frac{1}{d-4} + \ln \mu + \cdots\right] , \tag{17.5.43}$$

where $\cdots$ denotes finite μ-independent terms. Here we have

$$L_A = \frac{g^2}{2\pi^2}\left(\frac{11}{12}C_1 - \frac{1}{3}C_2\right)\left(\frac{1}{d-4} + \ln \mu + \cdots\right) + O(g^4)$$

and so

$$g_R = g\left[1 - \frac{g^2}{4\pi^2}\left(\frac{11}{12}C_1 - \frac{1}{3}C_2\right)\left(\frac{1}{d-4} + \ln \mu + \cdots\right) + O(g^4)\right] . \tag{17.5.44}$$

Note that the ultraviolet divergence here takes a different form, but the dependence on the infrared cut-off μ is the same. Eq. (17.5.44) will provide an important input to our discussion of asymptotic freedom in Section 18.7.

Problems

1. Carry out the proof of renormalizability given in Section 17.2, including elementary scalar fields in the Lagrangian.

2. Carry out the quantization of a non-Abelian gauge theory in background field gauge, using the BRST method of quantization discussed at the end of Section 17.2.

3. Derive the relation (17.5.44) between the renormalized and unrenormalized gauge couplings by calculating the terms in $\Gamma^{(1\text{ loop})}$ that are quadratic in a spacetime-dependent gauge field.

4. Calculate the one-loop relation between the renormalized and unrenormalized gauge couplings in a gauge theory containing elementary scalar fields.

References

1. J. Zinn-Justin, in *Trends in Elementary Particle Theory — International Summer Institute on Theoretical Physics in Bonn 1974* (Springer-Verlag, Berlin, 1975).

2. Before the advent of BRST symmetry, proofs of the renormalizability of non-Abelian gauge theories were based directly on the Slavnov–Taylor identities for gauge transformations; see B. W. Lee and J. Zinn-Justin, *Phys. Rev.* **D5**, 3121, 3137 (1972); *Phys. Rev.* **D7**, 1049 (1972); G. 't Hooft and M. Veltman, *Nucl. Phys.* **B50**, 318 (1972); B. W. Lee, *Phys. Rev.* **D9**, 933 (1974). The original proof of renormalizability based on BRST symmetry was given by C. Becchi, A. Rouet, and R. Stora, *Commun. Math. Phys.* **42**, 127 (1975); in *Renormalization Theory — Proceedings of the International School of Mathematical Physics at Erice, August 1975*, eds. G. Velo and A.S. Wightman (D. Reidel, Dordrecht, 1976): pp. 269–97, 299–343. The proof given here follows the general outline of J. Zinn-Justin, Ref. 1; B. W. Lee, in *Methods in Field Theory*, eds. R. Balian and J. Zinn-Justin (North-Holland, Amsterdam, 1976): pp. 79–139.

3. The point of view and presentation here is based on that of J. Gomis and S. Weinberg, *Nucl. Phys.* **B469**, 475 (1996). For earlier use of these methods, see B. L. Voronov and I. V. Tyutin, *Theor. Math. Phys.* **50**, 218 (1982); **52**, 628 (1982); B. L. Voronov, P. M. Lavrov, and I. V. Tyutin, *Sov. J. Nucl. Phys.* **36**, 292 (1982); P. M. Lavrov and I. V. Tyutin *Sov. J. Nucl. Phys.* **41**, 1049 (1985); D. Anselmi, *Class. and Quant. Grav.* **11**, 2181 (1994); **12**, 319 (1995); M. Harada, T. Kugo, and K. Yamawaki, *Prog. Theor. Phys.* **91**, 801 (1994).

4. G. Barnich and M. Henneaux, *Phys. Rev. Lett.* **72**, 1588 (1994); G. Barnich, F. Brandt, and M. Henneaux, *Phys. Rev.* **51**, R143 (1995); *Commun. Math. Phys.* **174**, 57, 93 (1995); *Nucl. Phys.* **B455**, 357 (1995).

5. The background field gauge was introduced by B. S. De Witt, *Phys. Rev.* **162**, 1195, 1239 (1967). For the treatment of multi-loop effects, see G 't Hooft, in *Functional and Probabilistic Methods in Quantum Field Theory: Proceedings of the 12th Karpacz Winter School of Theoretical Physics* (Acta Universitatis Wratislavensis no. 38, 1975); B. S. De Witt, in *Quantum Gravity* II, eds. C. Isham, R. Penrose, and D. Sciama (Oxford University Press, Oxford, 1982); L. F. Abbott, *Nucl. Phys.* **B185**, 189 (1981).

18
Renormalization Group Methods

The method of the renormalization group was originally introduced by Gell-Mann and Low[1] as a means of dealing with the failure of perturbation theory at very high energies in quantum electrodynamics. An n-loop contribution to an amplitude involving momenta of order q, such as the vacuum polarization $\Pi_{\mu\nu}(q)$, is found to contain up to n factors of $\ln(q^2/m_e^2)$ as well as a factor α^n, so perturbation theory will break down when $\alpha|\ln(q^2/m_e^2)|$ is large, even though the fine structure constant α is small. Even in a massless theory like a non-Abelian gauge theory we must introduce some scale μ to specify a renormalization point at which the renormalized coupling constants are to be defined, and in this case we encounter logarithms $\ln(E/\mu)$, so that perturbation theory may break down if $E \gg \mu$ or $E \ll \mu$, even if the coupling constant is small.

Fortunately, there is a modified version of perturbation theory that can often be used in such cases. The key idea of this approach consists in the introduction of coupling constants g_μ defined at a sliding renormalization scale μ — that is, a scale that is not related to particle masses in any fixed way. By then choosing μ to be of the same order of magnitude as the energy E that is typical of the process in question, the factors $\ln(E/\mu)$ are rendered harmless. We can then do perturbation theory as long as g_μ remains small. In particular, given the coupling constants defined at scale μ, we can use perturbation theory to calculate physical amplitudes at an energy $\mu + d\mu$, and use these to calculate the coupling constants defined at a renormalization scale $\mu + d\mu$. By integrating the resulting differential equation we can then relate the coupling constants at the scale of interest to the coupling constants as conventionally defined. (The name 'renormalization group' arose originally because one is concerned here with equations that describe how the appearance of a theory changes under a redefinition of the renormalized coupling constants, but it really has nothing to do with group theory.) The method of the renormalization group can also provide qualitative guidance regarding asymptotic behavior at very high or (in massless theories) at very low energy, even where the coupling constants at the scale of interest are too large to allow the use of perturbation theory.

Although the method of the renormalization group arose originally in connection with changes in the prescription used to define renormalized coupling constants, it has come to have a wider meaning. When we replace bare couplings and fields with renormalized couplings and fields defined in terms of matrix elements evaluated at a characteristic energy scale μ, the integrals over virtual momenta will be effectively cut off at energy and momentum scales of order μ. Thus as we change μ, we are in effect changing the scope of the degrees of freedom taken into account in our calculations. The lesson of the renormalization group, that in order to avoid large logarithms we should take μ to be of the order of the energy E typical of the process being studied, is a special case of a broader principle, that in order to do calculations at a given energy we should first get rid of the degrees of freedom of much higher energy.

There are various other ways to accomplish this. As we saw in Section 12.4, in the approach to the renormalization group pioneered by Wilson[2] one introduces a *finite* explicit cut-off accompanied by a change in the parameters of the theory designed to keep physical quantities cut-off-independent. This approach requires introduction of an infinite number of interaction types, all those allowed by the symmetries of the theory, and is therefore not particularly convenient in dealing with theories that are actually renormalizable, like quantum electrodynamics (although, as discussed in Section 12.3, quantum electrodynamics is today regarded as only a very good approximation to a non-renormalizable theory in which the higher-dimensional interactions are suppressed by negative powers of some very large mass.) Where the cut-off is imposed by quantizing a gauge theory on a finite spacetime lattice, the Wilson approach has the advantage that calculations can be done while maintaining manifest gauge invariance (the volume of the gauge group equaling the volume of the global symmetry group times the number of lattice sites), but it has the disadvantage of not maintaining manifest Lorentz or rotational invariance. In any case much of the formalism of the renormalization group remains the same whatever approach is used to eliminate the high-energy degrees of freedom.

18.1 Where do the Large Logarithms Come From?

Let us first consider how large logarithms can arise at very high energies. Consider a physical amplitude or cross section or other rate parameter $\Gamma(E, x, g, m)$, that depends on an over-all energy scale E, on various angles and energy ratios collectively called x, on various dimensionless coupling constants collectively called g, and on various masses collectively called m. If Γ has dimensionality [mass]^D (as, for instance, a cross section would have $D = -2$) then simple dimensional analysis tells us that

$$\Gamma(E, x, g, m) = E^D \Gamma\left(1, x, g, \frac{m}{E}\right) . \tag{18.1.1}$$

We might expect that in the limit $E \to \infty$ such an amplitude would behave as a simple power

$$\Gamma(E, x, g, m) \longrightarrow E^D \Gamma(1, x, g, 0) .$$

But this is not what is found. Instead, in perturbation theory calculations the factor E^D is found to be accompanied by powers of $\ln(E/m)$, which invalidate this simple power-law behavior.

Clearly, powers of $\ln(E/m)$ can enter as $E \to \infty$ with fixed m only if the amplitude Γ at *fixed energy* E becomes singular as $m \to 0$. There are two classes of such mass singularities, one which is simply eliminated by calculating the right sort of amplitude or rate constant, the other of which requires a change in our renormalization procedure.

Zero-mass singularities of the first sort arise from a confluence of poles of propagators on the mass shells of the corresponding particles. For instance, suppose that a Feynman diagram has an incoming line with total four-momentum p^μ, attached at a vertex to internal lines of mass $m_1, m_2, \ldots, m_n$. According to the arguments of Chapter 10, the corresponding Feynman diagram will have a cut running along the negative real p^2 axis, from $p^2 = -(m_1 + \ldots + m_n)^2$ to $-\infty$. This does not lead to singularities if the external line is for a stable particle, with a mass $M < m_1 + \ldots + m_n$, because then $p^2 = -M^2$ is off the cut. However, when $M, m_1, \ldots, m_n$ all go to zero, the value of $-p^2$ on the mass shell and the branch point at the tip of this cut move together to join at $p^2 = 0$, producing a singularity.

This suggests that we can avoid the infrared divergences at $m = 0$ by simply staying off the mass shell, as, for instance, by letting p^2 for all external lines go to $+\infty$ along with all energy variables. We would then have to employ dispersion relations or some other technique of analytic continuation to use the results for the behavior of Feynman amplitudes in this limit to tell us anything about S-matrix elements. Often this continuation is unnecessary because we are interested not in on-shell S-matrix elements but rather in the matrix elements of currents carrying momenta q unrelated to any masses. For instance, the vacuum polarization function $\pi(q^2)$ defined in Section 10.5 is free of zero-mass singularities of the first sort except for $q^2 < 0$.

Another approach to the elimination of mass singularities of the first type is suggested by the observation that zero-mass singularities typically occur if we try to calculate a cross section that becomes unmeasurable in the limit $m \to 0$. For instance, in quantum electrodynamics the cross section for any process involving definite numbers of electrons and photons becomes infrared-divergent in the limit $m_e \to 0$, even if we sum over unlimited numbers of soft photons, because for $m_e \to 0$ it is impossible to distinguish an electron from a jet of electrons, positrons, and photons with total charge $-e$, all moving in the same direction at the same speed. As shown in Chapter 13, such infrared divergences can be cured by considering only suitably integrated cross sections, which *would* be measurable for $m_e \to 0$. For instance, instead of trying to calculate the cross section for a specific Compton scattering process, we would calculate the cross section for the scattering of a jet with total charge $-e$ with another of total charge zero into two other such jets, plus soft photons. Such inclusive rates or cross sections, which remain finite when all masses vanish, are known as 'infrared safe.'

Our troubles are not over. Even where we avoid infrared divergences by integrating over cross sections or staying off the mass shell, the resulting integrated cross sections or off-shell amplitudes for energies E contain mass singularities of a second type, leading to factors $\ln(E/m)$ that invalidate the naive power-law behavior suggested by dimensional analysis. The reason can be traced to the fact that renormalized coupling constants are conventionally defined in terms of amplitudes that become infrared-divergent when all masses vanish. For instance, consider the theory of a real scalar field with Lagrangian density

$$\mathscr{L} = -\frac{1}{2}\,\partial_\lambda\phi\partial^\lambda\phi - \frac{1}{2}\,m^2\phi^2 - \frac{1}{24}\,g\,\phi^4. \tag{18.1.2}$$

To one-loop order, the invariant elastic scattering amplitude for a scattering process with initial four-momenta p_1, p_2, and final four-momenta p_1', p_2', is given by Eq. (12.2.24) as

$$A = g - \frac{g^2}{32\pi^2}\int_0^1 dx\left\{\ln\left(\frac{\Lambda^2}{m^2 - sx(1-x)}\right)\right.$$
$$\left. + \ln\left(\frac{\Lambda^2}{m^2 - tx(1-x)}\right) + \ln\left(\frac{\Lambda^2}{m^2 - ux(1-x)}\right) - 3\right\} + O(g^3), \tag{18.1.3}$$

where s, t, and u are the Mandelstam variables

$$s = -(p_1 + p_2)^2\,, \quad t = -(p_1 - p_1')^2\,, \quad u = -(p_1 - p_2')^2$$

and Λ is an ultraviolet cut-off. This has no zero-mass singularities as long as we keep s, t, and u away from the positive real axis, and in particular if they all go to $-\infty$ (which violates the mass shell condition $s+t+u = 4m^2$.) Of course, the amplitude depends on the cut-off Λ as well as on m, so even though there are no zero-mass singularities, we do not find the result $A \to$ constant that would be found on the basis of naive scaling arguments in the limit as s, t, and u go to $-\infty$. The dependence on the cut-off can be buried by renormalization; we replace the bare coupling g with a renormalized coupling g_R, defined as the value of A at some convenient renormalization point. For instance, we might take

$$\begin{aligned} g_R &\equiv A(s=t=u=0) \\ &= g - \frac{3g^2}{32\pi^2}\left\{\ln\frac{\Lambda^2}{m^2} - 1\right\} + O(g^3) . \end{aligned} \tag{18.1.4}$$

Then (18.1.3) becomes

$$\begin{aligned} A = g_R + \frac{g_R^2}{32\pi^2}\int_0^1 dx\Big\{\ln\left(1 - \frac{sx(1-x)}{m^2}\right) \\ + \ln\left(1 - \frac{tx(1-x)}{m^2}\right) + \ln\left(1 - \frac{ux(1-x)}{m^2}\right)\Big\} + O(g_R^3) . \end{aligned} \tag{18.1.5}$$

(We can freely replace g^2 with g_R^2 in the second term, because the difference is only of order g^3.) This is free of ultraviolet divergences, but it now has a singularity at $m = 0$, even where s, t, and u are all kept negative. In consequence, where s, t, and u all go to $-\infty$, we again find an asymptotic behavior in disagreement with expectations based on naive scaling

$$A \to g_R + \frac{g_R^2}{32\pi^2}\left\{\ln\left(\frac{-s}{m^2}\right) + \ln\left(\frac{-t}{m^2}\right) + \ln\left(\frac{-u}{m^2}\right) - 6\right\} . \tag{18.1.6}$$

(Much the same happens with any other 'natural' definition of the renormalized coupling; for instance, we might define g_R as the value of A at the on-shell symmetrical point $s = t = u = 4m^2/3$, and would recover the same asymptotic behavior as in Eq. (18.1.6), except that -6 would be replaced with some other numerical constant.) It is clear that the zero-mass singularity that is encountered when A is expressed in terms of g_R arises entirely from the $\ln m^2$ term in the formula (18.1.4) for the renormalized coupling g_R in terms of the bare coupling g.

There are in addition other zero-mass singularities that are encountered when we calculate matrix elements of operators (such as off-shell Feynman amplitudes) rather than integrals of cross sections. These are due to the necessity of renormalizing these operators as well as coupling constants. For instance, suppose that in the scalar field theory with Lagrangian (18.1.2), we wish to calculate some matrix element $\langle\beta|\mathcal{O}(p)|\alpha\rangle$ of the operator

$$\mathcal{O}(p) \equiv \int d^4x\, e^{-ip\cdot x}\phi^2(x) . \tag{18.1.7}$$

In Feynman diagram terms, this corresponds to inserting a vertex in which two internal ϕ-lines come together, and through which flows a total four-momentum p in diagrams for the transition $\alpha \to \beta$. (See Figure 18.1.)

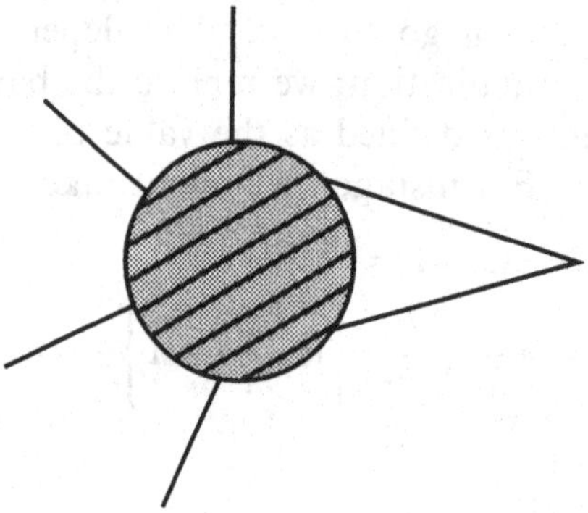

Figure 18.1. Momentum-space Feynman diagrams for the matrix element of the operator $\int d^4x \exp(-ip \cdot x)\, \phi^2(x)$ in the theory of an elementary scalar field $\phi(x)$. The cross-hatched disk represents the sum of diagrams with the indicated external lines. Apart from the pair of external lines that meet in a ϕ^2 vertex, the other external lines attached to the disk represent the particles in the initial and final states between which the matrix element is evaluated.

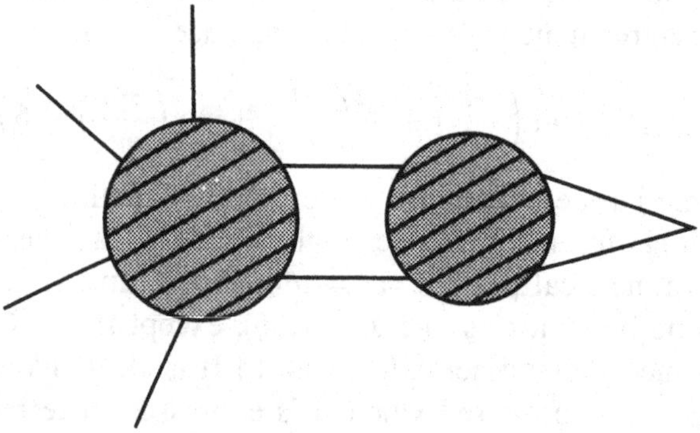

Figure 18.2. A class of Feynman diagrams for the matrix element of the operator $\int d^4x \exp(-ip \cdot x)\, \phi^2(x)$ that exhibit an ultraviolet divergence. The notation is the same as in Figure 18.1.

Ultraviolet divergences arise from a class of diagrams in which this new vertex is part of a subdiagram that is connected to the rest of the graph by just two ϕ-lines. (See Figure 18.2.) Dimensional analysis shows that the subgraph would be convergent if connected to the rest of the diagram by more than two ϕ-lines, because the interaction ϕ^2 has dimensionality +2, and hence a subdiagram in which this vertex is connected to the rest of the graph by $n > 2$ lines has dimensionality $4 - 2 - n < 0$. The divergent part of this subdiagram is just a logarithmically divergent constant, so

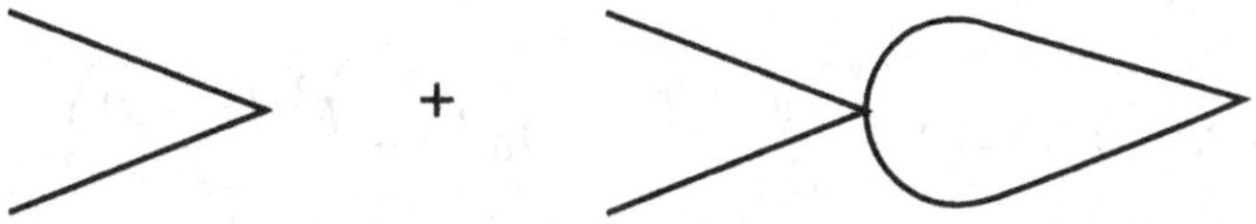

Figure 18.3. The divergent part of the diagram in Figure 18.2, to one-loop order.

the matrix elements of ϕ^2 can be made finite* by multiplying ϕ^2 with a suitable divergent constant Z_{ϕ^2}. To order g^2, the relevant subdiagram is given by the diagrams of Figure 18.3, and hence contributes to matrix elements of $\mathscr{O}(p)$ a divergent factor

$$F(p) = 1 + \frac{1}{2}\left[-i(2\pi)^4 g\right]\left[\frac{-i}{(2\pi)^4}\right]^2 \int \frac{d^4k}{\left[k^2+m^2-i\epsilon\right]\left[(p-k)^2+m^2-i\epsilon\right]} . \tag{18.1.8}$$

Combining denominators, rotating the k^0 integration contour, and imposing an ultraviolet cut-off Λ gives a result, for $\Lambda \to \infty$

$$F(p) = 1 - \frac{g}{32\pi^2}\int_0^1 dx \left[\ln\left(\frac{\Lambda^2}{m^2+p^2x(1-x)}\right) - 1\right] + O(g^2) . \tag{18.1.9}$$

This has no zero-mass singularity (as long as we keep p^2 positive), but of course it does depend on the cut-off. This logarithmic divergence is eliminated by defining a renormalized ϕ^2 operator

$$(\phi^2)_R = N^{(\phi^2)}\phi^2 \tag{18.1.10}$$

with $N^{(\phi^2)}$ chosen so that $N^{(\phi^2)}F(p)$ has some definite finite value at some definite renormalization point. For instance, we could define the renormalized ϕ^2 operator so that

$$N^{(\phi^2)}F(0) = 1 , \tag{18.1.11}$$

in which case

$$N^{(\phi^2)} = 1 + \frac{g}{32\pi^2}\left(\ln\frac{\Lambda^2}{m^2} - 1\right) + O(g^2) . \tag{18.1.12}$$

The matrix elements of the renormalized operator $(\phi^2)_R$ then contain a

* In making this argument it was assumed that any divergences arising from subsubdiagrams containing the ϕ^2 vertex which are attached to the rest of the subdiagram by just two ϕ-lines are eliminated in the same way.

factor

$$F_R(p) \equiv N^{(\phi^2)} F(p) = 1 + \frac{g}{32\pi^2} \int_0^1 dx \, \ln\left(1 + \frac{p^2 x(1-x)}{m^2}\right) + O(g^2). \tag{18.1.13}$$

This is finite for all $p^2 > 0$ and $m^2 > 0$, but it now contains an infrared singularity for $m \to 0$, corresponding to large logarithms in the asymptotic behavior when $p^2 \to +\infty$. Of course in order to eliminate the cut-off in higher-order calculations we would have to introduce a renormalized coupling constant as well as a renormalized ϕ^2 operator, and we would encounter logarithms arising from both sources.

Similar renormalization factors are needed for any sort of operator, not just $\phi^2(x)$. In particular, taking a matrix element of one of the elementary fields ψ of a theory introduces ultraviolet divergences that arise from radiative corrections to the corresponding propagator. As we saw in Chapter 12, these infinities can be cancelled by working with a renormalized field ψ_R:

$$\psi_R = N^{(\psi)} \psi \tag{18.1.14}$$

with $N^{(\psi)}$ chosen to make the matrix element of ψ_R between a one-particle state and the vacuum the same as for a conventionally normalized field in the absence of interactions. This is related to the usual Z-factor of renormalization theory by

$$Z^{(\psi)} = |N^{(\psi)}|^{-2}. \tag{18.1.15}$$

For instance, let's recall our earlier results for the renormalization of the photon field in spinor quantum electrodynamics.** In this case, the renormalized electromagnetic field A^μ_R is conventionally written in terms of the 'bare' field A^μ_B as

$$A^\mu_R = Z_3^{-1/2} A^\mu_B$$

with Z_3 given by Eq. (11.2.21) as

$$Z_3 = 1 - \frac{e^2}{12\pi^2} \ln\left(\frac{\Lambda^2}{m^2}\right) + O(e^4). \tag{18.1.16}$$

This has a zero-mass singularity, which will affect the asymptotic behavior of matrix elements of the renormalized photon field. In particular,

** In the scalar field theory with ϕ^4 interaction used as an example above, the lowest-order terms in $N^{(\phi)}$ arise from a two-loop graph, so it will not be convenient to use this theory to illustrate the calculation of the N-factors.

Eq. (11.2.22) gives the self-energy function of the renormalized electromagnetic field as

$$\pi(q^2) = \frac{e_R^2}{2\pi^2} \int_0^1 dx\, x(1-x) \ln\left[1 + \frac{q^2 x(1-x)}{m^2}\right] + O(e_R^4). \tag{18.1.17}$$

This has a singularity at $m = 0$, and so has large logarithms in its asymptotic behavior: for $q^2 \to +\infty$

$$\pi(q^2) \longrightarrow \frac{e_R^2}{2\pi^2}\left[\frac{1}{6}\ln\frac{q^2}{m^2} - \frac{5}{18}\right] + O(e_R^4). \tag{18.1.18}$$

Electrodynamics has the special feature that the constant Z_3 that appears in the renormalization of the electromagnetic field also appears in the renormalization of the electric charge:

$$e_R = Z_3^{1/2}\, e_{\text{BARE}}\,, \tag{18.1.19}$$

but this is not generally the case. Renormalization group techniques were first applied in quantum electrodynamics, but the scalar field theory discussed here gives a more typical illustration of these methods, with separate renormalizations for fields and couplings.

18.2 The Sliding Scale

We have seen in the previous section that the large logarithms that appear at high energy in suitably integrated cross sections or off-shell Feynman amplitudes can be traced to the prescription used to define renormalized coupling constants and operators. The central idea of the renormalization group method is to change this prescription.

Suppose we find some way of defining a new kind of renormalized coupling constant $g(\mu)$, that depends on a sliding energy scale μ, but that (at least for $\mu \gg m$) has no dependence on the scale m of the masses of the theory. Then suitably integrated cross sections or other infrared-safe rate parameters may be expressed as functions of g_μ and μ instead of g_R. By dimensional analysis such functions may be written as

$$\Gamma(E, x, g_\mu, m, \mu) = E^D\, \Gamma\left(1, x, g_\mu, \frac{m}{E}, \frac{\mu}{E}\right). \tag{18.2.1}$$

(Our notation here is the same as in Section 18.1; in particular, x stands for all dimensionless angles, energy ratios, etc. on which Γ may depend.) Since μ is a completely arbitrary renormalization scale, we can choose $\mu = E$, in which case Eq. (18.2.1) reads

$$\Gamma(E, x, g_\mu, m, \mu) = E^D\, \Gamma\left(1, x, g_E, \frac{m}{E}, 1\right). \tag{18.2.2}$$

This now has no zero-mass singularities because g_E does not depend on m for $m \ll E$, so there are no large logarithms, and we can use perturbation theory to calculate Γ in terms of g_E as long as g_E itself remains sufficiently small. In particular, in any finite order of perturbation theory Γ has the asymptotic behavior, for $E \gg m$,

$$\Gamma(E, x, g_\mu, m, \mu) \longrightarrow E^D \, \Gamma(1, x, g_E, 0, 1)\,. \tag{18.2.3}$$

(Non-perturbative corrections are considered in Section 18.4.)

It remains to calculate g_E. For instance, in the scalar field theory with Lagrangian (18.1.2), we may define g_μ in terms of the value of the scattering amplitude at a renormalization point $s = t = u = -\mu^2$:

$$\begin{aligned} g_\mu &\equiv A(s = t = u = -\mu^2) \\ &= g - \frac{3g^2}{32\pi^2} \int_0^1 dx \left\{ \ln\left(\frac{\Lambda^2}{m^2 + \mu^2 x(1-x)} \right) - 1 \right\} + O(g^3) \end{aligned} \tag{18.2.4}$$

or, in terms of the conventional renormalized coupling (18.1.4),

$$g_\mu = g_R + \frac{3g_R^2}{32\pi^2} \int_0^1 dx \ln\left(1 + \frac{\mu^2 x(1-x)}{m^2} \right) + O(g_R^3)\,. \tag{18.2.5}$$

But this formula is reliable only if the correction term is smaller than g_R; that is, only if $|g_R \ln(\mu/m)| \ll 1$. If this were the case for $\mu \simeq E$, then we would not need the methods of the renormalization group; ordinary perturbation theory would be good enough.

Instead of using formulas like (18.2.5) directly for large μ, we must instead proceed in stages: g_μ may be calculated in terms of g_R as long as μ/m is not much larger than unity; then $g_{\mu'}$ may be calculated in terms of g_μ as long as μ'/μ is not much larger than unity; and so on, up to g_E. Instead of discrete stages, this may also be done continuously. Dimensional analysis tells us that the relation between $g_{\mu'}$ and g_μ takes the form

$$g_{\mu'} = G(g_\mu, \mu'/\mu, m/\mu)\,. \tag{18.2.6}$$

Differentiating with respect to μ' and then setting $\mu' = \mu$ yields the differential equation

$$\mu \frac{d}{d\mu} g_\mu = \beta\left(g_\mu, \frac{m}{\mu} \right), \tag{18.2.7}$$

where

$$\beta\left(g_\mu, \frac{m}{\mu} \right) \equiv \left[\frac{\partial}{\partial z} G(g_\mu, z, m/\mu) \right]_{z=1}. \tag{18.2.8}$$

There are no zero-mass singularities here, so for $\mu \gg m$ the differential equation becomes simply

$$\mu \frac{d}{d\mu} g_\mu = \beta(g_\mu, 0) \equiv \beta(g_\mu)\,, \tag{18.2.9}$$

which is often known as the *Callan–Symanzik equation.*[1] We are to calculate g_E by integrating the differential equation (18.2.9), with an initial value g_M at some scale $\mu = M$, chosen in practice large enough so that for $\mu \geq M$ we can neglect the masses m compared with μ, but small enough so that large logarithms $\ln(M/m)$ do not prevent us from using perturbation theory to calculate g_M in terms of the conventional renormalized coupling constant g_R. The solution may be formally written

$$\ln(E/M) = \int_{g_M}^{g_E} dg/\beta(g) \tag{18.2.10}$$

as long as $\beta(g)$ does not vanish between g_M and g_E.

The results of the previous paragraph do not rely on perturbation theory, but we usually need to use perturbation theory to calculate the functions G and β. As an example, suppose we calculate $g_{\mu'}$ in the scalar field theory with interaction Hamiltonian density $g\phi^4/24$, renormalizing by expressing g in terms of g_μ rather than g_R. Following the same procedure that led to Eq. (18.2.5), this gives

$$g_{\mu'} = g_\mu - \frac{3g_\mu^2}{32\pi^2} \int_0^1 dx \ln\left(\frac{m^2 + \mu^2 x(1-x)}{m^2 + \mu'^2 x(1-x)}\right) + O(g_\mu^3)\,.$$

Then Eq. (18.2.8) gives

$$\beta\left(g_\mu, \frac{m}{\mu}\right) = +\frac{3g_\mu^2}{16\pi^2} \int_0^1 dx \frac{\mu^2 x(1-x)}{m^2 + \mu^2 x(1-x)} + O(g_\mu^3)\,. \tag{18.2.11}$$

For $\mu \gg m$, this is

$$\beta(g_\mu) = \frac{3g_\mu^2}{16\pi^2} + O(g_\mu^3)\,. \tag{18.2.12}$$

In next order, the beta-function for $\mu \gg m$ is[3]

$$\beta(g_\mu) = g_\mu \left[3\left(\frac{g_\mu}{16\pi^2}\right) - \frac{17}{3}\left(\frac{g_\mu}{16\pi^2}\right)^2 + \cdots\right]\,.$$

If we are content with the one-loop approximation the calculation of $\beta(g)$ can be done even more easily. In order to avoid large radiative corrections in matrix elements at energies of order μ, we must write the bare coupling g in terms of a finite renormalized coupling g_μ, as

$$g = g_\mu + B(g_\mu) \ln \frac{\Lambda}{\mu} + \cdots\,. \tag{18.2.13}$$

For instance, from Eq. (18.1.3) we could have immediately read off that the $\ln \Lambda$ terms in g have coefficient

$$B(g) = -\frac{3}{2}\left[-i(2\pi)^4 g^2\right]\left[\frac{-i}{(2\pi)^4}\right]^2 2\pi^2 i = \frac{3g^2}{16\pi^2}\,. \tag{18.2.14}$$

The unrenormalized coupling is of course independent of μ, so to lowest order

$$\mu \frac{d}{d\mu} g_\mu - B(g_\mu) = 0$$

and so to lowest order

$$\beta(g) = B(g)\,. \tag{18.2.15}$$

With $B(g)$ given by Eq. (18.2.14), this agrees with our previous result (18.2.12) for $\beta(g)$ in the ϕ^4 scalar field theory.

Instead of using a simple ultraviolet cut-off, which can interfere with gauge invariance, it is often more convenient to deal with ultraviolet divergences by use of dimensional regularization. For a spacetime dimensionality $d < 4$, we find in place of $\ln(\Lambda/\mu)$ the convergent integral

$$\int_\mu^\infty k^{d-4}\frac{dk}{k} = \frac{\mu^{d-4}}{4-d} \xrightarrow[d\to 4]{} \left[\frac{1}{4-d} - \ln\mu\right]\,.$$

Thus instead of eliminating the cut-off dependence by writing the unrenormalized coupling constant as in (18.2.13), we instead write

$$g = g_\mu + B(g_\mu)\left[\frac{1}{4-d} - \ln\mu\right] \tag{18.2.16}$$

with the same function $B(g_\mu)$ as before. Thus in order to calculate $\beta(g_\mu)$, all we need to do is to pick out the coefficient of the singular factor $1/(4-d)$ in the renormalized coupling. This argument is extended to all orders of perturbation theory in Section 18.6.

As long as g_μ is sufficiently small in the scalar field theory with Lagrangian (18.1.2), the solution of Eqs. (18.2.9) and (18.2.12) can be well approximated by

$$g_\mu = -\frac{16\pi^2}{3\ln(\mu/M)}\,, \tag{18.2.17}$$

where M is an integration constant. This expression illustrates a common aspect of renormalization group calculations, that dimensionless couplings like g_R become replaced with parameters like M that have the dimensionality of mass. The value of M may be related to g_R by comparing the solution (18.2.17) with the behavior of the coupling for values of μ that are large enough to allow us to use approximations based on $\mu \gg m$, but small enough so that $|g_R \ln(\mu/m)| \ll 1$, where (18.2.5) gives

$$g_\mu \simeq g_R + \frac{3g_R^2}{16\pi^2} \ln\left(\frac{\mu}{m}\right) . \tag{18.2.18}$$

In this way, we find

$$M \simeq m \exp\left(\frac{16\pi^2}{3g_R}\right) , \tag{18.2.19}$$

so that Eq. (18.2.17) may be put in a more conventional form

$$g_\mu = g_R \left[1 - \frac{3g_R}{16\pi^2} \ln \frac{\mu}{m}\right]^{-1} . \tag{18.2.20}$$

To repeat, this is valid provided g_μ is small, even if $|g_R \ln(\mu/m)|$ is of order unity, so it represents a significant improvement over the perturbative result (18.2.18). Of course, the condition that g_μ should be small will become violated when $g_R \ln(\mu/m)$ is sufficiently close to the critical value $16\pi^2/3$. But at least Eq. (18.2.20) makes the unequivocal prediction that g_E becomes large enough to invalidate perturbation theory at some energy E *below* the critical value (18.2.19).

In calculating off-shell matrix elements of operators instead of integrated cross sections, we also need to take into account the N-factors that appear in the definition of the renormalized operators whose matrix elements are finite. We saw in the previous section that if these N-factors are defined in a conventional way (say, so that the correction factors produced by the divergent subgraphs are cancelled when the operator carries zero four-momentum, or is a field on its mass shell) then the formula for the N-factor involves zero-mass singularities as in Eq. (18.1.12) or Eq. (18.1.16), resulting in large logarithms at energies $E \gg m$. The cure is to define renormalization constants $N_\mu^{(\mathcal{O})}$ at a sliding scale μ, so that in matrix elements of the renormalized operator

$$\mathcal{O}_\mu = N_\mu^{(\mathcal{O})} \mathcal{O} \tag{18.2.21}$$

the correction factor produced by divergent subgraphs containing operator $\mathcal{O}$ are cancelled at a renormalization point characterized by four-momenta of order μ. If M_R is a matrix element of operators that are conventionally renormalized, and M is one in which the operators are renormalized as in Eq. (18.2.21), then for any μ

$$M_R = \left[\prod_{\mathcal{O}} \left(N^{(\mathcal{O})}/N_\mu^{(\mathcal{O})}\right)\right] M(E, x, g_\mu, m, \mu) . \tag{18.2.22}$$

We can again use dimensional analysis (assuming M has dimensionality

D) and set $\mu = E$, to write this as

$$M_R = E^D \left[\prod_{\mathcal{O}} \left(N^{(\mathcal{O})}/N_E^{(\mathcal{O})}\right)\right] M\left(1, x, g_E, \frac{m}{E}, 1\right) . \tag{18.2.23}$$

Thus to find the high energy behavior of an off-shell amplitude M_R, we need to know how N_μ varies with the renormalization scale μ.

For any two renormalization scales μ and μ', the renormalized operators $N_\mu^{(\mathcal{O})}\mathcal{O}$ and $N_{\mu'}^{(\mathcal{O})}\mathcal{O}$ both have finite matrix elements, so the ratio $N_{\mu'}^{(\mathcal{O})}/N_\mu^{(\mathcal{O})}$ must be cut-off independent. On dimensional grounds, this ratio must take the form

$$N_{\mu'}^{(\mathcal{O})}\Big/N_\mu^{(\mathcal{O})} = G^{(\mathcal{O})}(g_\mu, \mu'/\mu, m/\mu) . \tag{18.2.24}$$

Differentiating with respect to μ' and then setting $\mu' = \mu$ gives

$$\mu \frac{d}{d\mu} N_\mu^{(\mathcal{O})} = \gamma^{(\mathcal{O})}(g_\mu, m/\mu)\, N_\mu^{(\mathcal{O})} , \tag{18.2.25}$$

where

$$\gamma^{(\mathcal{O})}(g_\mu, m/\mu) \equiv \left[\frac{\partial}{\partial z} G^{(\mathcal{O})}(g_\mu, z, m/\mu)\right]_{z=1} . \tag{18.2.26}$$

The solution is

$$N_E^{(\mathcal{O})} \propto \exp\left[\int^E \gamma^{(\mathcal{O})}\left(g_\mu, \frac{m}{\mu}\right) \frac{d\mu}{\mu}\right] . \tag{18.2.27}$$

This is a useful result because the introduction of the sliding scale prevents the appearance of zero-mass singularities in $N_\mu^{(\mathcal{O})}$ and $N_{\mu'}^{(\mathcal{O})}$, and hence also in $G^{(\mathcal{O})}$ and $\gamma^{(\mathcal{O})}$. Hence as long as g_μ is small, there are no large logarithms that prevent the application of perturbation theory to calculate $\gamma^{(\mathcal{O})}$. Also, for $\mu \gg m$, $\gamma^{(\mathcal{O})}(g_\mu, m/\mu)$ has a smooth limit

$$\gamma^{(\mathcal{O})}(g_\mu) \equiv \gamma^{(\mathcal{O})}(g_\mu, 0) . \tag{18.2.28}$$

As an example, consider the operator $\mathcal{O} = \phi^2$ in the scalar field theory with interaction $g\phi^4/24$. Instead of renormalizing it so that the correction factor (18.1.9) is cancelled at $p^2 = 0$, we cancel it at a sliding scale $p^2 = \mu^2$, by introducing a new renormalized ϕ^2 operator $N_\mu^{(\phi^2)}\phi^2$, with

$$N_\mu^{(\phi^2)} \equiv F^{(\phi^2)}(\mu^2)^{-1} = 1 + \frac{g}{32\pi^2} \int_0^1 dx \left[\ln\left[\frac{\Lambda^2}{m^2 + \mu^2 x(1-x)}\right] - 1\right] + O(g^2).$$

Then the function (18.2.24) for this operator is

$$G^{(\phi^2)}\left(g_\mu, \frac{\mu'}{\mu}, \frac{m}{\mu}\right) \equiv \frac{N_{\mu'}^{(\phi^2)}}{N_\mu^{(\phi^2)}} = 1 + \frac{g_\mu}{32\pi^2}\int_0^1 dx \ln\left[\frac{m^2+\mu^2 x(1-x)}{m^2+\mu'^2 x(1-x)}\right] + O(g_\mu^2)\,.$$

(We can use g_μ here instead of g or $g_{\mu'}$ because the difference only affects the terms of order g_μ^2 or higher.) From Eq. (18.2.26), we have then

$$\gamma^{(\phi^2)}\left(g_\mu, \frac{m}{\mu}\right) = -\frac{g_\mu}{16\pi^2}\int_0^1 \frac{\mu^2 x(1-x)}{m^2+\mu^2 x(1-x)}\,dx + O(g_\mu^2)$$

or for $\mu \gg m$,

$$\gamma^{(\phi^2)}(g_\mu) = -\frac{g_\mu}{16\pi^2} + O(g_\mu^2)\,. \tag{18.2.29}$$

Another good example is provided by the N-factor associated with the renormalization of the electromagnetic field in quantum electrodynamics. Recall that the photon propagator can be made finite for all momenta by evaluating it for renormalized electromagnetic fields, or equivalently by multiplying the propagator of the unrenormalized field by Z_3^{-1}:

$$\tilde{\Delta}_{\rho\sigma}(q) = Z_3^{-1}\Delta_{\rho\sigma}(q)\,. \tag{18.2.30}$$

Eq. (10.5.17) shows that this renormalized propagator may be written

$$\tilde{\Delta}_{\rho\sigma}(q) = \frac{\eta_{\rho\sigma}}{[q^2-i\epsilon][1-\pi(q^2)]} + q_\rho q_\sigma\text{-terms}\,. \tag{18.2.31}$$

Suppose we instead define a renormalized field $N_\mu^{(A)}A^\rho$, whose propagator has a term proportional to $\eta_{\rho\sigma}/[q^2-i\epsilon]$ with a coefficient that is equal to unity at a sliding renormalization scale $q^2=\mu^2$. For this purpose, we must clearly take

$$N_\mu^{(A)} = Z_3^{-1/2}[1-\pi(\mu^2)]^{1/2}\,. \tag{18.2.32}$$

Using Eq. (11.2.22), the function (18.2.24) is then

$$G^{(A)}(g_\mu, \mu'/\mu, m/\mu) = \left[\frac{1-\pi(\mu'^2)}{1-\pi(\mu^2)}\right]^{1/2}$$
$$= 1 - \frac{e_\mu^2}{4\pi^2}\int_0^1 dx\, x(1-x)\ln\left[\frac{m^2+\mu'^2 x(1-x)}{m^2+\mu^2 x(1-x)}\right] + O(e_\mu^4) \tag{18.2.33}$$

and so Eq. (18.2.26) gives

$$\gamma^{(A)}(e_\mu, m/\mu) = -\frac{e_\mu^2}{2\pi^2}\int_0^1 dx\,\frac{x^2(1-x)^2\mu^2}{m^2+\mu^2 x(1-x)} + O(e_\mu^4)\,. \tag{18.2.34}$$

As promised, this has a smooth limit for $\mu \gg m$

$$\gamma^{(A)}(e_\mu) \equiv \gamma^{(A)}(e_\mu, 0) = -\frac{e_\mu^2}{12\pi^2} + O(e_\mu^4)\,. \tag{18.2.35}$$

As already mentioned, electrodynamics is a special case, because the renormalization constant $Z_3^{-1/2}$ in the definition of the renormalized electromagnetic field is just the reciprocal of the constant used to define the renormalized electric charge of the electron: $e_R = Z_3^{1/2} e$. The natural definition of the renormalized electric charge at a sliding scale μ is then

$$e_\mu = N_\mu^{(A)-1} e = Z_3^{-1/2} N_\mu^{(A)-1} e_R\,, \tag{18.2.36}$$

so that e_μ times the field $N_\mu^{(A)} A^\rho$ renormalized at scale μ is independent of μ. From Eq. (18.2.25) we see that the function $\beta(e)$ which gives the μ dependence of e_μ according to Eq. (18.2.9) for $\mu \gg m$ has the value

$$\beta(e) = -e\gamma^{(A)}(e) = \frac{e^3}{12\pi^2} + O(e^5)\,. \tag{18.2.37}$$

Using an earlier calculation[3a] of the fourth-order term in the vacuum polarization function $\pi(q^2)$, Gell-Mann and Low were able also to give the term in $\beta(e)$ of next order in e:

$$\beta(e) = \frac{e^3}{12\pi^2} + \frac{e^5}{64\pi^2} + O(e^7)\,. \tag{18.2.38}$$

In other words, the electric charge at a sliding scale μ satisfies the renormalization group equation

$$\mu \frac{d}{d\mu} e_\mu = \frac{e_\mu^3}{12\pi^2} + \frac{e_\mu^5}{64\pi^2} + O(e_\mu^7)\,. \tag{18.2.39}$$

This shows, for small e_μ, that e_μ *increases* with increasing μ.

We also need an initial condition. This is provided by the known value of the conventionally renormalized charge $e_R = Z_3^{1/2} e$, for which $\alpha \equiv e_R^2/4\pi = 1/137.036\ldots$. Eqs. (18.2.32) and (18.2.36) yield

$$\begin{aligned} e_R/e_\mu &= Z_3^{1/2} N_\mu^{(A)} = \sqrt{1-\pi(\mu^2)} \\ &= 1 - \frac{e_R^2}{4\pi^2}\int_0^1 dx\, x(1-x)\ln\left[1 + \frac{\mu^2 x(1-x)}{m_e^2}\right] + O(e_R^4)\,. \end{aligned} \tag{18.2.40}$$

We need to match this with the solution of Eq. (18.2.39) at a value of μ which is large enough to justify the approximation $\mu \gg m_e$ in (18.2.39) but small enough so that the logarithm in Eq. (18.2.40) is still small enough compared with $4\pi^2/e_R^2$ to justify the use of perturbation theory. (For instance, we might take μ to be of order 100 MeV.) For such values of μ, Eq. (18.2.40) gives

$$e_\mu \simeq e_R + \frac{e_R^3}{12\pi^2}\left[\ln\frac{\mu}{m_e} - \frac{5}{6}\right]. \tag{18.2.41}$$

On the other hand, the solution of Eq. (18.2.39) for e_μ small (keeping only the leading term on the right-hand side) is

$$e_\mu = \left[\text{constant} - \frac{\ln\mu}{6\pi^2}\right]^{-1/2}. \tag{18.2.42}$$

Comparing Eqs. (18.2.41) and (18.2.42) gives the solution

$$e_\mu = e_R\left[1 - \frac{e_R^2}{6\pi^2}\left(\ln\left(\frac{\mu}{m_e}\right) - \frac{5}{6}\right)\right]^{-1/2}. \tag{18.2.43}$$

Unlike Eq. (18.2.41), Eq. (18.2.43) is valid as long as $e_\mu^2/6\pi^2$ is small, whether or not $(e_R^2/6\pi^2)\ln(\mu/m_e)$ is small.

For instance, we have already seen in Section 11.3 that the leading fourth-order radiative correction to the magnetic moment of the muon can be obtained by multiplying the second-order (Schwinger) term (11.3.16) by the vacuum polarization function $\pi_e(k^2)$ at $k^2 \approx m_\mu^2$, which according to Eq. (18.2.40) is the same (to this order) as using $e_{m_\mu}^2/4\pi$ in place of α in the Schwinger term.

Another example: Experiments at high energy electron–positron colliders such as LEP at CERN or SLC at SLAC now study physical processes at energies of the order of the mass of the Z^0 particle, or 91 GeV. Eq. (18.2.43) shows that at these energies, radiative corrections in pure quantum electrodynamics should be calculated using a value for the fine structure constant which is not $\alpha = 1/137.036$ but rather

$$\frac{e^2(91\,\text{GeV})}{4\pi} = \frac{\alpha}{1 - 2(11.25)\alpha/3\pi} = \frac{1}{134.6}. \tag{18.2.44}$$

This is for a theory in which electrons are the only charged particles with masses below m_Z. In the real world there are many such particle types, and the effective fine structure constant[4] at m_Z is $(128.87 \pm 0.12)^{-1}$.

* * *

The sliding scale at which the coupling parameters of a theory are calculated may be the value of an external *field* rather than the momentum of an external line. In one of the early applications of the renormalization group method, Coleman and E. Weinberg[4a] considered the effective potential $V(\phi)$ for a spacetime-independent external scalar field ϕ. In the simple case where the field interacts with itself alone, their one-loop result is given by Eq. (16.2.15). It is particularly interesting to consider this potential in the case where the renormalized mass m_R vanishes, where to one-loop order we may set $\mu^2(\phi)$ in the last term equal to $g_R\phi^2/2$, so that Eq. (16.2.15) here reads (with g a slightly redefined coupling):

$$V(\phi) = \lambda_R + \frac{g}{24}\phi^4 + \frac{g^2\phi^4 \ln \phi^2}{256\pi^2} . \tag{18.2.45}$$

This looks like at first sight as if for $g > 0$ the potential becomes less than λ_R for very small ϕ, so that the point $\phi = 0$ is a local maximum instead of a minimum, but for such small values of ϕ the third term is larger than the second, and the perturbation theory is obviously untrustworthy. Also, we would like to be able to argue that g must be positive in order for the potential to be bounded below at large fields, but Eq. (18.2.45) shows that however small g is, there is some sufficiently large ϕ where perturbation theory breaks down, and hence Eq. (18.2.45) cannot be trusted to tell us whether the potential is bounded below at large fields.

We can do much better by using a coupling constant defined at a sliding scale μ of field strength. Suppose we define a coupling g_μ by the condition that

$$V(\mu) = \lambda_R + \frac{g_\mu}{24}\mu^4 . \tag{18.2.46}$$

If we had used g_μ as the coupling parameter from the beginning, then in place of Eq. (18.2.45) we would have obtained

$$V(\phi) = \lambda_R + \frac{g_\mu}{24}\phi^4 + \frac{g_\mu^2\phi^4}{256\pi^2} \ln\left(\frac{\phi^2}{\mu^2}\right) , \tag{18.2.47}$$

which obviously satisfies Eq. (18.2.46).* The renormalization group equation for g_μ can be obtained from the condition that this effective potential

* Eq. (18.2.47) differs from what we would get from Eq. (18.2.45) by simply using Eq. (18.2.46) to express g in terms of g_μ, in that the last term is proportional to g_μ^2 rather than g^2. The difference is of higher order in g, but can become significant if $V(\phi)$ is evaluated at ϕ very different from μ, where large logarithms can compensate for powers of coupling. If we use g_μ as the coupling parameter from the beginning and take μ to be of order ϕ then no such large logarithms occur, and the approximation Eq. (18.2.47) is valid as long as g_μ remains small.

is independent of μ,**

$$\mu \frac{dg_\mu}{d\mu} = \frac{3g_\mu^2}{16\pi^2} \,. \tag{18.2.48}$$

(Terms involving the derivative of g_μ^2 are dropped here, because they are of higher order in g_μ, and are hence negligible as long as g_μ is sufficiently small.) It is not a coincidence that this takes the same form as the renormalization group equation (18.2.9), (18.2.12), where μ was a renormalization momentum, because as we shall see in the next section the first two terms in the renormalization group equation are always independent of the way that we define the sliding scale. The solution of this equation is given by Eq. (18.2.17), in general with a different integration constant M. Hence by taking $\mu = \phi$ in Eq. (18.2.17), we now have

$$V(\phi) = \lambda_R - \frac{4\pi^2\phi^4}{9\ln(\phi^2/M^2)} \,, \tag{18.2.49}$$

$$g_\phi = -\frac{32\pi^2}{3\ln(\phi^2/M^2)} \,. \tag{18.2.50}$$

This result should be used with some care, because it is only valid where the coupling constant g_ϕ is small. The problem is not just that Eq. (18.2.49) loses its validity for ϕ near M; we also cannot integrate the renormalization group equation through the singularity at $\phi = M$, so a knowledge of g_ϕ on one side of this singularity tells us nothing about its behavior on the other side.

If g_ϕ is found to have a small *positive* value for some ϕ_0 then from Eq. (18.2.50) we know that $M > |\phi_0|$, so g_ϕ remains small and Eq. (18.2.49) is valid for $|\phi| < |\phi_0|$. This shows that the point $\phi = 0$ is a local *minimum* of $V(\phi)$, contrary to what Eq. (18.2.45) might have led us to suppose. This means that the vacuum which is invariant under the symmetry transformation $\phi \rightarrow -\phi$ is stable in this model, aside from the possibility of quantum mechanical barrier penetration. On the other hand, we do not know that Eq. (18.2.49) becomes valid for $|\phi|$ sufficiently large compared with M. The coupling might remain too large to use perturbation theory for all $|\phi| > M$, and even if it becomes small for some $|\phi| > M$, the potential might be given for such ϕ by Eq. (18.2.49) with a renormalization

** To eliminate all cut-off dependence, ϕ should be written in terms of a renormalized field $\phi_\mu = N_\mu^\phi \phi$, which gives $V(\phi_\mu)$ a μ dependence arising from the μ dependence of the renormalization constant N_μ^ϕ. This point is ignored here, because in the scalar field theory with interaction $\propto \phi^4$ the lowest-order graphs contributing to the μ dependence of N_μ^ϕ have two loops, so that to the order of the calculations presented here we can take $N_\mu^\phi = 1$.

scale $M' > \phi$, producing a second singularity. Thus in this case we cannot conclude that $V(\phi) \to -\infty$ for $|\phi| \to \infty$.

Similarly, if g_ϕ is found to have a small *negative* value for some ϕ_0 then we know that $M < |\phi_0|$, so g_ϕ remains small and Eq. (18.2.49) is valid for $|\phi| > |\phi_0|$. In this case we cannot conclude anything about the behavior of the potential for $|\phi| < M$, but here we *can* use Eq. (18.2.49) to see that $V(\phi) \to -\infty$ for $|\phi| \to \infty$, ruling out the possibility of any stable vacuum. Because we are here considering the limit $|\phi| \to \infty$, this conclusion holds also for scalar fields of mass $m > 0$, provided $m \ll |\phi_0|$. This is why it is necessary to assume that the ϕ^4 coupling (renormalized at any scale much larger than the scalar mass) is positive.

18.3 Varieties of Asymptotic Behavior

The renormalization group method provides useful insight into the types of possible asymptotic behavior that are encountered in quantum field theories, even in cases where the running coupling g_μ does not remain small enough to allow the use of perturbation theory. We will distinguish four different ways that g_μ may behave for $\mu \to \infty$, that correspond to four different shapes of the function $\beta(g)$ in theories with a single coupling constant. In the next section we will take up the case of theories with several independent couplings.

Let's first recall the results for $\beta(g)$ obtained for the two examples considered in the previous section. One of these is the scalar field theory with interaction $g\phi^4/24$, for which the beta-function for small g is

$$\beta(g) = \frac{3g^2}{16\pi^2} - \frac{17}{3}\frac{g^3}{(16\pi^2)^2} + O(g^4)\,. \tag{18.3.1}$$

The other is quantum electrodynamics. Instead of writing the beta-function here in the form (18.2.38), we will emphasize the similarities between this theory and the scalar field theory by writing

$$g \equiv e^2\,, \tag{18.3.2}$$

with $\beta(g_\mu)$ now understood to be $\mu\, dg_\mu/d\mu$, so that for small g

$$\begin{aligned}\beta(g) &= 2e\left[\frac{e^3}{12\pi^2} + \frac{e^5}{64\pi^2} + O(e^7)\right] \\ &= \frac{g^2}{6\pi^2} + \frac{g^3}{32\pi^2} + O(g^4)\,.\end{aligned} \tag{18.3.3}$$

Note that in both cases the physically allowed coupling constants fall in the range $g \geq 0$, where $\beta(g) \geq 0$ for small g. In electrodynamics this is simply

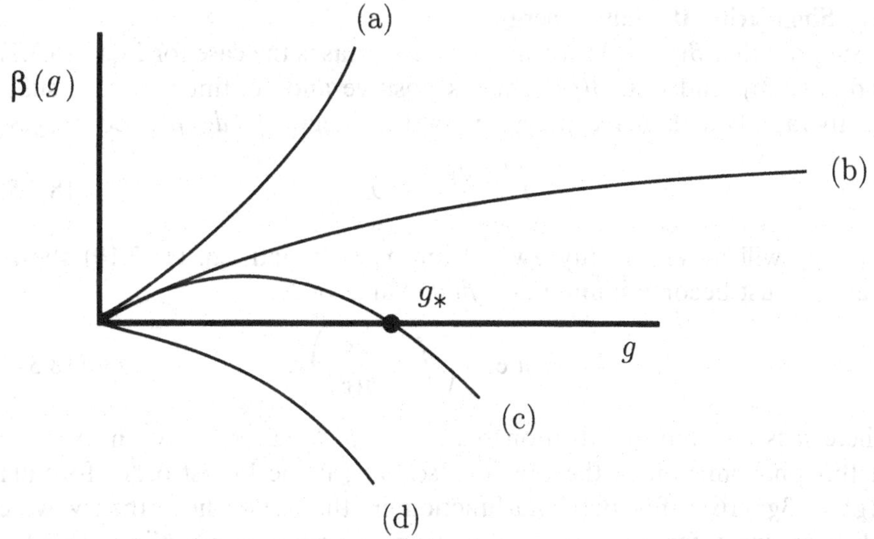

Figure 18.4. Schematic representation of four possible forms of the function $\beta(g)$. For such forms of $\beta(g)$, the running coupling g_μ would: (a) approach infinity at a finite value of μ; (b) continue to grow as μ increases; (c) approach a finite limit g_* for $\mu \to \infty$; (d) approach zero for $\mu \to \infty$.

because the reality of the Lagrangian requires e to be real. In the scalar field theory, as we saw at the end of the previous section, it is necessary to have $g > 0$ in order to have any stable vacuum state. However, there are other examples that have $\beta(g) \leq 0$ for $g \geq 0$. For instance, we can consider a scalar field theory with interaction Hamiltonian $-g\phi^4/24$, with g taken positive. This may be unphysical, but stability problems will not bother us as long as we stick to perturbation theory. Eq. (18.2.9) shows that if we redefine $g \to -g$ the beta-function undergoes the change $\beta(g) \to -\beta(-g)$, so our previous result that $\beta(g) = 3g^2/16\pi^2 + O(g^3)$ for an interaction $g\phi^4/24$ now gives

$$\beta(g) = -3g^2/16\pi^2 + O(g^3) \tag{18.3.4}$$

for an interaction Hamiltonian density $-g\phi^4/24$. Of greater relevance to physics, we shall see in Section 18.7 that non-Abelian gauge theories with not too many spinor fields have $\beta \leq 0$ for small positive gauge coupling constants. In what follows we will always define the coupling g so that $g \geq 0$, but we shall consider the cases of $\beta(g)$ either positive or negative for small g.

Now let us turn to our list of possibilities. (See Figure 18.4.)

(a) **Singularity at Finite Energy**

Suppose that $\beta(g) > 0$ for small positive g (as is the case for Eqs. (18.3.1) and (18.3.3)), and that $\beta(g)$ remains positive and continues to rise sufficiently rapidly with increasing g, so that the integral $\int^\infty dg/\beta(g)$ converges:

$$\int^\infty \frac{dg}{\beta(g)} < \infty . \tag{18.3.5}$$

Then g_μ will move steadily away from $g_\mu = 0$, and Eq. (18.2.10) shows that g_E must become infinite at a *finite* value of E:

$$E_\infty = \mu \exp\left(\int_{g_\mu}^\infty \frac{dg}{\beta(g)}\right) , \tag{18.3.6}$$

where μ is any renormalization scale with $\mu \gg m$. We saw an example of this phenomenon in the previous section; if the lowest-order formula $\beta(g) = 3g^2/16\pi^2$ for the beta-function in the scalar field theory were taken as exact for all values of g, then the running coupling (18.2.17) would become infinite at the energy (18.2.19). Similarly, if the lowest order formula $\beta(g) = g^2/6\pi^2$ (with $g \equiv e^2$) for the beta-function in spinor quantum electrodynamics were taken as exact for all g, then the energy (18.3.6) at which g_E and e_E become infinite would be

$$E_\infty = \mu \exp(6\pi^2/g_\mu) . \tag{18.3.7}$$

Using Eq. (18.2.43), we may express this in terms of the conventional renormalized charge:

$$E_\infty \simeq m_e \exp\left(\frac{6\pi^2}{e_R^2} + \frac{5}{6} + O(e_R^2)\right) = e^{646.6} m_e . \tag{18.3.8}$$

Of course, the approximation that $\beta(g) = g^2/6\pi^2$ will break down before this energy is reached, so all we can say with confidence is that e_E will become large enough to invalidate perturbation theory at some energy E below E_∞.

(b) **Continued Growth**

Now suppose that in some theory $\beta(g)$ remains positive definite for $g \to \infty$, but rises slowly enough (or decreases) so that $\int^\infty dg/\beta(g)$ is divergent. The coupling constant g_E then continues to increase as $E \to \infty$, but becomes infinite only for $E = \infty$. Furthermore, the leading term in the asymptotic behavior of g_E for $E \to \infty$ is independent of the conventionally renormalized coupling. For instance, if $\beta(g)$ behaves for large g like bg^k with $b > 0$ and $k < 1$, then the solution of Eq. (18.2.9) is

$$g_E = \left[1 + (1-k)b\, g_\mu^{k-1} \ln\frac{E}{\mu}\right]^{1/(1-k)} g_\mu . \tag{18.3.9}$$

If g_μ is small for some μ (say, of order m) then the growth of g_E is seen only at energies which are exponentially large compared with this μ. However, in the extreme high energy limit the coupling grows according to

$$g_E \to [(1-k)b \ln E]^{1/(1-k)} , \tag{18.3.10}$$

a limiting behavior independent of g_μ!

(c) **Fixed Point at Finite Coupling**

Suppose next that $\beta(g)$ remains positive-definite for $0 < g < g_*$, but drops to zero at $g = g_*$ and is negative thereafter. Then Eq. (18.2.9) dictates that as μ increases g_μ will increase for $g_\mu < g_*$ and decrease for $g_\mu > g_*$, in either case approaching the fixed point g_* for $\mu \to \infty$. If the zero of $\beta(g)$ at g_* is simple, then in the neighborhood of this point we have

$$\beta(g) \to a\,(g_* - g) \quad \text{for} \quad g \to g_* \tag{18.3.11}$$

with $a > 0$. The solution of Eq. (18.2.9) is then

$$g_* - g_\mu \propto \mu^{-a} . \tag{18.3.12}$$

(The behavior of type (b) described above may be regarded as the special case where the fixed point g_* is at infinity.) Also, $\gamma(g)$ for a general operator $\mathcal{O}$ may be expected to behave smoothly near g_*:

$$\gamma(g) = \gamma(g_*) + c(g_* - g) + O\left((g_* - g)^2\right) . \tag{18.3.13}$$

(We are here dropping the label $\mathcal{O}$ on γ and c.) Hence in matrix elements of this (and perhaps other) operators, we encounter a factor (see Eq. (18.2.27))

$$N_E^{-1} \propto \exp\left[-\int^E \gamma(g_\mu)\,\frac{d\mu}{\mu}\right] \propto E^{-\gamma(g_*)}\left[1 + O(E^{-a})\right] . \tag{18.3.14}$$

The product of the factors $E^{-\gamma(g_*)}$ can be lumped together with the factor E^D in Eq. (18.2.23), with the result that the whole matrix element goes as

$$M_R \propto E^{D_*} , \tag{18.3.15}$$

where the dimensionality D_* is calculated adding an 'anomalous dimension' $-\gamma(g_*)$ to the actual dimensionality of each operator appearing in the matrix element.

(d) **Asymptotic Freedom**

In the examples that have been discussed so far, $\beta(g)$ was positive for small positive g, so that g_μ is driven away from $g = 0$ as μ increases. Suppose that for some other theory $\beta(g)$ is negative for small positive g.

Then

$$\beta(g) \to -b\,g^n\,, \tag{18.3.16}$$

where $b > 0$. Here n is the order of the lowest-order diagrams that contribute to $\beta(g)$, and hence is always an integer greater than unity. (In the theories used as examples here, $n = 2$.) The solution of Eq. (18.2.9) here is

$$g_E = g_\mu \left[1 + b(n-1)g_\mu^{n-1} \ln \frac{E}{\mu}\right]^{-1/(n-1)}\,. \tag{18.3.17}$$

For $E \to \infty$, this has a limit independent of g_μ:

$$g_E \to [b(n-1)\ln E]^{-1/(n-1)}\,. \tag{18.3.18}$$

Since this gives a vanishing g_E for $E \to \infty$, we can trust perturbation theory in this limit, provided only that g_E for some finite E is within the region around $g = 0$ where g and $\beta(g)$ have opposite signs. The anomalous dimensions $\gamma^{(\mathcal{O})}$ of various operators $\mathcal{O}$ are expected to have the weak-coupling behavior (dropping the label $\mathcal{O}$)

$$\gamma(g) \to c\,g^m\,, \tag{18.3.19}$$

where m is the order of the lowest-order diagrams that contribute to the renormalization of the operator, and c is a real constant that can be positive or negative. The asymptotic behavior at high energies of the factor introduced into matrix elements by renormalization of this operator is then

$$\begin{aligned} N_E^{-1} &\propto \exp\left[-\int^E \gamma(g_\mu)\frac{d\mu}{\mu}\right] \\ &\to \exp\left[-c\int^E [b(n-1)\ln\mu]^{-m/(n-1)}\,\frac{d\mu}{\mu}\right] \\ &\propto \exp\left[-\frac{c[b(n-1)]^{-m/(n-1)}}{(1-m/(n-1))}(\ln E)^{1-m/(n-1)}\right] \end{aligned} \tag{18.3.20}$$

except that for $m = n-1$

$$N_E^{-1} \propto (\ln E)^{-c/b(n-1)}\,. \tag{18.3.21}$$

We see that in the case of asymptotic freedom there are no corrections to those effective dimensionalities that determine the powers of energy appearing in the asymptotic behavior of the matrix element, but this asymptotic behavior is instead modified by powers of $\ln E$.

For a toy model that exhibits asymptotic freedom, we can use the scalar field theory with interaction Hamiltonian density $-g\phi^4/24$, with g taken positive. Eq. (18.3.4) here gives the parameters of Eq. (18.3.16) as:

$$b = 3/16\pi^2 , \qquad n = 2 , \tag{18.3.22}$$

so Eq. (18.3.17) gives, for $E \to \infty$,

$$g_E \to \left[\frac{3 \ln E}{16\pi^2}\right]^{-1} . \tag{18.3.23}$$

Also, the operator ϕ^2 in this theory has an anomalous dimension given by Eq. (18.2.29) as

$$\gamma(g) = -\frac{g}{16\pi^2} + O(g^2) . \tag{18.3.24}$$

Hence Eq. (18.3.19) applies here, with

$$c = -1/16\pi^2 , \qquad m = 1 . \tag{18.3.25}$$

Each ϕ^2 operator in a matrix element therefore contributes a factor given by Eq. (18.3.21) as

$$N_E^{-1} \propto (\ln E)^{1/3} . \tag{18.3.26}$$

The scalar field ϕ itself in this theory has $\gamma(g) \propto g^2$, and hence $m = 2$, so each ϕ operator in a matrix element therefore contributes a factor given by Eq. (18.3.20) as

$$N_E^{-1} \propto 1 + O\left(\frac{1}{\ln E}\right) . \tag{18.3.27}$$

We will see another, more physical, example of asymptotic freedom when we take up quantum chromodynamics in Section 18.7.

In all cases where g_E can be extended to infinite energy, its behavior in this limit turns out to be independent of the renormalized coupling g_R. However, this does not necessarily mean that the theory involves no arbitrary dimensionless parameters. In all cases, in order to describe how g_E approaches its limit for $E \to \infty$, we need to specify a free parameter λ with the dimensions of energy. For case (b), Eq. (18.3.10) may be written

$$g_E \to \left[(1-k)b \ln(E/\lambda)\right]^{1/(1-k)} .$$

For case (c), Eq. (18.3.12) may be written

$$g_E \to g_* \left[1 - \left(\frac{\lambda}{E}\right)^b\right] .$$

Finally, for case (d), Eq. (18.3.18) may be written

$$g_E \to \left[b(n-1)\ln(E/\lambda)\right]^{-1/(n-1)} .$$

Such theories in general do have a free dimensionless parameter: the ratio of λ to the mass m. Coupling constants like e_R that are renormalized at scales tied to m may be expressed as functions of m/λ. It is only when all masses in a theory vanish that we can say that the theory has no free dimensionless parameters.

Of the four types of asymptotic behavior described here, types (a) and (b) lead to the apparently unphysical behavior that the running coupling g_E becomes infinite, either at a finite energy (case (a)) or for $E \to \infty$ (case (b)). This does not in itself mean disaster: we have to look at how the coupling is defined. For instance, if g_μ drops smoothly from a finite value g_m at $\mu = m$ to zero for $\mu \to \infty$, and we define a new coupling $\tilde{g}_\mu \equiv g_\mu/[1 - g_\mu/g_{2m}]$, then $\tilde{g}_\mu$ becomes infinite at $\mu = 2m$, but this is just an artifact of this particular choice of coupling parameter. However, the conventional renormalized couplings g_μ in both the ϕ^4 scalar field theory and quantum electrodynamics have been defined here in terms of the values of matrix elements at energies of order μ. Specifically, g_μ in ϕ^4 scalar field theory is defined as the invariant Feynman amplitude A for scalar–scalar scattering at $s = t = u = \mu^2$, where A is supposed to be analytic. Also, $g_\mu = e_\mu^2$ in spinor electrodynamics is given by

$$e_\mu^2/e_R^2 = Z_3\Big/N_\mu^{(A)2} = \Big[1 - \pi(\mu^2)\Big]^{-1} . \tag{18.3.28}$$

An infinity in e_μ^2 at a point μ_∞ would therefore produce a pole or other singularity in the renormalized photon propagator at a positive value of p^2, that is, at $p^2 = \mu_\infty^2$, where the propagator is supposed to be analytic. Thus, with g_μ as defined here, the type of asymptotic behavior described in case (a) *is* ruled out physically.

How then do our various quantum field theories behave? Years ago, Landau[4b] argued that in quantum electrodynamics the increasing powers of $\ln(E/M)$ encountered at each order of perturbation theory would add up to give singularities (so-called 'Landau ghosts') at finite values of E. In modern terms, Landau could be said to have discovered possibility (a) above, but he did not give any argument against cases (b) or (c).

Nevertheless, there is today a widespread view that interacting quantum field theories that are not asymptotically free, like quantum electrodynamics or the scalar field theory with ϕ^4 interaction, are not mathematically consistent. In quantum electrodynamics there is some evidence against case (c), the existence of a finite fixed point e_*. Such a fixed point would only be possible[4c] if non-perturbative effects changed the qualitative nature of the operator product expansion, the subject of Chapter 20, or if there were a non-perturbative renormalization of the triangle anomaly discussed in Chapter 22. But even if case (c) is indeed ruled out in quantum electrodynamics, there is still the possibility of case (b), a fixed point at infinite coupling.

Most of the evidence against the consistency of interacting non-asymptotically free quantum field theories comes from the study of the scalar field theory in four spacetime dimensions with ϕ^4 interaction, quantized on a finite spacetime lattice. There are rigorous theorems[4d] to the effect that this theory (with arbitrary dependence of the parameters of the theory on lattice spacing) does not have an interacting continuum theory as its limit for zero lattice spacing unless the theory is asymptotically free, which of course is contrary to what is found for this theory in perturbation theory. This argument also seems inconclusive. It is true that if there were a consistent continuum scalar field theory that is not asymptotically free, then it would be possible to construct a lattice theory by integrating out the values of the scalar field at all points except on a spacetime lattice. *But this would not be the lattice theory that is considered in these theorems.* It would be a lattice theory with every possible coupling allowed by symmetry principles — not just a term proportional to ϕ^4, but also terms proportional to ϕ^6, $\phi^3\Box\phi$, etc, with coefficients having a dependence on the cut-off (the inverse lattice spacing) governed by the Wilson renormalization group equations discussed in Section 12.4. No one has proved anything about the continuum limit of such a theory.

If it is really true that there is no interacting continuum scalar field theory in the limit of zero lattice spacing, then we must encounter some obstruction when we try to solve the Wilson renormalization group equations, which for weak renormalized couplings would have to be at very small lattice spacings. Such a theory would appear like an interacting continuum field theory unless examined at very short distances. The renormalized coupling constant in this approximate continuum theory presumably has a singularity at finite energy, as in case (a) above, so that it too breaks down at short distances. (But for strong couplings there is no direct connection between the forms of the Wilson and Gell-Mann–Low renormalization group equations, so the existence of a singularity in the bare couplings at finite lattice spacing does not necessarily imply a singularity in the renormalized coupling constants at a finite renormalization scale.)

Theories of this sort are sometimes called *trivial*, either because, under various assumptions about the bare couplings of the theory quantized on a lattice, the continuum limit turns out to be a free field theory, or because the only way to make a continuum theory of type (a) physically satisfactory at all energies is to adopt the solution $g_\mu = 0$ of the renormalization group equation (18.2.9). Even if a field theory is trivial in either sense, there is no reason not to include it as part of a realistic theory of physical phenomena. The existence of an obstruction to the solution of the Wilson renormalization group equations for a field theory at very small lattice spacings is not important if in the real world there are other fields that must also be taken into account at such short distances. Similarly, the fact that a given quantum field theory has unphysical singularities at some large energy E_∞ is not a physical problem if the theory in question is only a low-energy approximation to a larger theory, an approximation valid only at energies far below E_∞. In particular, long before we reach the energies near (18.3.8) where quantum electrodynamics could be expected to become singular, it becomes necessary to take even gravitation into account, and no one knows how to calculate the effects of strong gravitational forces at such energies.

Despite these reassuring remarks, it is possible that in order to avoid unphysical singularities, all our separate quantum field theories like spinor quantum electrodynamics will eventually have to be integrated into an asymptotically free theory. Fortunately, the question of whether a theory is asymptotically free for *some* finite range of coupling constants can be settled by perturbative calculations: if $\beta(g)$ is negative as $g \to 0+$, then the theory is asymptotically free for all renormalized couplings g_μ lying between zero and the first zero of $\beta(g)$.

* * *

In this connection, it is worth noting that although the detailed form of $\beta(g)$ depends on the gauge and on precisely how the running coupling is defined, the first two terms in the power series for $\beta(g)$ do not. Suppose we have two definitions g_μ and $\tilde{g}_\mu$ of the running coupling, perhaps employing different definitions of the renormalization scale μ or different gauges. Since both g_μ and $\tilde{g}_\mu$ are dimensionless and cut-off independent, there is no way that $\tilde{g}_\mu$ for $\mu \gg m$ can depend on anything but g_μ:

$$\tilde{g}_\mu = \tilde{g}(g_\mu)\,.$$

We then have

$$\tilde{\beta}(\tilde{g}_\mu) \equiv \mu \frac{d}{d\mu}\tilde{g}_\mu = \frac{d\tilde{g}(g_\mu)}{dg_\mu}\,\beta(g_\mu)$$

and so

$$\tilde{\beta}(\tilde{g}) = \frac{d\tilde{g}(g)}{dg}\,\beta(g)\,. \tag{18.3.29}$$

As long as we are sticking to the same definition of the unrenormalized coupling, all renormalized couplings are equal in lowest order, so the power series for $\tilde{g}$ in terms of g may be written

$$\tilde{g}(g) = g + a\,g^2 + O(g^3)$$

or, equivalently,

$$g = \tilde{g} - a\,\tilde{g}^2 + O(\tilde{g}^3)\,.$$

The derivative is

$$\frac{d\tilde{g}}{dg} = 1 + 2\,a\,g\, + O(g^2) = 1 + 2\,a\,\tilde{g} + O(\tilde{g}^2)\,.$$

Also, for the couplings we have been considering here (including $g = e^2$) the power series for $\beta(g)$ takes the form

$$\beta(g) = b\,g^2 + b'g^3 + O(g^4)$$

or in terms of $\tilde{g}$

$$\beta(\tilde{g}) = b\tilde{g}^2 + (b' - 2ab)\,\tilde{g}^3 + O(\tilde{g}^4)\,.$$

From (18.3.29), we have then

$$\begin{aligned}\tilde{\beta}(\tilde{g}) &= \left[1 + 2a\,\tilde{g} + O(\tilde{g}^2)\right]\left[b\tilde{g}^2 + (b' - 2ab)\tilde{g}^3 + O(\tilde{g}^4)\right]\\ &= b\tilde{g}^2 + b'\tilde{g}^3 + O(\tilde{g}^4)\,. \end{aligned} \tag{18.3.30}$$

We see that the first two terms in the power series for $\tilde{\beta}$ in terms of $\tilde{g}$ have the same coefficients as in the power series for β in terms of g. However, this is definitely not the case for the higher-order terms. In fact, it is always possible to choose the function $\tilde{g}(g)$ so that all terms in $\tilde{\beta}(\tilde{g})$ of higher than third order in $\tilde{g}$ vanish, so we can describe the asymptotic behavior of $\tilde{g}_E$ for $E \to \infty$ by inspection of the first two terms in the perturbation series for $\beta(g)$. But this is of little value, since we would need to carry our calculations to all orders to determine how g depends on $\tilde{g}$, and without this we cannot use our knowledge of the asymptotic behavior of $\tilde{g}$ to say anything about the asymptotic behavior of g, or of physical quantities.

The same argument that led to Eq. (18.3.30) shows that, *at small coupling*, the Wilson renormalization group equation for the bare coupling constant as a function of lattice spacing for inverse lattice spacings greater than particle masses is the same as the Gell-Mann–Low renormalization group equation for the renormalized coupling constant as a function of renormalization scale for scales greater than particle masses. Hence if a continuum theory is asymptotically free, then there will be no obstruction in passing to the continuum limit of the theory quantized on a lattice.

18.4 Multiple Couplings and Mass Effects

Up to now, we have considered theories with only one dimensionless coupling g. It is easy to extend the formalism to incorporate several such couplings g^ℓ : for each g^ℓ, we have a renormalization group equation that for $\mu \gg m$ takes the form

$$\mu \frac{d}{d\mu} g^\ell(\mu) = \beta^\ell(g(\mu)) \,, \tag{18.4.1}$$

with each β^ℓ depending in general on all the gs. There are now many more possibilities for the asymptotic behavior of the $g^\ell(\mu)$ as $\mu \to \infty$; in a given theory we may have some trajectories in g-space that go off to infinity, at finite or infinite values of μ, other trajectories that approach fixed points, and yet other trajectories that approach closed curves known as 'limit cycles'. To get a taste of some of the various possibilities, let's consider the behavior of $g^\ell(\mu)$ near a fixed point.

Eq. (18.4.1) has a fixed-point solution $g^\ell(\mu) = g_*^\ell$ if

$$\beta^\ell(g_*) = 0 \,. \tag{18.4.2}$$

In the neighborhood of this point, Eq. (18.4.1) becomes

$$\mu \frac{d}{d\mu}\left[g^\ell(\mu) - g_*^\ell\right] = \sum_k M^\ell{}_k \left[g^k(\mu) - g_*^k\right] \,, \tag{18.4.3}$$

where M is the matrix

$$M^\ell{}_k \equiv \left[\frac{\partial \beta^\ell(g)}{\partial g^k}\right]_{g=g_*} \,. \tag{18.4.4}$$

The solution can be expanded in eigenvectors of this matrix

$$g^\ell(\mu) = g_*^\ell + \sum_m c_m \, V_m{}^\ell \, \mu^{\lambda_m} \,, \tag{18.4.5}$$

where V_m is a eigenvector of M with eigenvalue λ_m (normalized in any convenient way):

$$\sum_k M^\ell{}_k V_m{}^k = \lambda_m \, V_m{}^\ell \,, \tag{18.4.6}$$

and the c_m are a set of expansion coefficients.* (The summation convention is suspended in this section.)

Eq. (18.4.5) shows that the coupling constants approach the fixed point as $\mu \to \infty$ if and only if $c_m = 0$ for all eigenvectors with $\lambda_m > 0$. (For simplicity we are assuming here that none of the eigenvalues vanish.) Thus

* We are assuming here that the eigenvectors V_m form a complete set. This is not always so, but it is the generic case; the eigenvectors of a finite matrix M will form a complete set if all of the roots of the secular equation $\text{Det}(M - \lambda 1) = 0$ are different. A matrix whose eigenvectors do not form a complete set can be regarded as a limiting case of a matrix with a complete set of eigenvectors when some of its eigenvalues become degenerate.

in general the trajectories that are attracted to the fixed point lie on an N_--dimensional surface, where N_- is the number of negative eigenvalues of M; the tangents to this surface at g_* are the eigenvectors with negative eigenvalues. Trajectories that are not on this surface may approach close to the fixed point, but are eventually repelled, predominantly in the direction of eigenvectors with the largest positive eigenvalues. Of course, if all eigenvalues are negative then there is a finite region around the fixed point within which all trajectories converge on this point.

Since the eigenvalues λ_m are evidently important in learning the asymptotic behavior of trajectories that approach a fixed point, it is useful to note that these eigenvalues are independent of the definition of the couplings. Suppose we introduce a new set of couplings $\tilde{g}^\ell$, defined as functions of the gs. These satisfy renormalization group equations

$$\mu \frac{d}{d\mu} \tilde{g}^\ell(\mu) = \sum_m \left. \frac{\partial \tilde{g}^\ell(g)}{\partial g^m} \right|_{g=g(\mu)} \beta^m(g(\mu)) \equiv \tilde{\beta}^\ell(\tilde{g}(\mu)) ,$$

so

$$\tilde{\beta}^\ell(\tilde{g}) = \sum_m \frac{\partial \tilde{g}^\ell(g)}{\partial g^m} \beta^m(g) . \tag{18.4.7}$$

(That is, β transforms as a contravariant vector in coupling-constant space.) Differentiating, we have

$$\sum_m \frac{\partial \tilde{\beta}^\ell(\tilde{g})}{\partial \tilde{g}^m} \frac{\partial \tilde{g}^m}{\partial g^k} = \sum_m \frac{\partial^2 \tilde{g}^\ell}{\partial g^m \partial g^k} \beta^m(g) + \sum_m \frac{\partial \tilde{g}^\ell}{\partial g^m} \frac{\partial \beta^m(g)}{\partial g^k} .$$

At a fixed point g_* the first term on the right vanishes, so this gives the matrix equation

$$\tilde{M} S = S M , \tag{18.4.8}$$

where

$$\tilde{M}^\ell{}_k \equiv \left[\frac{\partial \tilde{\beta}^\ell}{\partial \tilde{g}^k} \right]_{\tilde{g}=\tilde{g}(g_*)} , \tag{18.4.9}$$

$$S^\ell{}_k \equiv \left[\frac{\partial \tilde{g}^\ell}{\partial g^k} \right]_{g=g_*} . \tag{18.4.10}$$

As long as the transformation $g \to \tilde{g}$ is non-singular, Eq. (18.4.8) is a similarity transformation, and hence M and $\tilde{M}$ have the same eigenvalues λ_m.

The renormalization group formalism may be extended to non-renormalizable as well as renormalizable theories. As explained in Section 12.3, the infinities in non-renormalizable theories are eliminated by a suitable renormalization of coupling constants and masses, just as in renormalizable theories; the only difference is that in non-renormalizable theories the Lagrangian must be supposed to contain all possible interactions allowed by the symmetries of the theory. If g_B^ℓ is the unrenormalized coupling constant multiplying an operator of dimensionality D_ℓ in the Lagrangian (that is, a product of fields and spacetime derivatives of fields whose dimensionality in powers of mass or energy is D_ℓ), then g_B^ℓ will have a dimensionality $\Delta_\ell = 4 - D_\ell$. We may then re-express the bare couplings in terms of a set of *dimensionless* renormalized couplings $g^\ell(\mu)$ and a cut-off Λ, through relations of the general form

$$g_B^\ell \equiv \mu^{\Delta_\ell}\left[g^\ell(\mu) + \sum_{k,m} b^\ell{}_{km}\, g^k(\mu) g^m(\mu) \ln\left(\frac{\Lambda}{\mu}\right) + O(g(\mu)^3)\right] , \qquad (18.4.11)$$

with the dimensionless numerical coefficients $b^\ell{}_{km}$ and similar coefficients in higher-order terms chosen to cancel the cut-off dependence of physical quantities. (In some theories the leading term might be trilinear or even higher order in couplings; the modifications that would be needed here are obvious.) From the requirement that g_B be cut-off-independent, we obtain the renormalization group equation (18.4.1), with

$$\beta^\ell(g) = -\Delta_\ell g^\ell - \sum_{k,m} b^\ell_{km} g^k g^m + O(g^3) \,. \qquad (18.4.12)$$

Non-renormalizable interactions are those with $D_\ell > 4$, or $\Delta_\ell < 0$, so as long as the $g^\ell(\mu)$ all remain sufficiently small we expect the non-renormalizable renormalized couplings to have positive β^ℓ and hence to grow with μ, but no one knows what happens when the couplings become large enough to invalidate perturbation theory.

However, as explained in the next section, even theories with infinite number of independent parameters commonly have fixed points g_* at which the number N_- of negative eigenvalues of the matrix (18.4.6) is *finite*, just as it is at zero coupling. (In particular, often $N_- = 1$.) Where $N_- \neq 0$, the fixed point lies on an N_--dimensional critical surface, consisting of trajectories that are attracted into the fixed point as $\mu \to \infty$. A non-renormalizable theory with coupling parameters on such a critical surface, although not of course asymptotically free, is said to be asymptotically *safe*,[5] because the renormalized couplings remain finite for large values of μ. The condition of asymptotic safety in such a theory would play the role that used to be associated with the principle of renormalizability, of eliminating all but a finite number of free parameters, the coordinates of the critical surface.

In a renormalizable theory, all physical quantities are made cut-off-independent by adjusting the cut-off dependence of a *finite* number of bare couplings. These bare couplings may be expressed in terms of an equal number of μ-dependent renormalized couplings, and the condition that the bare couplings are μ-independent yields renormalization group equations relating only these renormalized couplings. From the broader point of view which allows non-renormalizable as well as renormalizable couplings, a renormalizable theory just corresponds to a finite-dimensional *invariant surface* in the infinite-dimensional space of all renormalizable and non-renormalizable theories; that is, it is a surface for which $\beta^\ell(g)$ at any point g on the surface is tangent to the surface at that point.

So far in this section we have tacitly assumed that $\mu \gg m$, so that we could neglect the dependence of β^ℓ on m/μ. However, this is not necessary; we can if we like treat a mass as just another coupling parameter.[6] That is, all renormalized couplings can be defined as before in terms of various Greens functions at off-mass-shell momenta of order μ, but now evaluated with all bare masses zero. The dimensionless renormalized mass parameters for Dirac fields ψ or scalar fields ϕ may be defined as

$$m_\psi(\mu) \equiv N^{(\bar{\psi}\psi)}(\Lambda/\mu)^{-1}\, m_{\psi,\text{BARE}}(\Lambda)/\mu\,, \tag{18.4.13}$$

$$m_\phi^2(\mu) \equiv N^{(\phi^2)}(\Lambda/\mu)^{-1}\, m^2_{\phi,\text{BARE}}(\Lambda)/\mu^2\,, \tag{18.4.14}$$

where $N^{(\mathcal{O})}(\Lambda/\mu)$ are the dimensionless constants which, when multiplied into corresponding operators $\mathcal{O}$, cancel the infinities in the matrix elements of these operators, also evaluated with all bare masses zero. (See Section 18.1.) These new renormalized masses and couplings have no direct physical significance, but the true physical masses and all physical matrix elements can be expressed in terms of them. These matrix elements take the form of sums of matrix elements for zero bare mass, with any number of insertions of the renormalized mass operators $N^{(\phi^2)}\phi^2$ and $N^{(\bar{\psi}\psi)}\bar{\psi}\psi$, times the corresponding renormalized mass parameters.

In this renormalization scheme the beta-functions for the various couplings are obviously mass-independent, and the beta-functions for the mass parameters are proportional to these parameters, with coefficients that depend on all the various couplings; using Eq. (18.2.25), we have

$$\mu\frac{d}{d\mu}m_\psi(\mu) = [-1 - \gamma_{\bar{\psi}\psi}(g_\mu)]\, m_\psi(\mu)\,, \tag{18.4.15}$$

$$\mu\frac{d}{d\mu}m_\phi^2(\mu) = [-2 - \gamma_{\phi^2}(g_\mu)]\, m_\phi^2(\mu)\,. \tag{18.4.16}$$

For instance, we noted in Section 18.2 that in the scalar field theory with Lagrangian (18.1.2), the mass operator ϕ^2 has anomalous dimension (18.2.29) for $m = 0$, so here

$$\mu\frac{d}{d\mu}m_\phi^2(\mu) = \left[-2 + \frac{g_\mu}{16\pi^2} + O(g_\mu^2)\right] m_\phi^2(\mu)\,. \tag{18.4.17}$$

Also, Eq. (11.4.3) shows that the effect of higher-order corrections to the electron propagator is to replace the electron mass by $m_e - \Sigma^*(p, m_e)$, so the effect of these corrections on matrix elements of the operator $\bar{\psi}_e\psi_e$ between one-electron states of four-momentum p^μ is to multiply them by a factor

$$F(p) = 1 - \left(\frac{\partial\Sigma^*(p, m_e)}{\partial m_e}\right)_{m_e=0} .$$

The renormalization constant $N^{\bar{\psi}\psi}$ for the operator $\bar{\psi}_e\psi_e$ is therefore equal to $F^{-1}(p)$, evaluated with p^2 equal to some renormalization scale, say $+\mu^2$. According to Eq. (11.4.8), to one-loop order this is

$$\begin{aligned} N^{(\bar{\psi}\psi)} &= 1 + \left(\frac{\partial\Sigma^*_{1\text{ loop}}(p, m_e)}{\partial m_e}\right)_{m_e=0,\, p^2=\mu^2} \\ &= 1 - \frac{4\pi^2 e^2}{(2\pi)^4}\int_0^1 dx\, \ln\left[1 + \left(\frac{\Lambda^2}{\mu^2}\right)(1-x)\right] \\ &\to 1 - \frac{e^2}{4\pi^2}\left[\ln\left(\frac{\Lambda^2}{\mu^2}\right) - 1\right] , \end{aligned} \tag{18.4.18}$$

where Λ is an ultraviolet cut-off,** and we take the limit $\Lambda \gg \mu$. The anomalous dimension of the operator $\bar{\psi}_e\psi_e$ is therefore given by (18.2.25) as

$$\gamma^{(\bar{\psi}\psi)} = \mu\frac{d}{d\mu}\ln N^{(\bar{\psi}\psi)} = \frac{e_\mu^2}{2\pi^2} + O(e_\mu^4) , \tag{18.4.19}$$

so Eq. (18.4.15) here reads

$$\mu\frac{d}{d\mu}m_e(\mu) = \left[-1 - \frac{e_\mu^2}{2\pi^2} + O(e_\mu^4)\right] m_e(\mu) . \tag{18.4.20}$$

The same formula holds in general gauge theories, with e^2 replaced with the value of $\sum_\alpha (t_\alpha)^2$ for the particular species of fermion in question.

The important difference between the $m(\mu)$ and the other renormalized parameters of the theory is of course that bare masses have positive dimensionality, so as long as the couplings remain small the $m(\mu)$ all decrease in magnitude. Our previous assumption that masses may be neglected as $\mu \to \infty$ is justified if in fact $m(\mu)$ does vanish for $\mu \to \infty$. However this is only known to be the case in asymptotically free theories, where the couplings all do remain small for $\mu \to \infty$; in all other cases this assumption is just an educated guess.

** This notation is different from that of Eq. (11.4.8), where the ultraviolet cut-off was called μ.

18.5 Critical Phenomena*

For some purposes we may be interested in the limit of very low rather than high energies or wave numbers. The arguments of Section 18.2 can be repeated to study this limit, except that here we must examine the case $\mu \to 0$ rather than $\mu \to \infty$. This limit is of course simplest if there are no masses in the theory, as, for instance, in quantum electrodynamics with a symmetry under the chiral transformation $\psi \to \gamma_5 \psi$ which forbids an electron mass. In this particular case the only renormalizable coupling $eA^\mu \bar{\psi}\gamma_\mu \psi$ as well as all non-renormalizable couplings have $\beta^\ell > 0$ for sufficiently small couplings, so all trajectories in at least a finite region around the origin are attracted into the point $g^\ell = 0$ as $\mu \to 0$.

The same considerations may be applied even to theories with very small but non-zero masses if we include these masses among the coupling parameters of the theory, as described in the previous section. The coefficient Δ in Eq. (18.4.12) is positive for a mass parameter, so in this case the trajectories can never reach the point $g = 0$, but they may come close if the masses are small.

Of course, even if we can regard some degree of freedom like the electron field as having zero or very small mass, in the real world there are many other degrees of freedom whose masses are not small. The renormalization group should properly be applied not to the true theory that encompasses all these heavy degrees of freedom, but to an 'effective' field theory, in which only massless or nearly massless degrees of freedom appear explicitly, with interactions that include the effects of internal heavy particle lines. (We shall have more to say about effective field theories in Chapter 19.)

The low wave number limit is of particular interest in the study of critical phenomena, such as long-range correlations at or near a second-order phase transition (a smooth phase transition, with no latent heat) in condensed matter. Because we are interested in the limit $\mu \to 0$, the important eigenvectors of the matrix (18.4.4) are those with eigenvalues $\lambda < 0$, which are called *relevant*. The eigenvectors with $\lambda = 0$ and $\lambda > 0$ are called *marginal* and *irrelevant*, respectively.

Suppose that there is a non-trivial fixed point g_* with just one negative eigenvalue λ_0, perhaps corresponding approximately to a mass operator. The set of trajectories of $g^\ell(\mu)$ that are attracted into this fixed point for $\mu \to 0$ therefore forms a critical surface of codimension one; that is, a surface defined by a single condition on the couplings, the condition that

* This section lies somewhat out of the book's main line of development, and may be omitted in a first reading.

for $g \to g_*$, the tangents $g^\ell - g_*^\ell$ have no components in the direction of the eigenvector with negative eigenvalue. There is a phase transition as the physical value of the couplings at any fixed characteristic scale approaches this surface. Because the critical surface has codimension one, the phase transition can be reached by adjusting any one parameter on which the couplings depend, such as the pressure or temperature. The fact that a wide variety of substances do exhibit phase transitions of this sort shows that it is common to encounter fixed points for which the matrix (18.4.4) has a single negative eigenvalue, as already mentioned in the previous section.

To be specific, as the temperature T approaches its critical value T_c we expect the coefficient c_0 of the growing term in Eq. (18.4.5) to become proportional to $T - T_c$, because there is no reason why it should be singular or why it should vanish faster than this. Hence for $\mu \to 0$ and then $T \to T_c$, the couplings go as

$$g^\ell(\mu) \to (T - T_c)\, V_0{}^\ell\, \mu^{\lambda_0}\,, \tag{18.5.1}$$

where λ_0 is the only negative eigenvalue at g_*, and $V_0{}^\ell$ is the corresponding eigenvector.** Applying our renormalization group arguments to wave numbers instead of energies, the N-point function (the Nth partial derivative of the effective action with respect to a field ϕ of dimensionality [wave number]$^{D_\phi}$) at a small characteristic wave number scale κ has the form†

$$\Gamma_N(\kappa) \to \kappa^{d-N(D_\phi+\gamma_\phi(g_*))} F_N((T-T_c)\kappa^{\lambda_0})\,, \tag{18.5.2}$$

where $\gamma_\phi(g)$ is the anomalous dimension associated with the field ϕ, and d is the spacetime dimensionality, or in classical statistical mechanics the spatial dimensionality. It is convenient to rewrite this in the equivalent form

$$\Gamma_N(\kappa) \to (T-T_c)^{-[d-N(D_\phi+\gamma_\phi(g_*))]/\lambda_0} G_N(\kappa(T-T_0)^{1/\lambda_0})\,. \tag{18.5.3}$$

This shows for one thing that the correlation length ξ (the characteristic length that determines the scale over which the Fourier transform of Γ_N varies) increases as T approaches T_c like

$$\xi \propto (T - T_c)^{-\nu} \tag{18.5.4}$$

where ν is a conventionally defined positive 'critical exponent', given by Eq. (18.5.3) as

$$\nu = -1/\lambda_0\,. \tag{18.5.5}$$

** Other contributions to the couplings will go as $(T - T_0)^0\mu^{\lambda_1}$ with $\lambda_1 > 0$. Thus Eq. (18.5.1) is valid here provided $T - T_0$ does not go to zero as fast as $\mu^{\lambda_1-\lambda_0}$.

† The function F_N also depends on dimensionless angles and ratios of wave numbers. Note that Γ_N has 'naive' dimensionality $d - ND_\phi$ because $\delta^d(\sum\kappa)\Gamma$ must be dimensionless.

Also, the zero-field effective action Γ_0 (or in statistical physics, the free energy) must be κ-independent because it corresponds to graphs with no external lines. It follows that Eq. (18.5.3) here becomes for $T \to T_c$:

$$\Gamma_0 - F_0 \propto (T - T_c)^{\nu d} , \tag{18.5.6}$$

where the constant F_0 is the effective action or free energy due to the heavy degrees of freedom which have been integrated out. Thus the exponent ν also governs the behavior of the part of the free energy which is not analytic in temperature for T near T_c.

In 1972 Wilson and Fisher[7] used an expansion in powers of $d-4$ both to show that the theory of a scalar field actually fits the above description, and also to carry out an approximate calculation of critical exponents such as ν. Consider a theory with a single 'light' degree of freedom, a scalar field ϕ, such as the magnetization in a ferromagnet, with a symmetry under $\phi \to -\phi$ that rules out interactions odd in ϕ. In addition to the 'mass' term $-g_2\phi^2/2$, the Lagrangian density of the effective field theory will contain interactions $-g_4\phi^4/4!$, $-g_6\phi^6/6!$, etc. The dimensionality of the field ϕ in powers of wave number is $(d-2)/2$ (so that $\int d^dx\,(\nabla\phi)^2$ should be dimensionless) so the dimensionalities of the couplings g_2, g_4, g_6, etc. in d dimensions are $+2$, $4-d$, $6-2d$, etc. For the fixed point at zero coupling in three dimensions, there are two relevant couplings, g_2 and g_4, but this conclusion is changed by interactions at non-trivial fixed points. Let's examine the surface in coupling constant space in which only g_2 and g_4 are non-zero, and take g_4 to be small.[††] Eq. (18.2.12) gives $\beta(g_4) = 3g_4^2/16\pi^2 + O(g_4^3)$ for $d = 4$, and Eq. (18.4.12) tells us that for $d = 4-\epsilon$ dimensions we must add to this a term $-\epsilon g_4$, so

$$\mu\frac{d}{d\mu}g_4(\mu) = -\epsilon\, g_4(\mu) + \frac{3g_4^2(\mu)}{16\pi^2} + O(g_4^3(\mu)) . \tag{18.5.7}$$

Also, Eq. (18.4.17) gives

$$\mu\frac{d}{d\mu}g_2(\mu) = \left[-2 + \frac{g_4(\mu)}{16\pi^2} + O(g_4(\mu))\right] g_2(\mu) . \tag{18.5.8}$$

Therefore for small ϵ there is a non-trivial fixed point at

$$g_{4*} = \frac{16\pi^2\epsilon}{3} , \qquad g_{2*} = 0 . \tag{18.5.9}$$

---

†† These are the only renormalizable couplings for $3 \le d \le 4$, so for such d this is an invariant surface. Note that we do not include the coefficient of $(\nabla\phi)^2$ among the couplings here, because this is a redundant coupling, in the sense described in Section 7.7.

The matrix (18.4.4) at this fixed point is diagonal, with eigenvalues

$$\lambda_4 = M^4{}_4 = -\epsilon + \frac{3g_{4*}}{8\pi^2} + O(g_{4*}^2) = +\epsilon + O(\epsilon^2)\,, \qquad (18.5.10)$$

$$\lambda_2 = M^2{}_2 = -2 + \frac{g_{4*}}{16\pi^2} + O(g_{4*}^2) = -2 + \frac{\epsilon}{3} + O(\epsilon^2)\,. \qquad (18.5.11)$$

From Eq. (18.5.10) we see that the coupling g_4 is actually irrelevant, so that there is just one relevant coupling here, signalling the presence of a second-order phase transition. From Eq. (18.5.11) we see that the anomalous exponent (18.5.5) is

$$\nu = -\frac{1}{\lambda_2} = \frac{1}{2} + \frac{\epsilon}{12} + O(\epsilon^2)\,. \qquad (18.5.12)$$

For the physical value $\epsilon = 1$ the first two terms give $\nu \simeq 0.58$. Three-loop calculations[8a] give this critical exponent to order ϵ^3 as

$$\nu = \frac{1}{2} + \frac{\epsilon}{12} + \frac{7\epsilon^2}{162} - 0.01904\epsilon^3\,, \qquad (18.5.13)$$

which for $\epsilon = 1$ gives $\nu = 0.61$.

In the calculation presented here nothing was assumed about the system under study except that there is a second-order phase transition, near which the only long-wavelength degree of freedom is a single scalar field. There are a number of different physical systems that fit this description, such as the spontaneous appearance of magnetization (represented here by ϕ) in ferromagnetic and antiferromagnetic materials, and also second-order phase transitions between liquids and gases and in binary fluids. All of these systems are therefore expected to have the same value of ν. This is confirmed by experiment, which gives a value[8] $\nu = 0.63 \pm 0.04$, in good agreement with the three-loop result (18.5.13), and in fair agreement even with the one-loop result (18.5.12). It is fortunate though still somewhat mysterious that an expansion in powers of 1 should work so well.

More generally, all the systems that are described by the same set of long-wavelength degrees of freedom near their second-order phase transitions are said to belong to the same *universality class*. All the critical exponents are the same for all systems in a given universality class.

18.6 Minimal Subtraction

We saw in Section 11.2 that dimensional regularization provides a particularly convenient method for calculating radiative corrections in quantum electrodynamics, because it preserves the conservation laws associated with gauge invariance. For the same reason, dimensional regularization also turns out to provide a very convenient alternative definition for the sliding scale of the renormalization group in general gauge theories.[9]

In calculations using dimensional regularization, ultraviolet divergences arise as poles in physical amplitudes when the spacetime dimensionality d approaches the physical value $d = 4$. (For an example, see Eq. (11.2.13).) To cancel these poles, the bare coupling constants $g_B^\ell(d)$ (including masses) must themselves have such poles, with residues fixed by the condition that physical amplitudes be regular as $d \to 4$. These bare couplings in general have non-zero dimensionalities $\Delta_\ell(d)$ that depend on the spacetime dimensionality d, so it is convenient to consider the dimensionless quantity $g_B^\ell(d)\mu^{-\Delta_\ell(d)}$, where μ is a sliding scale with the dimensions of energy or mass. This rescaled bare coupling may be expressed as a sum of terms proportional to positive-definite powers ν of $1/(d-4)$, with coefficients b_ν fixed by the requirement of cancellation of singularities as $d \to 4$ in physical amplitudes, plus a remainder that is analytic in d at $d = 4$. This remainder is identified as the dimensionless renormalized coupling constant $g^\ell(\mu, d)$, so

$$g_B^\ell(d)\mu^{-\Delta_\ell(d)} = g^\ell(\mu, d) + \sum_{\nu=1}^{\infty}(d-4)^{-\nu}\, b_\nu^\ell(g(\mu, d))\,. \tag{18.6.1}$$

We are free to give the bare couplings any d dependence we like, as long as the singularities at $d = 4$ in physical amplitudes are cancelled; we shall remove this ambiguity by requiring that $g^\ell(\mu, d)$ be analytic in d not only at $d = 4$, but for all d.

To calculate the renormalization group equation satisfied by $g^\ell(\mu, d)$, first differentiate Eq. (18.6.1) with respect to μ:

$$-\Delta^\ell(d)\left[g^\ell + \sum_{\nu=1}^{\infty}(d-4)^{-\nu} b_\nu^\ell(g)\right] = \beta^\ell(g, d) + \sum_{\nu=1}^{\infty}\sum_m b_{\nu m}^\ell(g)\,\beta^m(g, d)(d-4)^{-\nu} \tag{18.6.2}$$

where

$$b_{\nu m}^\ell(g) \equiv \frac{\partial}{\partial g^m} b_\nu^\ell(g) \tag{18.6.3}$$

and as before

$$\mu\frac{d}{d\mu}g^\ell(\mu, d) = \beta^\ell(g(\mu, d), d)\,. \tag{18.6.4}$$

Note that β^ℓ is a function of all of the $g^m(\mu, d)$ and also of d, but it cannot depend separately on μ because, with rescaled masses included among the dimensionless coupling parameters, there are no other dimensionful parameters besides μ.

As we have seen, the dimensionalities $\Delta_\ell(d)$ are always linear functions of d, which we shall now write as

$$\Delta_\ell(d) = \Delta_\ell + \rho_\ell(d-4)\,. \tag{18.6.5}$$

We rewrite the left-hand side of Eq. (18.6.2) as

$$-\rho_\ell g^\ell(d-4) - \left[\Delta_\ell g^\ell + b_1^\ell(g)\rho^\ell\right] - \sum_{\nu=1}^{\infty}(d-4)^{-\nu}\left[\rho_\ell\, b_{\nu+1}^\ell(g) + \Delta_\ell b_\nu^\ell(g)\right]\,.$$

The highest power of d in the analytic part here is of first order, so the same must be true on the right-hand side of Eq. (18.6.2), and therefore $\beta(g,d)$ must be linear in d:

$$\beta^\ell(g,d) = \beta^\ell(g) + (d-4)\alpha^\ell(g)\,. \tag{18.6.6}$$

Equating terms of first and zeroth order in (18.6.2) gives then

$$\alpha^\ell(g) = -\rho_\ell\, g^\ell \tag{18.6.7}$$

and, more importantly,

$$\beta^\ell(g) = -\Delta_\ell g^\ell - b_1^\ell(g)\rho_\ell + \sum_{\nu=1}^{\infty}\sum_m b_{1m}^\ell(g)\,\rho_m g^m\,. \tag{18.6.8}$$

It is noteworthy that the beta-function depends only on the coefficients of the simple pole in the bare couplings. In fact, these coefficients also determine the coefficients of all the higher poles; equating the pole terms on the right and left of Eq. (18.6.2) yields the recursion relation

$$\rho_\ell b_{\nu+1}^\ell(g) - \sum_m \rho_m g^m b_{\nu+1m}^\ell(g) = -\Delta_\ell b_\nu^\ell(g) - \sum_m b_{\nu m}^\ell(g)\beta^m(g)\,. \tag{18.6.9}$$

For instance, in order for $\int d^dx\, F_{\mu\nu}F^{\mu\nu}$ to be dimensionless, any gauge field A^μ must have dimensions (in powers of mass) $(d-2)/2$, and since $g_B A^\mu$ must have the same dimensions as $\partial/\partial x^\mu$, g_B must have dimensions $(4-d)/2$, so that for gauge couplings $\Delta = 0$ and $\rho = -1/2$. Eq. (18.6.8) gives then for a gauge theory with a single coupling constant:

$$\beta(g) = \tfrac{1}{2}\left[b_1(g) - g b_1'(g)\right]\,. \tag{18.6.10}$$

In particular, Eq. (11.2.20) shows that in quantum electrodynamics in one-loop order the bare electric charge has a pole at $d \to 4$ with

$$e_B = Z_3^{-1/2} e \to e - \frac{e^3}{12\pi^2}\frac{1}{d-4}\,. \tag{18.6.11}$$

Setting $b_1(e) = -e^3/12\pi^2$ in Eq. (18.6.10) gives then

$$\beta(e) = \frac{e^3}{12\pi^2}\,, \tag{18.6.12}$$

in agreement with the previous result (18.2.37).

The coupling constants $g^\ell(\mu)$ introduced in this section are said to be defined by *minimal subtraction.* There is a slightly different scheme that is somewhat more convenient. The simple poles $(d-4)^{-1}$ typically arise from functions $(4\pi)^{d/2-2}\Gamma(2-d/2)$ (as in Eq. (11.2.13)) which for $d \to 4$ have the limit

$$(4\pi)^{2-d/2}\Gamma\left(2-\frac{d}{2}\right) \to \frac{1}{2-d/2} - \gamma + \ln 4\pi\,, \tag{18.6.13}$$

where γ is the Euler constant, $\gamma = 0.5772157$. It is therefore convenient to make the replacement everywhere in Eq. (18.6.1):

$$\frac{1}{d-4} \to \frac{1}{d-4} + \frac{\gamma}{2} - \frac{1}{2}\ln 4\pi\,. \tag{18.6.14}$$

With this prescription, the coupling constants are said to be defined by *modified minimal subtraction.*

One of the distinguishing characteristics of the definition of couplings by minimal subtraction (or modified minimal subtraction) is that, since no factors of an ultraviolet cut-off ever appear in any calculation, loop diagrams have poles at $d = 4$ that correspond only to logarithmic ultraviolet divergences, not divergences that are linear, quadratic, etc. Hence a residue function $b_1^\ell(g)$ can contain a term of order $g^a g^b g^c \cdots$ only if at $d = 4$ the dimensionality of g_B^ℓ equals the total dimensionality of the couplings g_B^a, g_B^b, g_B^c, etc.:

$$\Delta_\ell = \Delta_a + \Delta_b + \Delta_c + \ldots\,, \tag{18.6.15}$$

and it follows then from (18.6.8) that the same is true of $\beta^\ell(g)$. In particular, in a theory with no superrenormalizable couplings (like a gauge theory with massless spinors and no scalars, such as quantum electrodynamics with vanishing electron mass) all couplings have $\Delta_\ell \leq 0$, so *the renormalization group equations for the renormalizable couplings (with $\Delta_\ell = 0$) are unaffected by the presence of any non-renormalizable interactions.*[5] Also, in such a theory the beta-functions for the non-renormalizable couplings are polynomials of finite order in the non-renormalizable couplings, with each coefficient in each polynomial given by an infinite series of powers of the renormalizable couplings. For instance, in the theory of photons and massless electrons (assuming invariance under $\psi \to \gamma_5\psi$ and P), there are no nonrenormalizable interactions of dimensionality +5, and several interactions of dimensionality +6 (four-fermion interactions as well as the purely photonic interaction $F_{\mu\nu}\Box F^{\mu\nu}$) with couplings f_i of dimensionality −2. The beta-function for f_i is of the form $\sum_j b_{ij}(e) f_j$, with the coefficients $b_{ij}(e)$ given by a power series in e.

18.7 Quantum Chromodynamics

Quantum chromodynamics is the modern theory of strong interactions. It is a non-Abelian gauge theory, based on the gauge group $SU(3)$. In addition to the gauge fields, quantum chromodynamics involves fields of spin $\frac{1}{2}$ particles known as quarks. There are quarks of six types, or 'flavors', the u, c, and t quarks having charge $2e/3$, and the d, s, and b quarks having charge $-e/3$. Quarks of each flavor come in three 'colors' which furnish the defining representation **3** of the $SU(3)$ gauge group.* Baryons like the protons and neutron may be approximately regarded as color-neutral bound states of three quarks, totally antisymmetric in quark colors, while mesons like the rho meson behave approximately like color-neutral bound states of quarks and antiquarks.[11]

In the approximation where the quark masses may be regarded as negligible compared with the energies of interest, the inversion of Eq. (17.5.44) shows that the bare coupling constant in a general gauge theory has a pole at spacetime dimensionality $d \to 4$ with residue given by

$$g_B \to \left[\frac{g^3}{4\pi^2}\left(\frac{11}{12}C_1 - \frac{1}{3}C_2\right) + O(g^5)\right]\frac{1}{d-4},$$

where C_1 and C_2 are defined by Eqs. (17.5.33) and (17.5.34). That is, in the notation of the previous section,

$$b_1(g) = \frac{g^3}{4\pi^2}\left(\frac{11}{12}C_1 - \frac{1}{3}C_2\right) + O(g^5). \tag{18.7.1}$$

Using this in Eq. (18.6.10) gives

$$\beta(g) = -\frac{g^3}{4\pi^2}\left(\frac{11}{12}C_1 - \frac{1}{3}C_2\right) + O(g^5). \tag{18.7.2}$$

For an $SU(3)$ theory with n_f massless quarks in the defining representation **3** of $SU(3)$, Eq. (17.5.35) gives

$$C_1 = 3, \qquad C_2 = n_f/2. \tag{18.7.3}$$

Because we are taking the quarks as massless here, this formula may be applied only in the effective field theory obtained by integrating out all quarks heavier than the typical energy E under consideration, so that n_f is the number of quark flavors with masses much less than E. With this

* Before the final formulation of quantum chromodynamics several authors had speculated that there might be three varieties of quarks of each flavor,[10] both in order to account for the rate of decays like $\pi^0 \to \gamma + \gamma$ (see Sections 22.1 and 22.2) and to introduce an additional degree of freedom that would explain how the wave function of fermionic quarks in a baryon could be symmetric in spin, space, and flavor coordinates.[11]

understanding, Eqs. (18.7.2) and (18.7.3) yield

$$\beta(g) = -\frac{g^3}{4\pi^2}\left(\frac{11}{4} - \frac{1}{6}n_f\right) + O(g^5)\,. \tag{18.7.4}$$

We see that the theory is asymptotically free as long as there are no more than 16 quark flavors with masses below the energy scale of interest. Since in fact there seem to be only six quark flavors of any mass, the theory of strong interactions based on the gauge group $SU(3)$ is asymptotically free.

It was the 1973 discovery of asymptotic freedom in non-Abelian gauge theories of this sort by Gross and Wilczek[12] and Politzer[13] that convinced theoretical physicists that this is the correct theory of strong interactions. Their calculation immediately explained the puzzling result of a famous 1968 experiment[14] at SLAC on deep-inelastic electron–nucleon scattering, that strong interactions seem to get weaker at high energies.** (This experiment will be discussed further in Section 20.6.) But the historical importance of the discovery of asymptotic freedom in Yang–Mills theories is not just that it explained an old experimental result; it for the first time opened up the prospect of doing reliable perturbative calculations of strong interaction processes, at least at high energy.

Asymptotic freedom was soon found to have another important implication. At first after the discovery of asymptotic freedom it was widely assumed that the gauge bosons in a realistic Yang–Mills theory of strong interactions would have to be quite heavy, to explain why these strongly-interacting bosons had not been discovered long before. Following the precedent of the theory of weak and electromagnetic interactions (discussed in Chapter 21), it was supposed that the masses of the gauge bosons arose from a spontaneous breakdown of the color $SU(3)$ gauge group, triggered by the vacuum expectation values of scalar fields in a non-trivial representation of this group. But these strongly interacting scalars would contribute positive terms to $\beta(g)$, which could destroy asymptotic freedom. Even worse, in a theory with strongly interacting scalar fields, radiative corrections involving weak interactions would introduce large violations of various symmetries like charge conjugation invariance and flavor conservation which, as we shall see, would not be violated without the scalars.[17] Then it was suggested to drop the strongly-interacting

---

** Zee[15] and perhaps other theorists had already understood that this experimental result could be understood in a theory with a beta-function that becomes negative for small positive coupling, but calculations of $\beta(g)$ in all renormalizable field theories *except* non-Abelian gauge theories gave $\beta(g) > 0$. On the other hand, by 1972 't Hooft had developed techniques for calculating $\beta(g)$ in Yang–Mills theories, and in June 1972 he announced at a conference on gauge theory at Marseilles[16] that $\beta(g) < 0$, but he waited to publish this result and work out its implications while he was doing other things, so his result did not attract much attention.

scalars, and accept the consequence that the gluons, the $SU(3)$ gauge bosons, have zero mass.[18] The decrease of the strong coupling constant at high energy or short distance of course implies an increase at low energy or large distance, and it was suggested that this might explain why massless gluons and quarks had not been detected. According to this hypothesis, only color-neutral particles like baryons or mesons will ever appear in isolation.[19] This is unfortunately still a hypothesis rather than a theorem, but after two decades there seems to be little doubt that it is correct.

Even though quarks cannot materialize as free particles, they are in a sense observed as jets produced in high energy collision processes. For instance, in many events in electron–positron annihilation, the final state consists of two narrowly collimated hadron jets, with a distribution in the angle θ between the colliding lepton momenta and the jet directions (in the center-of-mass system) given by $1 + \sin^2\theta$, just as expected from the tree graph for electron–positron annihilation into quark–antiquark final states.[20] This can be understood[21] in terms of the general analysis of infrared divergences in Section 13.4. At extremely high energies we would expect the rate for a physical process to be given by lowest-order perturbation theory, provided it is 'infrared-safe,' in the sense of not becoming infrared divergent when all masses are taken to zero. The total rate for electron-positron annihilation into hadrons is infrared-safe, since we sum over all hadronic final states. (We are ignoring higher-order electromagnetic effects here.) Therefore we can rely on perturbation theory, which immediately tells us that the ratio R of this rate to the rate for $e^+ + e^- \to \mu^+ + \mu^-$ is $R = 3\sum_q Q_q^2$, where the sum runs over all quark flavors, Q_q is their charge in units of e, and the factor 3 is the number of colors. (For instance, in the wide energy range between $m_b \approx 4.5$ GeV and $m_t \approx 180$ GeV, $R \simeq 3(2(2/3)^2 + 3(-1/3)^2) = 11/3$.) On the other hand, the rate for electron–positron annihilation into some definite state of quarks and gluons is not infrared-safe, and its rate therefore cannot be calculated in perturbation theory at all; in fact, it is zero. In between these two extremes is the rate for electron–positron annihilation into a definite number of jets, each jet carrying a definite total momentum and charge, together with a set of unobserved hadrons with limited total energy outside the jets. As discussed in Section 13.4, this rate is infrared-safe. It can therefore be calculated at high energy in the tree approximation of perturbation theory, identifying jets (in this order of perturbation theory) with the outgoing quarks, antiquarks, and gluons. We can even calculate the rate for three-jet events, arising from tree diagrams in which a gluon is emitted from the outgoing quark or antiquark, and use the comparison of the results with experiment to measure the value of $\alpha_s(\mu)$.[22] But we cannot use perturbation theory to predict the distribution of momenta within a jet, because such a differential rate is not infrared-safe. Similar remarks apply to the production of jets in deep inelastic lepton–hadron collisions, to be discussed in Section 20.6, but the presence of hadrons in the initial state makes the analysis more complicated.

Following the same reasoning as in Section 12.5, with no scalar fields the most general renormalizable Lagrangian for quantum chromodynamics can be put in the form

$$\mathscr{L} = -\tfrac{1}{4}F_\alpha^{\mu\nu}F_{\alpha\mu\nu} - \sum_n \overline{\psi}_n[\not{\partial} - ig\,\not{A}_\alpha t_\alpha + m_n]\psi_n\,, \tag{18.7.5}$$

where A_α^μ is the color gauge vector potential; $F_\alpha^{\mu\nu}$ is the color gauge-covariant field strength tensor; g is the strong coupling constant; t_α are a complete set of generators of color $SU(3)$ in the **3** representation (that is, Hermitian traceless 3×3 matrices with rows and columns labelled by the three quark colors), normalized so that $\mathrm{Tr}(t_\alpha t_\beta) = \frac{1}{2}\delta_{\alpha\beta}$; and the subscript n labels quark flavors, with quark color indices suppressed. Just as we found for electrodynamics in Section 12.5, this Lagrangian has important accidental symmetries: it conserves space parity,† charge conjugation parity, and the numbers of quarks of each flavor (minus the number of the corresponding antiquarks), including the long-established 'strangeness' quantum number, which counts the numbers of 's' quarks. Thus quantum chromodynamics immediately explained the mysterious fact that the strong interactions respect various symmetries that are not symmetries of all interactions. This argument also makes it clear why, as mentioned earlier, in this theory the weak interactions do not introduce large violations of parity, charge conjugation, strangeness, etc. Since all renormalizable interactions among quarks and gluons conserve these symmetries, at energies E much less than the masses m_W of the particles that carry the weak interactions these symmetries could be violated only by non-renormalizable terms in the effective field theory, such as $\bar{\psi}\psi\bar{\psi}\psi$ interactions, which as discussed in Section 12.3 would be suppressed by negative powers of m_W as well as by the coupling constants of the weak interactions.

Of course, it is possible that the quarks and gluons exhibit some new kind of *strong* interaction at an energy scale Λ' much larger than the scale Λ characteristic of quantum chromodynamics. For instance, as discussed in Section 22.5, the quarks might be bound states of more fundamental fermions, which interact with gauge fields whose asymptotically free couplings become strong at energies of order Λ', trapping them into the quarks. In that case the effective Lagrangian density for quarks at en-

† Although it was not known in 1973, we shall see in Section 23.6 that non-perturbative effects can violate parity in quantum chromodynamics. Various ways of avoiding strong parity violation have been suggested, but it is not yet clear which is correct.

ergies $E \ll \Lambda'$ would contain non-renormalizable interactions such as $\bar{\psi}\psi\bar{\psi}\psi$, which are suppressed only by powers of E/Λ'. These interactions could show up, not only in small violations of symmetries like parity and quark flavor conservation at ordinary energies, but also in departures[23] from the quantitative predictions of quantum chromodynamics at energies approaching Λ'.

Now let us consider the behaviour of the coupling constant of quantum chromodynamics in greater detail. In lowest order, the renormalization group equation is given by Eq. (18.7.4) as

$$\mu\frac{d}{d\mu}g(\mu) = -\frac{g^3(\mu)}{4\pi^2}\left(\frac{11}{4}-\frac{1}{6}n_f\right). \tag{18.7.6}$$

The solution is

$$\alpha_s(\mu) \equiv \frac{g^2(\mu)}{4\pi} = \frac{12\pi}{(33-2n_f)\ln(\mu^2/\Lambda^2)}, \tag{18.7.7}$$

where Λ is an integration constant. This formula exhibits a characteristic property of theories of massless (or, for the quarks, approximately massless) particles: in such theories one of the dimensionless couplings in the Lagrangian is exchanged for a free dimensionful parameter. Eq. (18.7.7) involves no free dimensionless parameters, but it does involve one free parameter with the dimensions of mass, the integration constant Λ.

These calculations have been carried to three-loop order. The renormalization group equations to this order are[24]

$$\mu\frac{d}{d\mu}g(\mu) = -\beta_0\frac{g^3(\mu)}{16\pi^2} - \beta_1\frac{g^5(\mu)}{128\pi^4} - \beta_2\frac{g^7(\mu)}{8192\pi^6}, \tag{18.7.8}$$

where β_n are the numerical coefficients:

$$\beta_0 = 11 - \frac{2}{3}n_f, \tag{18.7.9}$$

$$\beta_1 = 51 - \frac{19}{3}n_f, \tag{18.7.10}$$

$$\beta_2 = 2857 - \frac{5033}{9}n_f - \frac{325}{27}n_f^2. \tag{18.7.11}$$

The solution is

$$\begin{aligned}\alpha_s(\mu) &\equiv \frac{g^2(\mu)}{4\pi}\\ &= \frac{4\pi}{\beta_0\ln(\mu^2/\Lambda^2)}\Bigg[1-\frac{2\beta_1}{\beta_0^2}\frac{\ln[\ln(\mu^2/\Lambda^2)]}{\ln(\mu^2/\Lambda^2)}\\ &\quad+\frac{4\beta_1^2}{\beta_0^4\ln^2(\mu^2/\Lambda^2)}\left(\left(\ln[\ln(\mu^2/\Lambda^2)]-\tfrac{1}{2}\right)^2+\frac{8\beta_2\beta_0}{\beta_1^2}-\frac{5}{4}\right)\Bigg]\end{aligned} \tag{18.7.12}$$

It should be recalled that n_f in the above results is the number of quark flavors with masses below the energies of interest. In each energy range between any two successive quark masses we have a different value of n_f, and also a different Λ, chosen to make $g(\mu)$ continuous at each quark mass. In particular, experiments on the deep inelastic scattering of electrons typically involve energies above only the first four quark flavors (u, d, s, and c), so here we must take $n_f = 4$. On the other hand, experiments at electron–positron colliders like PEP, PETRA, TRISTRAN, and LEP are at energies well above the fifth (b) quark mass, so in these experiments we must take $n_f = 5$. But these results may be expressed in terms of those for $n_f = 4$ by matching the solutions of the renormalization group equations at the b quark mass. In this way it is found[25] (using the modified minimum subtraction prescription in calculating β_2) that the strong coupling extrapolated to $m_Z = 91.2$ GeV is $\alpha_s(m_Z) \equiv g_s^2(m_Z)/4\pi = 0.118 \pm 0.006$, corresponding to $\Lambda \approx 250$ MeV for energies μ with $m_b \ll \mu \ll m_t$, where $n_f = 5$. A more recent study[26] of hadronic production in e^+–e^- annihilation at the Z resonance has given a directly measured value $\alpha_s(m_Z) = 0.1200 \pm 0.0025$, with a theoretical uncertainty of ± 0.0078, corresponding to $\Lambda = 253^{+130}_{-96}$ MeV.

18.8 Improved Perturbation Theory*

The ground-breaking paper[1] of Gell-Mann and Low was in large part directed to the problem of 'improving' perturbation theory — that is, of using the ideas of the renormalization group and the results of perturbation theory to a given order to say something about the next order of perturbation theory. To illustrate this, let's return to the specific case studied by Gell-Mann and Low: vacuum polarization in quantum electrodynamics.

Recall that the renormalized electric charge e_μ at a sliding scale μ is given by Eq. (18.2.36) in terms of the bare charge e_B as

$$e_\mu = N_\mu^{(A)-1} e_B \,, \tag{18.8.1}$$

where $N_\mu^{(A)}$ is the constant which, when multiplied into the unrenormalized electromagnetic field, gives a field renormalized at scale μ. (See Eq. (18.2.21).) Thus we can define a renormalized (and hence cut-off-independent) complete photon propagator $\Delta'_{\rho\sigma}(q, \mu, e_\mu)$ in terms of the complete propagator $\Delta'_{B\rho\sigma}(q, e_B)$ of the unrenormalized field as

$$\Delta'_{\rho\sigma}(q, \mu, e_\mu) = N_\mu^{(A)2} \Delta'_{B\rho\sigma}(q, e_B) \tag{18.8.2}$$

* This section lies somewhat out of the book's main line of development, and may be omitted in a first reading.

in such a way that the function $e_\mu^2 \Delta'_{\rho\sigma}(q,\mu,e_\mu)$ is both independent of μ, because it equals $e_B^2 \Delta'_{B\rho\sigma}(q,e_B)$, and independent of the cut-off, because both e_μ and $\Delta'_{\rho\sigma}(q,\mu,e_\mu)$ are renormalized quantities. (We are not explicitly displaying cut-off dependence here.) But Lorentz invariance and dimensional analysis tell us that this function must take the form:

$$e_\mu^2 \Delta'_{\rho\sigma}(q,\mu,e_\mu) = \frac{\eta_{\rho\sigma} d(q^2/\mu^2, e_\mu)}{q^2} + q_\rho q_\sigma \text{ terms}. \tag{18.8.3}$$

Since Eq. (18.8.3) is μ-independent, we can set $\mu = \sqrt{q^2} \equiv q$ here, so that

$$d(q^2/\mu^2, e_\mu) = d(1, e_q). \tag{18.8.4}$$

Now let us see what this tells us about the structure of the perturbation series for $d(q^2/\mu^2, e_\mu)$. The beta-function for e has the expansion:

$$\beta(e) = b_1 e^3 + b_2 e^5 + b_3 e^7 + \dots. \tag{18.8.5}$$

The renormalization group equation for e_μ then has a power series solution

$$e_q^2 = e_\mu^2 - b_1 e_\mu^4 \ln\frac{q^2}{\mu^2} - b_2 e_\mu^6 \ln\frac{q^2}{\mu^2} - \left(\frac{b_1 b_2}{2}\ln^2\frac{q^2}{\mu^2} + b_3 \ln\frac{q^2}{\mu^2}\right) e_\mu^8 + \dots. \tag{18.8.6}$$

If we also expand d:

$$d(1,e) = e^2 + d_1 e^4 + d_2 e^6 + d_3 e^8 + \dots \tag{18.8.7}$$

then

$$d(q^2/\mu^2, e_\mu) = d(1,e_q) = e_\mu^2 - \left(b_1 \ln\frac{q^2}{\mu^2} - d_1\right) e_\mu^4 - \left(b_2 \ln\frac{q^2}{\mu^2} - d_2\right) e_\mu^6 - \left(\frac{1}{2} b_1 b_2 \ln^2\frac{q^2}{\mu^2} + (b_3 - b_1 d_2)\ln\frac{q^2}{\mu^2} - d_3\right) e_\mu^8 + \dots. \tag{18.8.8}$$

Note that the leading powers of $\ln(q^2/\mu^2)$ in each order of $d(q^2/\mu^2, e_\mu)$ are respectively $0, 1, 1, 2, 3, \dots$. Also, if we calculate $d(q^2/\mu^2, e_\mu)$ to order e_μ^6 and thus determine b_1 and b_2, we can immediately write down the coefficient of the leading logarithm in order e_μ^8, as $-\frac{1}{2} b_1 b_2$. None of this would be easy to infer without using the method of the renormalization group.

Problems

1. Consider an $SU(N)$ gauge theory with a scalar field in the defining representation of $SU(N)$. Calculate the beta-function for the gauge

coupling to one-loop order, including the contribution of a scalar loop. (Recommendation: Use the background field gauge, with a constant background field.)

2. Suppose that the beta-function $\beta(g)$ for a theory with positive coupling constant g has a simple zero at $g = g_*$, where $\beta(g) \to a(g_* - g)$ with $a > 0$. What is the asymptotic behavior of the correction to the leading term $\propto E^{-\gamma^{\mathcal{O}}(g_*)}$ in the factor $N_E^{\mathcal{O}-1}$ associated with the inclusion of an operator $\mathcal{O}$ in a vacuum expectation value.

3. Show that in a theory with $\beta(g) = bg^2 + b'g^3 + b''g^4 + \cdots$, it is possible by a redefinition of the coupling constant to make the coefficient b'' anything we want.

4. Calculate the effective electric charge that should be used in studying processes at energy 100 GeV, taking account of all known charged quarks and leptons with masses below 100 GeV.

5. Calculate the asymptotic behavior for large four-momentum of the electron propagator in quantum electrodynamics. (You may use the one-loop value of Z_2 calculated elsewhere, for instance in Section 11.4.)

6. Calculate the anomalous exponent ν to first order in the expansion in $\epsilon = 4 - d$ for an $O(N)$-invariant theory of scalar fields $\phi_n(x)$ with $n = 1, \cdots, N$ belonging to the vector representation of $O(N)$, and an interaction $\frac{1}{4}g(\sum_n \phi_n^2)^2$.

References

1. M. Gell-Mann and F. E. Low, *Phys. Rev.* **95**, 1300 (1954). The freedom of choosing the definition of renormalized coupling constants had been discussed a little earlier by E. C. G. Stueckelberg and A. Peterman, *Helv. Phys. Acta* **26**, 499 (1953) (who also introduced the unfortunate term 'renormalization group') but without an explanation of its relevance to the calculation of physical processes at very high energy or very low energy. After the work of Gell-Mann and Low renormalization group methods were further developed by N. N. Bogoliubov and D. V. Shirkov, *Introduction to the Theory of Quantized Fields* (Interscience, New York, 1959): Chapter VIII, and references quoted therein. Interest in renormalization group methods in particle physics was revived in 1970 by C. G. Callan, *Phys. Rev.* **D2**, 1541 (1970); K. Symanzik, *Commun. Math. Phys.* **18**, 227 (1970);

C. G. Callan, S. Coleman, and R. Jackiw, *Ann. of Phys.* (New York), **59**, 42 (1970).

2. K. G. Wilson, *Phys. Rev.* **B4**, 3174, 3184 (1971); *Rev. Mod. Phys.* **47**, 773 (1975).

3. J. C. Collins, *Phys. Rev.* **D10**, 1213 (1974).

3*a*. R. Jost and J. M. Luttinger, *Helv. Phys. Acta* **23**, 201 (1950).

4. This is the value quoted by the Particle Data Group in 1994. More recent calculations have been summarized by G. Altarelli, CERN preprint CERN-TH-95/203, to be published in *Proceedings of the Workshop on Physics at DAΦNE, April 1995.* These more recent values of $\alpha^{-1}(m_Z)$ range from 128.89 to 129.08.

4*a*. S. Coleman and E. Weinberg, *Phys. Rev.* **D7**, 1888 (1973).

4*b*. L. D. Landau, in *Niels Bohr and the Development of Physics* (Pergamon Press, New York, 1955): p. 52; and earlier works cited therein.

4*c*. S. L. Adler, C. G. Callan, D. J. Gross, and R. Jackiw, *Phys. Rev.* **D6**, 2982 (1972); M. Baker and K. Johnson, *Physica* **96A**, 120 (1979).

4*d*. For a discussion and references, see J. Glimm and A. Jaffe, *Quantum Physics – A Functional Integral Point of View*, Second Edition (Springer-Verlag, New York, 1987): Section 21.6; R. Fernandez, J. Fröhlich, and A. D. Sokal, *Random Walks, Critical Phenomena, and Triviality in Quantum Field Theory* (Springer-Verlag, Berlin, 1992): Chapter 15.

5. S. Weinberg, in *General Relativity*, eds. S. W. Hawking and W. Israel, eds. (Cambridge University Press, Cambridge, 1979): p. 790.

6. S. Weinberg, *Phys. Rev.* **D8**, 3497 (1973).

7. The original calculation is by K. G. Wilson and M. E. Fisher, *Phys. Rev. Lett.* **28**, 240 (1972); K. G. Wilson, *Phys. Rev. Lett.* **28**, 548 (1972). For reviews, see K. G. Wilson and J. Kogut, *Phys. Rep.* **12C**, No. 2 (1974); M. E. Fisher, *Rev. Mod. Phys.* **46**, 597 (1974); E. Brézin, J. C. Le Guillou, and J. Zinn-Justin, in *Phase Transitions and Critical Phenomena*, eds. C. Domb and M. S. Green (Academic Press, London, 1975).

8. See, e. g., P. M. Chaikin and T. C. Lubensky, *Principles of Condensed Matter Physics* (Cambridge University Press, Cambridge, 1995): p. 231.

8*a*. E. Brézin, J. C. Le Guillou, J. Zinn-Justin, and B. G. Nickel, *Phys. Lett.* **44A**, 227 (1973); K. G. Wilson and J. Kogut, Ref. 7, Section 8.

9. G. 't Hooft, *Nucl. Phys.* **B61**, 455 (1973); *Nucl. Phys.* **B82**, 444 (1973). The derivation presented here is a somewhat simplified version of 't Hooft's.

10. O. W. Greenberg, *Phys. Rev. Lett.*, **13**, 598 (1964); M. Y. Han and Y. Nambu, *Phys. Rev.* **139**, B1006 (1965); W. A. Bardeen, H. Fritzsch, and M. Gell-Mann, in *Scale and Conformal Invariance in Hadron Physics*, ed. R. Gatto (Wiley, New York, 1973).

11. M. Gell-Mann, *Phys. Lett.* **8**, 214 (1964); G. Zweig, CERN preprint TH401 (1964).

12. D. J. Gross and F. Wilczek, *Phys. Rev. Lett.*, **30**, 1343 (1973).

13. H. D. Politzer, *Phys. Rev. Lett.* **30**, 1346 (1973)

14. E. D. Bloom *et al.*, *Phys. Rev. Lett.* **23**, 930 (1969); M. Breidenbach *et al.*, *Phys. Rev. Lett.* **23**, 933 (1969); J. L. Friedman and H. W. Kendall, *Annual Reviews of Nuclear Science* **22**, 203 (1972).

15. A. Zee, unpublished.

16. G. 't Hooft, unpublished.

17. S. Weinberg, *Phys. Rev.* **D8**, 605 (1973).

18. D. J. Gross and F. Wilczek, *Phys. Rev.* **D8**, 3633 (1973); S. Weinberg, *Phys. Rev. Lett.* **31**, 494 (1973).

19. A similar idea had been suggested before the discovery of asymptotic freedom by H. Fritzsch, M. Gell-Mann, and H. Leutwyler, *Phys. Lett.* **47B**, 365 (1973).

20. G. Hanson *et al.*, *Phys. Rev. Lett.* **35**, 1609 (1975); R. F. Schwitters, in *Proceedings of the International Conference on Lepton and Photon Interactions at High Energy at Stanford, 1975*, ed. W. T. Kirk (Stanford Linear Accelerator Center, Stanford, 1975): p. 5; G. Hanson, Stanford Linear Accelerator Center Report SLAC–PUB–1814 (1976), unpublished.

21. G. Sterman and S. Weinberg, *Phys. Rev. Lett.* **39**, 1436 (1977).

22. J. Ellis, M. K. Gaillard, and G. G. Ross, *Nucl. Phys.* **B111**, 253 (1976).

23. E. Eichten, K. Lane, and M. Peskin, *Phys. Rev. Lett.* **50**, 811 (1983).

24. For a review, see I. Hinchliffe, in 'Review of Particle Properties,' *Phys. Rev.* **D50**, 1177 (1994): Section 25.

25. G. Altarelli, in *Proceedings of the Rencontres de Hanoi,* CERN preprint CERN-PPE/94-71 (1994).

26. K. Abe *et al.* (SLD collaboration), *Phys. Rev.* **D51**, 962 (1995). A summary of earlier data on Z^0 decay into hadrons by M. Shifman, Minnesota preprint hep-ph/9501222 (1995), gave a value $\alpha_s(m_Z) = 0.125 \pm 0.005$, corresponding to $\Lambda \approx 500$ MeV.

19

Spontaneously Broken Global Symmetries

Much of the physics of this century has been built on principles of symmetry: first the spacetime symmetries of Einstein's 1905 special theory of relativity, and then internal symmetries, such as the approximate $SU(2)$ isospin symmetry of the 1930s. It was therefore exciting when in the 1960s it was discovered that there are more internal symmetries than could be guessed by inspection of the spectrum of elementary particles. There are exact or approximate symmetries of the underlying theory that are 'spontaneously broken,' in the sense that they are not realized as symmetry transformations of the physical states of the theory, and in particular do not leave the vacuum state invariant. The breakthrough was the discovery of a broken approximate global $SU(2) \times SU(2)$ symmetry of the strong interactions, which will be discussed in detail in Section 19.3. This was soon followed by the discovery of an exact but spontaneously broken local $SU(2) \times U(1)$ symmetry of the weak and electromagnetic interactions, which will be taken up along with more general broken local symmetries in Chapter 21. In this chapter we shall begin with a general discussion of broken global symmetries, and then move on to physical examples.

19.1 Degenerate Vacua

We do not have to look far for examples of spontaneous symmetry breaking. Consider a chair. The equations governing the atoms of the chair are rotationally symmetric, but a solution of these equations, the actual chair, has a definite orientation in space. Here we will be concerned not so much with the breaking of symmetries by *objects* like chairs, but rather with the symmetry breaking in the ground state of any realistic quantum field theory, the vacuum.

A spontaneously broken symmetry in field theory is always associated with a degeneracy of vacuum states. For instance, consider a symmetry transformation of the action, and of the measure used in integrating over fields, that acts linearly on a set of scalar fields $\phi_n(x)$:

$$\phi_n(x) \to \phi'_n(x) = \sum_m L_{nm}\phi_m(x) . \tag{19.1.1}$$

(The ϕ_n need not be elementary fields; they can be composite objects, like $\bar{\psi}\Gamma_n\psi$.) As we saw in Section 16.4, the quantum effective action $\Gamma[\phi]$ will then have the same symmetry

$$\Gamma[\phi] = \Gamma[L\phi] . \tag{19.1.2}$$

For the vacuum, the expectation value of $\phi(x)$ must be at a minimum of the vacuum energy $-\Gamma[\phi]$, say at $\phi(x) = \bar{\phi}$ (a constant). But if $L\bar{\phi} \neq \bar{\phi}$, then this vacuum is not unique; $-\Gamma[\phi]$ has the same value at $\phi = L\bar{\phi}$ as it does at $\bar{\phi}$. In the simple special case where the symmetry transformation (19.1.1) is a reflection, $\phi \to -\phi$, if $-\Gamma(\phi)$ has a minimum at a non-zero value $\bar{\phi}$ of ϕ, then it has two minima, at $\bar{\phi}$ and $-\bar{\phi}$, each corresponding to a state of broken symmetry.

We are not yet ready to conclude that in such cases the symmetry is broken, because we have not yet ruled out the possibility that the true vacuum is a linear superposition of vacuum states in which ϕ_m has various expectation values, which would respect the assumed symmetry. For instance, in a theory with a symmetry $\phi \to -\phi$, even if $\Gamma(\phi)$ has a minimum for some non-zero value $\bar{\phi}$ of ϕ, how do we know that the true vacuum is one of the states $|\text{VAC},\pm\rangle$ for which Φ has expectation values $\bar{\phi}$ and $-\bar{\phi}$, and not some linear combination like $|\text{VAC},+\rangle + |\text{VAC},-\rangle$ that *would* respect the symmetry under $\phi \to -\phi$? The assumed symmetry under the transformation $\phi \to -\phi$ tells us that the vacuum matrix elements of the Hamiltonian are

$$\langle \text{VAC},+|H|\text{VAC},+\rangle = \langle \text{VAC},-|H|\text{VAC},-\rangle \equiv a$$

(with a real) and

$$\langle \text{VAC},+|H|\text{VAC},-\rangle = \langle \text{VAC},-|H|\text{VAC},+\rangle \equiv b ,$$

(with b real), so the eigenstates of the Hamiltonian are $|\text{VAC},+\rangle \pm |\text{VAC},-\rangle$, with energies $a \pm |b|$. These energy eigenstates are invariant (or invariant up to a sign) under the symmetry $\phi \to -\phi$. In fact, the same issue also arises for chairs. The quantum mechanical ground state of an isolated chair is actually rotationally invariant; it is a state with zero angular momentum quantum numbers, and hence with no definite orientation in space.

Spontaneous symmetry breaking actually occurs only for idealized systems that are infinitely large. The appearance of broken symmetry for a chair arises because it has a macroscopic moment of inertia I, so that its ground state is part of a tower of rotationally excited states whose energies are separated by only tiny amounts, of order $\hbar^2/I$. This gives the state vector of the chair an exquisite sensitivity to external perturbations; even very weak external fields will shift the energy by much more than the energy difference of these rotational levels. In consequence, any rotationally asymmetric external field will cause the ground state or any other state of the chair with definite angular momentum quantum numbers rapidly to develop components with other angular momentum quantum numbers. The states of the chair that are relatively stable with respect to small external perturbations are not those with definite angular momentum quantum numbers, but rather those with a definite orientation, in which the rotational symmetry of the underlying theory is broken.

For the vacuum also, the possibility of spontaneous symmetry breaking is again related to the large size of the system, specifically to the large volume of space. In the above example of a reflection symmetry, the off-diagonal matrix element b of the Hamiltonian involves an integration over field configurations that tunnel from the minimum at $\phi = \bar{\phi}$ to the one at $\phi = -\bar{\phi}$, so it is smaller than the diagonal matrix element a by a barrier penetration factor that for a spatial volume $\mathscr{V}$ is of the form $\exp(-C\mathscr{V})$, where C is a positive constant* depending on the microscopic parameters of the theory. The two energy eigenstates $|\text{VAC},+\rangle \pm |\text{VAC},-\rangle$ are thus essentially degenerate for any macroscopic volume, and so are strongly mixed by any perturbation that is an odd functional of ϕ. Even if such a perturbation H' is very weak, its diagonal elements $\langle\text{VAC},\pm|H'|\text{VAC},\pm\rangle$ will differ by much more than the exponentially suppressed off-diagonal elements of either H or the perturbation. Thus the vacuum eigenstates of the perturbed Hamiltonian will be very close to either one of the broken symmetry states $|\text{VAC},\pm\rangle$ which diagonalize the perturbation, and *not* to the invariant states $|\text{VAC},+\rangle \pm |\text{VAC},-\rangle$. Which one of the states $|\text{VAC},\pm\rangle$ is the true vacuum for very small perturbations? This depends on the perturbation, but since these two states are related by a symmetry transformation of the original Hamiltonian, it doesn't matter; if the perturbation is sufficiently small, no observer will be able to tell the difference.

The vanishing of matrix elements between vacuum states with different field expectation values becomes exact in a space of infinite volume.[1] For infinite volume, a general vacuum state $|v\rangle$ may be defined as a state with

---

* For instance, by analogy with the classic wave mechanical problem of barrier penetration, for a Lagrangian density of the form $-\frac{1}{2}\partial_\mu\phi\partial^\mu\phi - V(\phi)$, we have $C = \int_{-\bar{\phi}}^{+\bar{\phi}} \sqrt{2V(\phi)}\, d\phi$. We will not bother to calculate the off-diagonal matrix element b here, because we shall soon give a general argument that shows that it vanishes for infinite volume.

zero momentum

$$\mathbf{P}|v\rangle = 0 \tag{19.1.3}$$

for which this is a *discrete* momentum eigenvalue. (This excludes single-particle or multiparticle states, for which the momentum value zero is always part of a continuum of momentum values in a space of infinite volume.) In general there may be a number of such states. They can usually be expanded in a discrete set, and our notation will treat them as if they were discrete. They will be chosen to be orthonormal

$$\langle u|v\rangle = \delta_{uv} \,. \tag{19.1.4}$$

Any matrix element of a product of local Hermitian operators at equal times between these states may be expressed as a sum over states:

$$\langle u|A(\mathbf{x})\,B(0)|v\rangle = \sum_w \langle u|A(0)|w\rangle\,\langle w|B(0)|v\rangle + \int d^3p \sum_N \langle u|A(0)|N,\mathbf{p}\rangle\,\langle N,\mathbf{p}|B(0)|v\rangle e^{-i\mathbf{p}\cdot\mathbf{x}} \,, \tag{19.1.5}$$

where $|N,\mathbf{p}\rangle$ are a set of orthonormalized continuum states of definite three-momentum $\mathbf{p}$ that together with the $|v\rangle$ span the whole physical Hilbert space. (Here N may include continuous as well as discrete labels. Also, we are dropping time arguments.) We assume without proof that because the $|N,\mathbf{p}\rangle$ belong to the continuous spectrum of the momentum operator $\mathbf{P}$, the dependence of matrix elements on $\mathbf{p}$ is smooth enough (that is, Lebesgue integrable) to allow the use of the Riemann–Lebesgue theorem,[2] so that the integral over $\mathbf{p}$ vanishes as $|\mathbf{x}| \to \infty$. In this limit, we have then

$$\langle u|A(\mathbf{x})\,B(0)|v\rangle \xrightarrow[|\mathbf{x}|\to\infty]{} \sum_w \langle u|A(0)|w\rangle\,\langle w|B(0)|v\rangle \,. \tag{19.1.6}$$

Likewise,

$$\langle u|B(0)\,A(\mathbf{x})|v\rangle \xrightarrow[|\mathbf{x}|\to\infty]{} \sum_w \langle u|B(0)|w\rangle\,\langle w|A(0)|v\rangle \,. \tag{19.1.7}$$

But causality tells us that the equal-time commutator $[A(\mathbf{x}), B(0)]$ vanishes for $\mathbf{x} \neq 0$ (see Section 5.1), so the matrix elements (19.1.6) and (19.1.7) are equal, and thus the Hermitian matrices $\langle u|A(0)|v\rangle$, $\langle u|B(0)|v\rangle$, etc., must all commute with one another. It follows that they can all be simultaneously diagonalized. Changing if necessary to this basis, we have then for every Hermitian local operator $A(\mathbf{x})$ of the theory

$$\langle u|A(0)|v\rangle = \delta_{uv}\, a_v \tag{19.1.8}$$

with a_v a real number, the expectation value of A in the state $|v\rangle$. So for infinite volume any Hamiltonian constructed from local operators will have vanishing matrix elements between the different vacua $|v\rangle$. In the absence of off-diagonal terms in the Hamiltonian, any two $|v\rangle$s connected by a symmetry operation will be degenerate. A symmetry-breaking perturbation built out of such local operators will be diagonal in the same basis, and will therefore yield a ground state that is one of the $|v\rangle$s, rather than a linear combination of them.

It is reassuring that the vacuum states $|v\rangle$ which are stable against small field-dependent perturbations are also vacuum states in which the cluster decomposition condition (see Chapter 4) is satisfied. This principle requires that for the physical vacuum state $|\text{VAC}\rangle$

$$\langle\text{VAC}|A(\mathbf{x})B(0)|\text{VAC}\rangle \xrightarrow[\mathbf{x}\to\infty]{} \langle\text{VAC}|A(\mathbf{x})|\text{VAC}\rangle\;\langle\text{VAC}|B(0)|\text{VAC}\rangle\,. \quad (19.1.9)$$

This condition is satisfied if we take the vacuum state $|\text{VAC}\rangle$ to be any one of the states $|v\rangle$ in the basis defined by Eq. (19.1.8), but not if we take it to be a general linear combination of several of the $|v\rangle$s.

19.2 Goldstone Bosons

We now specialize to the case of a spontaneously broken *continuous* symmetry. In this case there is a theorem, that (with one important exception, to be considered in Chapter 21) the spectrum of physical particles must contain one particle of zero mass and spin for each broken symmetry. Such particles, known as Goldstone bosons (or Nambu–Goldstone bosons) were first encountered in specific models by Goldstone[3] and Nambu[4]; two general proofs of their existence were then given by Goldstone, Salam, and myself.[5] This section will present both of these proofs, and then go on to consider the properties of the Goldstone bosons.

Suppose that the action and measure are invariant under a continuous symmetry, under which a set of Hermitian scalar fields $\phi_n(x)$ (either elementary or composite) are subjected to the linear infinitesimal transformations

$$\phi_n(x) \to \phi_n(x) + i\epsilon \sum_m t_{nm}\phi_m(x)\,, \quad (19.2.1)$$

with it_{nm} a finite real matrix. As we found in Section 16.4, the effective action is then also invariant under this transformation

$$\sum_{n,m}\int \frac{\delta\Gamma[\phi]}{\delta\phi_n(x)}\, t_{nm}\phi_m(x)\, d^4x = 0\,. \quad (19.2.2)$$

We shall specialize to the case of a translationally invariant theory with constant fields ϕ_n, where as we saw in Section 16.2, the effective action takes the form

$$\Gamma[\phi] = -\mathscr{V} V(\phi)\,, \tag{19.2.3}$$

where $\mathscr{V}$ is the spacetime volume and $V(\phi)$ is known as the effective potential. Eq. (19.2.2) may then be written

$$\sum_{n,m} \frac{\partial V(\phi)}{\partial \phi_n}\, t_{nm}\, \phi_m = 0\,. \tag{19.2.4}$$

We will use this symmetry requirement in a form obtained by differentiating again with respect to ϕ_ℓ:

$$\sum_n \frac{\partial V(\phi)}{\partial \phi_n}\, t_{n\ell} + \sum_{n,m} \frac{\partial^2 V(\phi)}{\partial \phi_n \partial \phi_\ell}\, t_{nm}\phi_m = 0\,. \tag{19.2.5}$$

Now specialize to the case where ϕ_n is at the minimum of $V(\phi)$, that is, at the vacuum expectation value $\bar{\phi}_n$. Since $V(\phi)$ is stationary at its minimum, the first term in Eq. (19.2.5) vanishes, so

$$\sum_{n,m} \left.\frac{\partial^2 V(\phi)}{\partial \phi_n \partial \phi_\ell}\right|_{\phi=\bar{\phi}} t_{nm}\, \bar{\phi}_m = 0\,. \tag{19.2.6}$$

The general results of Section 16.1 show that the second derivative in Eq. (19.2.6) is just the sum of all connected one-particle-irreducible momentum-space Feynman diagrams with external lines labelled n and ℓ and carrying zero four-momentum. As shown at the end of Section 16.1, it is related to the reciprocal of the momentum-space propagator by

$$\frac{\partial^2 V(\phi)}{\partial \phi_n\, \partial \phi_\ell} = \Delta^{-1}_{n\ell}(0)\,, \tag{19.2.7}$$

so Eq. (19.2.6) gives

$$\sum_{nm} \Delta^{-1}_{n\ell}(0) t_{nm}\bar{\phi}_m = 0\,. \tag{19.2.8}$$

Thus, if the symmetry is broken, so that $\sum_m t_{nm}\bar{\phi}_m$ is non-zero, then this is an eigenvector of $\Delta^{-1}_{n\ell}(0)$ with eigenvalue zero. The existence of such an eigenvector means that $\Delta_{n\ell}(q)$ has a pole at $q^2 = 0$. The rank of the residue of the pole at $q^2 = 0$ is equal to the dimensionality of the space of the vectors $t\bar{\phi}$, with t running over all the generators of continuous symmetries of the theory. Roughly speaking, there is one massless boson for every independent broken symmetry.

In the classic example of a broken symmetry, the Lagrangian involves a set of N real scalar fields ϕ_n, and takes the form

$$\mathscr{L} = -\frac{1}{2}\sum_n \partial_\mu\phi_n\partial^\mu\phi_n - \frac{\mathscr{M}^2}{2}\sum_n \phi_n\phi_n - \frac{g}{4}\left(\sum_n \phi_n\phi_n\right)^2 . \tag{19.2.9}$$

This is invariant under the group $O(N)$, consisting of rotations of the N-vector with components ϕ_n. For constant fields the effective potential in the tree approximation is given simply by minus the non-derivative terms in the Lagrangian density

$$V(\phi) \simeq \frac{\mathcal{M}^2}{2} \sum_n \phi_n \phi_n + \frac{g}{4} \left(\sum_n \phi_n \phi_n \right)^2 . \tag{19.2.10}$$

As usual we suppose that g is positive-definite. (Otherwise the minimum, if any, of $V(\phi)$ lies outside the range of validity of perturbation theory.) If $\mathcal{M}^2$ is also positive, the minimum of $V(\phi)$ is at the point $\phi = 0$, which is invariant under $O(N)$. On the other hand, for $\mathcal{M}^2 < 0$ the minimum is at points $\bar{\phi}_n$ at which

$$\sum_n \bar{\phi}_n \bar{\phi}_n = -\mathcal{M}^2 / g . \tag{19.2.11}$$

The mass matrix in the tree approximation is then

$$\begin{aligned} M^2_{nm} &= \left. \frac{\partial^2 V(\phi)}{\partial \phi_n \partial \phi_m} \right|_{\phi = \bar{\phi}} \\ &= \mathcal{M}^2 \delta_{nm} + g \delta_{nm} \sum_\ell \bar{\phi}_\ell \bar{\phi}_\ell + 2 g \bar{\phi}_n \bar{\phi}_m \\ &= 2 g \, \bar{\phi}_n \, \bar{\phi}_m . \end{aligned} \tag{19.2.12}$$

This has one eigenvector $\bar{\phi}_n$ with a non-zero eigenvalue:

$$m^2 = 2 g \sum_n \bar{\phi}_n \bar{\phi}_n = 2 |\mathcal{M}^2| , \tag{19.2.13}$$

and $N - 1$ eigenvectors perpendicular to $\bar{\phi}$ with eigenvalue zero. The reason for the appearance of only $N - 1$ Goldstone bosons is just that $O(N)$ is broken down to $O(N-1)$ (the subgroup of $O(N)$ that leaves $\bar{\phi}$ invariant), and so the number of independent broken symmetries is the dimensionality of $O(N)$ minus the dimensionality of $O(N-1)$, or

$$\frac{1}{2} N(N-1) - \frac{1}{2}(N-1)(N-2) = N - 1 . \tag{19.2.14}$$

Here is another proof of the existence of Goldstone bosons, one that makes no use of the effective action formalism. As we saw in Chapter 7, any continuous symmetry of the action leads to the existence of a conserved current J^μ:

$$\frac{\partial J^\mu(x)}{\partial x^\mu} = 0 , \tag{19.2.15}$$

with a charge Q that induces the associated symmetry transformation

$$Q = \int d^3x \, J^0(\mathbf{x}, 0) , \tag{19.2.16}$$

$$\left[Q, \phi_n(x)\right] = -\sum_m t_{nm}\phi_m(x)\,. \tag{19.2.17}$$

Operator relations like Eqs. (19.2.15)–(19.2.17) are unaffected by spontaneous symmetry breaking, which is manifested in the properties of physical *states*. Now, consider the vacuum expectation value of the commutator of the current and field. Summing over intermediate states, this is

$$\left\langle\left[J^\lambda(y),\, \phi_n(x)\right]\right\rangle_{\text{VAC}} = (2\pi)^{-3}\int d^4p\left[\rho_n^\lambda(p)\,e^{ip\cdot(y-x)} - \tilde{\rho}_n^\lambda(p)\,e^{ip\cdot(x-y)}\right], \tag{19.2.18}$$

where, using translation invariance,

$$(2\pi)^{-3}i\rho_n^\lambda(p) = \sum_N \langle\text{VAC}|J^\lambda(0)|N\rangle\,\langle N|\phi_n(0)|\text{VAC}\rangle\delta^4(p-p_N)\,, \tag{19.2.19}$$

$$(2\pi)^{-3}i\tilde{\rho}_n^\lambda(p) = \sum_N \langle\text{VAC}|\phi_n(0)|N\rangle\,\langle N|J^\lambda(0)|\text{VAC}\rangle\delta^4(p-p_N)\,. \tag{19.2.20}$$

We usually take J^λ as well as ϕ_n to be Hermitian operators, in which case Eqs. (19.2.19) and (19.2.20) are complex conjugates

$$\rho_n^\lambda(p) = -\tilde{\rho}_n^{\lambda*}(p)\,, \tag{19.2.21}$$

but this will not be assumed here.

Lorentz invariance tells us that ρ and $\tilde{\rho}$ must take the forms

$$\rho_n^\lambda(p) = p^\lambda\rho_n(-p^2)\theta(p^0)\,, \tag{19.2.22}$$

$$\tilde{\rho}_n^\lambda(p) = p^\lambda\tilde{\rho}_n(-p^2)\theta(p^0)\,. \tag{19.2.23}$$

(The factor $\theta(p^0)$, which is $+1$ for $p^0 > 0$ and zero otherwise, is required by the fact that p_N is the four-momentum of a physical state.) This gives

$$\left\langle\left[J^\lambda(y), \phi_n(x)\right]\right\rangle_{\text{VAC}} = \frac{\partial}{\partial y_\lambda}\int d\mu^2\Big[\rho_n(\mu^2)\Delta_+(y-x;\mu^2) + \tilde{\rho}_n(\mu^2)\Delta_+(x-y;\mu^2)\Big]\,, \tag{19.2.24}$$

where Δ_+ is the familiar function

$$\Delta_+(z;\mu^2) = (2\pi)^{-3}\int d^4p\,\theta(p^0)\,\delta(p^2+\mu^2)\,e^{ip\cdot z}\,. \tag{19.2.25}$$

As remarked in Chapter 5, Lorentz invariance allows $\Delta_+(z;\mu^2)$ to depend only on z^2, μ^2, and $\theta(z^0)$, and for $z^2 > 0$ only on z^2 and μ^2. Hence $\Delta_+(x-y;\mu^2)$ and $\Delta_+(y-x;\mu^2)$ are equal for $x-y$ spacelike, and so in this case

$$\left\langle\left[J^\lambda(y),\, \phi_n(x)\right]\right\rangle_{\text{VAC}} = \frac{\partial}{\partial y_\lambda}\int d\mu^2\left[\rho_n(\mu^2)+\tilde{\rho}_n(\mu^2)\right]\Delta_+(y-x;\mu^2)\,. \tag{19.2.26}$$

But for $x - y$ spacelike all commutators must vanish, so

$$\rho_n(\mu^2) = -\tilde{\rho}_n(\mu^2)\,, \tag{19.2.27}$$

and therefore Eq. (19.2.24) gives for general x and y:

$$\left\langle \Big[J^\lambda(y),\, \phi_n(x)\Big] \right\rangle_{\rm VAC} = \frac{\partial}{\partial y_\lambda} \int d\mu^2 \rho_n(\mu^2) \Big[\Delta_+(y-x;\mu^2) - \Delta_+(x-y;\mu^2)\Big]\,. \tag{19.2.28}$$

Where Eq. (19.2.21) applies, Eq. (19.2.27) also shows that $\rho_n(\mu^2)$ is *real.*

Now let us use the fact that $J^\lambda(y)$ is conserved. Applying the derivative $\partial/\partial y^\lambda$ to both sides of Eq. (19.2.28), and using the familiar equation

$$(\Box_y - \mu^2)\,\Delta_+(y-x;\mu^2) = 0\,, \tag{19.2.29}$$

we find, for *all* x and y

$$0 = \int d\mu^2\, \mu^2 \rho_n(\mu^2)\, \Big[\Delta_+(y-x;\mu^2) - \Delta_+(x-y;\mu^2)\Big]\,, \tag{19.2.30}$$

and so (since $\Delta_+(x-y)$ is *not* even for $x - y$ timelike or lightlike)

$$\mu^2 \rho_n(\mu^2) = 0\,. \tag{19.2.31}$$

Normally we would conclude from this that $\rho_n(\mu^2)$ vanishes for all μ^2. However, this is not possible in the case of broken symmetry. Set $\lambda = 0$ and $x^0 = y^0 = t$ in Eq. (19.2.28):

$$\begin{aligned}\left\langle \Big[J^0(\mathbf{y},t),\, \phi_n(\mathbf{x},t)\Big] \right\rangle_{\rm VAC} &= 2i(2\pi)^{-3} \int d\mu^2 \rho_n(\mu^2) \\ &\quad \times \int d^4p \sqrt{\mathbf{p}^2 + \mu^2}\; e^{i\mathbf{p}\cdot(\mathbf{y}-\mathbf{x})} \delta(p^2 + \mu^2) \\ &= i\delta^3(\mathbf{y}-\mathbf{x}) \int d\mu^2 \rho_n(\mu^2)\,.\end{aligned}$$

Integrating and using Eqs. (19.2.16) and (19.2.17) gives

$$-\sum_m t_{nm} \langle \phi_m \rangle_{\rm VAC} = i \int d\mu^2 \rho_n(\mu^2)\,. \tag{19.2.32}$$

Eqs. (19.2.31) and (19.2.32) can be reconciled only if

$$\rho_n(\mu^2) = i\,\delta(\mu^2) \sum_m t_{nm} \langle \phi_m(0) \rangle_{\rm VAC}\,. \tag{19.2.33}$$

(This is real for Hermitian fields ϕ_n because in this case Eq. (19.2.1) requires t_{nm} to be imaginary.) Thus as long as the symmetry is broken, $\rho_n(\mu^2)$ cannot vanish, but rather consists entirely of a term proportional to $\delta(\mu^2)$. Such a term can obviously only arise in a theory that has massless particles, because otherwise the spectrum of center-of-mass squared energies $-p_N^2$ would not extend down to zero. Furthermore a delta-function $\delta(\mu^2)$ can only arise from single particle states of zero mass; multiparticle states would contribute a continuum extending down to $\mu^2 = 0$. The state $\phi_n(0)|\text{VAC}\rangle$ is rotationally invariant, so $\langle N|\phi_n(0)|\text{VAC}\rangle$ must vanish for any state N of non-zero helicity. Also $\langle\text{VAC}|J^0|N\rangle$ vanishes for any state N that has different intrinsic parity or (unbroken) internal quantum numbers from J^0. We conclude then that *a broken symmetry with $t_{nm}\langle\phi_n(0)\rangle_{\text{VAC}} \neq 0$ requires the existence of a massless particle of spin zero and the same parity and internal quantum numbers as J^0*. These are our Goldstone bosons.

The above argument breaks down when the spontaneously broken symmetry is a local rather than a global symmetry. Either we choose a Lorentz-invariant gauge, such as the Landau gauge with $\partial_\mu A^\mu = 0$, in which case as shown in Section 15.7 the positivity assumptions of quantum mechanics are violated, or we adopt a gauge like the axial gauge with $A_3^\mu = 0$, in which case the ordinary rules of quantum mechanics apply but manifest Lorentz invariance is lost. As we shall see in Chapter 21, this exception is not just a technicality; spontaneously broken local symmetries do not lead to Goldstone bosons.

It will be useful to look in a little more detail at how the coefficient of the delta-function in $\rho_n(\mu^2)$ is related to the properties of the Goldstone boson. For a spin zero boson B of four-momentum p_B^μ, Lorentz invariance requires the matrix element of the current between the vacuum and one-particle states to take the form

$$\langle\text{VAC}|J^\lambda(x)|B\rangle = i\,\frac{F\,p_B^\lambda\,e^{ip_B\cdot x}}{(2\pi)^{3/2}\sqrt{2p_B^0}}\,, \tag{19.2.34}$$

where $\mathbf{p}_B$ is the momentum of B, $p_B^0 \equiv |\mathbf{p}_B|$ and F is a constant coefficient with the dimensions of energy. (This is consistent with current conservation because $p_{B\lambda}p_B^\lambda = 0$.) Also, the matrix element of the scalar field $\phi_n(y)$ between a single-particle state and the vacuum is of the form

$$\langle B|\phi_n(y)|\text{VAC}\rangle = \frac{Z_n\,e^{-ip_B\cdot y}}{(2\pi)^{3/2}\sqrt{2p_B^0}}\,, \tag{19.2.35}$$

where Z_n is a dimensionless constant. From Eqs. (19.2.34) and (19.2.35), we have then

$$\begin{aligned}(2\pi)^{-3} i\rho_n(-p^2)p^\lambda\theta(p^0) &\equiv \int d^3p_B\langle\text{VAC}|J^\lambda(0)|B\rangle\,\langle B|\phi_n(0)|\text{VAC}\rangle\,\delta^4(p-p_B)\\ &= \delta(p^0-|\mathbf{p}|)\,(2\pi)^{-3}(2p^0)^{-1}p^\lambda iFZ_n\\ &= \theta(p^0)\,\delta(-p^2)\,(2\pi)^{-3}p^\lambda iFZ_n\,,\end{aligned}$$

so

$$\rho_n(\mu^2) = FZ_n\,\delta(\mu^2)\,. \tag{19.2.36}$$

Comparing with Eq. (19.2.33), this gives

$$iFZ_n = -\sum_m t_{nm}\langle\phi_m(0)\rangle_{\rm VAC}\ . \tag{19.2.37}$$

More generally, we may have several broken symmetries with generators t_a and currents J^μ_a, which we can take independent in the sense that no linear combination of the t_a is unbroken. For each of these, there is a Goldstone boson $|B_a\rangle$, and we define Z_{an} and F_{ab} by

$$\langle{\rm VAC}|J^\lambda_a(x)|B_b\rangle = i\,\frac{F_{ab}p^\lambda_B e^{ip_B\cdot x}}{(2\pi)^{3/2}\sqrt{2p^0_B}}\ , \tag{19.2.38}$$

$$\langle B_a|\phi_n(y)|{\rm VAC}\rangle = \frac{Z_{an}\,e^{-ip_B\cdot y}}{(2\pi)^{3/2}\sqrt{2p^0_B}}\ . \tag{19.2.39}$$

Eq. (19.2.37) applies for each a, so

$$i\sum_b F_{ab}Z_{bn} = -\sum_m [t_a]_{nm}\langle\phi_m(0)\rangle_{\rm VAC}\ . \tag{19.2.40}$$

For instance, in the $O(N)$ example discussed earlier, we can adapt our basis so that the vacuum expectation value points in the one–direction

$$\bar\phi_m \equiv \langle\phi_m(0)\rangle_{\rm VAC} = v\delta_{m1}\ . \tag{19.2.41}$$

The $N-1$ broken symmetry generators t_a (with $a = 2\cdots N$) can be defined as those for infinitesimal rotations in the 1–a plane. With a convenient choice of normalization, these have the non-zero elements

$$[t_a]_{1a} = -[t_a]_{a1} = i \tag{19.2.42}$$

(with no summation over a). The unbroken $O(N-1)$ symmetry, under which the $N-1$ Goldstone bosons transform according to the vector representation, tells us that

$$F_{ab} = \delta_{ab}F\ , \qquad Z_{a1} = 0\ , \qquad Z_{ab} = Z\delta_{ab}\ . \tag{19.2.43}$$

Then Eq. (19.2.40) requires

$$FZ = v\ . \tag{19.2.44}$$

It is conventional to adopt the field renormalization prescription that $Z = 1$, so $F = v$. Thus F is a measure of the strength of the symmetry breaking. As we shall now see, the parameter $1/F$ determines the strength with which the Goldstone bosons interact with each other and with other particles.

A broken symmetry tells us more about the Goldstone bosons than just that they have zero mass; it also tightly constrains their interactions at low energy. To see this in the simplest case, consider the matrix element of

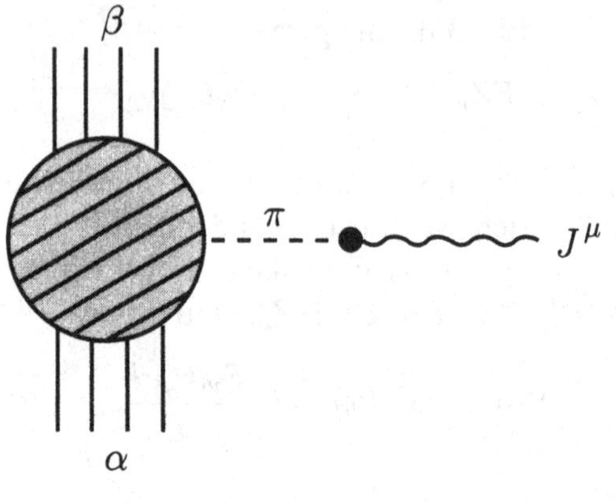

Figure 19.1. Feynman diagrams for pole terms in the matrix element of a symmetry current $J^\mu(x)$ between general states α and β, due to a Goldstone boson internal line, denoted π.

the current $J^\mu(x)$ associated with a broken symmetry. Between arbitrary states α, β:

$$\left\langle \beta | J^\mu(x) | \alpha \right\rangle = e^{iq\cdot x} \left\langle \beta | J^\mu(0) | \alpha \right\rangle , \tag{19.2.45}$$

with

$$q^\mu \equiv p^\mu_\alpha - p^\mu_\beta \, . \tag{19.2.46}$$

We know that $J^\mu(x)$ has a non-zero matrix element between the vacuum and a one-Goldstone-boson state $|B, \mathbf{q}\rangle$, given by Eq. (19.2.34). It follows by the usual rules of polology (see Chapter 10) that the matrix element (19.2.45) has a pole at $q^2 \to 0$, with*

$$\langle \beta | J^\mu(0) | \alpha \rangle \to \frac{iF\, q^\mu}{q^2} \, M_{\beta\alpha} \, , \tag{19.2.47}$$

where $i(2\pi)^4 \delta^4(p_\alpha - p_\beta - q) M_{\beta\alpha}/(2\pi)^{3/2}(2q^0)^{1/2}$ is the S-matrix element for emitting a Goldstone boson of four-momentum q in the transition $\alpha \to \beta$. Let us therefore write

$$\langle \beta | J^\mu(0) | \alpha \rangle \equiv N^\mu_{\beta\alpha} + \frac{iFq^\mu}{q^2} \, M_{\beta\alpha} \, , \tag{19.2.48}$$

where $N^\mu_{\beta\alpha}$ is defined as the non-pole contributions to the matrix element of the current. From Eq. (19.2.45), we see that the conservation law

* Where the Goldstone boson corresponds to an elementary field, Eq. (19.2.47) may be obtained by inspection of the class of Feynman diagrams shown in Figure 19.1. The factor $i(2\pi)^4$ in the S-matrix element for $\alpha \to \beta + B$ is cancelled by the factor $-i(2\pi)^{-4}$ associated with the B propagator. The general rules of polology tell us that the same holds even where the Goldstone boson is a composite particle.

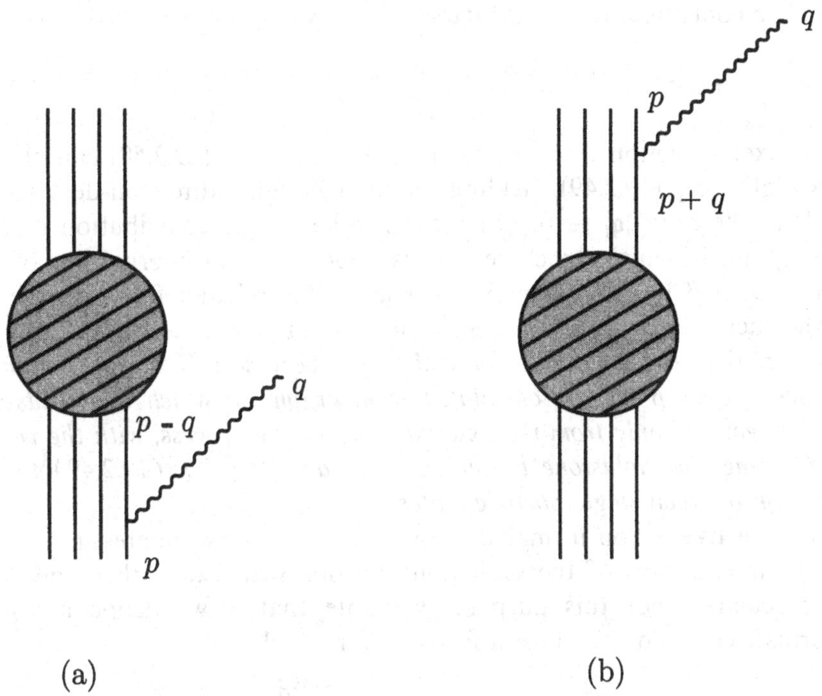

Figure 19.2. Feynman diagrams for pole terms in the matrix element of a symmetry current $J^\mu(x)$ between general states α and β, due to internal lines close to their mass shell. Solid lines indicate 'hard' external particles; wavy lines indicate insertions of the current $J^\mu(x)$; the cross-hatched disk represents the sum of diagrams with the indicated external lines.

$\partial_\mu J^\mu = 0$ for the current J^μ requires (19.2.48) to vanish when contracted with q^μ, and so

$$M_{\beta\alpha} = \frac{i}{F}\, q_\mu N^\mu_{\beta\alpha} \,. \tag{19.2.49}$$

One immediate consequence is that unless $N^\mu_{\beta\alpha}$ has a pole at $q \to 0$, the matrix element $M_{\beta\alpha}$ for emitting a Goldstone boson in a transition $\alpha \to \beta$ vanishes as $q \to 0$. This is called an 'Adler zero.'[6]

In fact, it often happens that $N^\mu_{\beta\alpha}$ *does* have a pole at $q = 0$. This is because the vertex for the current $J^\mu(x)$ might be attached to an external line of the process $\alpha \to \beta$. (See Figure 19.2.) For instance, if a four-momentum q^μ is carried away by a current J^μ inserted into an outgoing or incoming particle line of four-momentum p and mass m, then the internal line that connects this vertex to the rest of the diagram will carry a four-momentum $p^\mu + q^\mu$ or $p^\mu - q^\mu$ respectively, and its propagator will therefore contribute to $N^\mu_{\beta\alpha}$ a factor

$$\left[(p\pm q)^2+m^2\right]^{-1}=\left[\pm 2p\cdot q+q^2\right]^{-1}\to\pm\frac{1}{2p\cdot q}\,. \qquad (19.2.50)$$

For a fixed direction of $\mathbf{q}$, the factor $1/|\mathbf{q}|$ from Eq. (19.2.50) cancels the factor $|\mathbf{q}|$ in Eq. (19.2.49), yielding a finite (though $\mathbf{q}$-direction-dependent) result in the limit $|\mathbf{q}|\to 0$. On the other hand, the contribution to $N^\mu_{\beta\alpha}$ of diagrams in which the current J^μ is attached to an internal line of the process $\alpha\to\beta$ has no singularity for $|\mathbf{q}|\to 0$, and therefore is cancelled by the factor $|\mathbf{q}|$ in Eq. (19.2.49). Thus Eq. (19.2.49) can be interpreted to mean that *the amplitude for emitting a very soft Goldstone boson in a process $\alpha\to\beta$ can be calculated from graphs in which the Goldstone boson is emitted only from the external lines of the process, with the vertex for emitting the Goldstone boson given by applying Eq. (19.2.49) to the transition between single-particle states.*

The effective action formalism can be used to derive interesting results for the interactions of the Goldstone bosons with each other and with other scalars. For this purpose, we note that if we define a set of renormalized Goldstone boson fields π_a for which

$$\langle B_a|\pi_b(x)|\mathrm{VAC}\rangle=\frac{e^{-ip_B\cdot x}\delta_{ab}}{(2\pi)^{3/2}\sqrt{2p_B^0}}\,, \qquad (19.2.51)$$

then Eq. (19.2.39) tells us that

$$\phi_n(x)=\sum_a Z_{an}\pi_a(x)+\dots\,, \qquad (19.2.52)$$

where '...' indicates fields that do not create Goldstone bosons. But Eq. (19.2.40) gives $Z_{an}=\sum_b F^{-1}_{ab}(it_b\bar\phi)_n$. Thus the amplitude for any reaction among N zero-four-momentum Goldstone bosons $\pi_{a_1},\dots,\pi_{a_N}$, is the same as would be calculated in the tree approximation from the effective interaction

$$\mathscr{H}_{\mathrm{eff}}=\frac{1}{N!}g_{a_1\cdots a_N}\pi_{a_1}\cdots\pi_{a_N}\,, \qquad (19.2.53)$$

with

$$g_{a_1\cdots a_N}=\sum_{b_1\dots b_N}F^{-1}_{a_1b_1}\cdots F^{-1}_{a_Nb_N}(it_{b_1}\bar\phi)_{n_1}\cdots(it_{b_N}\bar\phi)_{n_N}\left.\frac{\partial^N V(\phi)}{\partial\phi_{n_1}\cdots\partial\phi_{n_N}}\right|_{\phi=\bar\phi}\,. \qquad (19.2.54)$$

Eq. (19.2.4) at $\phi=\bar\phi$ then implies that the sum of all 'tadpole' graphs for a single Goldstone boson line disappearing into the vacuum vanishes, and Eq. (19.2.6) tells us that the amplitudes vanish for Goldstone bosons to make transitions at zero four-momentum into any other scalars. To go beyond these results, we can continue to differentiate with respect to the scalar fields. For instance, the derivative of Eq. (19.2.5) gives for any symmetry generator t:

$$\sum_n \frac{\partial^2 V(\phi)}{\partial\phi_n \partial\phi_m} t_{n\ell} + \sum_n \frac{\partial^2 V(\phi)}{\partial\phi_n \partial\phi_\ell} t_{nm} + \sum_{n,k} \frac{\partial^3 V(\phi)}{\partial\phi_n \partial\phi_m \partial\phi_\ell} t_{nk}\phi_k = 0\,. \quad (19.2.55)$$

Taking t as one of the broken symmetry generators t_a, setting $\phi = \bar{\phi}$, contracting with $(t_b\bar{\phi})_m(t_c\bar{\phi})_\ell$, and using Eq. (19.2.6) gives

$$\sum_{nm\ell} \left. \frac{\partial^3 V(\phi)}{\partial\phi_n \partial\phi_m \partial\phi_\ell} \right|_{\phi=\bar{\phi}} (t_a\bar{\phi})_n(t_b\bar{\phi})_m(t_c\bar{\phi})_\ell = 0\,, \quad (19.2.56)$$

so the sum of all graphs with three external zero-four-momentum Goldstone boson lines vanishes. In particular, this means that in general processes, to leading order in small Goldstone boson energies, low energy Goldstone bosons are not emitted from external low energy Goldstone boson lines.

19.3 Spontaneously Broken Approximate Symmetries

In the previous section we dealt with exact symmetries of the action that do not leave the vacuum invariant, and are said to be spontaneously broken. We shall now consider the effect of adding small symmetry-breaking terms to the action in such a theory. Such spontaneously broken approximate symmetries are important in the theory of strong interactions, and in some areas of condensed matter physics. As we shall see, the spontaneous breakdown of an approximate symmetry does not lead to the appearance of massless Goldstone bosons, but of low-mass spinless particles, often called *pseudo-Goldstone bosons.*[7]

We continue here to treat translationally invariant theories, in which the effective action is expressed as in Eq. (19.2.3) in terms of an effective potential $V(\phi)$ that depends on a set of spacetime-independent scalar field expectation values ϕ_n. For an action that obeys some set of approximate continuous symmetries with generators t_α, the effective potential may be written

$$V(\phi) = V_0(\phi) + V_1(\phi)\,, \quad (19.3.1)$$

where $V_0(\phi)$ satisfies the invariance condition*

$$\sum_{nm} \frac{\partial V_0(\phi)}{\partial\phi_n} (t_\alpha)_{nm}\phi_m = 0 \quad (19.3.2)$$

* We denote general symmetry generators as t_α, t_β, etc., in contrast with the independent broken symmetry generators, for which the subscript α takes values a, b, etc.

and $V_1(\phi)$ is a small correction due to the symmetry breaking in the action. Suppose that this perturbation shifts the minimum of the potential from ϕ_0, the minimum of $V_0(\phi)$, to $\bar{\phi} = \phi_0 + \phi_1$, where ϕ_1 is small, of first order in the symmetry-breaking perturbation. The equilibrium condition for the vacuum is then

$$\left.\frac{\partial V(\phi)}{\partial \phi_n}\right|_{\phi=\phi_0+\phi_1} = 0 . \tag{19.3.3}$$

The zeroth-order term on the left is just $[\partial V_0(\phi)/\partial \phi_n]_{\phi=\phi_0}$, which vanishes because ϕ_0 is defined as the minimum of $V_0(\phi)$. Thus the first-order terms must also vanish, and so:

$$\sum_m \left.\frac{\partial^2 V_0(\phi)}{\partial \phi_n \partial \phi_m}\right|_{\phi=\phi_0} \phi_{1m} + \left.\frac{\partial V_1(\phi)}{\partial \phi_n}\right|_{\phi=\phi_0} = 0 . \tag{19.3.4}$$

Eq. (19.2.6) holds here with V replaced with the invariant term V_0 and $\bar{\phi}$ replaced with ϕ_0:

$$\sum_{nl} \left.\frac{\partial^2 V_0(\phi)}{\partial \phi_n \partial \phi_m}\right|_{\phi=\phi_0} (t_\alpha)_{nl}\, \phi_{0l} = 0 . \tag{19.3.5}$$

Hence multiplying Eq. (19.3.4) with $(t_\alpha \phi_0)_n$ and summing over n gives

$$\sum_n (t_\alpha \phi_0)_n \left.\frac{\partial V_1(\phi)}{\partial \phi_n}\right|_{\phi=\phi_0} = 0 . \tag{19.3.6}$$

Recalling the interpretation of $V(\phi)$ as the generating function for one-particle-irreducible graphs, and noting that in the absence of the perturbation V_1 the Goldstone components of ϕ_n are those in the direction of $t_\alpha \phi_0$ (see Eq. (19.2.33)), the left-hand side of Eq. (19.3.6) is proportional to the sum of all 'tadpole' graphs, in which a pseudo-Goldstone boson disappears into the vacuum. Eq. (19.3.6) may thus be paraphrased as the condition that, to first order in V_1, *pseudo-Goldstone bosons have no tadpoles.*

The moral of this calculation is that if we do not start with a zeroth-order vacuum expectation value that satisfies Eq. (19.3.6), then even a small perturbation will produce a large change in $\bar{\phi}$, invalidating the expansion of $\bar{\phi}$ around ϕ_0. Fortunately, for compact Lie groups it is always possible to choose ϕ_0 to satisfy Eq. (19.3.6). To see this, note that the invariance of the potential $V_0(\phi)$ under a group of real linear transformations $\phi \to L\phi$ implies that if ϕ_* is one minimum of the potential, then so is $L\phi_*$. For continuous groups of transformations, we can always parameterize the transformations as $L(\theta)$, in such a way that

$$\frac{\partial L(\theta)}{\partial \theta^\alpha} L^{-1}(\theta) = i N_{\alpha\beta}(\theta) t_\beta , \tag{19.3.7}$$

where $N_{\alpha\beta}$ is a non-singular matrix depending on the group parameters θ_α. (Recall that in a real representation, it is it_α rather than t_α that is real.) Now consider the function $V_1(L(\theta)\phi_*)$. Where the group is compact, $L(\theta)\phi_*$ maps out a compact manifold as θ runs over the group volume, and as long as $V_1(\phi)$ is continuous, it must have a minimum on any such compact surface, say at a point $L(\theta_*)\phi_*$. The derivative of $V_1(L(\theta)\phi_*)$ with respect to θ_α is

$$\frac{\partial V_1(L(\theta)\phi_*)}{\partial\theta_\alpha} = \sum_n \left.\frac{\partial V_1(\phi)}{\partial\phi_n}\right|_{\phi=L(\theta)\phi_*} N_{\alpha\beta}(\theta)\,(it_\beta L(\theta)\phi_*)_n\,. \tag{19.3.8}$$

This must vanish at the minimum θ_*, and since N_{ab} is non-singular, this implies that

$$0 = \sum_n \left.\frac{\partial V_1(\phi)}{\partial\phi_n}\right|_{\phi=L(\theta_*)\phi_*} (t_\beta L(\theta_*)\phi_*)_n\,. \tag{19.3.9}$$

But then Eq. (19.3.6) is satisfied if we make the choice $\phi_0 = L(\theta_*)\phi_*$.

Eq. (19.3.6) is known as a *vacuum alignment* condition,[8] because it generally has the effect of forcing the direction of the symmetry breaking by the vacuum into some sort of alignment with the symmetry-breaking terms in the Hamiltonian. For instance, consider the case of $SO(N)$ spontaneously broken to $SO(N-1)$, introduced in the previous section. In the absence of any symmetry-breaking perturbation, there is no way to tell *which* $SO(N-1)$ subgroup is left unbroken; if the dynamics of the theory leads to a ground state that is invariant under the $SO(N-1)$ subgroup of $SO(N)$ that leaves some N-vector ϕ_{0n} invariant, then by performing an $SO(N)$ rotation we can find a ground state that is invariant under the $SO(N-1)$ subgroup that leaves any other N-vector invariant. If we add a perturbation that transforms under $SO(N)$ like, say, the component $\sum_n u_n\phi_n$ of an N-vector ϕ_n (not necessarily consisting of elementary scalars), then the Hamiltonian is invariant under a specific $SO(N-1)$ subgroup of $SO(N)$, consisting of rotations that leave the vector u invariant. Without a vacuum alignment condition, we might think that the remaining exact symmetry is $SO(N-2)$, consisting of those rotations that leave invariant *both* u and the vector ϕ_0 characterizing the vacuum symmetry. But with $V_1(\phi) = \sum_n u_n\phi_n$, the condition (19.3.6) tells us that at the true vacuum, $\sum_n(t_\alpha\phi_0)_n u_n = 0$ for all $SO(N)$ generators t_α. The $SO(N)$ generators t_α span the space of all antisymmetric $N\times N$ matrices, so this condition requires that ϕ_0 must be in the same direction as u, and so the unbroken symmetry is $SO(N-1)$, not $SO(N-2)$.

According to the general results of Section 16.1, the mass matrix M_{ab}

of the pseudo-Goldstone bosons is given to first order by

$$M^2_{ab} = \sum_{mn} Z_{an} Z_{bm} \left. \frac{\partial^2 V(\phi)}{\partial \phi_m \partial \phi_n} \right|_{\phi=\phi_0+\phi_1} , \tag{19.3.10}$$

where Z_{an} is the field renormalization constant defined by Eq. (19.2.39). Since the mass matrix (19.3.10) vanishes in zeroth order, the first-order terms give

$$M^2_{ab} = \sum_{mn} Z_{an} Z_{bm} \left[\sum_{\ell} \left. \frac{\partial^3 V_0(\phi)}{\partial \phi_\ell \partial \phi_m \partial \phi_n} \right|_{\phi=\phi_0} \phi_{1\ell} + \left. \frac{\partial^2 V_1(\phi)}{\partial \phi_m \partial \phi_n} \right|_{\phi=\phi_0} \right] , \tag{19.3.11}$$

where Z_{an} is here given by the zeroth-order approximation to (19.2.40):

$$Z_{an} = \sum_b F^{-1}_{ab} \, (it_b \phi_0)_n \, . \tag{19.3.12}$$

To calculate the mass matrix (19.3.11), we take t in Eq. (19.2.55) to be one of the broken symmetry generators t_a, set $\phi = \phi_0$, and contract with $(t_b\phi_0)_m \phi_{1\ell}$:

$$0 = \sum_{nm\ell} \left. \frac{\partial^3 V_0(\phi)}{\partial \phi_\ell \partial \phi_m \partial \phi_n} \right|_{\phi=\phi_0} \phi_{1\ell} (t_a\phi_0)_n (t_b\phi_0)_m$$
$$+ \sum_{nm} \left. \frac{\partial^2 V_0(\phi)}{\partial \phi_n \partial \phi_m} \right|_{\phi=\phi_0} (t_a\phi_1)_n (t_b\phi_0)_m + \sum_{n\ell} \left. \frac{\partial^2 V_0(\phi)}{\partial \phi_n \partial \phi_\ell} \right|_{\phi=\phi_0} (t_a t_b \phi_0)_n \phi_{1\ell} \, .$$

The second term on the right-hand side vanishes according to Eq. (19.2.6), while the third may be rewritten using Eq. (19.3.4), leaving us with

$$\sum_{nm\ell} \left. \frac{\partial^3 V_0(\phi)}{\partial \phi_\ell \partial \phi_m \partial \phi_n} \right|_{\phi=\phi_0} \phi_{1\ell} (t_a\phi_0)_n (t_b\phi_0)_m = \left. \frac{\partial V_1(\phi)}{\partial \phi_n} \right|_{\phi=\phi_0} (t_a t_b \phi_0)_n \, . \tag{19.3.13}$$

Using this in Eq. (19.3.11) then yields a formula for the pseudo-Goldstone boson mass matrix in terms of V_1:

$$M^2_{cd} = -\sum_{ab} F^{-1}_{ca} F^{-1}_{db} \left[(t_a\phi_0)_n (t_b\phi_0)_m \left. \frac{\partial^2 V_1(\phi)}{\partial \phi_m \partial \phi_n} \right|_{\phi=\phi_0} + (t_a t_b \phi_0)_n \left. \frac{\partial V_1(\phi)}{\partial \phi_n} \right|_{\phi=\phi_0} \right] . \tag{19.3.14}$$

For this to be a sensible mass matrix, it had better be positive. To see that it is, it is convenient to rewrite this result in terms of derivatives with respect to the group parameters θ_a. Differentiating Eq. (19.3.8) with respect to θ_β, setting $\theta = \theta_*$, and using Eq. (19.3.9) and $\phi_0 = L(\theta_*)\phi_*$

gives

$$M_{ab}^2 = \sum_{cd\alpha\beta} N_{a\alpha}^{-1}(\theta_*) N_{b\beta}^{-1}(\theta_*) F_{ac}^{-1} F_{bd}^{-1} \frac{\partial^2 V_1(L(\theta)\phi_*)}{\partial\theta_\alpha \partial\theta_\beta}\bigg|_{\theta=\theta_*}. \qquad (19.3.15)$$

The matrix on the right is positive, because θ_* is the *minimum* of the function $V_1(L(\theta)\phi_*)$.

This formula has a somewhat more familiar version, in terms of the vacuum expectation value of a double commutator of symmetry generators with the symmetry-breaking perturbation. Suppose that the symmetry-breaking perturbation H_1 in the Hamiltonian is a linear combination

$$H_1 = \sum_n u_n \Phi_n \qquad (19.3.16)$$

of operators Φ_n (not necessarily elementary scalar fields), which furnish a representation of the symmetry group with generators t_α, in the sense that

$$[T_\alpha, \Phi_n] = -(t_\alpha)_{nm}\Phi_m, \qquad (19.3.17)$$

where T_α are the quantum mechanical generators of the symmetry group. According to the results of Section 16.3, the symmetry-breaking part of the potential is

$$V_1(\phi) = \langle H_1\rangle_{\langle\Phi\rangle=\phi} = \sum_n u_n \phi_n, \qquad (19.3.18)$$

the subscript in the middle expression indicating that the expectation value is to be taken in the state of minimum energy in which Φ_n has expectation value ϕ_n. The vacuum alignment condition (19.3.6) then reads

$$0 = \sum_n u_n (t_\alpha \phi_0)_n,$$

or using Eq. (19.3.17)

$$0 = \langle[T_\alpha, H_1]\rangle_0, \qquad (19.3.19)$$

the subscript 0 now indicating that the expectation value is to be taken in the vacuum state, in which Φ_n has expectation value ϕ_{0n}. Also, Eq. (19.3.14) gives the mass matrix here as

$$M_{cd}^2 = -\sum_{ab} F_{ca}^{-1} F_{db}^{-1} \sum_n u_n (t_a t_b \phi_0)_n,$$

and, using Eq. (19.3.17), this is

$$M_{cd}^2 = -\sum_{ab} F_{ca}^{-1} F_{db}^{-1} \langle[T_a, [T_b, H_1]]\rangle_0. \qquad (19.3.20)$$

This is symmetric in c and d. To see this, note that the Jacobi identity and group commutation relations may be used to write the difference of Eq. (19.3.20) and the same with c and d interchanged as a linear combination of terms $\langle [T_\alpha, H_1]\rangle_0$, which vanish according to the vacuum alignment condition (19.3.19). The mass matrix (19.3.20) is also positive, because the point $\theta = 0$ is at the *minimum* of the vacuum energy $\langle \exp(-i\theta_a T_a) H_1 \exp(i\theta_a T_a)\rangle_0$ for rotated vacuum states $\exp(i\theta_a T_a)|0\rangle$.

19.4 Pions as Goldstone Bosons

The classic example of a broken symmetry in elementary particle physics is the approximate symmetry of strong interactions known as chiral $SU(2) \times SU(2)$. According to our present understanding, this symmetry arises because there are two quark fields, u and d, that happen to have relatively small masses. (An estimate is given in Section 19.7.) In the approximation that the u and d are massless, the Lagrangian (18.7.5) of quantum chromodynamics is

$$\mathscr{L} = -\bar{u}\gamma^\mu D_\mu u - \bar{d}\gamma^\mu D_\mu d - \cdots \,, \tag{19.4.1}$$

where D_μ is a color-gauge-covariant derivative (see Eq. (15.1.10)) and '$\cdots$' refers to terms involving only gluon fields and/or other quark flavors, but not u or d. This Lagrangian is invariant under the transformations

$$\begin{pmatrix} u \\ d \end{pmatrix} \to \exp\left(i\vec{\theta}^V \cdot \vec{t} + i\gamma_5 \vec{\theta}^A \cdot \vec{t}\right) \begin{pmatrix} u \\ d \end{pmatrix} , \tag{19.4.2}$$

where $\vec{t}$ is the three-vector* of isospin matrices

$$t_1 = \frac{1}{2}\begin{pmatrix} 0 & 1 \\ 1 & 0 \end{pmatrix} , \qquad t_2 = \frac{1}{2}\begin{pmatrix} 0 & -i \\ i & 0 \end{pmatrix} , \qquad t_3 = \frac{1}{2}\begin{pmatrix} 1 & 0 \\ 0 & -1 \end{pmatrix} ,$$

and $\vec{\theta}^V$ and $\vec{\theta}^A$ are independent real three-vectors.** This Lie algebra may be written in terms of two commuting $SU(2)$ subalgebras that act respectively only on the left- and right-handed parts of the quark fields, with generators

$$\vec{t}_L = \frac{1}{2}(1+\gamma_5)\vec{t} \,, \qquad \vec{t}_R = \frac{1}{2}(1-\gamma_5)\vec{t} \tag{19.4.3}$$

* We use arrows for three-vectors in isotopic spin space to distinguish them from ordinary three-vectors, which will continue to be indicated by boldface letters.

** The Lagrangian (19.4.1) has this symmetry because $\overline{\gamma_5\psi}\gamma^\mu = -\bar{\psi}\gamma_5\gamma^\mu = +\bar{\psi}\gamma^\mu\gamma_5$. The Lagrangian also has two other continuous internal symmetries. One is baryon conservation, the invariance under a common phase transformation of the u and d quark fields. This is unbroken and commutes with the other symmetries, so it does not affect our discussion in this section. The other symmetry is invariance under multiplication of the quark doublet with $\exp(i\alpha\gamma_5)$. As discussed in Section 23.5, this $U(1)$ symmetry is strongly intrinsically broken by non-perturbative effects associated with instantons.

satisfying the commutation relations

$$[t_{Li}, t_{Lj}] = i\epsilon_{ijk}\, t_{Lk}\,, \tag{19.4.4}$$

$$[t_{Ri}, t_{Rj}] = i\epsilon_{ijk}\, t_{Rk}\,, \tag{19.4.5}$$

$$[t_{Li}, t_{Rj}] = 0\,. \tag{19.4.6}$$

The underlying symmetry group is therefore identified as $SU(2) \times SU(2)$. It has another obvious $SU(2)$ subgroup, consisting of ordinary isospin transformations with $\vec{\theta}^A = 0$, and generators

$$\vec{t} = \vec{t}_L + \vec{t}_R\,. \tag{19.4.7}$$

The algebra of $SU(2) \times SU(2)$ may be written in terms of $\vec{t}$ and another triplet of generators:

$$\vec{x} = \vec{t}_L - \vec{t}_R = \gamma_5 \vec{t} \tag{19.4.8}$$

with commutation relations

$$[t_i, t_j] = i\epsilon_{ijk}\, t_k\,, \tag{19.4.9}$$

$$[t_i, x_j] = i\epsilon_{ijk}\, x_k\,, \tag{19.4.10}$$

$$[x_i, x_j] = i\epsilon_{ijk}\, t_k\,. \tag{19.4.11}$$

We will see that the $SU(2) \times SU(2)$ symmetry is spontaneously broken, while its isotopic spin subgroup generated by the $\vec{t}$ is an ordinary unbroken (though approximate) symmetry.

By Noether's method (see, e.g., Section 7.3) we may derive from the Lagrangian (19.4.1) the conserved vector and axial-vector currents

$$\vec{V}^\mu = i\bar{q}\gamma^\mu \vec{t} q\,, \qquad \vec{A}^\mu = i\bar{q}\gamma^\mu\gamma_5 \vec{t} q\,, \tag{19.4.12}$$

$$\partial_\mu \vec{V}^\mu = \partial_\mu \vec{A}^\mu = 0\,, \tag{19.4.13}$$

where q is the quark doublet,

$$q \equiv \begin{pmatrix} u \\ d \end{pmatrix}. \tag{19.4.14}$$

Their associated charges are the generators respectively of isospin and of the remaining symmetries

$$\vec{T} = \int d^3x\, \vec{V}^0\,, \tag{19.4.15}$$

$$\vec{X} = \int d^3x\, \vec{A}^0\,. \tag{19.4.16}$$

The currents (19.4.12) are normalized so that the quantum operators $\vec{T}$ and $\vec{X}$ satisfy the same commutation relations as the matrices $\vec{t}$ and $\vec{x}$:

$$[T_i, T_j] = i\epsilon_{ijk}\, T_k\,, \tag{19.4.17}$$

$$[T_i, X_j] = i\epsilon_{ijk} X_k\,, \tag{19.4.18}$$
$$[X_i, X_j] = i\epsilon_{ijk} T_k\,. \tag{19.4.19}$$

Acting on the quark fields, these operators induce the transformation (19.4.2), in the sense that

$$\left[\vec{T},\, q\right] = -\vec{t}\,q\,, \tag{19.4.20}$$
$$\left[\vec{X},\, q\right] = -\vec{x}\,q\,. \tag{19.4.21}$$

This symmetry if exact and unbroken would require any one-hadron state $|h\rangle$ to be degenerate with another state $\vec{X}|h\rangle$ of opposite parity and equal spin, baryon number, and strangeness.† No such parity doubling is seen in the hadron spectrum, so we are forced to conclude that if the chiral symmetry $SU(2) \times SU(2)$ is a good approximation at all, then it must be spontaneously broken to its isotopic spin $SU(2)$ subgroup. In this case the operator $\vec{X}$ takes a one-hadron state $|h\rangle$ into a hadron h plus a massless pseudoscalar Goldstone boson, so there is no need for parity doubling of the hadron spectrum.

The question of whether quantum chromodynamics actually exhibits such a pattern of symmetry breaking involves all the complications of strong interaction dynamics. As we shall see in Section 19.9, there are general grounds for believing that the isospin $SU(2)$ is not spontaneously broken in quantum chromodynamics, but it is much more difficult to show that the chiral part of $SU(2) \times SU(2)$ *is* spontaneously broken. (But according to an argument given in Section 22.5, the $SU(3) \times SU(3)$ symmetry of quantum chromodynamics with *three* massless quark flavors must be spontaneously broken.) It was something of a breakthrough in the 1960s to realize that one does not have to have a detailed understanding of the mechanism of the breaking of chiral symmetry; we derive the most interesting consequences of this symmetry breaking by simply *assuming* that $SU(2) \times SU(2)$ is spontaneously broken to $SU(2)$.

The u and d quarks have small but nonzero masses, so the $SU(2)\times SU(2)$ symmetry is not exact. A broken *approximate* chiral symmetry entails the existence of an approximately massless Goldstone boson with the same quantum numbers as the broken symmetry generator $\vec{X}$: it must be a state of negative parity, zero spin, unit isospin, and zero baryon number

† One way to satisfy this condition is for the hadron to have zero mass, with two states $|\pm\rangle$ of helicity $\pm\frac{1}{2}$ and equal spin, baryon number, and strangeness; the states $|+\rangle + |-\rangle$ and $|+\rangle - |-\rangle$ would then have opposite parity. It is not true that an unbroken chiral symmetry necessarily implies a zero nucleon mass, unless we make further assumptions about the matrix elements of the axial-vector current. But as we shall see in Section 22.5, an exact unbroken chiral symmetry would in fact require that *some* baryons be massless.

and strangeness. In fact the lightest of all hadrons is the pion, which has precisely these quantum numbers, so we are led to identify the pion as the Goldstone boson associated with the spontaneous breaking of approximate chiral symmetry. As we shall see below, it is m_π^2 rather than m_π that is proportional to a linear combination of m_u and m_d, and $m_\pi^2/m_N^2 \simeq 0.022$ is very small, so the consequences we derive from spontaneously broken $SU(2) \times SU(2)$ should be reasonably accurate.

In exploring the consequences of chiral symmetry for pion interactions, it is very useful to note that although the chiral symmetry of the strong interactions does not depend in any way on the existence of weak interactions, the symmetry currents $\vec{V}^\mu$ and $\vec{A}^\mu$ happen to be the currents entering into strangeness-conserving semileptonic weak interactions like nuclear beta decay. As we shall see in Section 21.3, the standard model of electroweak interactions requires that the effective Lagrangian for these interactions at low energy must take the form:

$$\mathscr{L}_{\rm wk} = -\frac{iG_{\rm wk}}{\sqrt{2}}\left(V_+^\lambda + A_+^\lambda\right)\sum_\ell \bar{\ell}\gamma_\lambda(1+\gamma_5)\nu_\ell + \text{H.c.} \tag{19.4.22}$$

where ℓ runs over the renormalized fields of the three charged leptons e, μ and τ; ν_ℓ runs over the renormalized fields of the associated neutrinos; and $V_\pm^\lambda$ and $A_\pm^\lambda$ are the charge changing currents

$$V_\pm^\lambda = V_1^\lambda \pm i\,V_2^\lambda\,, \qquad A_\pm^\lambda = A_1^\lambda \pm i\,A_2^\lambda\,. \tag{19.4.23}$$

The constant†† $G_{\rm wk}$ may be measured from the rates of beta transitions between states of zero spin within the same isotopic multiplet, such as the decays $\pi^+ \to \pi^0 + e^+ + \nu_e$ and $^{14}O \to {}^{14}N^* + e^+ + \nu_e$. The momentum transfer in these transitions is very small, so parity conservation (in strong interactions) and rotational invariance tell us that only the matrix elements of $\int d^3x\, V_-^0 = T_1 - iT_2$ enter in the S-matrix elements for these decays. This operator has matrix elements between states in a given isospin multiplet that are just known Clebsch–Gordan coefficients, so from the rates for these '0 → 0' processes we may calculate a value[9] for the coupling in Eq. (19.4.22): $G_{\rm wk} \simeq 1.14959(38) \times 10^{-5}$ GeV^{-2}. On the other hand, in the process of pion decay, $\pi^+ \to \mu^+ + \nu_\mu$, the only current matrix element we need is the matrix element of A_-^λ between a one-pion state and the vacuum:

$$\langle {\rm VAC}|A_i^\mu(x)|\pi_j\rangle = \frac{iF_\pi\,\delta_{ij}\,p_\pi^\mu\,e^{ip_\pi\cdot x}}{2(2\pi)^{3/2}\sqrt{2p_\pi^0}}\,, \tag{19.4.24}$$

†† $G_{\rm wk}$ is related to the conventional Fermi coupling constant G_F and the Cabibbo angle θ_C by $G_{\rm wk} = G_F\cos\theta_C$; see Section 21.3.

which is completely known except for the factor F_π. The rate for pion decay turns out to be

$$\Gamma(\pi \to \mu + \nu) = \frac{G_{\rm wk}^2 F_\pi^2 m_\mu^2 (m_\pi^2 - m_\mu^2)^2}{16\pi m_\pi^3} . \tag{19.4.25}$$

From the known rate for $\pi^+ \to \mu^+ + \nu_\mu$, $\Gamma = (2.6033(24) \times 10^{-8}\ {\rm s})^{-1}$, and the above value of $G_{\rm wk}$, one finds that‡

$$F_\pi \simeq 184\ {\rm MeV} . \tag{19.4.26}$$

Now let us consider the matrix element of $\vec{A}^\mu$ between one-nucleon states. This is of interest in its own right, and as discussed in Section 19.2, it provides information we need to calculate the emission of low energy pions in collisions of nucleons. Following the same reasoning as used for the electromagnetic current in Section 10.6, one finds that Lorentz invariance and parity conservation require this matrix element to take the form‡‡

$$\langle p|A_+^\mu(x)|n\rangle = (2\pi)^{-3}\, e^{iq\cdot x}$$
$$\times\ \bar{u}_p\Big[- i\gamma^\mu \gamma_5 f(q^2) + q^\mu \gamma_5 g(q^2) + i q_\nu [\gamma^\mu, \gamma^\nu]\gamma_5 h(q^2)\Big] u_n\,, \tag{19.4.27}$$

where $q \equiv p_n - p_p$. In the approximation in which the $SU(2) \times SU(2)$ symmetry is exact, the conservation of current requires that

$$q_\mu \langle p|A_+^\mu(x)|n\rangle = 0 . \tag{19.4.28}$$

Using the defining equations for the Dirac spinors u_p and u_n,

$$\bar{u}_p(i\, \not{p}_p + m_N) = (i\, \not{p}_n + m_N)u_n = 0\,,$$

we see that

$$q_\mu \bar{u}_p[-i\gamma^\mu\gamma_5]u_n = -2m_N \bar{u}_p \gamma_5 u_n$$

and so Eq. (19.4.28) requires that

$$2m_N f(q^2) = q^2 g(q^2) . \tag{19.4.29}$$

If $g(q^2)$ had no singularity at $q^2 = 0$, then (19.4.29) would require that either $m_N = 0$, which is certainly not true, or that $f(0) = 0$, which is not true either. In fact, the quantity $f(0)$ is measured in low energy nuclear

‡ It is common to encounter a pion decay constant f_π, which in terms of the F_π used here is variously defined as F_π, $F_\pi/\sqrt{2}$, or $F_\pi/2$.

‡‡ The currents $A_\pm^\mu$ in the standard model have charge conjugation properties that make the coefficient $h(q^2)$ vanish. It is possible that there are 'second-class' terms[10] in the weak currents with opposite charge conjugation properties that give a non-vanishing $h(q^2)$, but there is no evidence for such terms. As we shall see, keeping the $h(q^2)$ term here has no effect on the inferences drawn from chiral symmetry.

beta decays, like neutron decay, where it is usually called g_A; it is found to have the value

$$f(0) \equiv g_A = 1.2573(28)\,. \tag{19.4.30}$$

The fact that neither m_N nor $f(0) = g_A$ is small requires that in the limit of exact $SU(2) \times SU(2)$ symmetry, $g(q^2)$ must have a pole as $q^2 \to 0$:

$$g(q^2) \to \frac{2m_N g_A}{q^2}\,. \tag{19.4.31}$$

Such a pole is naturally provided by the massless pion that would be required by the spontaneous breakdown of an exact $SU(2) \times SU(2)$ symmetry. Suppose that the pion couples to the one-nucleon state as if the interaction Lagrangian were♮ $-2iG_{\pi N}\vec{\pi}\bar{N}\gamma_5\vec{t}N$. Eq. (19.4.24) tells us that the matrix element of the current between a one-pion state and the vacuum is the same as if there were a term in $\vec{A}^\mu(x)$ of the form $F_\pi\partial^\mu\vec{\pi}/2$. Therefore in the limit $q^2 \to 0$ the matrix element (19.4.27) has the pole

$$\langle p|A^\mu_+(x)|n\rangle \to \left[\frac{e^{iq\cdot x}}{(2\pi)^3}\right]\left[i(2\pi)^4\,2G_{\pi N}\,\bar{u}_p(i\gamma_5)u_n\right]\left[\frac{-i}{(2\pi)^4q^2}\right]\left[iq^\mu F_\pi/2\right].$$

Comparing with Eq. (19.4.27), we see that one-pion exchange gives the function $g(q^2)$ a pole

$$g(q^2) \to \frac{G_{\pi N}F_\pi}{q^2} \tag{19.4.32}$$

for $q^2 \to 0$. Putting together Eqs. (19.4.31) and (19.4.32), we find

$$G_{\pi N} = \frac{2m_N g_A}{F_\pi}\,. \tag{19.4.33}$$

This is the famous Goldberger–Treiman relation.[11] It works reasonably well; taking $m_N = (m_p + m_n)/2 = 938.9$ MeV, $g_A = 1.257$, and $F_\pi = 184$ MeV gives $G_{\pi N} \simeq 12.7$, in fair agreement with the value♮♮ $G_{\pi N} = 13.5$ measured in various ways (including the effects of the one-pion pole in nucleon–nuclear scattering and the one-nucleon pole in pion–nucleon scattering.)

In the real world the pion is not massless and the $SU(2) \times SU(2)$ symmetry is not exact (even before being spontaneously broken.) This circumstance may be analyzed using the general formalism presented in

♮ This is the conventional definition of the pseudoscalar pion–nucleon coupling $G_{\pi N}$. The factor 2 is introduced here to cancel the 1/2 in the isospin matrices.

♮♮ The textbook[12] value is $G^2_{\pi N}/4\pi = 14.3$, or $G_{\pi N} = 13.4$. More recently, a high precision study[13] of neutron–proton charge-exchange scattering at 162 MeV has given a value $G^2_{\pi N}/4\pi = 14.6 \pm 0.3$, or $G_{\pi N} = 13.5$.

the previous section. The Lagrangian (19.4.1) yields a symmetry-breaking term in the Hamiltonian

$$H_1 = m_u \bar{u}u + m_d \bar{d}d = (m_u + m_d)\Phi_4^+ + (m_u - m_d)\Phi_3^- \,, \tag{19.4.34}$$

where

$$\Phi_4^+ \equiv \tfrac{1}{2}(\bar{u}u + \bar{d}d) \,, \qquad \Phi_3^- = \tfrac{1}{2}(\bar{u}u - \bar{d}d) \,. \tag{19.4.35}$$

The operators Φ_4^+ and Φ_3^- are spatial scalars and, as this notation is meant to suggest, they transform under $SU(2) \times SU(2)$ as components of independent chiral four-vectors $\Phi_\alpha^\pm$:

$$\vec{\Phi}^+ = i\bar{q}\gamma_5 \vec{t} q \,, \qquad \Phi_4^+ = \tfrac{1}{2}\bar{q}q \,, \tag{19.4.36}$$

$$\vec{\Phi}^- = \bar{q}\vec{t} q \,, \qquad \Phi_4^- = -\tfrac{1}{2}i\bar{q}\gamma_5 q \,. \tag{19.4.37}$$

These are chiral four-vectors, in the sense that

$$\left[\vec{T},\, \Phi_n^\pm\right] = -\sum_m (\vec{\mathscr{T}})_{nm}\Phi_m^\pm \,, \tag{19.4.38}$$

$$\left[\vec{X},\, \Phi_n^\pm\right] = -\sum_m (\vec{\mathscr{X}})_{nm}\Phi_m^\pm \,, \tag{19.4.39}$$

where $\vec{\mathscr{T}}$ and $\vec{\mathscr{X}}$ are Hermitian 4×4 matrices that furnish the four-vector representation of the algebra of $SU(2) \times SU(2) \equiv SO(4)$:

$$(\mathscr{T}_a)_{bc} = -\, i\epsilon_{abc} \,, \qquad (\mathscr{T}_a)_{b4} = (\mathscr{T}_a)_{4b} = (\mathscr{T}_a)_{44} = 0 \,, \tag{19.4.40}$$

$$(\mathscr{X}_a)_{b4} = -\, (\mathscr{X}_a)_{4b} = -i\delta_{ab} \,, \qquad (\mathscr{X}_c)_{ab} = (\mathscr{X}_c)_{44} = 0 \,. \tag{19.4.41}$$

This notation makes it easy to see that the vacuum alignment conditions for generators T_1, T_2, X_1, X_2, and X_3 respectively take the form

$$\begin{aligned} 0 &= \langle \Phi_2^- \rangle_0 = \langle \Phi_1^- \rangle_0 = \langle \Phi_1^+ \rangle_0 = \langle \Phi_2^+ \rangle_0 \\ &= (m_u + m_d)\langle \Phi_3^+ \rangle_0 + (m_d - m_u)\langle \Phi_4^- \rangle_0 \,. \end{aligned} \tag{19.4.42}$$

We have been assuming that in the absence of u and d quark masses, the symmetry $SU(2) \times SU(2)$ is spontaneously broken in such a way as to preserve unbroken the $SU(2)$ symmetry generated by $\vec{T}$ as well as parity, in which case $\langle \Phi_n^+ \rangle_0$ points in the four-direction and $\langle \Phi_n^- \rangle_0 = 0$ vanishes, so that the conditions (19.4.42) are satisfied. But with $m_u = m_d = 0$, we can find other symmetry-breaking solutions with other definitions of parity by subjecting this one to an arbitrary $SU(2) \times SU(2)$ transformation. Thus in the absence of u and d quark masses, there would be no way to tell in what direction $\langle \Phi_n^+ \rangle_0 = 0$ should point, or which $SU(2)$ subgroup of $SU(2) \times SU(2)$ is left unbroken, though in all cases $\langle \Phi_n^- \rangle_0 = 0$. The vacuum alignment condition (19.4.42) tells us that, with the symmetry broken intrinsically by the perturbation (19.4.34) and spontaneously in such a way that $\langle \Phi_n^- \rangle_0 = 0$, the vacuum must 'line up' in such a way that $\langle\Phi_n^+\rangle_0 = 0$ points in the four-direction, so that the unbroken $SU(2)$ symmetry is ordinary isospin.

This formalism may be used to calculate the pion mass. From Eqs. (19.4.39)–(19.4.41) we find

$$[X_a, [X_b, \Phi_4^+]] = \delta_{ab}\Phi_4^+ \,, \qquad [X_a, [X_b, \Phi_3^-]] = \Phi_a^- \delta_{b3} \,. \tag{19.4.43}$$

Also, isospin invariance tells us that the symmetry-breaking parameter F_{ab} introduced in Section 19.2 is proportional to δ_{ab}, with a proportionality factor that according to (19.4.24) is just $F_\pi/2$, so

$$F_{ab} = \delta_{ab}F_\pi/2 \,. \tag{19.4.44}$$

Eq. (19.3.20) thus gives the pion mass matrix as

$$m_{ab}^2 = \delta_{ab}m_\pi^2 \,, \tag{19.4.45}$$

where

$$m_\pi^2 = -4(m_u + m_d)\langle\Phi_4^+\rangle_0/F_\pi^2 \,. \tag{19.4.46}$$

It is striking that the masses of the charged and neutral pions turn out to be equal, even though we have made no assumption about the ratio of m_u and m_d. We shall see below that this ratio is not near unity; isospin is a good quantum number not because the u and d quark masses are nearly equal, but because they are *small*. Also, as promised, we see that it is the *square* of the pion mass that is proportional to the quark masses, so the quark masses should be quite small. (See Section 19.7 for an estimate.) The observed pion mass difference arises not from the u–d quark mass difference, but from electromagnetism. Indeed, the pion isospin multiplet is the only one whose mass difference has been successfully calculated on the basis of one-photon exchange alone.[14]

Of course, even with quark masses taken into account, Eq. (19.4.24) can still be used to define F_π; its divergence gives

$$\langle \text{VAC}|\partial_\mu A_i^\mu(x)|\pi_j\rangle = \frac{F_\pi \delta_{ij} m_\pi^2 e^{ip_\pi\cdot x}}{2(2\pi)^{3/2}\sqrt{2p_\pi^0}} \,. \tag{19.4.47}$$

Instead of assuming that $\partial_\mu \vec{A}^\mu$ vanishes, we can now assume that it is *small*, of order m_π^2, except where a pion pole compensates for the smallness of m_π^2. According to Eq. (19.4.47), the behavior of matrix elements of $\partial_\mu \vec{A}^\mu$ near a one-pion pole is the same as if $\partial_\mu \vec{A}^\mu$ were $F_\pi m_\pi^2$ times a properly renormalized pion field. For instance, the one-nucleon matrix element of $\partial_\mu A_+^\mu$ should be

$$\langle p|\partial_\mu A_+^\mu(0)|n\rangle \simeq \frac{F_\pi m_\pi^2}{2}\left[\frac{-i}{(2\pi)^4(q^2+m_\pi^2)}\right]\left[i(2\pi)^4 2G_{\pi N}\bar{u}_p(-i\gamma_5)u_n\right] , \tag{19.4.48}$$

so in terms of the form-factors in (19.4.27)

$$q^2 g(q^2) - 2m_N f(q^2) \simeq -\frac{G_{\pi N} F_\pi m_\pi^2}{q^2 + m_\pi^2} . \tag{19.4.49}$$

This is expected to be valid when q^2 is of order m_π^2, and not only in the limit $q^2 \to -m_\pi^2$, because the pion pole dominates matrix elements of $\partial_\mu A^\mu$ for all such small q^2. Also, for such q^2, in place of Eq. (19.4.32) we have

$$g(q^2) \simeq \frac{G_{\pi N} F_\pi}{q^2 + m_\pi^2} . \tag{19.4.50}$$

From Eqs. (19.4.49) and (19.4.50) we find that for q^2 of order m_π^2,

$$f(q^2) \simeq G_{\pi N} F_\pi \Big/ 2m_N . \tag{19.4.51}$$

It should be no surprise that this function is roughly constant over the range of q^2 from zero to of order m_π^2, because it has no one-pion pole, and there is nothing else that could give it a substantial variation in such a small range of q^2. This constant value is approximately $f(0) \equiv g_A$, and Eq. (19.4.51) thus again yields the Goldberger–Treiman relation.

We can now use the results of Section 19.2 to calculate the amplitude for emission of a single low energy pion in an arbitrary process $\alpha \to \beta$. We found that the amplitude is to be calculated from Feynman diagrams in which the pion is emitted only from the external lines of the process, and Eq. (19.2.49) shows that these contributions are to be calculated as if the pion field interaction were $\partial_\mu \vec{\pi} \cdot \vec{A}_N / F_\pi$, in which the subscript N indicates that we are to drop the one-pion pole term in matrix elements of the axial-vector current. From Eq. (19.4.27) (and using isospin invariance) we conclude that for soft pions emitted from a nucleon line this interaction is effectively

$$-\frac{i g_A}{F_\pi} \partial_\mu \vec{\pi}\, \bar{N} \gamma^\mu \gamma_5 \vec{t} N .$$

Using the free-particle Dirac equation, we can see that for nucleons on the mass shell (that is, at the one-nucleon poles in Figure 19.2) this is equivalent to the pseudoscalar interaction $-2i\, m_N\, g_A\, \vec{\pi} \cdot \bar{N} \gamma_5 \vec{t} N / F_\pi$. This provides yet another demonstration of the Goldberger–Treiman relation (19.4.33).

* * *

Our discussion here has not paralleled the historical development of these ideas. In fact, the course of historical development was chronologically almost exactly opposite to the line of argument presented here. Broken symmetries in particle physics started with the Goldberger–Treiman relation (19.4.33), which was derived[11] in 1957 on the basis of a dynamical calculation of pion decay. In order to explain the surprising success of this very approximate calculation, several theorists[15] introduced the idea of a 'partial conservation of the axial-vector current' (PCAC), the idea that although the axial-vector current is not conserved (as shown by the fact that pions do decay) its divergence $\partial_\mu A^\mu_\pm$ is proportional to the pion field. In itself, this assumption is meaningless — we saw in Chapter 10 that *any* field with a non-vanishing matrix element between the vacuum and one-pion states may be regarded as a pion field. Although it was not clear at the time, what was really being assumed was that the divergence of the axial-vector current is *small*, of order m_π^2, except where a pion pole gives it a large matrix element. The problem was greatly clarified by a 1960 paper of Nambu,[16] who pointed out that the axial-vector current could be regarded as exactly conserved in the limit of zero pion mass, in which case the Goldberger–Treiman relation could be derived as we have done here. In this and a subsequent paper with Jona-Lasinio,[17] Nambu recognized that the appearance of this massless or nearly massless pion was a symptom of a broken exact or approximate symmetry. With other collaborators,[18] Nambu also showed how to calculate the rates of emission of a single soft pion in various processes. Subsequently Goldstone[3] remarked that broken symmetries always entail massless bosons, and this was proved in 1962 by Goldstone, Salam, and myself,[5] using the arguments presented here in Section 19.2.

None of this early work on pions as Goldstone bosons depended on any specific assumptions about the nature of the broken symmetry group; for instance it might have been a direct product of three commuting $U(1)$ groups, whose generators form an isotopic spin vector, or it might have been the non-compact group $SO(3,1)$, for which a minus sign appears on the right-hand side of the commutation relation (19.4.19). The nature of the broken symmetry group became important only with the consideration of processes involving more than one pion, starting with the Adler–Weisberger sum rule in 1965,[19] whose success showed that the broken symmetry is indeed $SU(2) \times SU(2)$. (Such processes are discussed in the following section.) The identification of the symmetry group $SU(2)\times SU(2)$ led to a shift of emphasis,[20] toward the implications of this symmetry within strong interaction physics, and away from an earlier concentration[21] on the currents themselves.

All of this work was done without a specific theory of the strong interactions. One of the reasons for the rapid acceptance of quantum chromodynamics in 1973 as the correct theory of strong interactions was that it explained the $SU(2) \times SU(2)$ symmetry as a simple consequence of the smallness of the u and d quark masses.

19.5 Effective Field Theories: Pions and Nucleons

In Section 19.2 we learned how to calculate the amplitude for emission of a single low energy Goldstone boson B in a transition $\alpha \to \beta + B$ by applying the condition of current conservation to the matrix element of the symmetry current between the states α and β. In this calculation we never had to use any information about the details of the broken symmetry group; current conservation was all we needed. A new element enters if we wish to calculate the matrix element for the emission and/or absorption of *two* Goldstone bosons, as for instance in a Goldstone boson scattering process. Here we must apply the condition of current conservation to a matrix element like

$$\langle\beta|T\{J_1^{\lambda_1}(x_1), J_2^{\lambda_2}(x_2)\}|\alpha\rangle\,,$$

where the states α and β contain the other particles besides the two Goldstone bosons participating in the reaction. But when we let the divergence operator $\partial/\partial x_1^{\lambda_1}$ act on this matrix element we encounter a non-zero contribution from the derivative of the functions $\theta(x_1^0 - x_2^0)$ and $\theta(x_2^0 - x_1^0)$ that appear in the definition of the time-ordered product $T\{\cdots\}$. This contribution is equal to the matrix element of an equal-time commutator $\delta(x_1^0 - x_2^0)[J_1^0(x_1), J_2^{\lambda_2}(x_2)]$, whose value depends on the commutation relations of the group algebra. This makes such multi-Goldstone-boson processes specially interesting, because they can be used to decide experimentally on the nature of the broken symmetry group, in a way that is not possible for processes involving just a single Goldstone boson. Because of the appearance of such current commutators, this approach is known as the method of *current algebra.*[21]

The current algebra method was used in early calculations of multi-Goldstone-boson amplitudes.[22] Unfortunately, such calculations are tedious, especially when three or more Goldstone bosons are involved, and it was also difficult to see how to deal with symmetries like the chiral symmetry of quantum chromodynamics that are not exact. For this reason a different and more physical calculational technique was introduced,[23] based on the use of effective Lagrangians: We simply calculate the Goldstone boson amplitudes by the methods of perturbation theory, using some Lagrangian for the Goldstone bosons and the other particles in the states α and β that obeys the assumed broken symmetry.

Originally the justification for the effective Lagrangian procedure was based on current algebra. By using current algebra one could see that the amplitude for emission of a set of low energy Goldstone bosons in a process $\alpha \to \beta + B_1 + B_2 + \cdots$ is fixed once one knows the equal-time commutation relations of the currents associated with the broken symmetries as well as the matrix element for the process $\alpha \to \beta$ and the matrix elements of the currents between various one-particle states. We know that a Lagrangian that respects the broken symmetry will allow the construction of conserved currents with the appropriate equal-time commutators by Noether's technique, so if we simply calculate the low energy Goldstone boson amplitudes with such a Lagrangian and insert the correct values for $M_{\beta\alpha}$ and the one-particle matrix elements of the currents, we must get the same answer as provided by current algebra. In the case of interactions among Goldstone bosons alone, the states α and β may both be taken as the vacuum, and we do not need any extra information beyond the matrix element F between a Goldstone boson state and the vacuum.

In the first example of this sort,[23] the starting point was the Lagrangian of the σ-model.[24] Restricting our attention for the moment to the bosonic sector of this model, its Lagrangian is the $SO(4)$-invariant one used as an example in Section 19.2:

$$\mathscr{L} = -\frac{1}{2}\partial_\mu \phi_n \partial^\mu \phi_n - \frac{\mathscr{M}^2}{2}\phi_n \phi_n - \frac{\lambda}{4}(\phi_n \phi_n)^2 \,, \tag{19.5.1}$$

where n is understood to be summed over the values 1, 2, 3, 4, with $\vec{\phi}$ an isovector pseudoscalar field and ϕ_4 an isoscalar scalar field.

The immediate problem faced by any sort of effective Lagrangian is that in order to use it to calculate scattering amplitudes, we must either include all Feynman diagrams of all orders of perturbation theory, or else find some rationale for dropping higher-order diagrams. We can find no such rationale with Lagrangians like (19.5.1), in which the broken symmetry is realized through linear transformations on the various fields. Fortunately any such Lagrangian can be recast in a form that allows the use of Feynman diagrams to generate an expansion for scattering amplitudes in powers of Goldstone boson energies. To do this, we perform a symmetry transformation at every point in spacetime that eliminates the fields corresponding to the Goldstone bosons of the theory. The Goldstone boson degrees of freedom reappear in the transformed theory as the parameters of this symmetry transformation. However, since the Lagrangian is invariant under spacetime-independent symmetry transformations, it cannot have any dependence on the new Goldstone boson fields when they are constant, and so every term in the Lagrangian that involves these new Goldstone boson fields must contain at least one spacetime derivative of the field. These derivatives introduce factors of the Goldstone boson energy when we calculate S-matrix elements for Goldstone boson reactions, and as we shall see we can use the Lagrangian to construct a series for the S-matrix elements in powers of these energies.

For instance, to recast the Lagrangian (19.5.1) in a useful form, we write the four-vector ϕ_n as a chiral rotation R acting on a four-vector $\{0, 0, 0, \sigma\}$

whose first three components (the Goldstone part of ϕ_n) vanish:

$$\phi_n(x) = R_{n4}(x)\sigma(x) \tag{19.5.2}$$

with $R_{nm}(x)$ an orthogonal matrix

$$R^T(x)\,R(x) = 1 \tag{19.5.3}$$

and therefore

$$\sigma(x) = \sqrt{\sum_n \phi_n(x)^2}\,. \tag{19.5.4}$$

The Lagrangian (19.5.1) then becomes

$$\mathscr{L} = -\frac{1}{2}\sum_{n=1}^{4}\Big(R_{n4}\partial_\mu\sigma + \sigma\partial_\mu R_{n4}\Big)^2 - \frac{1}{2}\,\mathscr{M}^2\sigma^2 - \frac{\lambda}{4}\,\sigma^4\,. \tag{19.5.5}$$

Because R is an orthogonal matrix, the $\partial_\mu\sigma\partial^\mu\sigma$ term is R-independent, and the cross-term vanishes

$$\sum_n R_{n4}^2 = 1\,, \qquad \sum_n R_{n4}\partial_\mu R_{n4} = \frac{1}{2}\,\partial_\mu \sum_n R_{n4}^2 = 0\,,$$

so $\mathscr{L}$ becomes

$$\mathscr{L} = -\frac{1}{2}\,\partial_\mu\sigma\partial^\mu\sigma - \frac{1}{2}\,\sigma^2\sum_{n=1}^{4}\partial^\mu R_{n4}\,\partial_\mu R_{n4} - \frac{1}{2}\mathscr{M}^2\sigma^2 - \frac{\lambda}{4}\sigma^4\,. \tag{19.5.6}$$

If $\mathscr{M}^2$ is negative then σ has a non-vanishing vacuum expectation value, given in lowest order by the position of the minimum of the sum of the last two terms, at $\sigma = |\mathscr{M}|/\sqrt{\lambda}$.

In place of the field variables ϕ_n, our variables now are $\sigma - |\mathscr{M}|/\sqrt{\lambda}$ and whatever other variables are needed to parameterize the rotation R. For instance, these parameters could be simply chosen as the R_{a4} themselves (where $a, b, \cdots$ from now are isovector indices running over the values 1,2,3), with R_{44} given by the condition that R is orthogonal. A different parameterization will give simpler final results, and was historically the first to be used for these purposes. It is to define

$$\zeta_a \equiv \frac{\phi_a}{\phi_4 + \sigma} \tag{19.5.7}$$

and take

$$R_{a4} = \frac{2\zeta_a}{1+\vec{\zeta}^2} = -R_{4a}\,, \quad R_{44} = \frac{1-\vec{\zeta}^2}{1+\vec{\zeta}^2}\,, \quad R_{ab} = \delta_{ab} - \frac{2\zeta_a\zeta_b}{1+\vec{\zeta}^2}\,, \tag{19.5.8}$$

so that

$$\phi_a/\sigma = R_{a4} = \frac{2\zeta_a}{1+\vec{\zeta}^{\,2}}\,, \qquad \phi_4/\sigma = R_{44} = \frac{1-\vec{\zeta}^{\,2}}{1+\vec{\zeta}^{\,2}}\,. \tag{19.5.9}$$

Then the Lagrangian (19.5.6) becomes

$$\mathscr{L} = -\frac{1}{2}\partial_\mu\sigma\partial^\mu\sigma - 2\,\sigma^2\vec{D}_\mu\cdot\vec{D}^\mu - \frac{1}{2}\mathscr{M}^2\sigma^2 - \frac{\lambda}{4}\sigma^4\,, \tag{19.5.10}$$

where

$$\vec{D}_\mu \equiv \frac{\partial_\mu\vec{\zeta}}{1+\vec{\zeta}^{\,2}}\,. \tag{19.5.11}$$

Whatever parameterization we use, it is clear that the fields $\vec{\zeta}$ describe particles of zero mass, whose interactions all involve field derivatives. These are (up to a normalization) our new pion fields.

Despite appearances, this Lagrangian is still invariant under $SO(4)$, but with $SO(4)$ now realized non-linearly. Under an isospin transformation with infinitesimal parameters $\vec{\theta}$, the field $\vec{\zeta}$ simply rotates like an ordinary isovector, and σ is an isoscalar:

$$\delta\vec{\zeta} = \vec{\theta}\times\vec{\zeta}\,, \qquad \delta\sigma = 0\,, \tag{19.5.12}$$

so the Lagrangian (19.5.10) is manifestly isospin-invariant. On the other hand, under the broken symmetry transformations parameterized by an infinitesimal vector $\vec{\epsilon}$, the original fields transform according to

$$\delta\vec{\phi} = 2\vec{\epsilon}\phi_4\,, \qquad \delta\phi_4 = -2\vec{\epsilon}\cdot\vec{\phi}\,. \tag{19.5.13}$$

From Eq. (19.5.7), we then find

$$\delta\vec{\zeta} = \vec{\epsilon}(1-\vec{\zeta}^{\,2}) + 2\vec{\zeta}(\vec{\epsilon}\cdot\vec{\zeta})\,, \qquad \delta\sigma = 0\,. \tag{19.5.14}$$

The Lagrangian (19.5.10) is invariant under the broken symmetry transformation (19.5.14) because $\vec{D}_\mu$ undergoes a linear (though field-dependent) isospin rotation:[25]

$$\delta\vec{D}_\mu = 2(\vec{\zeta}\times\vec{\epsilon})\times\vec{D}_\mu \tag{19.5.15}$$

and Eq. (19.5.10) is isospin-invariant. Because of the transformation rule (19.5.15), $\vec{D}_\mu$ is often called the *covariant derivative* of the pion field.

The transformation rules (19.5.12) and (19.5.14) specify what is called a *non-linear realization* of the group $SU(2)\times SU(2)$.[25] The general theory of non-linear realizations of Lie groups is given in the next section; we shall show there that, up to field redefinitions, the transformation rules (19.5.12) and (19.5.14) provide the most general realization of $SU(2)\times SU(2)$ in which the isospin $SU(2)$ subgroup is realized linearly on $\vec{\zeta}$.

We see that each interaction of these new pion fields is accompanied with a spacetime derivative, so that the effective coupling is weak for low pion energies. (This remark will be made more precise below.) Therefore for sufficiently small pion energies we may use this Lagrangian in the tree approximation to reproduce the soft-pion theorems of current algebra. For this purpose, it is only necessary that the Lagrangian be $SO(4)$-invariant. But since the σ field is an $SO(4)$ scalar, it plays no role in maintaining the $SO(4)$ invariance of the Lagrangian, and may be simply discarded.* Of course, this procedure changes the physical content of the theory, but it does not change the amplitudes given by soft-pion theorems. The Lagrangian (19.5.10) then simplifies to

$$\mathscr{L} = -\frac{F^2}{2}\vec{D}_\mu \vec{D}^\mu = -\frac{F^2}{2}\frac{\partial_\mu \vec{\zeta} \cdot \partial^\mu \vec{\zeta}}{(1+\vec{\zeta}^{\,2})^2}\,, \tag{19.5.16}$$

where $F = 2\langle\sigma\rangle = 2|\mathscr{M}|/\sqrt{\lambda}$. (As we shall soon see, this F is the same as the constant F_π discussed in the previous section.) For many purposes it is more convenient to work with a conventionally normalized pion field

$$\vec{\pi} \equiv F\vec{\zeta} \tag{19.5.17}$$

for which the Lagrangian (19.5.16) reads

$$\mathscr{L} = -\frac{1}{2}\frac{\partial_\mu \vec{\pi} \cdot \partial^\mu \vec{\pi}}{(1+\vec{\pi}^2/F^2)^2}\,. \tag{19.5.18}$$

The factor $1/F$ acts as a coupling parameter that accompanies the interaction of each additional pion. Eq. (19.5.18) describes what is often called the 'non-linear σ-model,' for the special case of $SU(2) \times SU(2)$ spontaneously broken to $SU(2)$.

An important point: to derive Eq. (19.5.18) it was not really necessary to start with the 'linear σ-model' Lagrangian (19.5.1). Indeed, we did not need to start with *any* specific theory. Eq. (19.5.18) can be used simply because it is invariant under the $SO(4)$ transformation (19.5.12), (19.5.14), and current algebra tells us that this is all we need to get the right results for low energy pion reaction amplitudes.

Some years after the introduction of effective Lagrangians for soft pions, there emerged a different justification for the effective field theory technique,[26] one that does not rely on current algebra and allows calculations that are not limited to the limit of vanishingly small Goldstone boson energies. It is based on the realization (not yet formally embodied in a theorem) that when we calculate a physical amplitude from Feynman diagrams using the most general Lagrangian that involves the relevant degrees of freedom and satisfies the assumed symmetries of the theory, we are simply constructing the most general amplitude that is consistent with general principles of relativity, quantum mechanics, and the assumed symmetries. This was the point of view underlying Volume I of this book. In the present context, the 'relevant' degrees of freedom are the Goldstone

* Alternatively, we can pass to the limit where $\mathscr{M}$ and λ go to infinity together, keeping the expectation value of σ constant.

bosons themselves, together with the particles in the states α and β and any other particle states that can be produced from them by interactions with low energy Goldstone bosons. By invoking this justification for effective field theories, we are freed from any need to wrestle with the complications of current algebra. More importantly, the modern effective field theory approach yields results that take us beyond the extreme low energy limit and allows a systematic treatment of any intrinsic symmetry breaking.

According to this approach, in order to calculate pion interaction amplitudes to any desired order in pion energies, we must use the most general Lagrangian involving a pion field $\vec{\zeta}$ that transforms according to the rules (19.5.12) and (19.5.14):

$$\mathscr{L}_{\text{eff}} = -\frac{F^2}{2}\vec{D}_\mu\vec{D}^\mu - \frac{c_4}{4}(\vec{D}_\mu\vec{D}^\mu)^2 - \frac{c'_4}{4}(\vec{D}_\mu\cdot\vec{D}_\nu)(\vec{D}^\mu\cdot\vec{D}^\nu) - \ldots . \tag{19.5.19}$$

The terms indicated by ... will contain higher powers of the covariant derivative $\vec{D}_\mu$, or higher covariant derivatives, whose general structure is described in the next section. The coefficients c_4 and c'_4 are dimensionless, and all higher terms have coefficients with the dimensionality of negative powers of mass.

Consider a general process involving arbitrary numbers of incoming and outgoing pions. We suppose that their energies and momenta are all at most of some order Q, which is small compared with a typical quantum chromodynamics energy scale (say, the nucleon or ρ mass). Even though Lagrangians like (19.5.19) are not renormalizable in the usual sense, we saw in Section 12.3 that such Lagrangians can yield finite results as long as they contain all possible terms allowed by symmetries, for then there will be a counterterm available to cancel every infinity. If we define the renormalized values of the constants F^2, c_4, c'_4, ... by specifying the values of various Goldstone boson scattering amplitudes at energies of order Q, then the integrals in momentum-space Feynman diagrams will be dominated by contributions from virtual momenta which are also of order Q (because renormalization makes them finite, and there is no other possible effective cut-off in the theory.) We can then develop perturbation theory as a power series expansion in Q.

Each derivative and hence each $\vec{D}_\mu$ in each interaction vertex contributes one factor of Q to the order of magnitude of the diagram; each internal pion propagator contributes a factor Q^{-2}; and each integration volume d^4q associated with the loops of the diagram contributes a factor Q^4; so a general connected diagram makes a contribution of order Q^ν, where

$$\nu = \sum_i V_i d_i - 2I + 4L . \tag{19.5.20}$$

Here d_i is the number of derivatives in an interaction of type i, V_i is the number of interaction vertices of type i in the diagram, I is the number of internal pion lines, and L is the number of loops. These quantities are related by the familiar topological identity (see Eq. (4.4.7)):

$$L = I - \sum_i V_i + 1\,, \tag{19.5.21}$$

so we can eliminate I and write

$$\nu = \sum_i V_i(d_i - 2) + 2L + 2\,. \tag{19.5.22}$$

The point of this is that each term here is positive; every interaction in Eq. (19.5.19) has at least two derivatives, and of course $L \geq 0$. Therefore the leading term of each process is of order Q^2, and arises solely from *tree* graphs (that is, $L = 0$) constructed solely from the term in Eq. (19.5.19) with only two derivatives (that is, $V_i = 0$ for $d_i \neq 2$):

$$\mathscr{L}_2 = -\frac{F^2}{2}\,\vec{D}_\mu \vec{D}^\mu = -\frac{1}{2}\,\frac{\partial_\mu \vec{\pi}\,\partial^\mu \vec{\pi}}{(1 + \vec{\pi}^{\,2}/F^2)^2}\,. \tag{19.5.23}$$

For instance, the invariant amplitude M that appears in the pion–pion scattering S-matrix element

$$S = i(2\pi)^4 \delta^4(p_A + p_B - p_C - p_D) M (2\pi)^{-6} (16 E_A E_B E_C E_D)^{-1/2} \tag{19.5.24}$$

is given to this order by

$$M^{(\nu=2)}_{abcd} = 4F^{-2}\Big[\delta_{ab}\delta_{cd}(-p_A \cdot p_B - p_C \cdot p_D) + \delta_{ac}\delta_{bd}(p_A \cdot p_C + p_B \cdot p_D) + \delta_{ad}\delta_{bc}(p_A \cdot p_D + p_B \cdot p_C)\Big]\,, \tag{19.5.25}$$

where a, b, c, d are the isovector indices of the pions A, B, C, D, respectively. (The effect of the finite pion mass will be taken up a little later in this section.) Using Eq. (19.5.25) as the leading term does not depend on any assumption of the smallness of the coupling constant λ in the original Lagrangian (19.5.1), or even on the validity of this formula for the Lagrangian, but only on the assumed smallness of the typical pion energy Q.

The next term in the amplitude for any Goldstone boson reaction will be of order Q^4, and arises both from one-loop graphs involving only the Lagrangian (19.5.16), and also from tree graphs constructed solely from the interaction (19.5.16) plus a single vertex arising from the $d = 4$ terms in Eq. (19.5.19):

$$M^{(v=4)}_{abcd} = \frac{\delta_{ab}\delta_{cd}}{F^4}\Bigg[-\frac{1}{2\pi^2}\, s^2 \ln(-s) - \frac{1}{12\pi^2}\,(u^2 - s^2 + 3t^2)\ln(-t)$$
$$- \frac{1}{12\pi^2}\,(t^2 - s^2 + 3u^2)\ln(-u) + \frac{1}{3\pi^2}(s^2 + t^2 + u^2)\ln\Lambda^2$$
$$- \frac{1}{2}\, c_4 s^2 - \frac{1}{4}\, c_4'(t^2 + u^2)\Bigg] + \text{crossed terms}\,, \tag{19.5.26}$$

where 'crossed terms' denotes terms given by the interchanges $B \leftrightarrow C$ and $B \leftrightarrow D$, and s, t, and u are the Mandelstam variables

$$s = -(p_A + p_B)^2,\ t = -(p_A - p_C)^2,\ u = -(p_A - p_D)^2\,.$$

The dependence of this result on the cut-off Λ can be eliminated by a redefinition of the constants c_4 and c_4'. The renormalized couplings are

$$c_{4R} = c_4 - \frac{2}{3\pi^2}\,\ln\left(\frac{\Lambda^2}{\mu^2}\right)\,, \tag{19.5.27}$$

$$c_{4R}' = c_4' - \frac{4}{3\pi^2}\,\ln\left(\frac{\Lambda^2}{\mu^2}\right)\,, \tag{19.5.28}$$

where μ is an arbitrary renormalization scale of order Q inserted in order to make the logarithms well-defined. In terms of these renormalized couplings, the amplitude (19.5.26) takes the form

$$M^{(v=4)}_{abcd} = \frac{\delta_{ab}\delta_{cd}}{F^4}\Bigg[-\frac{1}{2\pi^2}\, s^2 \ln\left(\frac{-s}{\mu^2}\right) - \frac{1}{12\pi^2}\,(u^2 - s^2 + 3t^2)\ln\left(\frac{-t}{\mu^2}\right)$$
$$- \frac{1}{12\pi^2}\,(t^2 - s^2 + 3u^2)\ln\left(\frac{-u}{\mu^2}\right) - \frac{1}{2}\, c_{4R} s^2 - \frac{1}{4}\, c_{4R}'(t^2 + u^2)\Bigg]$$
$$+ \text{crossed terms}\,. \tag{19.5.29}$$

This sort of calculation can be carried to arbitrary orders in Q, always with the result that in each order we encounter a finite number of new couplings whose renormalization serves to eliminate the cut-off dependence of physical amplitudes. Notice that the ratio of the leading $v = 2$ terms and the $v = 4$ corrections is of order $Q^2/8\pi^2F^2$, so this is likely to be a useful expansion as long as the pion energies are all much less than an amount of order $2\pi F$.

This perturbative expansion may also be used to relate the ubiquitous parameter F to the measured pion decay amplitude F_π. Recalling the transformation rule (19.5.14) for $\vec{\zeta}$, the axial-vector current is given by

$$\vec{\epsilon}\cdot\vec{A}^\mu = -\frac{\partial\mathscr{L}}{\partial(\partial_\mu\vec{\zeta})}\cdot\delta\vec{\zeta}$$

and so

$$\vec{A}^\mu = -(1-\vec{\zeta}^{\,2})\,\frac{\partial \mathscr{L}}{\partial(\partial_\mu \vec{\zeta})} - 2\vec{\zeta}\,\vec{\zeta}\cdot\left(\frac{\partial \mathscr{L}}{\partial(\partial_\mu \vec{\zeta})}\right). \tag{19.5.30}$$

(This $\vec{A}^\mu$ is the Noether current of the symmetry generated by $2\vec{x}$, which on the nucleon doublet is represented by $2\gamma_5\vec{t} = \gamma_5\vec{\tau}$.) After integrating out all other fields, and using Eq. (19.5.17) to express $\vec{\zeta}$ in terms of the canonically normalized pion field $\vec{\pi}$, we find

$$\vec{A}^\mu = F\left[\partial^\mu\vec{\pi}\,\frac{(1-\vec{\pi}^2/F^2)}{(1+\vec{\pi}^2/F^2)} + \frac{2\vec{\pi}(\vec{\pi}\cdot\partial^\mu\vec{\pi})}{F^2(1+\vec{\pi}^2/F^2)^2}\right] + \cdots . \tag{19.5.31}$$

Using our perturbative expansion in powers of the pion energy to calculate $\langle\mathrm{VAC}|\vec{A}^\mu|\pi\rangle$, we see that in lowest order the pion decay amplitude here is

$$F_\pi = F . \tag{19.5.32}$$

Furthermore Lorentz invariance and Eq. (19.5.22) tell us that the higher-order corrections must be proportional to powers of p_π^2/F_π^2, which vanishes for a massless pion, so Eq. (19.5.32) is actually exact in the limit $m_\pi \to 0$. We may therefore guess that our perturbation expansion will be useful for pion energies that are less than an amount of order $2\pi F_\pi = 1200$ MeV.

In order to make contact with experiment, we must deal with the fact that the pion mass is not zero. On the mass shell it is not possible for the time component of a pion four-momentum to be less than m_π, so in counting powers of the typical pion energy and momentum Q, we must regard m_π as being of order Q. But we saw in the previous section that m_π^2 is proportional to a linear combination of quark masses, so our formula (19.5.22) for the order Q^ν of a given Feynman diagram should be rewritten to read

$$\nu = \sum_i V_i(d_i + 2m_i - 2) + 2L + 2 , \tag{19.5.33}$$

where m_i is the number of factors of quark masses in the interaction of type i.

The interactions involving quark masses may be distinguished by their transformation properties under $SU(2)\times SU(2)$, or equivalently, under $SO(4)$. Eq. (19.4.34) shows that the terms of first order in quark masses transform as a linear combination of two scalars, the fourth component of a chiral four-vector Φ_n^+, with coefficient $m_u + m_d$, and the third component of a different chiral four-vector Φ_n^-, with coefficient $m_u - m_d$. Thus the terms $\Delta\mathscr{L}^+$ and $\Delta\mathscr{L}^-$ in the effective Lagrangian that are of first order in $m_u + m_d$ and $m_u - m_d$ must have the chiral transformation properties of the fourth and third components of chiral four-vectors, respectively, and of course be Lorentz scalars. One obvious candidate for a chiral four-vector whose fourth component is a scalar is the ϕ_n with which we started in this section. According to Eq. (19.5.9), its fourth component is just $\sigma(1-\vec{\zeta}^2)/(1+\vec{\zeta}^2)$. The factor σ may be dropped, as it is a chiral scalar, and therefore has no effect on the chiral transformation properties of this quantity. We can fix the normalization of this term by requiring that the coefficient of $\vec{\pi}^2 = F_\pi^2\vec{\zeta}^2$ be $-m_\pi^2/2$, so that, apart from an additive constant,

$$\Delta\mathscr{L}^+ = -\frac{m_\pi^2 F_\pi^2}{2}\frac{\vec{\zeta}^2}{1+\vec{\zeta}^2} = -\frac{m_\pi^2}{2}\frac{\vec{\pi}^2}{1+\vec{\pi}^2/F_\pi^2}\,. \tag{19.5.34}$$

We shall see in the next section that this is the unique scalar function of the pion field without derivatives that transforms as the fourth component of a chiral four-vector. On the other hand, there is *no* scalar function of the pion field without derivatives that transforms as the third component of a chiral four-vector because such a function, if the third component of a chiral four-vector, would have to be odd in the pion field, and therefore pseudoscalar rather than scalar. Thus (19.5.34) is the only interaction with $d_i = 0$ and $m_i = 1$. It is striking that the isospin violating difference in the u and d quark masses has not only no effect on the pion masses, as we saw in the previous section, but also has no effect on any non-derivative multipion interaction.

We are now in a position to do a realistic calculation of the leading $\nu = 2$ terms in the pion–pion scattering amplitude. According to Eq. (19.5.33), these terms arise only from tree graphs constructed from the $d = 2$, $m = 0$ pion interaction in Eq. (19.5.23) or from the $d = 0$, $m = 1$ pion interaction in Eq. (19.5.34). To this order, the invariant amplitude M defined by Eq. (19.5.24) is now

$$M_{abcd}^{(\nu=2)} = 4F_\pi^{-2}\Big[\delta_{ab}\delta_{cd}(s-m_\pi^2)+\delta_{ac}\delta_{bd}(t-m_\pi^2)+\delta_{ad}\delta_{bc}(u-m_\pi^2)\Big]\,. \tag{19.5.35}$$

In particular, at threshold $s = 4m_\pi^2$, $t = u = 0$, so

$$\begin{aligned} M_{abcd}^{(\nu=2)}(\text{threshold}) &= 4m_\pi^2 F_\pi^{-2}[3\delta_{ab}\delta_{cd} - \delta_{ac}\delta_{bd} - \delta_{ad}\delta_{bc}] \\ &= 4m_\pi^2 F_\pi^{-2}\left[7\,M_{ab,cd}^{(0)} - 2\,M_{ab,cd}^{(2)}\right]\,, \end{aligned} \tag{19.5.36}$$

where $M^{(0)}$ and $M^{(2)}$ are the appropriate tensors representing two-pion states with isospin $T = 0$ or $T = 2$:

$$M_{ab,cd}^{(0)} = \tfrac{1}{3}\delta_{ab}\delta_{cd}\,, \tag{19.5.37}$$

$$M_{ab,cd}^{(2)} = \tfrac{1}{2}\left(\delta_{ac}\delta_{bd} + \delta_{ad}\delta_{bc} - \tfrac{2}{3}\,\delta_{ab}\delta_{cd}\right)\,, \tag{19.5.38}$$

normalized so that $\mathrm{Tr}M^{(T)} = 2T + 1$. This result is usually expressed in terms of the scattering lengths. According to Sections 3.6 and 3.7, the scattering length a_T for two pions in a state of isospin T is given[27] by $1/32\pi m_\pi$ times the coefficient of $M^{(T)}$ in Eq. (19.5.36):

$$a_0 = \frac{7m_\pi}{8\pi F_\pi^2} = 0.16\, m_\pi^{-1} \,, \qquad a_2 = -\frac{m_\pi}{4\pi F_\pi^2} = -0.046\, m_\pi^{-1} \,.$$

The pion scattering lengths are difficult to measure, but careful study of processes like $\pi+N \to \pi+\pi+N$ and $K \to \pi+\pi+e+\nu$ has given the results[28] $(0.26\pm0.05)\, m_\pi^{-1}$ and $(-0.028\pm0.012)\, m_\pi^{-1}$ for a_0 and a_2, respectively, which are consistent with these theoretical values. Corrections of higher order in $m_\pi/2\pi F_\pi$ seem to improve the agreement.

This formalism can be extended to describe the interactions of pions with nucleons or other particles. The easiest approach is to suppose that we add a term involving the nucleon doublet field N to the Lagrangian with which we started:

$$\mathscr{L}_N = -\bar{N}\Big(\not\partial + g[\phi_4 + 2i\vec{t}\cdot\vec{\phi}\gamma_5]\Big)\, N \,, \tag{19.5.39}$$

where $\vec{t}$ is the isospin matrix vector for isospin 1/2 (that is, half the Pauli spinor $\vec{\tau}$.) This is invariant under chiral $SU(2) \times SU(2)$ transformations with

$$\delta\vec{\phi} = 2\vec{\epsilon}\,\phi_4 \,, \qquad \delta\phi_4 = -2\vec{\epsilon}\cdot\vec{\phi} \,, \tag{19.5.40}$$

$$\delta N = -2i\gamma_5\vec{\epsilon}\cdot\vec{t}\, N \,. \tag{19.5.41}$$

Now in eliminating the non-derivative pion couplings we must express the nucleon field N as the $SO(4)$ rotation R in the representation (19.5.41) acting upon a new nucleon field $\tilde{N}$

$$N \equiv \frac{(1 - 2i\gamma_5\vec{t}\cdot\vec{\zeta})}{\sqrt{1+\vec{\zeta}^{\,2}}}\, \tilde{N} \tag{19.5.42}$$

with $\vec{\zeta}$ given again by Eq. (19.5.7). With this transformation the non-derivative term in Eq. (19.5.39) now depends only on $\tilde{N}$ and σ:

$$\bar{N}[\phi_4 + 2i\vec{t}\cdot\vec{\phi}\gamma_5]N = \sigma\bar{\tilde{N}}\tilde{N} \,.$$

On the other hand, the derivative term involves derivatives of the matrix in Eq. (19.5.41), yielding a nucleonic Lagrangian:

$$\mathscr{L}_N = -\bar{\tilde{N}}\left[\not\partial + g\sigma + 2i\,\frac{\vec{t}\cdot(\vec{\zeta}\times\not\partial\vec{\zeta})}{1+\vec{\zeta}^{\,2}} + 2i\gamma_5\vec{t}\cdot\vec{\not{D}}\right]\tilde{N} \tag{19.5.43}$$

or in terms of the canonically normalized pion field (19.5.17):

$$\mathscr{L}_N = -\bar{\tilde{N}}\left[\not\partial + g\sigma + \frac{2i\vec{t}\cdot(\vec{\pi}\times\not\partial\vec{\pi})}{F_\pi^2[1+\vec{\pi}^{\,2}/F_\pi^2]} + \frac{2i\gamma_5\vec{t}\cdot\not\partial\vec{\pi}}{F_\pi[1+\vec{\pi}^{\,2}/F_\pi^2]}\right]\tilde{N} \,. \tag{19.5.44}$$

Since σ has a non-vanishing vacuum expectation value, we see that the nucleon here has a non-zero rest mass, a result that would be prohibited if the symmetry under the transformation (19.5.40), (19.5.41) were unbroken.

By its construction, the Lagrangian (19.5.44) is chiral-invariant. This can also be seen directly, using the previously obtained chiral transformation properties of σ and $\vec{\pi}$, and now also noting that under the chiral transformation (19.5.40), (19.5.41), the new nucleon field defined by Eq. (19.5.42) transforms as

$$\delta\tilde{N} = 2i\vec{t}\cdot[\vec{\zeta}\times\vec{\epsilon}]\,\tilde{N}\,. \tag{19.5.45}$$

That is, under a chiral transformation $\tilde{N}$ simply undergoes the same isospin rotation (19.5.15) as the quantity $\vec{D}_\mu$, but of course in the $T = 1/2$ representation. The isospin rotation parameter $\vec{\zeta}\times\vec{\epsilon}$ is spacetime-dependent, so derivatives of $\tilde{N}$ do not have the same chiral transformation property, but it is straightforward to check that the combination of the first and third terms in Eq. (19.5.43) behaves as a chiral-covariant derivative: that is,

$$\delta\mathscr{D}_\mu\tilde{N} = 2i\vec{t}\cdot[\vec{\zeta}\times\vec{\epsilon}]\mathscr{D}_\mu\tilde{N}\,, \tag{19.5.46}$$

where

$$\mathscr{D}_\mu\tilde{N} \equiv \left[\partial_\mu + 2i\,\frac{\vec{t}\cdot(\vec{\zeta}\times\partial_\mu\vec{\zeta})}{1+\vec{\zeta}^{\,2}}\right]\tilde{N}\,. \tag{19.5.47}$$

Thus the Lagrangian (19.5.43) (and hence (19.5.44)) is obviously chiral-invariant, because it is isospin-invariant, and constructed solely from the ingredients $\tilde{N}$, $\mathscr{D}_\mu\tilde{N}$, σ, and $\vec{D}_\mu$, all of which transform under chiral transformations with the same isospin rotation.

Again, the specific Lagrangian with which we started here is not important. As in the case of the pure pion theory considered earlier, the important thing is the chiral invariance of the Lagrangian. For chiral invariance, the Lagrangian must conserve isospin, and be constructed only from the ingredients $\tilde{N}$, $\mathscr{D}_\mu\tilde{N}$, and $\vec{D}_\mu$ (together with higher covariant derivatives of these objects). The most general chiral-invariant Lagrangian that is bilinear in the new nucleon field and involves no more than one derivative therefore takes the form

$$\mathscr{L}_{N,0} = -\bar{\tilde{N}}\left[\not{\mathscr{D}} + m_N + 2ig_A\gamma_5\vec{t}\cdot\vec{\not{D}}\right]\tilde{N} \tag{19.5.48}$$

or in terms of the pion field (19.5.17):

$$\mathscr{L}_{N,0} = -\bar{\tilde{N}}\left[\not{\partial} + m_N + \frac{2i\vec{t}\cdot(\vec{\pi}\times\not{\partial}\vec{\pi})}{F_\pi^2[1+\vec{\pi}^{\,2}/F_\pi^2]} + \frac{2ig_A\gamma_5\vec{t}\cdot\not{\partial}\vec{\pi}}{F_\pi[1+\vec{\pi}^{\,2}/F_\pi^2]}\right]\tilde{N}\,. \tag{19.5.49}$$

Note that we have inserted an arbitrary constant g_A in the last term in Eq. (19.5.48) because this term is chiral symmetric by itself, and so chiral symmetry cannot dictate its coefficient. (This is in contrast with the third term in Eq. (19.5.43), whose coefficient is fixed by the condition that it and the first term together give a chiral invariant.) We can check that the constant g_A inserted here is indeed the axial-vector coupling of beta decay by constructing the extra term in the axial-vector current that arises from the nucleonic Lagrangian (19.5.48). Alternatively, integrating by parts and using the Dirac equation, we find that the pion–nucleon interaction $-2ig_A\bar{\tilde{N}}\gamma_5\vec{t}\cdot(\not{\partial}\vec{\pi})\tilde{N}/F_\pi$ is equivalent on the nucleon mass shell to an interaction $-4im_N g_A\bar{\tilde{N}}\gamma_5\vec{t}\cdot\vec{\pi}\tilde{N}/F_\pi$, corresponding to a pion–nucleon coupling constant

$$G_{\pi N} = 2m_N g_A/F_\pi \,,$$

which is just the Goldberger–Treiman relation (19.4.33).

Incidentally, if we had used the Lagrangian (19.5.44) in this calculation, we would have obtained the Goldberger–Treiman relation with $g_A = 1$. However, this result is in fact an artifact of the particular form (19.5.39) of the interaction with which we started. We might have included a non-renormalizable derivative coupling term**

$$\mathscr{L}'_N = ig'\bar{N}\left[(\vec{t}\cdot\vec{\phi}\not{\partial}\phi_4 - \phi_4\vec{t}\cdot\not{\partial}\cdot\vec{\phi})\gamma_5 + \vec{t}\cdot(\vec{\phi}\times\not{\partial}\vec{\phi})\right]N \,. \tag{19.5.50}$$

It is straightforward to show that in terms of the transformed fields $\sigma, \tilde{N}$, and $\vec{\zeta}$, this takes the form

$$\mathscr{L}'_N = -8i\sigma^2 g'\bar{\tilde{N}}\vec{t}\cdot\not{D}\vec{\zeta}\gamma_5\tilde{N} \,.$$

Since σ has a non-vanishing vacuum expectation value, this makes a contribution to the pion–nucleon coupling constant and to g_A proportional to g'. Hence the value $g_A = 1$ is *not* dictated by the broken $SU(2)\times SU(2)$ symmetry alone; by including the interaction (19.5.50) and adjusting g', we can make g_A anything we like.

Now let us consider how to use this Lagrangian to calculate amplitudes for reactions involving both pions and nucleons. We will have to give special attention to the nucleon propagators, since a nucleon can never be a 'soft' particle like a pion. A nucleon line that enters a diagram with an on-shell four-momentum p of order m_N, and then from interactions with

** The left- and right-handed parts of the nucleon doublet transform according to the representations $(\frac{1}{2},0)$ and $(0,\frac{1}{2})$ of $SU(2)\times SU(2)$, respectively. The bilinear $\bar{N}\gamma_\mu\gamma_5 N$ is thus a sum of terms quadratic in $(\frac{1}{2},0)$ or $(0,\frac{1}{2})$ terms, so it transforms like a direct sum of the (1,0), (0,0), and (0,1) representations. In Eq. (19.5.50) we have coupled the $(1,0)+(0,1)$ terms in $\bar{N}\gamma_\mu\gamma_5 N$ to the antisymmetric tensor formed from the $SU(2)\times SU(2)$ four-vectors ϕ_n and $\partial_\mu\phi_n$. Its invariance under the transformation (19.5.40), (19.5.41) can of course be checked directly.

soft pions absorbs a net four-momentum q with components much less than m_N, will have a propagator:

$$\frac{-i(\not{p}+\not{q})+m_N}{(p+q)^2+m_N^2} \xrightarrow[q\to 0]{} \frac{-i\not{p}+m_N}{2p\cdot q} . \tag{19.5.51}$$

(The neglected terms may be taken into account by including higher derivative terms in the nucleon Lagrangian.) Suppose again that all external pion four-momenta have components at most of order Q and define all renormalized couplings at renormalization points of order Q, so that integrals converge in such a way that internal pion lines also have four-momenta Q. Then Eq. (19.5.51) shows that internal nucleon lines make a contribution of order $1/Q$. A general Feynman diagram for such a process will make a contribution to the invariant amplitude of order Q^ν, where now

$$\nu = \sum_i V_i(d_i + 2m_i) - 2I_\pi - I_N + 4L . \tag{19.5.52}$$

Here V_i is the number of vertices associated with interactions of type i, d_i is the number of derivatives in each such interaction, m_i is the number of quark mass factors in each interaction, I_π and I_N are the numbers of internal pion and nucleon lines, respectively, and L is the number of loops in the graph. We use the familiar topological relations for connected graphs:

$$L = I_\pi + I_N - \sum_i V_i + 1 \tag{19.5.53}$$

and

$$2I_N + E_N = \sum_i V_i n_i , \tag{19.5.54}$$

where n_i is the number of nucleon fields in interactions of type i, and E_N is the number of external nucleon lines. Eliminating the quantities I_N and I_π, this gives

$$\nu = \sum_i V_i \left(d_i + 2m_i + \frac{n_i}{2} - 2 \right) + 2L - \frac{E_N}{2} + 2 . \tag{19.5.55}$$

The important point here is that the coefficient $d_i + 2m_i + \frac{1}{2}n_i - 2$ in the first term is always positive or zero. We have already seen that $d_i + 2m_i \geq 2$ for the purely pionic interactions with $n_i = 0$, and inspection of (19.5.49) shows that $d_i \geq 1$ for the pion–nucleon interactions with $n_i = 2$ and $m_i = 0$. Any interaction with $n_i = 2$ and $m_i \geq 1$ or $n_i \geq 4$ clearly also has $d_i \geq 2$. Hence the leading terms for $Q \ll 2\pi F_\pi$ are tree graphs (that is,

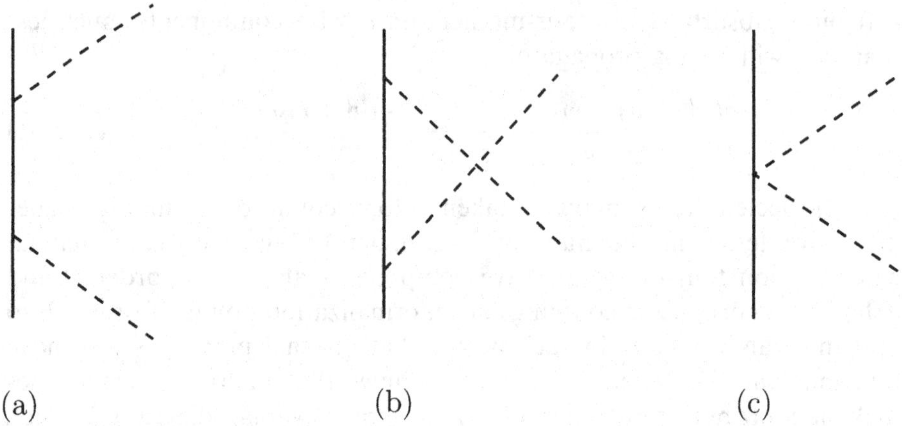

Figure 19.3. Feynman diagrams used with an effective chiral Langrangian to calculate the scattering of a soft pion from a nucleon. Dashed lines are pions; solid lines are nucleons.

$L = 0$) for which all interactions have

$$d_i + 2m_i + \frac{n_i}{2} - 2 = 0\,.$$

The interactions that satisfy this condition are just those shown explicitly in Eqs. (19.5.23), (19.5.34), and (19.5.49), plus possible interactions with $d_i = 0$ and $n_i = 4$:

$$\left(\bar{\tilde{N}}\,\Gamma_\alpha \tilde{N}\right)\left(\bar{\tilde{N}}\,\Gamma^\alpha \tilde{N}\right), \tag{19.5.56}$$

where Γ_α and Γ^α are any matrices in spin and isospin space that yield Lorentz, space inversion, and isospin-invariant four-fermion interactions. These last interactions are important for multinucleon processes,[29] which will not be considered here.

Let us now apply this method to pion–nucleon scattering. For pion energies roughly of order m_π, the pion–nucleon scattering amplitude is given by the Feynman graphs of Figure 19.3, all of which make contributions of order m_π to the invariant amplitude. However, at threshold the leading contribution is from Figure 19.3(c), the others being suppressed by an extra factor of m_π/m_N. This is because at threshold in the rest frame the incoming and outgoing pion four-momenta q, q' and nucleon four-momenta p, p' are given by

$$q = q' = (0, 0, 0, m_\pi) = \left(\frac{m_\pi}{m_N}\right) p = \left(\frac{m_\pi}{m_N}\right) p'\,. \tag{19.5.57}$$

Thus the invariant amplitude at threshold from either Figure 19.3(a) or

19.3(b) is proportional to

$$\bar{u}\gamma_5\, \not{q}\, \frac{(-i(\not{p}\pm\not{q})+m_N)}{(p\pm q)^2+m_N^2}\gamma_5\, \not{q} u$$
$$=\frac{m_\pi^2\,\bar{u}\gamma_5(-i\not{p}(1\pm m_\pi/m_N)+m_N)\gamma_5 u}{(\mp 2m_\pi m_N-m_\pi^2)}$$
$$=\frac{\mp m_\pi}{2m_N\pm m_\pi}\,\bar{u}(i\not{p}(1\pm m_\pi/m_N)+m_N)u=\frac{m_\pi^2}{2m_N\pm m_\pi}$$

where u is a Dirac spinor with $\bar{u}u=1$. In contrast, the diagram of Figure 19.3(c) gives a contribution to M of order m_π. We may write this amplitude as a 2×2 matrix in the nucleon isospin indices:

$$M_{ba}=\frac{-2i}{F_\pi^2}\,t_c\epsilon_{abc}\,\bar{u}(-i\not{q}-i\not{q}')u\,,$$

where q' and q are the final and initial pion four-momenta and b and a are the corresponding isovector indices. Using Eq. (19.5.57) and the momentum-space Dirac equation $(i\not{p}+m_N)u=0$, this becomes

$$M_{ba}=\frac{-4im_\pi}{F_\pi^2}\,t_c\,\epsilon_{abc}=\frac{-4m_\pi}{F_\pi^2}\,\vec{t}\cdot[\vec{t}^{(\pi)}]_{ba}\,, \tag{19.5.58}$$

where $[t_c^{(\pi)}]_{ba}\equiv -i\epsilon_{bac}$ is the pion isovector matrix. The matrix $\vec{t}\cdot\vec{t}^{(\pi)}$ has eigenvalues in states of total isospin T equal to $\frac{1}{2}[T(T+1)-2-\frac{3}{4}]$, so in the two isospin states, $T=1/2$ and $T=3/2$, the invariant amplitudes are[30]

$$M_{1/2}=4m_\pi/F_\pi^2\,,\qquad\qquad M_{3/2}=-2m_\pi/F_\pi^2\,,$$

These results are usually expressed in terms of the scattering lengths, which are defined as the invariant amplitudes divided by $4\pi(1+m_\pi/m_N)$:

$$a_{1/2}=\frac{m_\pi}{\pi F_\pi^2(1+m_\pi/m_N)}=0.15\,m_\pi^{-1}\,, \tag{19.5.59}$$
$$a_{3/2}=-\frac{m_\pi}{2\pi F_\pi^2(1+m_\pi/m_N)}=-0.075\,m_\pi^{-1}\,. \tag{19.5.60}$$

These are in reasonable agreement with the experimental values[12] $a_{1/2}=(0.173\pm0.003)\,m_\pi^{-1}$ and $a_{3/2}=(-0.101\pm0.004)\,m_\pi^{-1}$. (The results (19.5.59) and (19.5.60) are only supposed to be valid to lowest order in m_π/m_N, but we retain the factor $1+m_\pi/m_N$, because it arises just from the definition of the scattering lengths.)

The corrections to the scattering lengths of next order in m_π arise from several sources. There are the Born diagram graphs of Figures 19.3(a) and 19.3(b), which are nominally of leading order but as we have seen at threshold are suppressed by an extra factor of m_π/m_N. There are additional tree graphs containing a vertex with two derivatives. Of special interest are tree graphs containing a vertex with no derivatives that arises from a symmetry-breaking interaction proportional to m_π^2. These interactions must have the chiral and spacetime transformation properties of the operators in Eq. (19.4.34): the fourth and third components of two different chiral four-vectors Φ_n^+ and Φ_n^-. There are two obvious candidates for such operators that are bilinear in the nucleon fields: a Φ_4^+ term,

$$\bar{N}N = \left(\frac{1-\vec{\zeta}^2}{1+\vec{\zeta}^2}\right)\bar{\tilde{N}}\tilde{N} - 4i\left(\frac{\vec{\zeta}}{1+\vec{\zeta}^2}\right)\cdot\bar{\tilde{N}}\gamma_5\vec{t}\tilde{N}\,,$$

and a Φ_3^- term,

$$\bar{N}t_3N = \bar{\tilde{N}}t_3\tilde{N} - 2\left(\frac{\zeta_3}{1+\vec{\zeta}^2}\right)\bar{\tilde{N}}\vec{t}\cdot\vec{\zeta}\tilde{N} - i\left(\frac{\zeta_3}{1+\vec{\zeta}^2}\right)\bar{\tilde{N}}\gamma_5\tilde{N}\,.$$

Chiral symmetries act on $\tilde{N}$ like ordinary isospin rotations, so they cannot mix up scalar and pseudoscalar nucleon bilinears. Thus there are really two independent Φ_4^+ operators:

$$\left(\frac{1-\vec{\zeta}^2}{1+\vec{\zeta}^2}\right)\bar{\tilde{N}}\tilde{N} \tag{19.5.61}$$

and

$$i\left(\frac{\vec{\zeta}}{1+\vec{\zeta}^2}\right)\cdot\bar{\tilde{N}}\gamma_5\vec{t}\tilde{N}\,, \tag{19.5.62}$$

and two independent Φ_3^- operators:

$$\bar{\tilde{N}}t_3\tilde{N} - 2\left(\frac{\zeta_3}{1+\vec{\zeta}^2}\right)\bar{\tilde{N}}\vec{t}\cdot\vec{\zeta}\tilde{N} \tag{19.5.63}$$

and

$$i\left(\frac{\zeta_3}{1+\vec{\zeta}^2}\right)\bar{\tilde{N}}\gamma_5\tilde{N}\,. \tag{19.5.64}$$

We shall show in the next section that these are the only operators with the transformation properties of the terms in Eq. (19.4.34) that are bilinear in the nucleon fields. The operators (19.5.62) and (19.5.64) evidently provide isospin-conserving and isospin-violating corrections to the Goldberger–Treiman formula for the pion-nucleon couplings. The other two operators, (19.5.61) and (19.5.63), contribute directly to both the nucleon mass and to low energy pion–nucleon scattering. From their contribution to the nucleon mass, we see that these latter terms enter into the effective Lagrangian in the form (now replacing $\vec{\zeta}$ with the conventionally normalized pion field):

$$\delta\mathscr{L}_{eff} = -\frac{\delta m_p + \delta m_n}{2}\left(\frac{1-\vec{\pi}^2/F_\pi^2}{1+\vec{\pi}^2/F_\pi^2}\right)\bar{\tilde{N}}\tilde{N}$$
$$-(\delta m_p - \delta m_n)\left[\bar{\tilde{N}}t_3\tilde{N} - \frac{2}{F_\pi^2}\left(\frac{\pi_3}{1+\vec{\pi}^2/F_\pi^2}\right)\bar{\tilde{N}}\vec{t}\cdot\vec{\pi}\tilde{N}\right], \qquad (19.5.65)$$

where δm_p and δm_n are the contributions of the quark mass terms (19.4.34) to the proton and neutron masses. This makes a contribution to the pion–nucleon scattering amplitude (again written as a matrix in the isospin space of the nucleon):

$$\delta M_{ba} = \frac{2[\delta m_p + \delta m_n]}{F_\pi^2}\delta_{ab} + \frac{2[\delta m_p - \delta m_n]}{F_\pi^2}(t_a\delta_{3b} + t_b\delta_{3a}) . \qquad (19.5.66)$$

The first term in Eq. (19.5.65) is often known as the 'σ-term'. The second term is an isospin violating correction to the σ-term, which shows up only in processes involving neutral pions, such as the charge-exchange processes $\pi^+ + n \to \pi^0 + p$ and $\pi^- + p \to \pi^0 + n$. Even though it is not possible to measure the nucleon mass shifts $\delta m_{p,n}$ directly, we shall see in Section 19.7 that an $SU(3)$ symmetry allows their difference to be calculated from hyperon mass differences. This yields the result $\delta m_p - \delta m_n = -2.5$ MeV. Unfortunately, we still do not have a firm theoretical estimate of the coefficient $\delta m_p + \delta m_n$ of the first term in Eq. (19.5.65).

Eq. (19.4.34) shows that in quantum chromodynamics the coefficients $\delta m_p + \delta m_n$ and $\delta m_p - \delta m_n$ in Eq. (19.5.66) are respectively proportional to $m_u + m_d$ and $m_u - m_d$. We shall see in Section 19.7 that m_u and m_d are not at all degenerate, so these coefficients are roughly of the same order of magnitude; the isospin-violating corrections to the σ-term are not much smaller than the σ-term itself. We see again that the reason that isospin conservation is such a good approximation in hadronic physics is not that the u and d quark masses are nearly equal, but just that they are small.

The use here of Lagrangians like (19.5.19) or (19.5.49) is one example of the method of *effective field theories*, already introduced in Section 12.3. Similar techniques have been employed to deal with meson and baryon interactions including strange particles (see Section 19.7), with quark and lepton interactions at energies below the scale of electroweak symmetry breaking (see Section 21.3), and even with superconductivity (see Section 21.6). In all these cases effective field theories provide the most convenient method for working out the consequences of symmetries and the general principles underlying quantum field theory.

* * *

The low energy theorems provided by broken symmetries when combined with dispersion relations yield useful sum rules. Let's see how this works for chiral symmetry, ignoring the small u and d quark masses and the pion mass. Consider the forward scattering of a massless pion with isovector index a and four-momentum q on a nucleon of four-momentum p, yielding a pion with isovector index b. The scattering amplitude M, defined by $S_{fi} = -2\pi i\delta^4(P_f - P_i)M_{fi}$, is given to first order in q by the Feynman diagrams of Figure 19.3 as

$$
\begin{aligned}
-2\pi i M &= \frac{1}{(2\pi)^6(2q^0)} \\
&\times \bar{u}\left[\left(-i\frac{(2\pi)^4 2g_A\gamma_5 t_b \not{q}}{F_\pi}\right)\left(\frac{-i}{(2\pi)^4}\frac{-i(\not{p}+\not{q})+m_N}{(p+q)^2+M^2}\right)\left(i\frac{(2\pi)^4 2g_A\gamma_5 t_a \not{q}}{F_\pi}\right)\right. \\
&+\left(i\frac{(2\pi)^4 2g_A\gamma_5 t_a \not{q}}{F_\pi}\right)\left(\frac{-i}{(2\pi)^4}\frac{-i(\not{p}-\not{q})+m_N}{(p-q)^2+M^2}\right)\left(-i\frac{(2\pi)^4 2g_A\gamma_5 t_b \not{q}}{F_\pi}\right) \\
&\left.-\,4i\frac{t_c\epsilon_{abc}\not{q}}{F_\pi^2}\right]u\,.
\end{aligned} \tag{19.5.67}
$$

To this order in q, the nucleon propagator in the first two terms inside the square brackets may by approximated by

$$
\frac{-i(\not{p}\pm\not{q})+m_N}{(p\pm q)^2+M^2} \approx \frac{-i\not{p}+m_N}{\pm 2p\cdot q}\,.
$$

Eq. (19.5.67) can then be further simplified, using the relations $\not{q}\,\not{q} = q^2 = 0$, $\bar{u}\gamma^\mu u = -ip\cdot q/m_N$, and $[t_a, t_b] = i\epsilon_{abc}t_c$. Also, in the laboratory frame (which for small q is the same as the center-of-mass frame) we have $p\cdot q = -m_N\omega$, where $\omega \equiv q^0$ is the pion energy in the nucleon rest frame. Finally, the conventional forward scattering amplitude (whose absolute square is the differential cross section for forward scattering) is given by Eq. (3.6.9) as $f \equiv -4\pi^2\omega M$. Putting this all together, we find the forward scattering amplitude at pion energy ω is given for $\omega \ll m_N$ by

$$
f_{ba}(\omega) \to -i\frac{\omega}{\pi F_\pi^2}(1-g_A^2)\,\epsilon_{abc}t_c\,. \tag{19.5.68}
$$

In particular, for π^+–proton scattering, we must contract Eq. (19.5.68) with $v_b^* v_a$, where $\vec{v}$ is the normalized isovector $(1, i, 0)/\sqrt{2}$, and set $t_3 = +\frac{1}{2}$. The forward low energy π^+–proton scattering amplitude is then

$$
f_{\pi^+ p}(\omega) \to -\frac{\omega}{2\pi F_\pi^2}(1-g_A^2)\,. \tag{19.5.69}
$$

Now, the dispersion relation for the forward scattering of a massless π^+ on a proton is given by Eq. (10.8.24) (with subtraction polynomial

$P(E) \propto E$) as

$$f_{\pi^+p}(\omega) = R + \frac{i\omega}{4\pi}\sigma_{\pi^+p}(\omega) + \frac{\omega}{4\pi^2}\int_0^\infty \left[\frac{\sigma_{\pi^+p}(E)}{E-\omega} - \frac{\sigma_{\pi^-p}(E)}{E+\omega}\right] dE\,, \tag{19.5.70}$$

where R is a possible subtraction constant, and ω and E are pion energies in the proton rest frame. Comparison of this dispersion relation with the low energy limit (19.5.69) shows that $R = 0$, and that

$$g_A^2 = 1 + \frac{F_\pi^2}{2\pi}\int_0^\infty \left[\sigma_{\pi^+p}(E) - \sigma_{\pi^-p}(E)\right] dE/E\,. \tag{19.5.71}$$

This is the celebrated Adler–Weisberger sum rule,[19] for the case of exact chiral symmetry, with $m_\pi = 0$. A similar sum rule may be derived for the scattering of pions on any baryon or meson, including the pion itself.[30a]

19.6 Effective Field Theories: General Broken Symmetries

The techniques described in the previous section for constructing effective Lagrangians for the case of $SU(2) \times SU(2)$ broken to $SU(2)$ were soon generalized[31] to the case of a general group G broken to an arbitrary subgroup H. Consider a quantum field theory whose Lagrangian is invariant under an arbitrary compact Lie group G of ordinary linear spacetime-independent transformations g of the fields $\psi_n(x)$:

$$\psi_n(x) \to \sum_m g_{nm}\psi_m(x)\,. \tag{19.6.1}$$

We suppose that this symmetry group is spontaneously broken to some subgroup $H \subset G$ of symmetry transformations that leave all vacuum expectation values invariant: for $h \in H$,

$$\sum_m h_{nm}\langle\psi_m(x)\rangle_{\rm VAC} = \langle\psi_n(x)\rangle_{\rm VAC}\,. \tag{19.6.2}$$

(Not all of the ψs need be scalars, but of course only the scalars will have non-vanishing vacuum expectation values.) In the example at the beginning of the previous section, G was the group $SO(4)$, the fields ψ_n furnished a four-vector representation of this group, and their vacuum expectation values broke $SO(4)$ down to the subgroup $H = SO(3)$ of rotations that leave the four-axis invariant. In the general case the fields ψ_n may furnish a reducible rather than an irreducible representation of G, as for instance became the case when we introduced the nucleon fields in the previous section.

We next express a general 'point' ψ_n in field space as a G transformation acting on fields $\tilde{\psi}_n$ from which the massless Goldstone mode has been eliminated:

$$\psi_n(x) = \sum_m \gamma_{nm}(x)\tilde{\psi}_m(x) . \qquad (19.6.3)$$

For instance, in the previous section γ was the $SO(4)$ rotation we called R, and the condition satisfied by $\tilde{\psi}_n$ was that the first three components of the four-vector should vanish.

In order to formulate this condition more generally, let us suppose that we are working with a real representation of G. (There is no loss of generality here, because we are not insisting on an irreducible representation, so if we start with a set of fields forming a complex representation of G we can always take the ψ_n to be the real and imaginary parts of these complex fields.) As shown in Section 19.2, the massless eigenvectors of the mass matrix are just the independent linear combinations of the vectors $\sum_m [t^\alpha]_{nm}\langle\psi_m(0)\rangle_{\rm VAC}$, where t^α are the generators of G. (The reality of the representation means that it^α is real, and the fact that G is compact means that we can choose the representation to be unitary and hence orthogonal, so that the t^α are imaginary and antisymmetric.) The condition that $\tilde{\psi}_n$ does not contain Goldstone modes may thus be formulated as

$$\sum_{nm} \tilde{\psi}_n(x)[t^\alpha]_{nm}\langle\psi_m(0)\rangle_{\rm VAC} = 0 . \qquad (19.6.4)$$

The number of independent conditions here is the dimensionality of the group G minus the dimensionality of the subgroup H whose generators annihilate $\langle\psi_m(0)\rangle_{\rm VAC}$. As in the previous section, the Goldstone bosons whose fields are eliminated in this way will reappear in the $\dim(G) - \dim(H)$ fields needed to parameterize the transformation $\gamma_{nm}(x)$. In general the $\tilde{\psi}_n$ include all the heavy fields of the theory, including those (like nucleons in the previous section) that have different spacetime symmetry properties from the Goldstone bosons.

It is necessary to show that we may always choose the transformation $\gamma_{nm}(x)$ so as to satisfy Eq. (19.6.4). For this purpose, consider the quantity

$$V_\psi(g) \equiv \sum_{nm} \psi_n\, g_{nm}\langle\psi_m(0)\rangle_{\rm VAC} , \qquad (19.6.5)$$

where g runs over the whole group G in the real orthogonal representation furnished by the ψ_n. This is obviously a continuous real function of g, and since the group is compact $V_\psi(g)$ is also a bounded function of g. At each spacetime point x, $V_{\psi(x)}(g)$ therefore reaches a maximum value for some group element, which we shall call $\gamma(x)$. At $g = \gamma(x)$, $V_{\psi(x)}(g)$ must be stationary with respect to arbitrary variations in g. But any infinitesimal shift in a group element g may always be written as a linear combination

$$\delta g = i \sum_\alpha \epsilon_\alpha g t^\alpha ,$$

with real infinitesimal coefficients ϵ_α that may depend on g. Hence the condition that $V_{\psi(x)}(g)$ be stationary with respect to the variation δg at $g = \gamma(x)$ reads

$$0 = \delta V_{\psi(x)}\Big(\gamma(x)\Big) = i \sum_\alpha \epsilon_\alpha \sum_{nml} \psi_n(x) \gamma_{nl}(x)\, t^\alpha_{lm} \langle \psi_m(0) \rangle_{\rm VAC}$$
$$= i \sum_\alpha \epsilon_\alpha \sum_{nml} [\gamma^{-1}(x)]_{ln} \psi_n(x)\, t^\alpha_{lm} \langle \psi_m(0) \rangle_{\rm VAC} \,.$$

This must be satisfied for all variations, and thus for all ϵ_α, so we see that Eq. (19.6.4) is satisfied for $\tilde{\psi}(x) = \gamma^{-1}(x)\psi(x)$, as was to be shown. In the purely bosonic theory with Lagrangian (19.4.1), $\tilde{\psi}_n$ was the four-vector field $(0, 0, 0, \sigma)$. This points in the direction of the vacuum expectation value $\langle \psi_n \rangle$, but this is not always the case.

Incidentally, it is not necessary in dealing with complex fields to formulate the condition on $\tilde{\psi}(x)$ explicitly in terms of its real and imaginary parts. If a set of fields $\chi(x)$ transform according to some complex representation of G with Hermitian generators T^α, then in the real representation

$$\psi(x) = \begin{pmatrix} {\rm Re}\, \chi(x) \\ {\rm Im}\, \chi(x) \end{pmatrix}$$

the generators are

$$i\, t^\alpha = \begin{pmatrix} -{\rm Im}\, T^\alpha & -{\rm Re}\, T^\alpha \\ {\rm Re}\, T^\alpha & -{\rm Im}\, T^\alpha \end{pmatrix} .$$

The condition (19.6.4) may then be expressed directly in terms of T^α and $\chi(x)$ as

$${\rm Im}\Big(\tilde{\chi}(x)^\dagger\, T^\alpha \,\langle \chi(0) \rangle_{\rm VAC} \Big) = 0 \,.$$

Because the Lagrangian is assumed to be only invariant under spacetime-independent G transformations, it will be found after the transformation (19.6.3) to depend on $\gamma(x)$ as well as on $\tilde{\psi}(x)$, though always with at least one derivative in each γ-dependent term. As already remarked, the spacetime-dependent parameters needed to specify $\gamma(x)$ will play the role of Goldstone boson fields. We now have to consider how to parameterize $\gamma(x)$.

It must be recognized from the outset that the choice of γ in Eq. (19.6.3) is generally not unique. Because $\langle \psi_m(0) \rangle_{\rm VAC}$ is H-invariant in the sense of Eq. (19.6.2), the quantity $V_\psi(g)$ defined by Eq. (19.6.5) is invariant under right multiplication of g by any element h of the unbroken subgroup H:

$$V_\psi(g) = V_\psi(gh) \qquad \text{for} \qquad h \in H\,. \tag{19.6.6}$$

It follows that if γ maximizes $V_\psi(g)$, then so does γh, so that condition (19.6.4) is satisfied for $\tilde{\psi} = h^{-1}\gamma^{-1}\psi$ as well as $\tilde{\psi} = \gamma^{-1}\psi$. Hence γ is only defined so far up to right multiplication by an element of H. (For instance in the example of the previous section, the four-dimensional rotation R could have been multiplied on the right with any rotation acting only on the first three components of four-vectors.) We may think of two group elements γ_1 and γ_2 as equivalent if $\gamma_1 = \gamma_2 h$ with $h \in H$. This is an equivalence relation because it is reflexive (if γ_1 is equivalent to γ_2 then γ_2 is equivalent to γ_1), symmetric (γ is equivalent to itself), and transitive (if γ_1 is equivalent to γ_2 and γ_2 is equivalent to γ_3 then γ_1 is equivalent to γ_3). The elements of the group G can therefore be sorted into disjoint 'equivalence classes', each consisting of elements γ that differ only by right multiplication by an element of H. These are known as the *right cosets* of G with respect to H. What we need is a parameterization of the space (known as G/H) of right cosets.

For this purpose we need only choose one representative group element from each right coset. For the $SO(4)$ symmetry broken to $SO(3)$ of the previous section, it was convenient to choose these representative elements as the rotations R parameterized by the three-vector $\vec{\zeta}$. There is another choice that is available for any compact group G broken to any subgroup H. Let us first adapt our notation for the generators of G to the pattern of symmetry breaking. The complete set of independent generators of H will be called t_i. According to Eq. (19.6.2) these satisfy

$$\sum_m (t_i)_{nm}\,\langle\psi_m\rangle_{\rm VAC} = 0\,. \tag{19.6.7}$$

Since H is a subgroup, the t_i form a subalgebra

$$[t_i, t_j] = i\sum_k C_{ijk} t_k\,. \tag{19.6.8}$$

We will take the x_a to be the *other* independent generators of G, in any basis with totally antisymmetric structure constants. (Such a basis always exists for compact groups; see Section 15.2.) Since x_a does not appear on the right-hand side of Eq. (19.6.8), the structure constants C_{ija} all vanish, and since they are totally antisymmetric, it follows that $C_{iaj} = 0$, so

$$[t_i, x_a] = i\sum_b C_{iab} x_b\,. \tag{19.6.9}$$

However, it is *not* necessarily the case that commutators of xs with each other are linear combinations of ts; this depends on the nature of G and H, and also on how H is embedded in G. (Where C_{abc} vanishes, the coset space G/H is known as a *symmetric space*). In general we may write

$$[x_a, x_b] = i \sum_i C_{abi} t_i + i \sum_c C_{abc} x_c \,. \tag{19.6.10}$$

Any set of generators with commutation relations of the form (19.6.8)–(19.6.10) is called a *Cartan decomposition* of the Lie algebra. An example is provided by the chiral $SU(2) \times SU(2)$ Lie algebra (19.4.9)–(19.4.11), for which the C_{abc} did happen to vanish.

Because t_i and x_a span the Lie algebra of G, any finite element of G may be expressed in the form

$$g = \exp\left[i \sum_a \xi_a x_a\right] \exp\left[i \sum_i \theta_i t_i\right] , \tag{19.6.11}$$

where ξ_a and θ_i are a set of real parameters. But the transformation $\gamma(x)$ in Eq. (19.6.3) is only defined so far up to right-multiplication with an element of H, so we may standardize our definition of γ by taking it in the form

$$\gamma(x) = \exp\left[i \sum_a \xi_a(x) x_a\right] \,. \tag{19.6.12}$$

The $\xi_a(x)$ may be identified (apart from normalization) with the Goldstone boson fields.

In what follows we will simply assume that representative elements have been chosen from each right coset, and expressed as continuous functions $\gamma(\xi)$ of some parameters ξ_a. Eq. (19.6.12) in general provides one explicit example, but we will not limit ourselves to this parameterization.

Now, suppose we use Eq. (19.6.3) to replace all fields $\psi(x)$ in the Lagrangian with $\gamma(x)\tilde{\psi}(x)$. The derivatives of the fields are given by

$$\partial_\mu \psi(x) = \gamma(x)\left[\partial_\mu \tilde{\psi}(x) + (\gamma^{-1}(x)\partial_\mu \gamma(x))\tilde{\psi}(x)\right] . \tag{19.6.13}$$

Therefore when we express the Lagrangian in terms of $\tilde{\psi}$ rather than ψ, the Goldstone boson fields appear through the dependence of $\gamma^{-1}(x)\partial_\mu\gamma(x)$ on $\xi_a(x)$ and its derivatives.

Any variation of a group element like $\gamma(x)$ may be written as the group element times a linear combination of the generators of the group. In our case, we may write this as

$$\gamma^{-1}(x)\partial_\mu \gamma(x) = i \sum_a x_a D_{a\mu}(x) + i \sum_i t_i E_{i\mu}(x) , \tag{19.6.14}$$

where $D_{a\mu}$ and $E_{i\mu}$ take the forms

$$D_{a\mu}(x) = \sum_b D_{ab}\Big(\xi(x)\Big)\partial_\mu \xi_b(x) , \tag{19.6.15}$$

$$E_{i\mu}(x) = \sum_b E_{ib}\Big(\xi(x)\Big)\partial_\mu \xi_b(x)\,. \tag{19.6.16}$$

The Goldstone boson fields will thus enter the Lagrangian through the appearance of the quantities $D_{a\mu}(x)$ and $E_{i\mu}(x)$ (and their derivatives). Note that for exact broken symmetries every interaction of the ξs must involve at least one derivative, so mass terms $m^2_{ab}\xi_a\xi_b$ cannot appear in the Lagrangian, and all interactions vanish when any Goldstone boson four-momentum vanishes.

Even where we do not know the details of the underlying Lagrangian, we can learn a great deal about the way that the quantities $D_{a\mu}$ and $E_{i\mu}$ appear in the transformed Lagrangian from their transformation properties. Under an arbitrary element g of the group G, the original field ψ transforms according to Eq. (19.6.1):

$$\psi(x) \to \psi'(x) = g\psi(x) = g\gamma\Big(\xi(x)\Big)\tilde{\psi}(x)\,. \tag{19.6.17}$$

Now, $g\gamma(\xi)$ is an element of G, so it must be in the same right coset as some $\gamma(\xi')$, and may therefore be written in the form

$$g\gamma\Big(\xi(x)\Big) = \gamma\Big(\xi'(x)\Big)h\Big(\xi(x), g\Big)\,, \tag{19.6.18}$$

where h is some element of the unbroken subgroup H. Using this in Eq. (19.6.17), we find that $\psi'(x)$ is of the form (19.6.3)

$$\psi'(x) = \gamma\Big(\xi'(x)\Big)\tilde{\psi}'(x)\,, \tag{19.6.19}$$

with $\xi'(x)$ defined by Eq. (19.6.18), and

$$\tilde{\psi}'(x) \equiv h\Big(\xi(x), g\Big)\tilde{\psi}(x)\,. \tag{19.6.20}$$

In the example at the beginning of the previous section, $\tilde{\psi}$ consisted of a single isoscalar σ, and since this was invariant under the unbroken isospin subgroup, it was also chiral invariant. More generally, we find here that $\tilde{\psi}(x)$ is not necessarily invariant under general G transformations, but that its G transformation depends only on its transformation under the unbroken subgroup H. We saw this in the previous section when we introduced the nucleon fields; under a general infinitesimal broken chiral symmetry, the fields $\tilde{N}$ transformed according to the field-dependent isospin rotation (19.5.45).

The transformation rules $\xi \to \xi'$ and $\tilde{\psi} \to \tilde{\psi}'$ specified by Eqs. (19.6.18) and (19.6.20) make no reference to the particular linear G transformation properties of the original fields ψ_n. Indeed, we did not need to start with a Lagrangian that was invariant under linear G transformations in order to deduce these transformation rules. Given any set of fields on which a group G acts linearly or non-linearly, with a subgroup H that leaves one special set of field values (their vacuum expectation values) invariant, we can always express these fields (in at least a finite neighborhood of the special field values) in terms of a set ξ_a and $\tilde{\psi}_n$, which transform under G according to the standard realization $\xi_a \to \xi'_a$, $\tilde{\psi} \to \tilde{\psi}'$ defined by Eqs. (19.6.18) and (19.6.20). The essential uniqueness of this realization was first proved[25] for $SU(2) \times SU(2)$ broken to $SU(2)$, and then for general Lie groups broken to any subgroup.[31] For an easier argument, we may note that for any set of fields that transforms according to a non-linear realization of an arbitrary compact Lie group G, it is always possible to define functions of these fields (and perhaps additional fields*) that transform linearly under G.[32] Starting from this linear realization, the arguments of this section allow us to construct fields ξ and $\tilde{\psi}$ with the transformation rules (19.6.18) and (19.6.20).

Although the transformation of $\xi(x)$ and $\tilde{\psi}(x)$ is complicated for general $g \in G$, it is much simpler when g is itself a member h of the unbroken subgroup H. It is usual and always possible to choose the Goldstone boson fields $\xi_a(x)$ to transform according to some *linear* representation $\mathscr{D}_{ab}(h)$ of the unbroken subgroup H:

$$h\,\gamma(\xi)\,h^{-1} = \gamma\Big(\mathscr{D}(h)\xi\Big)\,. \tag{19.6.21}$$

For instance, in the parameterization of the coset space $SO(4)/SO(3)$ used in the previous section, the Goldstone boson fields ζ_a formed an isospin three-vector. For the more generally applicable parameterization based on Eq. (19.6.12), the commutation relations (19.6.9) show that the x_a transform linearly under H:

$$hx_bh^{-1} = \sum_a \mathscr{D}_{ba}(h)x_a\,, \tag{19.6.22}$$

from which Eq. (19.6.21) follows immediately. Comparing Eq. (19.6.21) with Eq. (19.6.18), we see that for $g = h \in H$,

$$\xi'_a(x) = \sum_b \mathscr{D}_{ab}(h)\xi_b(x)\,, \tag{19.6.23}$$

$$\tilde{\psi}'_n(x) = \sum_m h_{nm}\tilde{\psi}_m(x)\,. \tag{19.6.24}$$

In other words, ξ_a and $\tilde{\psi}_n$ transform under H according to the representations $\mathscr{D}_{ab}(h)$ and h_{nm} itself, respectively.

We must now consider how to construct the most general Lagrangian that is invariant under the transformations (19.6.18) and (19.6.20). The transformation $\xi \to \xi'$ for general $g \in G$ is non-linear and rather complicated, and rules out the appearance of $\xi_a(x)$ in the Lagrangian except in

* For instance, the polar and azimuthal angles θ and φ furnish a non-linear realization of $SO(3)$; by adding an additional variable r we can construct quantities x_1, x_2, and x_3 that transform linearly under $SO(3)$.

quantities like $D_{a\mu}(x)$ and $E_{i\mu}(x)$. Fortunately, these quantities obey very simple transformation rules. Differentiate Eq. (19.6.18) with respect to x^μ and multiply on the left with its inverse. This gives

$$\begin{aligned}\gamma^{-1}\Big(\xi(x)\Big)\partial_\mu\gamma\Big(\xi(x)\Big) &= \Big[g\gamma\Big(\xi(x)\Big)\Big]^{-1}\partial_\mu\Big[g\gamma\Big(\xi(x)\Big)\Big]\\ &= h^{-1}\Big(\xi(x),g\Big)\gamma^{-1}\Big(\xi'(x)\Big)\partial_\mu\Big[\gamma\Big(\xi'(x)\Big)h\Big(\xi(x),g\Big)\Big]\\ &= h^{-1}\Big(\xi(x),g\Big)\Big[\gamma^{-1}\Big(\xi'(x)\Big)\partial_\mu\gamma\Big(\xi'(x)\Big)\Big]\;h\Big(\xi(x),g\Big)\\ &\quad + h^{-1}\Big(\xi(x),g\Big)\partial_\mu h\Big(\xi(x),g\Big)\end{aligned}$$

and so

$$\begin{aligned}\gamma^{-1}\Big(\xi'(x)\Big)\partial_\mu\gamma\Big(\xi'(x)\Big) &= h\Big(\xi(x),g\Big)\Big[\gamma^{-1}\Big(\xi(x)\Big)\partial_\mu\gamma\Big(\xi(x)\Big)\Big]\;h^{-1}\Big(\xi(x),g\Big)\\ &\quad - \Big[\partial_\mu h\Big(\xi(x),g\Big)\Big]\;h^{-1}\Big(\xi(x),g\Big)\,. \qquad (19.6.25)\end{aligned}$$

Eq. (19.6.14) gives the quantity $\gamma^{-1}\partial_\mu\gamma$ as a linear combinations of xs and ts. Also, the second term on the right-hand side of Eq. (19.6.25) is a linear combination of ts alone. Since the x_a and t_i are all independent, the coefficients of each generator on both sides of Eq. (19.6.25) must all be equal:

$$\sum_a x_a D'_{a\mu} = h(\xi,g)\left(\sum_a x_a D_{a\mu}\right)\,h^{-1}(\xi,g)\,, \qquad (19.6.26)$$

$$\begin{aligned}\sum_i t_i E'_{i\mu} &= h(\xi,g)\left(\sum_i t_i E_{i\mu}\right)\,h^{-1}(\xi,g)\\ &\quad + i\,\Big[\partial_\mu h(\xi,g)\Big]\,h^{-1}(\xi,g)\,, \qquad (19.6.27)\end{aligned}$$

or in more detail

$$\begin{aligned}D'_{a\mu}(x) &= \sum_b \mathscr{D}_{ab}\left(h\Big(\xi(x),g\Big)\right)\,D_{b\mu}(x)\,,\\ E'_{i\mu}(x) &= \sum_j \mathscr{E}_{ij}\left(h\Big(\xi(x),g\Big)\right)E_{j\mu}(x)\\ &\quad -\sum_b \mathscr{H}_{ib}\left(h\Big(\xi(x),g\Big)\right)\partial_\mu\xi_b(x)\,, \qquad (19.6.28)\end{aligned}$$

where $D_{a\mu}$ and $E_{i\mu}$ are defined by Eqs. (19.6.15) and (19.6.16), and

$$\Big[\partial_\mu h\Big(\xi(x),g\Big)\Big]\,h^{-1}\Big(\xi(x),g\Big) = i\sum_{ib}\mathscr{H}_{ib}\Big(\xi(x)\Big)t_i\partial_\mu\xi_b(x)\,,$$

$$h\,t_j\,h^{-1} = \sum_i \mathscr{E}_{ij}(h)t_i\,,$$

while $\mathscr{D}_{ab}$ is defined by Eq. (19.6.22). We see that the quantities $D_{a\mu}(x)$ are 'covariant derivatives' of the Goldstone boson fields, transforming under a general element $g \in G$ much like the fields $\tilde{\psi}$: both are subjected to the H transformation $h(\xi(x), g)$, though in different representations. For instance, in the $SO(4)$ theory of the previous section, the covariant derivative of the pion field $\vec{\zeta}$ was the quantity $\partial_\mu \vec{\zeta}/(1+\vec{\zeta}^2)$, which is transformed under an infinitesimal chiral transformation by an isospin rotation (19.5.15).

On the other hand, $E_{j\mu}(x)$ transforms inhomogeneously, much like a gauge field. The extra term in Eq. (19.6.28) allows a cancellation of the inhomogeneous term in derivatives of $\tilde{\psi}$. Differentiating Eq. (19.6.20), we have

$$\partial_\mu \tilde{\psi}'(x) = h\Big(\xi(x), g\Big)\left[\partial_\mu \tilde{\psi}(x) + h^{-1}\Big(\xi(x), g\Big)\partial_\mu h\Big(\xi(x), g\Big)\tilde{\psi}(x)\right] .$$

Combining this with Eq. (19.6.27) gives

$$\Big(\mathscr{D}_\mu \tilde{\psi}(x)\Big)' = h\Big(\xi(x), g\Big)\mathscr{D}_\mu \tilde{\psi}(x) , \tag{19.6.29}$$

where $\mathscr{D}_\mu \tilde{\psi}$ is a covariant derivative of the heavy particle fields:

$$\mathscr{D}_\mu \tilde{\psi}(x) \equiv \partial_\mu \tilde{\psi}(x) + i \sum_i t_i E_{i\mu}(x) \tilde{\psi}(x) . \tag{19.6.30}$$

(Eq. (19.5.47) provides an example of such a covariant derivative in the case of $SO(4)$ broken to $SO(3)$.) *Any Lagrangian that is invariant under the unbroken subgroup H and is constructed from $\tilde{\psi}$, $\mathscr{D}_\mu \tilde{\psi}$, and $D_{a\mu}$ will thus also be invariant under the full group G.*

Renormalizability is not an issue here, so we can also consider quantities involving more than one spacetime derivative. One such is $\mathscr{D}_\nu \mathscr{D}_\mu \tilde{\psi}$, obtained by just repeating the operation (19.6.30):

$$\mathscr{D}_\nu \mathscr{D}_\mu \tilde{\psi} = \left[\partial_\nu + i \sum_j t_j E_{j\nu}\right]\left[\partial_\mu + i \sum_i t_i E_{i\mu}\right]\tilde{\psi} . \tag{19.6.31}$$

This transforms just like $\mathscr{D}_\mu \tilde{\psi}$

$$(\mathscr{D}_\nu \mathscr{D}_\mu \tilde{\psi})' = h(\xi, g)\mathscr{D}_\nu \mathscr{D}_\mu \tilde{\psi} . \tag{19.6.32}$$

Another covariant quantity can be constructed from the covariant derivative of the covariant Goldstone boson derivative (19.6.15), which is defined just like (19.6.30), but with t_i replaced with the corresponding matrix in the representation of H furnished by the x_a:

$$\mathscr{D}_\nu D_{a\mu} = \partial_\nu D_{a\mu} + \sum_{ib} C_{iab}\, E_{i\nu} D_{b\mu} , \tag{19.6.33}$$

which transforms just like $D_{a\mu}$:

$$(\mathscr{D}_\nu D_{a\mu})' = \sum_b \mathscr{D}_{ab}(h, g)\mathscr{D}_\nu D_{b\mu} . \tag{19.6.34}$$

Also, by antisymmetrizing the derivative of $E_{i\mu}$ we can eliminate the inhomogeneous term in (19.6.27). This yields a 'curvature'

$$R_{i\mu\nu} \equiv \partial_\nu E_{i\mu} - \partial_\mu E_{i\nu} - i\sum_{j,k} C_{ijk} E_{j\mu} E_{k\nu} \,. \tag{19.6.35}$$

This is not really new; it is easy to see that the antisymmetric covariant derivative of the 'matter' field $\tilde{\psi}$ is proportional to this curvature

$$[\mathscr{D}_\nu, \mathscr{D}_\mu]\tilde{\psi} = \sum_i t_i\, R_{i\mu\nu}\tilde{\psi} \,. \tag{19.6.36}$$

Of course, we can continue this process and construct yet more covariants by taking higher and higher covariant derivatives of $\mathscr{D}_\mu\tilde{\psi}$ and $D_{a\mu}$.

It is useful to note that the quantities $D_{a\mu}$ and $E_{i\mu}$ may be calculated once and for all for given groups G and H and a given parameterization of the coset space G/H, and do not depend on the assumed transformation properties of the fields ψ_n or the specific matrices x_a, t_i used to represent the Lie algebra of the broken symmetry group. For instance, in the exponential parameterization (19.6.12), the first few terms in the power series for these covariants are easily calculated from Eqs. (19.6.12) and (19.6.14) as

$$\begin{aligned} D_{a\mu} = {} & \partial_\mu \xi_a + \tfrac{1}{2}\sum_{bc} C_{abc}\xi_b \partial_\mu \xi_c \,, \\ & + \tfrac{1}{6}\sum_{bcd}\left[\sum_e C_{cde}C_{bea} + \sum_i C_{cdi}C_{bia}\right]\xi_b\xi_c\partial_\mu\xi_d \\ & + O(\xi^3\partial_\mu\xi)\,, \end{aligned} \tag{19.6.37}$$

$$\begin{aligned} E_{i\mu} = {} & \tfrac{1}{2}\sum_{ab} C_{abi}\xi_a\partial_\mu\xi_b + \tfrac{1}{6}\sum_{abcd} C_{acd}\,C_{bdi}\,\xi_a\xi_b\partial_\mu\xi_c \\ & + O(\xi^3\partial_\mu\xi)\,. \end{aligned} \tag{19.6.38}$$

All we need to know in order to construct the most general Lagrangian involving the Goldstone boson fields ξ_a and a set of heavy particle fields is the pattern of spontaneous symmetry breaking, of G to H, and the transformation of the heavy particle fields under H, not G: the Lagrangian is taken to be the most general H-invariant function of $\tilde{\psi}$, $\mathscr{D}_\mu\,\tilde{\psi}$, $D_{a\mu}$, and higher covariant derivatives.

Now consider the case where the symmetry under G is not only spontaneously broken, but not even exact to begin with. Suppose that there is a term $\Delta\mathscr{L}$ in the initial Lagrangian that is *not* invariant under the group G of linear transformations $\psi \to g\psi$, but transforms as a linear combination of the components of some representation (reducible or irreducible) of G.

That is, we take

$$\Delta\mathscr{L} = \sum_A c_A \mathcal{O}_A \,, \tag{19.6.39}$$

where $\mathcal{O}_A$ transforms under G according to some representation $D[g]_{AB}$:

$$\mathcal{O}_A \to \sum_B D[g]_{AB} \mathcal{O}_B \,. \tag{19.6.40}$$

If we now replace the fields ψ with the set $\xi_a, \tilde{\psi}$, this term in the Lagrangian will still take the form (19.6.39), but Eq. (19.6.40) now reads

$$\mathcal{O}_A\Big[f_a(\xi, g),\ h(\xi, g)\tilde{\psi}\Big] = \sum_B D[g]_{AB} \mathcal{O}_B\Big[\xi, \tilde{\psi}\Big] \,, \tag{19.6.41}$$

where $f_a(\xi, g)$ is the result of applying the transformation g to ξ_a, defined by Eq. (19.6.18):

$$g\gamma\Big(\xi\Big) = \gamma\Big(f(\xi, g)\Big)h\Big(\xi, g\Big) \,. \tag{19.6.42}$$

As we shall see, it is easy to find sets of operators satisfying (19.6.41), and the solution is unique up to a specification of certain H-invariant functions of the $\tilde{\psi}$.

First, consider the case $\xi_a = 0$, with $g = \gamma(\xi')$. In this case $g\gamma(\xi) = \gamma(\xi')$, so

$$f_a\Big(0, \gamma(\xi')\Big) = \xi'_a \,, \qquad h\Big(0, \gamma(\xi')\Big) = 1 \,.$$

Applying this in Eq. (19.6.41), we find (now dropping the prime):

$$\mathcal{O}_A\Big[\xi_a, \tilde{\psi}\Big] = \sum_B D[\gamma(\xi)]_{AB}\, \mathcal{O}_B\Big[0, \tilde{\psi}\Big] \,. \tag{19.6.43}$$

This gives the operators $\mathcal{O}_A[\xi_a, \tilde{\psi}]$ for all ξ_a if we know them for $\xi_a = 0$.

Now take $\xi_a = 0$, with g an element h of the unbroken subgroup H. Here we have $g\gamma(\xi) = h$, so

$$f_a(0, h) = 0 \,, \qquad h(0, h) = h \,.$$

Applying this in Eq. (19.6.41), we find here

$$\mathcal{O}_A\Big[0, h\tilde{\psi}\Big] = \sum_B D[h]_{AB}\, \mathcal{O}_B\Big[0, \tilde{\psi}\Big] \,. \tag{19.6.44}$$

In other words, the $\mathcal{O}_A$ for $\xi = 0$ must have the same transformation under linear H transformations as are found in the representation (19.6.40) of G. (However, there is nothing in this that fixes the normalization of the different irreducible H representations found in the G representation (19.6.40); some of them may even be absent altogether.)

Finally, we note that any set of $\mathcal{O}$ operators satisfying (19.6.44) allows the construction using Eq. (19.6.43) of operators satisfying the general G-transformation rules (19.6.40). Using Eqs. (19.6.43) and (19.6.44), the left-hand side of Eq. (19.6.41) becomes

$$\begin{aligned}\mathcal{O}_A\Big[f_a(\xi,g),\ h(\xi,g)\tilde{\psi}\Big] &= \sum_B D[\gamma\big(f(\xi,g)\big)]_{AB}\mathcal{O}_B\Big[0, h(\xi,g)\tilde{\psi}\Big] \\ &= \sum_{BC} D[\gamma\big(f(\xi,g)\big)]_{AB}D[h(\xi,g)]_{BC}\ \mathcal{O}_C\Big[0,\tilde{\psi}\Big] \\ &= \sum_B D[\gamma\big(f(\xi,g)\big)h(\xi,g)]_{AB}\ \mathcal{O}_B\Big[0,\tilde{\psi}\Big] \\ &= \sum_B D[g\gamma(\xi)]_{AB}\ \mathcal{O}_B\Big[0,\tilde{\psi}\Big] \\ &= \sum_B D[g]_{AB}\ \mathcal{O}_B\Big[\xi,\tilde{\psi}\Big]\,,\end{aligned}$$

as was to be shown.

As an example, consider the group $SO(4)$ spontaneously broken to $SO(3)$ as in the previous section, and suppose we want to construct operators that transform according to the four-vector representation of $SO(4)$ out of the Goldstone boson field $\vec{\zeta}(x)$ and the other fields $\tilde{\psi}(x)$. According to (19.6.43), these must take the form

$$\mathcal{O}_n\Big[\vec{\zeta},\tilde{\psi}\Big] = R_{nm}(\vec{\zeta})\,\mathcal{O}_m\Big[0,\tilde{\psi}\Big]\,,$$

where R is the $SO(4)$ rotation defined by Eqs. (19.5.8) and (19.5.9):

$$R_{a4} = \frac{2\zeta_a}{1+\vec{\zeta}^{\,2}}\,, \qquad R_{44} = \frac{1-\vec{\zeta}^{\,2}}{1+\vec{\zeta}^{\,2}}\,.$$

The condition that R is an orthogonal matrix is satisfied if we choose its other components as

$$R_{ab} = \delta_{ab} - \frac{2\zeta_a\zeta_b}{1+\vec{\zeta}^{\,2}}\,, \qquad R_{4a} = -\frac{2\zeta_a}{1+\vec{\zeta}^{\,2}}\,.$$

Thus any scalar (as opposed to pseudoscalar) operators that transform as the fourth and third components of chiral four-vectors must appear in the effective Lagrangian in the forms

$$\Phi_4^+\Big[\vec{\zeta},\tilde{\psi}\Big] = -\left(\frac{2\vec{\zeta}}{1+\vec{\zeta}^{\,2}}\right)\cdot\vec{\Phi}^+\Big[0,\tilde{\psi}\Big] + \left(\frac{1-\vec{\zeta}^{\,2}}{1+\vec{\zeta}^{\,2}}\right)\Phi_4^+\Big[0,\tilde{\psi}\Big]$$

and

$$\Phi_3^-\Big[\vec{\zeta},\tilde{\psi}\Big] = \left[\Phi_3^-\Big[0,\tilde{\psi}\Big] - \left(\frac{2\zeta_3}{1+\vec{\zeta}^{\,2}}\right)\vec{\zeta}\cdot\vec{\Phi}^-\ \Big[0,\tilde{\psi}\Big]\right] + \left(\frac{2\zeta_3}{1+\vec{\zeta}^{\,2}}\right)\Phi_4^-\Big[0,\tilde{\psi}\Big]\,.$$

The only condition on the operators $\vec{\Phi}^+[0,\tilde{\psi}]$, $\Phi_4^+[0,\tilde{\psi}]$, $\vec{\Phi}^-[0,\tilde{\psi}]$, and $\Phi_4^-[0,\tilde{\psi}]$ is that they should transform under Lorentz and isospin transformations respectively as a pseudoscalar isovector, a scalar isoscalar, a scalar isovector, and a pseudoscalar isoscalar, but they do not have to be related in any way. For processes involving only pions we use an effective Lagrangian that involves no fields except the pion field $\vec{\zeta}$, so the operators $\vec{\Phi}^+[0,\tilde{\psi}]$, $\vec{\Phi}^-[0,\tilde{\psi}]$, and $\Phi_4^-[0,\tilde{\psi}]$ must all vanish, while $\Phi_4^+[0,\tilde{\psi}]$ is just a numerical constant. We see as promised in the previous section that the only non-derivative chiral symmetry breaking operator in the effective Lagrangian is proportional to $(1-\vec{\zeta}^2)/(1+\vec{\zeta}^2)$, and hence to the operator (19.5.34). In order to include symmetry-breaking operators bilinear in the nucleon field, parity and isospin conservation require that we take

$$\vec{\Phi}^+\left[0,\tilde{N}\right] \propto \bar{\tilde{N}}\gamma_5\vec{t}\tilde{N}\,, \qquad \Phi_4^+\left[0,\tilde{N}\right] \propto \bar{\tilde{N}}\tilde{N}\,,$$

$$\vec{\Phi}^-\left[0,\tilde{N}\right] \propto \bar{\tilde{N}}\vec{t}\tilde{N}\,, \qquad \Phi_4^-\left[0,\tilde{N}\right] \propto \bar{\tilde{N}}\gamma_5\tilde{N}\,.$$

Thus as claimed in the previous section the only non-derivative symmetry-breaking terms involving nucleon bilinears are of the form (19.5.61)–(19.5.64).

The phenomenological Lagrangian may be used to calculate the amplitudes for processes involving particles of small three-momentum in much the same way as was done for the chiral theory of pions and nucleons in the previous section. In counting powers of a characteristic small momentum Q, Goldstone boson internal lines always contribute factors of order Q^{-2}. Also, an internal line for one of the heavy particles (of any spin) described by the fields $\tilde{\psi}$ that has absorbed a net four-momentum q from the Goldstone boson field will have a propagator

$$\frac{N(p+q)}{(p+q)^2+M^2} \to \frac{N(p)}{2p\cdot q}$$

(where N is some polynomial, depending on the particle spin and field type, and M is the particle mass) so it will contribute a factor of order Q^{-1}. The same argument that led to (19.5.55) applies here, and gives the number of powers of Q in a given Feynman diagram as

$$\nu = \sum_i V_i\,(d_i + h_i/2 - 2) + 2L - E_h/2 + 2\,, \tag{19.6.45}$$

where V_i is the number of vertices of type i, d_i is the number of derivatives (or for approximate broken symmetries, derivatives or Goldstone boson masses) at a vertex of type i, h_i is the number of heavy particle lines at a vertex of type i, L is the number of loops, and E_h is the number of external lines of heavy particles. In general the coefficient $d_i + h_i/2 - 2$ is non-negative for all interactions allowed by chiral symmetries. The interactions among Goldstone bosons alone always have $d_i \geq 2$, because they must either be at least bilinear in the covariant derivative (19.6.15), or proportional to symmetry-breaking parameters (like quark masses in the previous section) which are of order of the squares of the Goldstone boson masses. Interactions between Goldstone bosons and heavy particles must involve the covariant derivatives (19.6.15) or (19.6.30), and so must have $d_i \geq 1$ as well as $h_i \geq 2$. The only interactions for which $d_i + h_i/2 < 2$ would be trilinear interactions among the heavy particles, which we exclude because we are only considering graphs in which all heavy hadrons remain non-relativistic. With $d_i + h_i/2 \geq 2$ for all interactions, the leading graphs are those constructed solely from interactions with $d_i + h_i/2 = 2$ and no loops. Corrections are again provided by interactions with more derivatives and/or more heavy-particle fields, and/or one or more loops.

In the absence of intrinsic symmetry breaking, the part of the Lagrangian that involves only the Goldstone boson fields has a unique $d_i = 2$ term

$$\mathscr{L}_{GB} = -\frac{1}{2} \sum_{ab} F_{ab}^2 D_{a\mu} D_b^\mu \,, \tag{19.6.46}$$

with F_{ab}^2 some real positive-definite matrix. For (19.6.12) and a large class of other parameterizations of the cosets, $\gamma(\xi)$ approaches $1 + i\xi_a x_a$ for $\xi_a \to 0$, so Eq. (19.6.14) shows that the linear term in $D_{a\mu}$ is just $\partial_\mu \xi_a$. The canonically normalized Goldstone boson fields π_a with a kinematic Lagrangian $-\frac{1}{2}\sum_a \partial_\mu \pi_a \partial^\mu \pi_a$ are then

$$\pi_a = \sum_b F_{ab} \xi_b \,. \tag{19.6.47}$$

In the $SO(4)$ theory of the previous section, the generators x_a and the fields ξ_a transform according to an irreducible representation of the unbroken subgroup $H = SO(3)$, so in this case $F_{ab}^2 = F_\pi^2 \delta_{ab}$, with F_π a single constant that characterizes the energy scale of the spontaneous symmetry breakdown. In the general case we can choose the generators (without changing the structure constants) so that F_{ab}^2 is diagonal, with diagonal elements all equal within each irreducible representation of H, but otherwise independent.

* * *

We have seen how to construct theories of Goldstone boson fields by starting with fields belonging to linear representations of the broken symmetry group, such as the $SU(2) \times SU(2)$ four-vector $\phi_n(x)$ with which we started in Section 19.5. At very low energy the only significant degrees of freedom are the Goldstone bosons, so we generally throw away the non-Goldstone parts of the field, such as the field $\sigma(x)$ given by Eq. (19.5.4). But there are circumstances in which it is necessary to return to full linear representations of the broken symmetry group, in which the Goldstone boson fields are just one part.

This happens when, by varying the temperature or turning on an external field of some sort, a system is brought close to a second-order transition, in which it smoothly goes from broken to unbroken symmetry. On one side of this transition the symmetry is broken, so we have massless Goldstone bosons, and various other massive excitations, not generally forming complete multiplets that would furnish linear representations of the broken symmetry group. On the other side of the transition the symmetry is unbroken, so here we have complete linear multiplets, not generally massless. If this transition is continuous, then very near the phase transition the Goldstone boson must form part of a complete linear multiplet of nearly massless excitations. Barring accidents or other symmetries, this multiplet will form an *irreducible* representation of the broken symmetry group. This irreducible multiplet of fields, which become massless only just at the phase transition, is known as the *order parameter.*

This is a more precise definition of order parameters than usual. Often one speaks of an order parameter as any set of fields whose expectation values break the symmetry, but this is too vague — there is no one set of fields whose expectation values can be blamed for a broken symmetry. For instance, the $SU(2) \times SU(2)$ symmetry of quantum chromodynamics with massless u and d quarks is broken by the expectation value of $\bar{u}u+\bar{d}d$, which is the fourth component of a chiral four-vector, but quartic and higher powers of quark fields also have non-vanishing vacuum expectation values, and these belong to other representations of $SU(2) \times SU(2)$. In contrast, at a smooth phase transition at which $SU(2) \times SU(2)$ becomes unbroken the Goldstone boson becomes a member of just one massless representation of $SU(2) \times SU(2)$, and this provides an unambiguous definition of the order parameter.

It is important to identify the correct order parameters, both in order to calculate the critical exponents discussed in Section 18.5, and to deal with configurations like vortex lines or magnetic monopoles, where, as we shall see in Sections 21.6 and 23.3, there is a singularity at which the broken symmetry becomes unbroken. It is often assumed that the order parameter for the chiral $SU(2) \times SU(2)$ symmetry of quantum chromodynamics is a chiral four-vector, but this is not known to be the case.[32a]

19.7 Effective Field Theories: $SU(3) \times SU(3)$

The approximate $SU(2)$ isospin symmetry of nuclear physics was extended in the early 1960s by Gell-Mann[33] and Ne'eman[34] to an even less exact $SU(3)$ symmetry, which grouped the known baryons and mesons in various irreducible representations: an octet of $1/2^+$ baryons p, n, Λ^0, $\Sigma^{\pm,0}$, $\Xi^{-,0}$, an octet of 0^- mesons $K^{+,0}$, $\pi^{\pm,0}$, η^0, $\overline{K}^{-,0}$, an octet of 1^- mesons $K^{*+,0}$, $\rho^{\pm,0}$, ω, $\overline{K^*}^{-,0}$, and a decuplet of $3/2^+$ baryons $\Delta^{++,+,0,-}$, $\Sigma^{*+,0,-}$,

$\Xi^{*0,-}$, Ω^-. (The η and Ω were not discovered until later.) After the successes of the chiral $SU(2)\times SU(2)$ symmetry in the mid-1960s, it became natural to suppose that the strong interactions also respect an approximate $SU(3)\times SU(3)$ symmetry, which like $SU(2)\times SU(2)$ is spontaneously broken to its diagonal subgroup, the $SU(3)$ of Gell-Mann and Ne'eman. Then after the advent of quantum chromodynamics it became clear that this symmetry arises because there are not two but three fairly light quarks; the u and d, and a third quark s which like d has charge $-1/3$. In this case the $SU(3)\times SU(3)$ symmetry consists of independent $SU(3)$ transformations (analogous to Eq. (19.4.2)) on the left- and right-handed parts of the u, d, s quark fields:

$$\begin{pmatrix} u \\ d \\ s \end{pmatrix} \to \exp\left[i\sum_a(\theta_a^V\lambda_a + \theta_a^A\lambda_a\gamma_5)\right]\begin{pmatrix} u \\ d \\ s \end{pmatrix}, \tag{19.7.1}$$

where λ_a are a complete set of traceless Hermitian matrices:

$$\lambda_1 = \begin{pmatrix} 0 & 1 & 0 \\ 1 & 0 & 0 \\ 0 & 0 & 0 \end{pmatrix}, \quad \lambda_2 = \begin{pmatrix} 0 & -i & 0 \\ i & 0 & 0 \\ 0 & 0 & 0 \end{pmatrix}, \quad \lambda_3 = \begin{pmatrix} 1 & 0 & 0 \\ 0 & -1 & 0 \\ 0 & 0 & 0 \end{pmatrix},$$
$$\lambda_4 = \begin{pmatrix} 0 & 0 & 1 \\ 0 & 0 & 0 \\ 1 & 0 & 0 \end{pmatrix}, \quad \lambda_5 = \begin{pmatrix} 0 & 0 & -i \\ 0 & 0 & 0 \\ i & 0 & 0 \end{pmatrix}, \quad \lambda_6 = \begin{pmatrix} 0 & 0 & 0 \\ 0 & 0 & 1 \\ 0 & 1 & 0 \end{pmatrix},$$
$$\lambda_7 = \begin{pmatrix} 0 & 0 & 0 \\ 0 & 0 & -i \\ 0 & i & 0 \end{pmatrix}, \quad \lambda_8 = \frac{1}{\sqrt{3}}\begin{pmatrix} 1 & 0 & 0 \\ 0 & 1 & 0 \\ 0 & 0 & -2 \end{pmatrix}, \tag{19.7.2}$$

normalized so that $\mathrm{Tr}\,(\lambda_a\lambda_b) = 2\delta_{ab}$. The generators of $SU(3)\times SU(3)$ are thus represented on the quark fields by the generators $t_a = \lambda_a$ of the unbroken $SU(3)$ symmetry and the broken symmetry generators $x_a = \lambda_a\gamma_5$.

To define the Goldstone boson fields and work out their transformation properties, we note that in the representation furnished by the quark fields, any $SU(3)\times SU(3)$ transformation may be written as $\exp(-i\gamma_5\sum_a \xi_a\lambda_a)$ times a transformation $\exp(i\sum_a \theta_a\lambda_a)$ belonging to the unbroken $SU(3)$ subgroup of $SU(3)\times SU(3)$. (The minus sign accompanying ξ is inserted for convenience in later comparing components of this field with the pion fields introduced in Section 19.5.) Hence in this representation, each right coset of $SU(3)\times SU(3)/SU(3)$ is represented by the matrix of the form $\gamma(\xi) = \exp(-i\gamma_5\sum_a \xi_a\lambda_a)$ that it contains, and aside from normalization, the parameters ξ_a of these right cosets may be taken as our Goldstone boson fields. According to Eq. (19.6.18), the transformation rule of these fields is dictated by

$$\exp\left[i\sum_a(\theta_a^V\lambda_a + \theta_a^A\lambda_a\gamma_5)\right]\exp\left(-i\gamma_5\sum_a \xi_a(x)\lambda_a\right)$$
$$= \exp\left(-i\gamma_5\sum_a \xi_a'(x)\lambda_a\right)\exp\left(i\sum_a \theta_a(x)\lambda_a\right) \quad (19.7.3)$$

with $\theta_a(x)$ some function of θ^V, θ^A, and $\xi(x)$. Also, according to Eq. (19.6.3), the Goldstone-free quark fields $\tilde{q}(x)$ are here defined by

$$q(x) \equiv \begin{pmatrix} u(x) \\ d(x) \\ s(x) \end{pmatrix} = \exp\left(-i\gamma_5\sum_a \xi_a(x)\lambda_a\right)\tilde{q}(x) \quad (19.7.4)$$

and have the transformation rule given by Eq. (19.6.24):

$$\tilde{q}'(x) = \exp\left(i\sum_a \theta_a(x)\lambda_a\right)\tilde{q}(x)\,. \quad (19.7.5)$$

We could proceed to introduce covariant derivatives (19.6.15) and (19.6.30), and use them to construct chiral-invariant Lagrangian densities, but for chiral symmetries there is a simpler approach available.

Note that the parts of Eq. (19.7.3) that are proportional to $(1+\gamma_5)$ and $(1-\gamma_5)$ read

$$\exp\left(i\sum_a \theta_a^L\lambda_a\right)\exp\left(-i\sum_a \xi_a(x)\lambda_a\right)$$
$$= \exp\left(-i\sum_a \xi_a'(x)\lambda_a\right)\exp\left(i\sum_a \theta_a(x)\lambda_a\right) \quad (19.7.6)$$

and

$$\exp\left(i\sum_a \theta_a^R\lambda_a\right)\exp\left(i\sum_a \xi_a(x)\lambda_a\right)$$
$$= \exp\left(i\sum_a \xi_a'(x)\lambda_a\right)\exp\left(i\sum_a \theta_a(x)\lambda_a\right)\,, \quad (19.7.7)$$

where

$$\theta_a^L \equiv \theta_a^V + \theta_a^A\,, \qquad \theta_a^R \equiv \theta_a^V - \theta_a^A\,.$$

By multiplying Eq. (19.7.7) on the right by the inverse of Eq. (19.7.6), we find the simple transformation rule

$$U'(x) = \exp\left(i\sum_a \lambda_a\theta_a^R\right)U(x)\exp\left(-i\sum_a \lambda_a\theta_a^L\right)\,, \quad (19.7.8)$$

where $U(x)$ is the unitary unimodular matrix

$$U(x) \equiv \exp\left(2i \sum_a \xi_a(x)\lambda_a\right) . \tag{19.7.9}$$

In other words, $U(x)$ transforms according to the $(\bar{\mathbf{3}}, \mathbf{3})$ representation of $SU(3) \times SU(3)$. This is still a non-linear realization of $SU(3) \times SU(3)$ because the components of $U(x)$ are not independent; they are subject to the nonlinear constraints $U^\dagger U = 1$ and $\text{Det}\, U = 1$.

The unique $(SU(3) \times SU(3))$-invariant term in the Goldstone boson Lagrangian that is of second order in spacetime derivatives is

$$\mathscr{L}_{2\,\text{deriv}} = -\tfrac{1}{16} F^2 \,\text{Tr}\,\{\partial_\mu U\, \partial^\mu U^\dagger\}\,, \tag{19.7.10}$$

with F^2 a positive constant to be determined. We can express the ξ_a in terms of conventionally normalized pseudoscalar meson fields by writing

$$\sum_a \lambda_a \xi_a = \frac{\sqrt{2}}{F}\begin{bmatrix} \frac{1}{\sqrt{2}}\pi^0 + \frac{1}{\sqrt{6}}\eta^0 & \pi^+ & K^+ \\ \pi^- & -\frac{1}{\sqrt{2}}\pi^0 + \frac{1}{\sqrt{6}}\eta^0 & K^0 \\ \bar{K}^- & \bar{K}^0 & -\sqrt{\frac{2}{3}}\eta^0 \end{bmatrix} \equiv \frac{\sqrt{2}B}{F}\,, \tag{19.7.11}$$

so that the kinematic term in Eq. (19.7.10) is of the usual form

$$\mathscr{L}_{\text{kin}} = -\tfrac{1}{2}\partial_\mu\pi^0\partial^\mu\pi^0 - \partial_\mu\pi^+\partial^\mu\pi^- - \partial_\mu K^+\partial^\mu\bar{K}^- - \partial_\mu K^0\partial^\mu\bar{K}^0 - \tfrac{1}{2}\partial_\mu\eta^0\partial^\mu\eta^0\,.$$

To determine the constant F, we note by comparing Eq. (19.7.4) with Eq. (19.5.42), that for ξ infinitesimal, the components ξ_1, ξ_2, and ξ_3 are the same as the Goldstone boson fields ζ_1, ζ_2, and ζ_3 introduced in Section 19.5. Then, comparing (19.7.11) with (19.5.17) and (19.5.32), we see that the constant F defined by Eq. (19.7.10) is the same as the $F_\pi = 184$ MeV introduced in Section 19.4.

The $SU(3) \times SU(3)$ symmetry of quantum chromodynamics is broken by the quark mass term, which in terms of the quark field $\tilde{q}(x)$ defined by Eq. (19.7.4) is

$$\mathscr{L}_{\text{mass}} = -\bar{q}\, M_q\, q = -\bar{\tilde{q}}\, e^{-i\sqrt{2}\gamma_5 B/F_\pi}\, M_q\, e^{-i\sqrt{2}\gamma_5 B/F_\pi}\, \tilde{q} \tag{19.7.12}$$

where

$$M_q = \begin{bmatrix} m_u & 0 & 0 \\ 0 & m_d & 0 \\ 0 & 0 & m_s \end{bmatrix} . \tag{19.7.13}$$

Eq. (19.7.12) contains a purely bosonic part, obtained by replacing the quark bilinear with its vacuum expectation value, given by the unbroken

$SU(3)$ and parity symmetries as*

$$< \bar{\tilde{q}}_n \gamma_5 \tilde{q}_m >_0 = 0\,, \qquad < \bar{\tilde{q}}_n \tilde{q}_m >_0 = -v\delta_{nm}\,. \tag{19.7.14}$$

The Goldstone boson mass term in the Lagrangian is then

$$-\frac{v}{F_\pi^2}\,\mathrm{Tr}\,\{B,\{B,M_q\}\} = -\frac{v}{F_\pi^2}\Bigg[4m_u\left(\frac{1}{\sqrt{2}}\pi^0+\frac{1}{\sqrt{6}}\eta^0\right)^2+4(m_u+m_d)\pi^+\pi^-$$
$$+4(m_u+m_s)K^+\bar{K}^-+4m_d\left(-\frac{1}{\sqrt{2}}\pi^0+\frac{1}{\sqrt{6}}\eta^0\right)^2$$
$$+4(m_d+m_s)K^0\bar{K}^0+\frac{8}{3}m_s(\eta^0)^2\Bigg]\,. \tag{19.7.15}$$

From this, we can read off that[35]

$$\begin{aligned} m^2_{\pi^+} &= m^2_{\pi^0} = \frac{4v}{F_\pi^2}[m_u+m_d]\,,\\ m^2_{K^+} &= \frac{4v}{F_\pi^2}[m_u+m_s]\,,\\ m^2_{K^0} &= \frac{4v}{F_\pi^2}[m_d+m_s]\,,\\ m^2_{\eta^0} &= \frac{4v}{F_\pi^2}\left[\frac{4m_s+m_d+m_u}{3}\right]\,, \end{aligned} \tag{19.7.16}$$

and there is also a term that mixes the π^0 and η^0:

$$m^2_{\pi\eta} = \frac{4v}{\sqrt{3}F_\pi^2}[m_u-m_d]\,. \tag{19.7.17}$$

The same results can also be obtained by operator methods. Noether's theorem allows us to construct generators T_a and X_a, for which

$$[T_a,q] = -\lambda_a q\,, \qquad [X_a,q] = -\gamma_5\lambda_a q\,. \tag{19.7.18}$$

We can write the Goldstone boson fields in a real basis as π_a, with $\pi^\pm = (\pi_1 \pm i\pi_2)/\sqrt{2}$, $\pi^0 = \pi_3$, $K^\pm = (\pi_4 \pm i\pi_5)/\sqrt{2}$, $K^0 = (\pi_6 + i\pi_7)/\sqrt{2}$, $\bar{K}^0 = (\pi_6 - i\pi_7)/\sqrt{2}$, and $\eta^0 = \pi_8$. The $SU(3)$ symmetry that survives the spontaneous symmetry breaking tells us that the F-matrix

* The fact that the conservation laws of parity, charge, and strangeness respected by the vacuum are the same as those respected by the quark mass matrix is a result of the vacuum alignment condition discussed in Section 19.3. Likewise, if the quark fields are defined so that the quark masses are all positive, then with $v > 0$ the minus sign in Eq. (19.7.3) is required by the condition that the true vacuum be at a minimum rather than a maximum of the vacuum energy for vacuum states rotated by $SU(3) \times SU(3)$ transformations, for as we shall see, it is this sign that will yield positive masses for the pseudo-Goldstone boson octet.

for these bosons takes the form $F_{ab} = F_\pi \delta_{ab}$. The $SU(3) \times SU(3)$ symmetry is also intrinsically broken by the mass term in the Hamiltonian $H_1 = m_u \bar{u}u + m_d \bar{d}d + m_s \bar{s}s = \bar{q}M_q q$, where M_q is the quark mass matrix (19.7.13). The mass matrix of the pseudo-Goldstone bosons is given here by (19.3.20) as

$$M^2_{ab} = -F_\pi^{-2} \langle \bar{q}\, \{\lambda_a, \{\lambda_b, M_q\}\}\, q \rangle_0 \,. \tag{19.7.19}$$

Since this is already of first order in quark masses, to this order we may use the unbroken $SU(3)$ relations $\langle \bar{u}u \rangle_0 = \langle \bar{d}d \rangle_0 = \langle \bar{s}s \rangle_0 \equiv -v$, and find the results (19.7.16) and (19.7.17), as before.

The masses $m_{K^+} = 493.65$ MeV and $m_{K^0} = 497.7$ MeV of the kaon doublet are quite close, and considerably larger than the pion mass, so we can see from (19.7.16) that the m_u and m_d must be considerably smaller than m_s. In calculating quantities that are sensitive to m_u and m_d, like the kaon mass difference or the pion masses, we should also take into account another small correction, the effect of electromagnetism. The electromagnetic current is $J^0 = ie\bar{q}\gamma^\mu Q q$, where Q is a diagonal matrix with elements 2/3, −1/3, and −1/3. Its commutators with the $SU(3) \times SU(3)$ generators are

$$[T_a, J^\mu] = -ie\,\tfrac{1}{2}\bar{q}\gamma^\mu [Q, \lambda_a] q \,, \qquad [X_a, J^\mu] = -ie\,\tfrac{1}{2}\bar{q}\gamma^\mu\gamma_5 [Q, \lambda_a] q \,.$$

We see that J^μ commutes with X_3, X_6, X_7, and X_8 as well as T_3, T_6, T_7, and T_8, so the electromagnetic part of the Hamiltonian is invariant under the $SU(2) \times SU(2) \times U(1) \times U(1)$ subgroup with these generators. Therefore in the limit of zero quark masses, *electromagnetic effects give no mass to the neutral pseudo-Goldstone bosons*[36] π^0, K^0, $\bar{K}^0$, and η^0, the Goldstone bosons associated with the spontaneous breakdown of the symmetries with generators X_3, X_6, X_7 and X_8. Also, in the zero-quark-mass limit there is an unbroken $SU(2)$ symmetry generated by T_6, T_7, and $\sqrt{3}T_8 - T_3$, under which the K^+ and π^+ transform as a doublet, so that for zero quark masses *the electromagnetic corrections to the K^+ and π^+ masses are equal.*[36] Since the quark mass terms and electromagnetic corrections are all small, it is reasonable to treat these effects as additive corrections to the effective quark Hamiltonian. We see then that with electromagnetism taken into account, the mass formulas (19.7.16) should be corrected to read[37]

$$\begin{aligned} m^2_{K^\pm} &= 4v(m_u + m_s)/F_\pi^2 + \Delta \,, \\ m^2_{K^0} = m^2_{\bar{K}^0} &= 4v(m_d + m_s)/F_\pi^2 \,, \\ m^2_{\pi^\pm} &= 4v(m_d + m_u)/F_\pi^2 + \Delta \,, \\ m^2_{\pi^0} &= 4v(m_d + m_u)/F_\pi^2 \,, \\ m^2_\eta &= 4v(m_u + m_d + 4m_s)/3F_\pi^2 \,, \end{aligned} \tag{19.7.20}$$

where Δ is the common electromagnetic correction to the K^+ and π^+ squared masses.

These formulas impose one linear relation on the five pseudo-Goldstone masses, a version of the Gell-Mann–Okubo relation:**

$$3m_\eta^2 + 2m_{\pi^+}^2 - m_{\pi^0}^2 = 2m_{K^+}^2 + 2m_{K^0}^2 . \tag{19.7.21}$$

Taking the kaon and pion masses from experiment, this relation yields an η mass of 566 MeV, in comparison with the experimental value of 547 MeV. The small discrepancy is usually attributed to the mixing of the η state with a heavier pseudoscalar particle, the η' at 958 MeV.

From the mass formulas (19.7.20) we may derive formulas for the quark mass ratios in terms of the pion and kaon masses:[37]

$$\frac{m_d}{m_s} = \frac{m_{K^0}^2 + m_{\pi^+}^2 - m_{K^+}^2}{m_{K^0}^2 + m_{K^+}^2 - m_{\pi^+}^2} , \qquad \frac{m_u}{m_s} = \frac{2m_{\pi^0}^2 - m_{K^0}^2 - m_{\pi^+}^2 + m_{K^+}^2}{m_{K^0}^2 + m_{K^+}^2 - m_{\pi^+}^2} . \tag{19.7.22}$$

Using the mass values $m_{\pi^+} = 139.57$ MeV, $m_{\pi^0} = 134.974$ MeV, $m_{K^+} = 493.65$ MeV and $m_{K^0} = 497.7$ MeV gives the ratios $m_d/m_s = 0.050$ and $m_u/m_s = 0.027$. Thus the ratio of the d and u quark masses is closer to 2 than 1. (A 1996 calculation[38a] using various other pieces of information as well as pseudoscalar meson masses has given values $m_d/m_s = 0.053 \pm 0.002$ and $m_u/m_s = 0.029 \pm 0.003$.)

None of this gives definite values for the individual quark masses. Indeed, these masses are not well defined until we define a renormalization prescription for the quark bilinears. Often this prescription is fashioned so that m_s equals the mass difference between isomultiplets differing by one unit of strangeness in some $SU(3)$ multiplet. For instance, the lightest vector meson octet consists of an isotopic doublet K^* at 892 MeV, interpreted as a bound state of an $\bar{s}$ antiquark and a u or d; its antidoublet, consisting of an s and a $\bar{u}$ or $\bar{d}$; and a $T = 1$ ρ at 770

** This relation was originally derived[38] on the basis of the approximate $SU(3)$ symmetry of Gell-Mann and Ne'eman generated by the T_a, ignoring mass differences within isotopic multiplets. From this derivation, one cannot tell whether this relation should be applied to the pseudoscalar meson masses themselves, or as here to their squares. Applying this relation to the pseudoscalar masses themselves would not in fact work very well; it gives an η mass of 613 MeV. The fact that the Gell-Mann–Okubo relation works well for the squares of the meson masses, but not for their first powers, supports the interpretation of these particles as Goldstone bosons of a spontaneously broken approximate $SU(3) \times SU(3)$ symmetry. Gell-Mann and Okubo also used the approximate $SU(3)$ symmetry generated by the T_a to derive relations among the masses of other particles, such as (ignoring isospin violating effects) the relation $2m_N + 2m_\Xi = 3m_\Lambda + m_\Sigma$ among the masses of the lightest baryon octet. For such multiplets, the average mass is so much larger than the mass differences within the multiplet that it makes little difference whether one applies the relation to the masses or their squares.

MeV and a $T = 0$ ω at 783 MeV, both interpreted as bound states of a $\bar{u}$ or $\bar{d}$ and a u or d. If we wish to attribute the difference between the K^* mass and the average of the ρ and ω masses to the relatively large s quark mass, then we must renormalize the quark bilinear so that $m_s - \frac{1}{2}(m_u + m_d) = m_{K^*} - \frac{1}{2}(m_\rho + m_\omega) = 120$ MeV, yielding $m_s = 125$ MeV. With the above quark mass ratios, this gives $m_d = 6.0$ MeV and $m_u = 3.3$ MeV. However, these estimates of mass values are much less reliable than the values of the mass ratios given in Eq. (19.7.22). Often m_s is estimated[38b] as 180 MeV rather than 125 MeV.

The mass term (19.7.12) includes meson–meson interactions. Using Eqs. (19.7.14), (19.7.11), and (19.7.9), we may write the purely bosonic part of this term in the Lagrangian as

$$\mathscr{L}_{\text{mass, bosonic}} = \tfrac{1}{2}v \,\text{Tr}\left\{M_q(U^\dagger + U)\right\} . \tag{19.7.23}$$

The form of this term could have been deduced from general symmetry considerations, which also allow us to find the allowed terms of higher order in M_q. Suppose we invent an external 3×3 field χ, and replace the mass term (19.7.12) in the underlying quantum chromodynamics Lagrangian with the χ–quark coupling term

$$\mathscr{L}_\chi = -\bar{q}\left[\tfrac{1}{2}(1+\gamma_5)\chi + \tfrac{1}{2}(1-\gamma_5)\chi^\dagger\right]q\,, \tag{19.7.24}$$

which becomes the same as Eq. (19.7.12) when we make the replacements $\chi = \chi^\dagger = M_q$. The point of this procedure is that the Lagrangian (19.7.24) becomes formally invariant under $SU(3) \times SU(3)$ if we give χ the formal transformation rule

$$\chi \to \exp\left(i\sum_a \lambda_a\theta_a^R\right)\,\chi\,\exp\left(-i\sum_a \lambda_a\theta_a^L\right) . \tag{19.7.25}$$

Thus we can work out the allowed bosonic terms involving quark masses by writing the most general $SU(3) \times SU(3)$ Lagrangian (up to some given order in derivatives and M_q) involving U and χ, requiring also invariance under the parity transformation

$$U(\mathbf{x},t) \leftrightarrow U^\dagger(-\mathbf{x},t)\,, \qquad \chi \leftrightarrow \chi^\dagger\,, \tag{19.7.26}$$

and then making the replacements

$$\chi = \chi^\dagger = M_q\,. \tag{19.7.27}$$

For instance, the interaction $\text{Tr}\,(U^\dagger\chi + U\chi^\dagger)$ is invariant under $SU(3) \times SU(3)$ and parity, and becomes of the same form as Eq. (19.7.23) when we give χ the values (19.7.27).

Using this technique, Gasser and Leutwyler[39] have given the complete effective Lagrangian for the pseudoscalar octet of fourth order in momenta or meson masses (with quark masses counted as being of second order in meson masses) as

$$\begin{aligned}\mathcal{L}_4 &= L_1 \,\mathrm{Tr}\left\{\partial_\mu U^\dagger \partial^\mu U\right\}^2 + L_2 \,\mathrm{Tr}\left\{\partial_\mu U \,\partial_\nu U^\dagger\right\} \mathrm{Tr}\left\{\partial^\mu U \,\partial^\nu U^\dagger\right\} \\ &+L_3 \,\mathrm{Tr}\left\{\partial_\mu U \,\partial^\mu U^\dagger \,\partial_\nu U \,\partial^\nu U^\dagger\right\} + L_4 \,\mathrm{Tr}\left\{\partial_\mu U \,\partial^\mu U^\dagger\right\} \mathrm{Tr}\left\{M_q(U+U^\dagger)\right\} \\ &+L_5 \,\mathrm{Tr}\left\{\partial_\mu U \,\partial^\mu U^\dagger \,(M_q U + U^\dagger M_q)\right\} + L_6 \left[\mathrm{Tr}\left\{M_q(U+U^\dagger)\right\}\right]^2 \\ &+L_7 \left[\mathrm{Tr}\left\{(U^\dagger - U)M_q\right\}\right]^2 + L_8 \,\mathrm{Tr}\left\{((UM_q)^2 + (U^\dagger M_q)^2)\right\}\,, \qquad (19.7.28)\end{aligned}$$

where $L_1, \ldots, L_8$ are constants to be determined by comparison with experiment. The complete effective Lagrangian up to fourth order in meson masses and momenta is

$$\mathcal{L}_{\mathrm{eff}} = \mathcal{L}_2 + \mathcal{L}_4\,, \qquad (19.7.29)$$

where $\mathcal{L}_2$ is the sum of the terms (19.7.10) and (19.7.23):

$$\mathcal{L}_2 = -\,\tfrac{1}{16}F^2\,\mathrm{Tr}\left\{\partial_\mu U \,\partial^\mu U^\dagger\right\} + \tfrac{1}{2}v\,\mathrm{Tr}\left\{M_q(U^\dagger + U)\right\}\,. \qquad (19.7.30)$$

Electroweak interactions may be included by replacing the derivatives ∂_μ with suitable gauge-covariant derivatives D_μ, and adding a few additional terms to $\mathcal{L}_{\mathrm{eff}}$.

Following the same power-counting arguments as in Sections 19.5 and 19.6, to calculate S-matrix elements to fourth order in meson masses and momenta we must include both tree graphs to all orders in $\mathcal{L}_2$ and up to first order in $\mathcal{L}_4$, and also one-loop graphs constructed from $\mathcal{L}_2$ alone. Gasser and Leutwyler and others have carried out a comprehensive study of meson dynamics (and associated electroweak interactions) using this effective Lagrangian.[40]

* * *

The quark mass term (19.7.12) naturally affects other $SU(3)$ multiplets. For any multiplet other than the pseudoscalar octet this can be treated as a first-order perturbation, and therefore gives a shift in the mass matrix of a generic multiplet $|i\rangle$ equal to

$$\delta m_{ij} = \langle i|\bar{\tilde{q}} M_q \tilde{q}|j\rangle \qquad (19.7.31)$$

and $SU(3)$ may be used[41] to relate the various matrix elements $\langle i|\bar{\tilde{q}}_r \tilde{q}_s|j\rangle$. In this way, it is straightforward to show that

$$\frac{\delta m_p - \delta m_n}{m_u - m_d} = \frac{m_\Xi - m_\Sigma}{m_s}\,. \qquad (19.7.32)$$

Using this together with Eq. (19.7.22) gives the quark mass contribution to the nucleon mass splitting $\delta m_p - \delta m_n \approx -2.5$ MeV. This result may be used in Eqs. (19.5.65) and (19.5.66) to calculate the leading violations of isospin symmetry in low energy pion–nucleon interactions. Of course, as defined here $\delta m_p - \delta m_n$ is not the full proton–neutron mass difference, which receives an important contribution also from photon emission and absorption. Because the neutron is electrically neutral, this electromagnetic term is almost certainly positive, in agreement with the fact that the observed proton–neutron mass difference is -1.3 MeV, leaving $+1.2$ MeV to be accounted for by electromagnetism. Unfortunately, accurate calculation of this electromagnetic mass difference has proved difficult.

19.8 Anomalous Terms in Effective Field Theories*

In implementing the effective field theory program described in the previous three sections, it is necessary that the action should contain all possible terms allowed by the assumed symmetries of the theory. The methods described in these sections allow us to identify all *manifestly* invariant terms, either by using the general formalism of covariant derivatives described in Section 19.6, or for chiral symmetries by using a linearly transforming field subject to non-linear constraints, like the $U(x)$ of Section 19.7. But it is possible that there may be other terms in the action of the effective field theory that are anomalous in the sense of being given by integrals of four-dimensional Lagrangian densities that are *not* invariant, but whose variation under the broken symmetry is a spacetime derivative, preserving the invariance of the corresponding term in the action.

As we shall see in Section 22.7, such a term was discovered for $SU(3) \times SU(3)$ by Wess and Zumino,[42] in a study of 'anomalies' due to quark loop graphs. However, the structure of this term can be understood without knowing anything about the underlying theory of quarks and gluons.

The easiest way to describe the Wess–Zumino term is by an extension of spacetime to five dimensions, introduced for this purpose by Witten.[43] As long as we require the fields in an effective field theory to approach a common limit as $x^\mu \to \infty$ in any direction, we can think of spacetime as having the topology of a sphere S_4, with the point at infinity included as an ordinary point. As remarked in Section 19.6, when a group G is broken to a subgroup H, the possible values of the Goldstone boson fields ξ_a at any spacetime point may be regarded as defining a point in the coset space G/H (the space of elements of G, with two elements identified if they differ by multiplication on the right by elements of H), so a set of functions $\xi_a(x)$ represents a mapping of the spacetime S_4 sphere into G/H. Depending on the topology of G/H, it may be possible smoothly to

* This section lies somewhat out of the book's main line of development, and may be omitted in a first reading.

deform any four-sphere in G/H to a point; that is, it may be possible to extend any $\xi_a(x)$ to a continuous function $\xi_a(x;s)$ defined for $0 \geq s \geq 1$, for which $\xi_a(x;0) = \xi_a(x)$ and $\xi_a(x;1)$ is any fixed point (say, $\xi_a = 0$) on the original sphere. Where this is true, it is expressed mathematically as the statement that the homotopy group $\pi_4(G/H)$ is trivial. (For a discussion of homotopy groups, see Section 23.2.) It is known that this is the case for $SU(N) \times SU(N)$ spontaneously broken to $SU(N)$, where the coset space $SU(N) \times SU(N)/SU(N)$ has the topology of $SU(N)$ itself.** Hence in the case of physical interest, where $G = SU(3) \times SU(3)$ and $H = SU(3)$, we may extend the Goldstone boson fields, or equivalently $U(x)$, (see Eq. (19.7.9)) to a unitary unimodular matrix $U(y)$ defined in a five-dimensional ball B_5 with coordinates x^μ and s, whose surface is the four-dimensional sphere of spacetime.

Now consider the following function formed from $U(y)$:

$$\omega(y) \equiv -\frac{i}{240\pi^2}\epsilon^{ijklm}\mathrm{Tr}\left\{U^{-1}\frac{\partial U}{\partial y^i}U^{-1}\frac{\partial U}{\partial y^j}U^{-1}\frac{\partial U}{\partial y^k}U^{-1}\frac{\partial U}{\partial y^l}U^{-1}\frac{\partial U}{\partial y^m}\right\}, \tag{19.8.1}$$

where indices i, j, etc. run over the five coordinate directions for the coordinates x^μ and s. (The phase and numerical coefficient are chosen for later convenience.) This is manifestly invariant under the chiral transformations (19.7.8). Also, because ϵ^{ijklm} is a tensor density, the integral of $\omega(y)$ over the five-ball is manifestly independent (up to a sign) of the choice of five-dimensional coordinates y^i. Furthermore, this integral depends only on the values taken by $U(y)$ on the ball's surface; that is, in spacetime. To check this last point, note that when we make an infinitesimal variation $\delta U(y)$ in $U(y)$ in the interior of the ball, $\omega(y)$ changes by a derivative:

$$\delta\omega(y) = -\frac{i}{48\pi^2}\epsilon^{ijklm}\frac{\partial}{\partial y^m}\mathrm{Tr}\left\{U^{-1}\frac{\partial U}{\partial y^i}U^{-1}\frac{\partial U}{\partial y^j}U^{-1}\frac{\partial U}{\partial y^k}U^{-1}\frac{\partial U}{\partial y^l}U^{-1}\delta U\right\}, \tag{19.8.2}$$

(for this calculation, see Section 23.4), so a change in $U(y)$ that does not affect its value in spacetime also does not affect the integral of (19.8.1) over the five-ball B_5 whose surface is spacetime. We can therefore include this integral as a term in the action:

$$I_{\mathrm{WZW}}[U] = n\int_{B_5} d^5y\,\omega(y) \tag{19.8.3}$$

with n a coefficient that so far is arbitrary.

** This is shown by the fact that we can use $SU(N)$ elements $U(x)$ to represent the right cosets of $SU(N)$ in $SU(N) \times SU(N)$.

This term may be written as the four-dimensional integral of a Lagrangian density, but not of an $(SU(3) \times SU(3))$-invariant Lagrangian density. Using Eq. (19.7.9), the leading term in $\omega(x)$ in the limit of small meson fields is

$$\omega(x) \to \frac{8\sqrt{2}}{15\pi^2 F_\pi^5}\epsilon^{ijklm}\,\mathrm{Tr}\left\{\frac{\partial B}{\partial y^i}\frac{\partial B}{\partial y^j}\frac{\partial B}{\partial y^k}\frac{\partial B}{\partial y^l}\frac{\partial B}{\partial y^m}\right\}, \tag{19.8.4}$$

where B is the matrix (19.7.11) of Goldstone boson fields. It follows then from Gauss's theorem that

$$I_{\rm WZW}[U] = \frac{8\sqrt{2}n}{15\pi^2 F_\pi^5}\,\epsilon^{\mu\nu\rho\sigma}\int_{S_4} d^4x\,\mathrm{Tr}\left\{B\frac{\partial B}{\partial x^\mu}\frac{\partial B}{\partial x^\nu}\frac{\partial B}{\partial x^\rho}\frac{\partial B}{\partial x^\sigma}\right\} + O\left(\frac{B^6}{F_\pi^6}\right). \tag{19.8.5}$$

Although chiral invariant, this cannot be written as the integral of a chiral-invariant density over spacetime, because any chiral invariant density would have to be constructed out of the first and higher covariant derivatives of the Goldstone boson fields, and so, when expanded in powers of the Goldstone boson fields, such an invariant density would begin with a term involving only derivatives of B, not B itself.

As Witten noted, the inclusion of this term in the effective action resolved what otherwise would have been a conflict between the $SU(3) \times SU(3)$ effective field theory and experiment. Because no $\epsilon^{\mu\nu\rho\sigma}$ terms appear in the effective Lagrangians (19.7.28) and (19.7.30) (or in higher-order terms of this sort) parity conservation imposes the requirement that these terms are *even* in the Goldstone boson fields, ruling out processes like $K + \bar{K} \to 3\pi$. Not only is there no symmetry in the underlying theory of quantum chromodynamics that would account for such a selection rule — there is even experimental evidence against it, for as Witten pointed out, the ϕ meson is observed to decay both into $K + \bar{K}$ and 3π final states. Eq. (19.8.5) shows that this unwanted selection rule is removed by the Wess–Zumino–Witten term in the action.

Remarkably, the coefficient of the Wess–Zumino–Witten term is not a freely adjustable parameter. As Witten showed,[43] this is because, although this term is not changed by smooth deformations of the function $U(y)$ in B_5 that do not affect $U(x)$ on the spacetime boundary S_4, the Wess–Zumino–Witten term *can* be changed by a discontinuous change in $U(y)$ that leaves $U(x)$ unchanged on the boundary. We may think of the five-ball B_5 as half of a five-sphere S_5, with the spacetime S_4 as the border between B_5 and the other half B'_5. (Think of S_5 as analogous to the surface of the Earth, with the spacetime S_4 as the equator and B_5 and B'_5 as the northern and southern hemispheres.) Because S_4 is also the boundary of the other half of S_5, we could just as well have written a

Wess–Zumino–Witten term as

$$I'_{WZW}[U] = -n \int_{B'_5} d^5y\, \omega(y) \qquad (19.8.6)$$

with a minus sign inserted because the boundary of B'_5 is the four-sphere with opposite orientation. It is not possible to require that the terms (19.8.3) and (19.8.6) should be equal for arbitrary Goldstone boson fields without setting $n = 0$, but in order that the weighting factor $\exp(iI)$ in path integrals should be unaffected by the difference between the terms (19.8.3) and (19.8.6), it is only necessary to require that this difference should be 2π times an integer. That is,

$$I_{WZW}[U] - I'_{WZW}[U] = n \int_{S_5} d^5y\, \omega(y) = 2\pi \times \text{integer}\,. \qquad (19.8.7)$$

With the normalization factors we have inserted in the definition (19.8.1) of $\omega(y)$, the integral of $\omega(y)$ over any five-sphere turns out to have the value[44] 2π. It follows that *the coefficient n must be an integer.*

The example of the Wess–Zumino–Witten term raises the question whether there may be other anomalous terms in the action, not necessarily related to quark loops, that also are invariant under $SU(3) \times SU(3)$, despite not being four-dimensional integrals of $(SU(3) \times SU(3))$-invariant Lagrangian densities. Fortunately, the answer is no. It has been shown[45] that for a general group G broken to an arbitrary subgroup H (with $\pi_4(G/H) = 0$), any term $F[\xi]$ in the action of the Goldstone boson fields $\xi_a(x)$ may always be written as the integral of a G-invariant five-form Ω over a five-ball B_5 whose boundary is the spacetime four-sphere S_4:

$$F[\xi] = \int_{B_5} d^5y\, \epsilon^{ijklm} \frac{\partial \xi_a}{\partial y^i} \frac{\partial \xi_b}{\partial y^j} \frac{\partial \xi_c}{\partial y^k} \frac{\partial \xi_d}{\partial y^l} \frac{\partial \xi_e}{\partial y^m}\, \Omega_{abcde}(\xi(y))\,. \qquad (19.8.8)$$

In order that this should be independent of the particular way that $\xi_a(x)$ is extended into the interior of the five-ball, it is necessary that Ω should be exact, in the sense that there it is the exterior derivative of a four-form

$$\Omega_{abcde}(\xi) = (\partial/\partial \xi_{[a})\mathscr{L}_{bcde]}(\xi) \qquad (19.8.9)$$

(with square brackets as usual indicating antisymmetrization with respect to the enclosed indices), so that

$$F[\xi] = \int_{S_4} d^4x\, \epsilon^{\mu\nu\rho\sigma} \frac{\partial \xi^a}{\partial x^\mu} \frac{\partial \xi^b}{\partial x^\nu} \frac{\partial \xi^c}{\partial x^\rho} \frac{\partial \xi^d}{\partial x^\sigma} \mathscr{L}_{abcd}(\xi)\,. \qquad (19.8.10)$$

It follows that Ω is also closed; that is, it has a vanishing exterior derivative

$$(\partial/\partial \xi_{[f})\, \Omega_{abcde]}(\xi) = 0\,. \qquad (19.8.11)$$

Where the four-form $\mathscr{L}_{abcd}(\xi)$ is also G-invariant, the functional (19.8.10) is just one of the ordinary manifestly G-invariant terms in the action, discussed in the previous three sections. The anomalous terms in the action arise from the possibility that, although every term in $\Omega(y)$ is G-invariant and the exterior derivative of a four-form, some terms may not be the exterior derivatives of G-invariant four-forms. Thus the new terms in the action may be identified with the closed G-invariant five-forms that are independent, in the sense that no real linear combination of them is the exterior derivative of a G-invariant four-form. These are known in mathematics as the generators of the *de Rham cohomology group* $H^5(G/H;\mathbf{R})$. (The group multiplication rule here is just simple addition.) The de Rham cohomology groups have been calculated for manifolds of various topologies.[46] In particular, $H^5(SU(N) \times SU(N)/SU(N);\mathbf{R})$ has a single generator, given by Eq. (19.8.1). Thus without knowing anything about the underlying theory of quarks and gluons, we can learn everything about the anomalous terms in the Goldstone boson action, with the single exception of the value of the integer n. We will see in Section 22.7 that in $SU(N_c)$ gauge theories this integer equals the number N_c of colors, which in quantum chromodynamics is $n = 3$.

19.9 Unbroken Symmetries

We have seen how the properties of Goldstone bosons and their low energy interactions may be deduced from an assumption that the theory is invariant (or approximately invariant) under a group G spontaneously broken to a subgroup H. But in applying these methods to the cases where G is $SU(2) \times SU(2)$ or $SU(3) \times SU(3)$, we had to take the pattern of symmetry breaking, of $SU(2) \times SU(2)$ or $SU(3) \times SU(3)$ to their non-chiral $SU(2)$ or $SU(3)$ subgroups, from experiment. In Section 22.5 we shall show that the $SU(3) \times SU(3)$ symmetry for massless u, d, and s quarks must in fact be spontaneously broken in quantum chromodynamics, but it is more difficult to show on the basis of quantum chromodynamics that the $SU(2) \times SU(2)$ symmetry with only u and d massless is also spontaneously broken.* On the other hand, there is an intuitive argument that their non-chiral $SU(2)$ or $SU(3)$ subgroups are *not* broken, based on a conjecture known as the *persistent mass condition*[48], which states that composite particles will not be massless if the particles of which they are

* Weingarten[47] has used lattice methods to show that whether or not chiral symmetry is broken, the lightest particle in quantum chromodynamics with massless u and d quarks must have the quantum numbers of the pion. As we will see in Section 22.5, the existence of anomalies due to fermion loops in quantum chromodynamics together with the assumption of quark trapping requires that some hadron be massless, so it follows that the pion is massless, which strongly suggests that chiral symmetry is spontaneously broken.

composed are massive. Non-chiral symmetries like isospin conservation are not violated if we give the quarks equal masses, and then if they were spontaneouly broken we would have the massless Goldstone bosons formed as composites of massive quarks, in contradiction with the persistent mass condition. In what follows we shall present a proof by Vafa and Witten[49] that in gauge theories like quantum chromodynamics those non-chiral symmetries that are not violated by quark masses can not be spontaneously broken. This result is of more than academic interest; as we shall see in Section 21.4, it is possible that the spontaneous breakdown of electroweak gauge symmetries is described by a 'technicolor' theory similar to quantum chromodynamics, and in testing this idea it is important to know what symmetries are left unbroken in this theory.

Consider a gauge theory like quantum chromodynamics, with a number of fermion 'flavors' in identical representations of the gauge group. If all fermions have masses, then the theory will be invariant under all unitary global non-chiral transformations on the fermion flavors that commute with the fermion mass matrix. For instance, if n_1 fermions are degenerate with a common mass m_1, n_2 fermions are degenerate with some other common mass m_2, and so on, then this global symmetry group is $U(n_1)\times U(n_2)\times\ldots$. (As a special case, if there are no degeneracies we have a global symmetry under $U(1)\times U(1)\times\ldots$; an example is the conservation of baryon number, strangeness, etc. in quantum chromodynamics.) *These symmetries cannot be spontaneously broken.*

To prove this, let us consider a general Greens function for r fermion and antifermion fields.** In the path-integral formalism this is†

$$\langle T\left\{\Psi_{u_1k_1}(x_1)\cdots\Psi_{u_rk_r}(x_r)\Psi^\dagger_{v_1l_1}(y_1)\cdots\Psi^\dagger_{v_rl_r}(y_r)\right\}\rangle_{\rm VAC}$$
$$=\frac{1}{Z}\int[dA][d\psi][d\psi^\dagger]\,\psi_{u_1k_1}(x_1)\cdots\psi_{u_rk_r}(x_r)\,\psi^\dagger_{v_1l_1}(y_1)\cdots\psi^\dagger_{v_rl_r}(y_r)$$
$$\times\exp\Big(iI_{\rm gauge}[A]+iI_{\rm Dirac}[\psi,\psi^\dagger\,;A]\Big)\,, \qquad (19.9.1)$$

where $k_1\ldots k_r$ and $l_1\ldots l_r$ are Dirac spin indices, $u_1\ldots u_r$ and $v_1\ldots v_r$ are fla-

** In the original work of Vafa and Witten,[49] they first proved the absence of symmetry breaking in the case $r=1$ with $x=y$, then observed that the absence of symmetry breaking in this vacuum expectation value did not rule out the possibility of a spontaneous symmetry breakdown occurring in other Greens functions, and so went on to different methods of proof.

† As we saw in Section 15.5, both numerator and denominator are proportional to the infinite volume of the gauge group, which cancels in the ratio (19.9.1). The presence of this infinite factor detracts from the rigor of the following arguments, but if we were to remove it by introducing ghosts then some of the steps below that depend on the positivity of the action would raise difficulties. One way to deal with this problem is to replace the spacetime continuum with a finite lattice of points, in which case the gauge group has a finite volume, and no gauge fixing or ghosts are needed.

vor indices, $I_{\rm gauge}[A]$ is the action for a pure gauge theory, $I_{\rm Dirac}[\psi, \psi^\dagger; A]$ is the action for the Dirac fields in the presence of a gauge field $A^\mu_\alpha(x)$, and Z is the vacuum–vacuum amplitude

$$Z \equiv \int [dA][d\psi][d\psi^\dagger] \exp\Big(iI_{\rm gauge}[A] + iI_{\rm Dirac}[\psi, \psi^\dagger; A]\Big) . \qquad (19.9.2)$$

It will be necessary here to work in a Euclidean spacetime, with $x^4 = x_4 = ix^0$, $y^4 = y_4 = iy^0$, and $A^4_\alpha = A_{4\alpha} = iA_{0\alpha}$ all real. (See Appendix A of Chapter 23.) In this case, the Dirac action is

$$I_{\rm Dirac}[\psi, \psi^\dagger; A] = i \int d^3x \int dx^4 \, \psi^\dagger \, [\not{D} + M] \, \psi \, , \qquad (19.9.3)$$

where M is the fermion mass matrix, and $\not{D}$ is the Euclidean gauge-covariant derivative contracted with the Euclidean Dirac matrices

$$\not{D} = \sum_{i=1}^{4} (\partial_i - it^\alpha A_{i\alpha})\gamma_i \, , \qquad (19.9.4)$$

where as usual $\gamma_4 = i\gamma^0$. Because the action is quadratic in fermion fields, we can explicitly perform the integral over these fields

$$\begin{aligned} &\langle T\left\{\Psi_{u_1k_1}(x_1)\cdots\Psi_{u_rk_r}(x_r)\Psi^\dagger_{v_1l_1}(y_1)\cdots\Psi^\dagger_{v_rl_r}(y_r)\right\}\rangle_{\rm VAC} \\ &= \frac{1}{Z}\int [dA] \, {\rm Det}\,(\not{D} + M) \, \exp\left(iI_{\rm gauge}[A]\right) \\ &\times \left[(\not{D} + M)^{-1}_{x_1\,u_1\,k_1,\,y_1\,v_1\,l_1} \cdots (\not{D} + M)^{-1}_{x_r\,u_r\,k_r,\,y_r\,v_r\,l_r} \pm \text{permutations}\right] , \end{aligned} \qquad (19.9.5)$$

where '$\pm$ permutations' indicates that we must sum over all $r!$ permutations of the ψ fields, with a minus sign for odd permutations, and

$$Z = \int [dA] \, {\rm Det}\,(\not{D} + M) \exp\left(iI_{\rm gauge}[A]\right) . \qquad (19.9.6)$$

The expression (19.9.5) is manifestly invariant under any unitary transformation on flavor indices that commutes with the mass matrix M. It does not matter what non-perturbative effects are produced by the functional integral over gauge fields and ghosts; these fields are inert under the symmetries in question here, so that the remaining functional integral in Eq. (19.9.5) cannot break these symmetries.†† But for this argument to be convincing, we must show that the expression (19.9.5) is well defined.

†† It is more difficult to show that P, C, and T are not spontaneously broken in quantum chromodynamics with massive quarks, because these symmetries act non-trivially on the gauge fields. Vafa and Witten[50] have applied the methods of Ref. 49 to show that P is not spontaneously broken in quantum chromodynamics.

This is not merely an academic question of mathematical rigor. As we saw in Section 19.1, the sign that a symmetry is not spontaneously broken is not just that the ground state is invariant under the symmetry transformations, but also that the symmetry of the ground state is stable under small perturbations. If we break the symmetries of M by adding a small perturbation δM, and if the expression (19.9.5) becomes singular as $\delta M \to 0$, then factors of δM in symmetry-breaking matrix elements may be cancelled by the singularities in the matrix elements for $\delta M = 0$. Precisely this happens for chiral symmetries, which arise in a symmetry limit where some of the eigenvalues of M vanish. In this case, as we approach the symmetry limit of zero mass, the factors of mass in the numerators of symmetry-breaking expectation values are cancelled by factors of mass in the denominators of the propagators $(\not{D}+M)^{-1}$ wherever $\not{D}$ has zero eigenvalues. For this reason, the arguments here will apply only to the non-chiral symmetries of theories with mass matrices M whose eigenvalues are all non-zero.

To set a bound on the matrix element (19.9.5) in this case, note first that the differential operator (19.9.4) is here antihermitian, so as long as the hermitian matrix M has no null eigenvalues, $\not{D}+M$ has a well-defined inverse. It is still necessary to show that the remaining integration over the gauge fields in Eq. (19.9.5), including that in Z, does not make this expression singular when M satisfies the conditions for a symmetry, for instance that n of its eigenvalues are equal for a $U(n)$ symmetry. As we shall see, this will insure that when a fermion mass matrix M that is invariant under some global symmetry transformation is perturbed by adding a small term δM that breaks this symmetry, the change in the expectation value (19.9.5) under this symmetry transformation vanishes in the limit $\delta M \to 0$.

It is difficult to show that the coordinate-space Greens functions Eq. (19.9.5) are non-singular, so let us consider instead the matrix element for smeared fields

$$\Psi_u[f] \equiv \int d^4x\, f^{k\dagger}(x)\, \Psi_{uk}(x)\,, \tag{19.9.7}$$

where $f^k(x)$ are arbitrary smooth square-integrable functions. In these terms, Eq. (19.9.5) reads

$$\begin{aligned}\Big\langle T\Big\{\Psi_{u_1}[f_1]\cdots\Psi_{u_r}[f_r]\,\Psi^\dagger_{v_1}[g_1]\cdots\Psi^\dagger_{v_r}[g_r]\Big\}\Big\rangle_{\rm VAC}\\ = \frac{1}{Z}\int[dA]\,{\rm Det}\,(\not{D}+M)\,\exp\left(iI_{\rm gauge}[A]\right)\\ \times\Big[(\not{D}+M)^{-1}_{f_1u_1,\,g_1v_1}\cdots(\not{D}+M)^{-1}_{f_ru_r,\,g_rv_r}\pm\text{permutations}\Big]\,,\end{aligned} \tag{19.9.8}$$

where

$$(\not{D}+M)^{-1}_{fu,\,gv} \equiv \int d^4x \int d^4y\; f^{k\dagger}(x)\,(\not{D}+M)^{-1}_{xuk,\,yvl}\; g^l(y)\,. \qquad (19.9.9)$$

In this basis, the fermion propagators for a given gauge field are not only well defined, but bounded uniformly in $A^\mu_\alpha(x)$:‡

$$\left|(\not{D}+M)^{-1}_{fu,\,gv}\right| \leq \sum_a \frac{1}{|m_a|}\,, \qquad (19.9.10)$$

where m_a are the eigenvalues of M, and for convenience we are now normalizing the functions $f_i(x)$ and $g_j(y)$ so that

$$\int f_i^\dagger(x) f_i(x)\, d^4x = \int g_j^\dagger(y) g_j(y)\, d^4y = 1\,. \qquad (19.9.11)$$

(The summation convention is suspended here.) Furthermore, the weight function for the average over gauge fields is positive:‡‡

$$\text{Det}\,(\not{D}+M)\exp\left(iI_{\text{gauge}}[A]\right) = \exp\left(-\tfrac{1}{4}\int d^3x \int dx^4 \sum_{i=1}^{4}\sum_{j=1}^{4} F_{ij}{}^2\right) \times \sqrt{\text{Det}\left[(iD)^2+M^2\right]}\,. \qquad (19.9.12)$$

With a positive weight function, the average of any function is bounded by the bound of that function. Using the bound (19.9.10) in Eq. (19.9.8),

---

‡ To see this, we may expand in the eigenvectors of M

$$M_{uv}c^a_v = m_a\, c^a_u\,, \qquad c^{a*}_u c^b_u = \delta_{ab}$$

and, adapting a trick of Vafa and Witten,[49] write Eq. (19.9.9) as

$$(\not{D}+M)^{-1}_{fu,\,gv} = \sum_a c^{a*}_u c^a_v \int d^4x \int d^4y\; f^{k\dagger}(x)\,(\not{D}+m_a)^{-1}_{xk,\,yl}\; g^l(y)$$
$$= \sum_a \pm c^{a*}_u c^a_v \int_0^\infty d\tau\, \exp(-|m_a|\tau) \int d^4x \int d^4y\; f^{k\dagger}(x)\,(\exp(\mp\tau\,\not{D}))_{xk,\,yl}\; g^l(y)\,,$$

where $\pm$ is the sign of m_a. The matrix $c^{a*}_u c^a_v$ and the operator $(\exp(\mp\tau\,\not{D}))_{xk,\,yl}$ are both unitary, so

$$\left|c^{a*}_u c^a_v \int d^4x \int d^4y\; f^{k\dagger}(x)\,(\exp(\mp\tau\,\not{D}))_{xk,\,yl}\, g^l(y)\right|^2 \leq \int d^4x\, |f(x)|^2 \int d^4y\, |g(y)|^2 = 1\,.$$

Using this inside the integral yields Eq. (19.9.10).

‡‡ In the fermion determinant, we use the fact that $\text{Det}\,(\not{D}+M) = \text{Det}\,\gamma_5\,(\not{D}+M)\,\gamma_5 = \text{Det}\,(-\not{D}+M)$.

the matrix element is bounded by

$$\left|\langle T\left\{\Psi_{u_1}[f_1]\cdots\Psi_{u_r}[f_r]\,\Psi^\dagger_{v_1}[g_1]\cdots\Psi^\dagger_{v_r}[g_r]\right\}\rangle_{\rm VAC}\right| \leq r!\left[\sum_a \frac{1}{|m_a|}\right]^r . \tag{19.9.13}$$

Thus there are no singularities as the symmetry-breaking terms δM in M go to zero that could cancel factors of δM. To see this more explicitly, we can use the same methods to show that if we perturb M by a small symmetry-breaking term δM, then the term in Eq. (19.9.8) of first order in δM is bounded by

$$\left|\delta\langle T\left\{\Psi_{u_1}[f_1]\cdots\Psi_{u_r}[f_r]\,\Psi^\dagger_{v_1}[g_1]\cdots\Psi^\dagger_{v_r}[g_r]\right\}\rangle_{\rm VAC}\right|$$
$$\leq r\,r!\sum_{a,b}\frac{|(\delta M)_{ab}|}{|m_a m_b|}\left[\sum_a \frac{1}{|m_a|}\right]^{r-1} , \tag{19.9.14}$$

so this vanishes when $\delta M \to 0$.

In the real world none of the quark masses are zero or degenerate, so it follows immediately from the above arguments that the $U(1)$ symmetries like conservation of baryon number, strangeness, etc., are neither spontaneously nor intrinsically broken. The other strong-interaction symmetries like isospin or $SU(3)$ are more problematic. These symmetries would be completely unbroken if the u and d quarks, or u, d, and s quarks, had equal non-zero masses, but as we saw in Section 19.7, these masses are not at all degenerate; the isospin and $SU(3)$ symmetries arise because the quark masses are small, not equal. These symmetries do remain unbroken if we give two or three quarks equal masses and let the masses become arbitrarily small, so isospin or $SU(3)$ will be good symmetries in any process that is not sensitive to the small quark masses. But not all processes are insensitive to these masses, because for two or three massless quarks, the pion or the pseudoscalar meson octet to which it belongs becomes massless. This is not a problem for isospin, because as we saw in Section 19.5, the pion triplet remains degenerate to first order in quark masses even though $m_u \neq m_d$. For $SU(3)$ the quark mass differences produce first-order mass differences among the pion, kaon, and eta, so that processes dominated by one-meson poles *can* show large departures from $SU(3)$.

19.10 The $U(1)$ Problem

The success of quantum chromodynamics in explaining the pattern of strong-interaction symmetries seemed at first to be marred by one failure. As we saw in Section 19.5, the broken $SU(2)\times SU(2)$ symmetry of the strong interactions is a natural consequence of the smallness of the u and

d quark masses. But the Lagrangian of quantum chromodynamics with small u and d quark masses also has another chiral symmetry,[51] the $U(1)_A$ symmetry under transformations

$$u \to \exp(i\gamma_5\theta)u\,, \qquad d \to \exp(i\gamma_5\theta)d\,. \tag{19.10.1}$$

Such a symmetry if unbroken would, like $SU(2) \times SU(2)$, impose a parity doubling on the hadron spectrum, but no such parity doubling is observed. On the other hand, a broken $U(1)_A$ symmetry would imply the existence of an isoscalar 0^- Goldstone boson with a mass comparable to that of the pion, and this Goldstone boson is also not observed. It is true that the η meson is an isoscalar 0^- boson, but it is considerably heavier than the pion, and as we saw in Section 19.7, it is well understood as one of the Goldstone bosons of $SU(3) \times SU(3)$. With the s as well as the u and d quarks regarded as relatively light, the Lagrangian of quantum chromodynamics would have a $U(1)$ chiral symmetry in addition to $SU(3) \times SU(3)$, under the transformations

$$u \to \exp(i\gamma_5\theta)u\,, \qquad d \to \exp(i\gamma_5\theta)d\,, \qquad s \to \exp(i\gamma_5\theta)s\,. \tag{19.10.2}$$

The spontaneous breakdown of such a symmetry would require the existence of *two* isoscalar 0^- mesons: one the η, and the other with a mass comparable to that of the pion.

This prediction can be made more explicit.[52] If we include the transformation (19.10.2) among the spontaneously broken symmetries of massless quantum chromodynamics, then in the presence of quark masses we encounter a term (19.7.12):

$$\mathscr{L}_{\text{mass}} = -\bar{q}\, M_q\, q = -\bar{\tilde{q}}\, e^{-i\sqrt{2}\gamma_5 B/F_\pi}\, M_q\, e^{-i\sqrt{2}\gamma_5 B/F_\pi}\, \tilde{q}\,, \tag{19.10.3}$$

where again

$$M_q = \begin{bmatrix} m_u & 0 & 0 \\ 0 & m_d & 0 \\ 0 & 0 & m_s \end{bmatrix}, \tag{19.10.4}$$

but now B includes a Goldstone boson field ζ for the broken symmetry (19.10.2):

$$B = \begin{bmatrix} \frac{1}{\sqrt{2}}\pi^0 + \frac{1}{\sqrt{6}}\eta^0 & \pi^+ & K^+ \\ \pi^- & -\frac{1}{\sqrt{2}}\pi^0 + \frac{1}{\sqrt{6}}\eta^0 & K^0 \\ \bar{K}^- & \bar{K}^0 & -\sqrt{\frac{2}{3}}\pi^0 \end{bmatrix}$$
$$+\frac{F_\pi}{\sqrt{3}F_\zeta}\begin{bmatrix} \zeta & 0 & 0 \\ 0 & \zeta & 0 \\ 0 & 0 & \zeta \end{bmatrix}, \tag{19.10.5}$$

with F_ζ the unknown coupling of the $U(1)_A$ Goldstone boson to the corresponding current (and the factor $\sqrt{3}$ inserted for future convenience.) We can again use the unbroken $SU(3)$ symmetry to write the vacuum expectation value of the quark bilinear in the form (19.7.14), and expanding in powers of the boson fields, we find a Goldstone boson mass term in the Lagrangian of the form

$$\begin{aligned}
-\frac{v}{F_\pi^2}\,\mathrm{Tr}\,\{B,\{B,M_q\}\} = -\frac{v}{F_\pi^2}\Bigg[&4m_u\left(\frac{1}{\sqrt{2}}\pi^0+\frac{1}{\sqrt{6}}\eta^0+\frac{F_\pi}{\sqrt{3}F_\zeta}\zeta\right)^2 \\
&+4(m_u+m_d)\pi^+\pi^- + 4(m_u+m_s)K^+\bar{K}^- \\
&+4m_d\left(-\frac{1}{\sqrt{2}}\pi^0+\frac{1}{\sqrt{6}}\eta^0+\frac{F_\pi}{\sqrt{3}F_\zeta}\zeta\right)^2 \\
&+4(m_d+m_s)K^0\bar{K}^0 + 4m_s\left(-\sqrt{\frac{2}{3}}\eta^0+\frac{F_\pi}{\sqrt{3}F_\zeta}\zeta\right)^2\Bigg].
\end{aligned} \tag{19.10.6}$$

The charged and strange meson masses are the same as before, but now the neutral non-strange mesons have a mass matrix

$$M_0^2 = 8v\begin{bmatrix}
\frac{m_u+m_d}{2F_\pi^2} & \frac{m_u-m_d}{2\sqrt{3}F_\pi^2} & \frac{m_u-m_d}{\sqrt{6}F_\pi F_\zeta} \\
\frac{m_u-m_d}{2\sqrt{3}F_\pi^2} & \frac{m_u+m_d+4m_s}{6F_\pi^2} & \frac{m_u+m_d-2m_s}{3\sqrt{2}F_\pi F_\zeta} \\
\frac{m_u-m_d}{\sqrt{6}F_\pi F_\zeta} & \frac{m_u+m_d-2m_s}{3\sqrt{2}F_\pi F_\zeta} & \frac{m_u+m_d+m_s}{3F_\zeta^2}
\end{bmatrix} \tag{19.10.7}$$

with rows and columns listed in the order π^0, η^0, and ζ.

In the limit where m_u and m_d vanish, this has two eigenvectors with eigenvalue zero:

$$u_a = \begin{pmatrix}1\\0\\0\end{pmatrix}, \quad u_b = \frac{1}{\sqrt{F_\pi^2+2F_\zeta^2}}\begin{pmatrix}0\\F_\pi\\\sqrt{2}F_\zeta\end{pmatrix}. \tag{19.10.8}$$

In this orthonormal basis, the mass-squared matrix to first order in m_u and m_d is

$$m_{aa}^2 \simeq u_a^{\mathrm{T}} M_0^2 u_a = \frac{4v\,(m_u+m_d)}{F_\pi^2}, \tag{19.10.9}$$

$$m_{bb}^2 \simeq u_b^{\mathrm{T}} M_0^2 u_b = \frac{12v\,(m_u + m_d)}{F_\pi^2 + 2F_\zeta^2}\,, \tag{19.10.10}$$

$$m_{ab}^2 = m_{ba}^2 \simeq \frac{\sqrt{3}v(m_u - m_d)}{2F_\pi\sqrt{F_\pi^2 + 2F_\zeta^2}}\,. \tag{19.10.11}$$

The effect of the off-diagonal term (19.10.11) is to decrease the product of eigenvalues by a negligible fractional amount $(m_u - m_d)^2/64(m_u + m_d)^2$, while of course leaving the sum of the eigenvalues unchanged, so the eigenvalues are given to a good approximation by the diagonal elements (19.10.9) and (19.10.10). Comparing Eq. (19.10.9) with Eq. (19.7.16), we see that the particle corresponding to the eigenvector u_a is the π^0. The other particle, corresponding to u_b, has a mass

$$m_b \simeq \sqrt{m_{bb}^2} \simeq \frac{\sqrt{3}m_\pi F_\pi}{\sqrt{F_\pi^2 + 2F_\zeta^2}} \leq \sqrt{3}m_\pi\,. \tag{19.10.12}$$

Thus a broken $U(1)_A$ symmetry would require a neutral pseudoscalar Goldstone boson with mass less than $\sqrt{3}m_\pi$, in addition to the pion itself. It hardly needs to be said that no such strongly-interacting particle exists.* We shall see in Section 23.5 that this problem was eventually solved by the discovery of non-perturbative effects that violate the extra $U(1)_A$ symmetry.

Problems

1. Apply the general theory of broken global symmetries to the case where an $SO(3)$ symmetry group (with generators t_1, t_2, t_3) is spontaneously broken to its $SO(2)$ subgroup (with generator t_3). How do the Goldstone boson fields transform under infinitesimal $SO(3)$ transformations? (Use the exponential parameterization of the coset space $SO(3)/SO(2)$.) Evaluate the covariant derivative $D_{a\mu}$ of the Goldstone boson field, and of a general field with a non-vanishing value q for the unbroken symmetry generator t_3. What is the most general $SO(3)$-invariant Lagrangian involving Goldstone bosons alone with no more than two derivatives? Use this Lagrangian to calculate the terms in the invariant amplitude for elastic scattering of

* Note that taking $F_\zeta \gg F_\pi$ would give the extra neutral scalar a very light mass, but its interactions would be so weak that it might have escaped detection. It seems unlikely that a very large ratio of F_π to F_ζ could arise in quantum chromodynamics, but something like this happens in theories with extra field variables that have been proposed to avoid parity violation by instantons; see Section 23.6.

the Goldstone bosons to lowest order in their energy. What is the most general $SO(3)$-invariant Lagrangian with two non-Goldstone field factors and at most one derivative? What is the most general function of the Goldstone boson field alone (with no derivatives) that transforms like the three-component of an $SO(3)$ three-vector? Add this term to the Lagrangian, with a coefficient chosen to give the Goldstone boson a mass m, and recalculate the lowest order amplitude for Goldstone boson scattering.

2. Consider a theory with an $SU(N)$ global symmetry, spontaneously broken to $SU(N-1)$. Suppose we add a small symmetry-breaking perturbation, belonging to the defining representation N of $SU(N)$. Taking vacuum alignment into account, what symmetry group is completely unbroken? What about the case where the symmetry-breaking perturbation belongs to the adjoint representation of $SU(N)$?

3. Calculate the pion–pion scattering amplitude to one-loop order, taking account of a finite pion mass.

4. Derive the 'Adler sum rule,' the analog for the case of pion–pion scattering of the Adler–Weisberger sum rule.

5. Calculate the $SU(2) \times SU(2)$ transformation properties of the pion field ξ_a in the case where the cosets of $SU(2) \times SU(2)/SU(2)$ are parameterized as $\exp(i\xi_a x_a)$.

6. Use Eq. (19.7.31) and $SU(3)$ symmetry to derive the relation (19.7.32).

References

1. S. Coleman, 'Secret symmetry: an introduction to spontaneous symmetry breakdown and gauge fields,' in *Aspects of Symmetry: Selected Erice Lectures of Sidney Coleman* (Cambridge University Press, Cambridge, 1985).

2. E. C. Titchmarsh, *Introduction to the Theory of Fourier Integrals* (Oxford University Press, Oxford, 1937): Section 1.4.

3. J. Goldstone. *Nuovo Cimento* **9**, 154 (1961).

4. Y. Nambu, *Phys. Rev. Lett.* **4**, 380 (1960).

5. J. Goldstone, A. Salam, and S. Weinberg, *Phys. Rev.* **127**, 965 (1962).

6. S. Adler, *Phys. Rev.* **137**, B1022 (1965).

7. S. Weinberg, *Phys. Rev. Lett.*, **29**, 1698 (1972).

8. R. Dashen, *Phys. Rev.* **183**, 1245 (1969). For corrections to this relation, see J. F. Donoghue, B. R. Holstein, and D. Wyler, *Phys. Rev.* **D47**, 2089 (1993); J. Bijnens, *Phys. Lett.* **B306**, 343 (1993); K. Maltman and D. Kotchan, *Mod. Phys. Lett.*, **A5**, 2457 (1990); R. Urech, *Nucl. Phys.* **B433**, 234 (1995); R. Baur and R. Urech, Zurich–Karlsruhe preprint ZU-TH 22/95, TTP95-31, hep-ph/9508393 (1995).

9. For a review, see I. S. Towner *et al.*, Chalk River preprint nucl-th/9507005 (1995).

10. S. Weinberg, *Phys. Rev.* **112**, 1375 (1958).

11. M. L. Goldberger and S. Treiman, *Phys. Rev.* **111**, 354 (1966).

12. T. Ericson and W. Weise, *Pions and Nuclei* (Clarendon Press, Oxford, 1988).

13. T. Ericson *et al.*, CERN preprint CERN-TH/95-50 (1995).

14. T. Das, G. S. Guralnik, V. S. Mathur, F. E. Low, and J. E. Young, *Phys. Rev. Lett.* **18**, 759 (1967).

15. J. Bernstein, S. Fubini, M. Gell-Mann, and W. Thirring, *Nuovo Cimento* **17**, 757 (1960); M. Gell-Mann and M. Lévy, Ref. 24; K-C. Chou, *Soviet Physics JETP* **12**, 492 (1961).

16. Y. Nambu, *Phys. Rev. Lett.* **4**, 380 (1960).

17. Y. Nambu and G. Jona-Lasinio, *Phys. Rev.* **122**, 345 (1961).

18. Y. Nambu and D. Lurie, *Phys. Rev.* **125**, 1429 (1962); Y. Nambu and E. Shrauner, *Phys. Rev.* **128**, 862 (1962).

19. S. L. Adler, *Phys. Rev. Lett.* **14**, 1051 (1965); *Phys. Rev.* **140**, B736 (1965); W. I. Weisberger, *Phys. Rev. Lett.* **14**, 1047 (1965); *Phys. Rev.* **143**, 1302 (1965). The effect of a finite pion mass was included in a different way in Ref. 27, by combining the current algebra results for the pion–nucleon scattering lengths with the sum rule for the scattering lengths derived from the forward scattering dispersion relations by M. L. Goldberger, H. Miyazawa, and R. Oehme, *Phys. Rev.* **99**, 986 (1955); M. L. Goldberger, in *Dispersion Relations and Elementary Particles* (Wiley, New York, 1960): p. 146.

20. S. Weinberg, 'Current Algebra – Rapporteur's Report' in *Proceedings of the International Conference on High-Energy Physics, Vienna, 1968* (CERN, Geneva, 1968): p. 253.

21. M. Gell-Mann, *Physics* **1**, 63 (1964).

22. S. Weinberg, *Phys. Rev. Lett.* **16**, 879 (1966).

23. S. Weinberg, *Phys. Rev. Lett.* **18**, 188 (1967).

24. M. Gell-Mann and M. Lévy, *Nuovo Cimento* **16**, 705 (1960).

25. S. Weinberg, *Phys. Rev.* **166**, 1568 (1968).

26. S. Weinberg, *Physica* **96A**, 327 (1979).

27. These scattering lengths were first calculated, using current algebra techniques, by S. Weinberg, *Phys. Rev. Lett.* **17**, 616 (1966).

28. For a review, see J. F. Donoghue, E. Golowich, and B. R. Holstein, *Dynamics of the Standard Model* (Cambridge University Press, Cambridge, 1992): Section VI-4. Measurements of the strong-interaction energy shift and width of the 1s state in $\pi^- p$ atoms have yielded results $a(\pi^- p \to \pi^- p) = \frac{2}{3}a_{1/2} + \frac{1}{3}a_{3/2} = 0.0885(9)m_\pi^{-1}$ and $a(\pi^- p \to \pi^0 n) = \frac{1}{3}\sqrt{2}(a_3 - a_1) = -0.136(10)m_\pi^{-1}$; see D. Sigg *et al.*, preprint ETHZ-IPP PR-95-4, July 1995, to be published in *Phys. Rev. Lett.*

29. S. Weinberg, *Phys. Lett.* **B251**, 288 (1990); *Nucl. Phys.* **B363**, 3 (1991); *Phys. Lett.* **B295**, 114 (1992). C. Ordóñez and U. van Kolck, *Phys. Lett.* **B291**, 459 (1992); C. Ordóñez, L. Ray, and U. van Kolck, *Phys. Rev. Lett.* **72**, 1982 (1994); U. van Kolck, *Phys. Rev.*, **C49**, 2932 (1994); U. van Kolck, J. Friar, and T. Goldman, to appear in *Phys. Lett. B*. This approach to nuclear forces is summarized in C. Ordóñez, L. Ray, and U. van Kolck, nucl-th/9511380, submitted to *Phys. Rev. C*; J. Friar, *Few-Body Systems Suppl.* **99**, 1 (1996). For application of these techniques to related nuclear processes, see T.-S. Park, D.-P. Min, and M. Rho, *Phys. Rep.* **233**, 341 (1993); nucl-th/9505017; S. R. Beane, C. Y. Lee, and U. van Kolck, *Phys. Rev.*, **C52**, 2915 (1995); T. Cohen, J. Friar, G. Miller, and U. van Kolck, nucl-th/9512036; D. B. Kaplan, M. Savage, and M. Wise, nucl-th/9605002.

30. S. Weinberg, Ref. 27. The pion–nucleon scattering lengths were independently calculated by Y. Tomozawa, *Nuovo Cimento* **46A**, 707 (1966).

30*a*. S. L. Adler, *Phys. Rev.* **140**, B736 (1965).

31. S. Coleman, J. Wess, and B. Zumino, *Phys. Rev.* **177**, 2239 (1969) ; C. G. Callan, S. Coleman, J. Wess, and B. Zumino, *Phys. Rev.* **177**, 2247 (1969).

32. R. S. Palais, *J. Math. Mech.* **6**, 673 (1957); G. D. Mostow, *Annals of Math.* **65**, 432 (1957).

32*a*. For a discussion of other possibilities, see S. Weinberg, *Physica Scripta* **21**, 773 (1980).

33. M. Gell-Mann, Cal. Tech. Synchotron Laboratory Report CTSL–20 (1961), unpublished. This is reproduced along with other articles on $SU(3)$ symmetry in M. Gell-Mann and Y. Ne'eman, *The Eightfold Way* (Benjamin, New York, 1964).

34. Y. Ne'eman, *Nucl. Phys.* **26**, 222 (1961).

35. M. Gell-Mann, R. J. Oakes, and B. Renner, *Phys. Rev.* **175**, 2195 (1968); S. Glashow and S. Weinberg, *Phys. Rev. Lett.* **20**, 224 (1968).

36. R. Dashen, *Phys. Rev.* **183**, 1245 (1969). Corrections were considered by P. Langacker and H. Pagels, *Phys. Rev.* **D8**, 4620 (1973). The absence of electromagnetic corrections to the π^0 mass had been proved in the model of ref. 17 by G. S. Guralnik, *Nuovo Cimento* **36**, 1002 (1965).

37. S. Weinberg, contribution to a *festschrift* for I. I. Rabi, *Trans. N. Y. Acad. Sci.* **38**, 185 (1977).

38. M. Gell-Mann, Ref. 33; S. Okubo, *Prog. Theor. Phys.*, **27**, 949 (1962).

38*a*. H. Leutwyler, Bern–CERN preprint CERN-TH/96-44, hep-ph/-9602366, to be published (1996).

38*b*. J. F. Donoghue, E. Golowich, and B. R. Holstein, Ref. 28.

39. J. Gasser and H. Leutwyler, *Nucl. Phys.* **B250**, 465 (1985). Also see J. Gasser and H. Leutwyler, *Ann. Phys.* **158**, 142 (1984).

40. For surveys, see H. Leutwyler, in *Proceedings of the XXVI International Conference on High Energy Nuclear Physics, Dallas, 1992*, ed. J. Sanford (American Institute of Physics, New York, 1993): 185; U. G. Meissner, *Rep. Prog. Phys.* **56**, 903 (1993); A. Pich, Valencia preprint FTUV/95-4, February 1995, to be published in *Reports on Progress in Physics*; J. Bijnens, G. Ecker, and J. Gasser, in *The Daphne Physics Handbook*, Vol. 1, eds. L. Maiani, G. Pancheri, and N. Paver (INFN, Frascati, 1995): Chapters 3 and 3.1; G. Ecker,

preprint hep-ph/9501357, to be published in *Progress in Particle and Nuclear Physics*, Vol. 35 (Pergamon Press, Oxford).

41. S. Coleman and S. Glashow, *Phys. Rev. Lett.* **6**, 423 (1961); S. Okubo, *Phys. Lett.* **4**, 14 (1963).
42. J. Wess and B. Zumino, *Phys. Lett.* **37B**, 95 (1971).
43. E. Witten, *Nucl. Phys.* **B223**, 422 (1983).
44. R. Bott and R. Seely, *Comm. Math. Phys.* **62**, 235 (1978).
45. E. D'Hoker and S. Weinberg, *Phys. Rev.* **D50**, R6050 (1994).
46. *Encyclopedic Dictionary of Mathematics*, eds. S. Iyanaga and Y. Kawada (MIT Press, Cambridge, 1980); W. Greub, S. Halperin, and R. Vanstone, *Connections, Curvature and Cohomology*, Vol III, (Academic Press, New York, 1976); A. Borel, *Ann. Math.* (2) **57**, 115 (1953); also in A. Borel, *Collected Papers*, Vol I (Springer Verlag, Berlin, 1983).
47. D. Weingarten, *Phys. Rev. Lett.* **51**, 1830 (1983).
48. J. Preskill and S. Weinberg, *Phys. Rev.* **D24**, 1059 (1981). This was a modified version of an earlier argument by 't Hooft, who had considered the limit of large quark masses; see G. 't Hooft, lecture given at the Cargèse Summer Institute, 1979, in *Recent Developments in Gauge Theories*, eds. G 't Hooft *et al.* (Plenum, New York, 1980), reprinted in *Dynamical Gauge Symmetry Breaking*, eds. E. Farhi and R. Jackiw (World Scientific, Singapore, 1982), and in G. 't Hooft, *Under the Spell of the Gauge Principle* (World Scientific, Singapore, 1994). Also see Section 22.5.
49. C. Vafa and E. Witten, *Nucl. Phys.* **B234**, 173 (1984). Also see C. Vafa and E. Witten, *Commun. Math. Phys.* **95**, 257 (1984).
50. C. Vafa and E. Witten, *Phys. Rev. Lett.* **53**, 535 (1984).
51. S. L. Glashow, in *Hadrons and Their Interactions*, ed. A. Zichichi (Academic Press, New York, 1968); S. L. Glashow, R. Jackiw, and S-S. Shei, *Phys. Rev.* **187**, 1916 (1969); M. Gell-Mann, in *Proc. Third Topical Conf. on Part. Phys.*, eds. W. A. Simonds and S. F. Tuan (Western Periodicals, Los Angeles, 1970); in *Elementary Particle Physics* (Springer-Verlag, Bonn, 1972); *Acta Phys. Austriaca Suppl.* **IX**, 1972 (1972); H. Fritzsch and M. Gell-Mann, in *Proceedings of the XVI International Conference on High Energy Physics*, eds. J. D. Jackson and A. Roberts (Fermi National Accelerator Laboratory, Batavia, IL, 1972); *Phys. Lett.* **47B**, 365 (1973).
52. S. Weinberg, *Phys. Rev.* **D11**, 3583 (1975).

20
Operator Product Expansions

We often find ourselves needing to know how an amplitude behaves when the four-momentum brought in by one operator and out by another goes to infinity, with all other external lines held at fixed four-momenta. For instance, we will see in Section 20.6 that the total cross section for the scattering of an electron by an initial hadron H, with arbitrary hadrons in the final state, is given by unitarity as a linear combination (with known coefficients) of the components of the amplitude

$$\int d^4x \, e^{-ik\cdot x} \, \langle H|J^\mu(x)J^\nu(0)|H\rangle \, ,$$

where k is the four-momentum transferred from the electron to the hadrons, and $J^\mu(x)$ is the electromagnetic current. In the case of deep inelastic electric scattering the momentum k that is carried in by one current operator and out by the other is allowed to go to infinity. Similarly, in studying the high momentum limit of various propagators and deriving corresponding spectral function sum rules in Section 20.5, we shall encounter the high momentum limit of similar Fourier transforms, but with the one-hadron state $|H\rangle$ replaced with the vacuum.

If an operator product such as $J^\mu(x)J^\nu(0)$ were analytic in x^μ, then its Fourier transform would decrease exponentially as the Fourier variable k goes to infinity. The leading terms in the high momentum limit of the Fourier transform arise from the singularities of the operator product as the spacetime arguments approach one another. The study of such operator products was initiated in 1969 by Wilson,[1] originally as an attempt to formulate a substitute for conventional quantum field theory. As has happened earlier (for instance with dispersion relations and Feynman's diagrammatic rules) the effort to bypass quantum field theory led to valuable general results, but results that can best be understood as general properties of quantum field theory.

The operator product expansion will be stated in Section 20.1. The standard proof of this expansion was given in perturbation theory in 1970 by Zimmerman.[2] In Section 20.1 we shall offer a non-perturbative and simpler though less rigorous derivation based on the path-integral formulation of field theory. Section 20.2 will present a different perspective on the operator product expansion, in terms of the flow of large momenta through Feynman graphs, which will lead us to a perturbative proof.

There are several aspects of the operator product expansion that make it particularly useful for drawing consequences from theories like quantum chromodynamics. One such property, discussed in Section 20.3, is that the functions describing the singularities in this expansion have a momentum dependence governed by renormalization group equations, so that in asymptotically free theories they can be calculated at large momenta using perturbation theory. Another aspect, shown in Section 20.4, is that these functions exhibit the full symmetry of the underlying theory, unaffected by possible spontaneous symmetry breaking. Applications will be considered in Sections 20.5 and 20.6.

20.1 The Expansion: Description and Derivation

Wilson[1] hypothesized that the singular part as $x \to y$ of the product $A(x)B(y)$ of two operators is given by a sum over other local operators

$$A(x)B(y) \to \sum_C F_C^{AB}(x-y)\, C(y)\,, \tag{20.1.1}$$

where $F_C^{AB}(x-y)$ are singular c-number functions. Dimensional analysis suggests that $F_C^{AB}(x-y)$ behaves for $x \to y$ like the power $d_C - d_A - d_B$ of $x-y$, where d_O is the dimensionality of the operator O in powers of mass or momentum. Since d_O increases as we add more fields or derivatives to an operator O, the strength of the singularity of $F_C^{AB}(x-y)$ decreases for operators C of increasing complexity. The remarkable thing about the operator product expansion is that it is an *operator* relation; that is, in applying it to any matrix element $\langle\beta|A(x)B(y)|\alpha\rangle$, we get the same functions $F_C^{AB}(x-y)$ for all states $|\alpha\rangle$ and $|\beta\rangle$.

It is the decrease of the singularity in Eq. (20.1.1) with operators $C(y)$ of increasing complexity that makes this expansion useful in drawing conclusions about the behavior of the product $A(x)B(y)$ for $x \to y$. The simple power-counting argument above is modified by renormalization effects; the expansion (20.1.1) must be formulated in terms of operators renormalized at some scale μ, and then μ appears along with $x-y$ in the coefficient function $F_C^{AB}(x-y)$. We shall see in Section 20.3 that in asymptotically free theories $F_C^{AB}(x-y)$ does behave like the power $d_C - d_A - d_B$ of $x-y$ suggested by dimensional analysis only up to a power of $\ln (x-y)^2$. Even in more general theories, it is plausible that the singularities associated with various operators $C(y)$ will decrease with the complexity of the operators.

The corresponding statement in momentum space is that for $k \to \infty$,

$$\int d^4x\, e^{-ik\cdot x} A(x)B(0) \to \sum_C V_C^{AB}(k)\, C(0) \tag{20.1.2}$$

and correspondingly

$$\int d^4x\, e^{-ik\cdot x} T\{A(x)B(0)\} \to \sum_C U_C^{AB}(k)\, C(0)\,, \tag{20.1.3}$$

where $V_C^{AB}(k)$ and $U_C^{AB}(k)$ are functions of k^μ that for large k decrease increasingly rapidly for more and more complicated terms in the series.

We are going to derive a generalized version of the Wilson expansion, in which the momenta carried by any number of operators go to infinity together. For this purpose, let us consider a Greens function for local operators $A_1(x_1)$, $A_2(x_2)$, etc. whose arguments approach a point x, as well as other local operators $B_1(y_1)$, $B_2(y_2)$, etc. with fixed arguments:

$$\begin{aligned}&\langle T\{A_1(x_1),\, A_2(x_2),\, \dots B_1(y_1),\, B_2(y_2),\, \dots\}\rangle_0\\ &= \int \left[\prod_{\ell, z} d\phi_\ell(z)\right]\, a_1(x_1)\, a_2(x_2) \cdots b_1(y_1)\, b_2(y_2) \cdots \exp(iI\,[\phi])\,,\end{aligned} \tag{20.1.4}$$

where the lower-case letters a and b indicate replacement of the field operators in the As and Bs with the c-number fields ϕ. Now surround the point x with a ball $B(R)$ of radius R which is much larger than the separations among the x_1, x_2, etc. but much smaller than the separations between x, y_1, y_2, etc. Since the action is local, it may be written as

$$I = \int_{z\in B(R)} d^4z\, \mathscr{L}(z) + \int_{z\notin B(R)} d^4z\, \mathscr{L}(z)\,. \tag{20.1.5}$$

Eq. (20.1.4) may then be put in the form

$$\begin{aligned}&\langle T\{A_1(x_1),\, A_2(x_2),\, \dots B_1(y_1),\, B_2(y_2),\, \dots\}\rangle_0\\ &= \int \left[\prod_{z\notin B(R),\, \ell} d\phi_\ell(z)\right]\, b_1(y_1)\, b_2(y_2) \dots \exp\left(i \int_{z\notin B(R)} d^4z\, \mathscr{L}(z)\right)\\ &\times \int \left[\prod_{z\in B(R),\, \ell} d\phi_\ell(z)\right]\, a_1(x_1)\, a_2(x_2) \dots \exp\left(i \int_{z\in B(R)} d^4z\, \mathscr{L}(z)\right)\,,\end{aligned} \tag{20.1.6}$$

in which the path integral over the fields inside the ball is constrained by the boundary condition that the fields merge smoothly at the ball's surface with the fields outside the ball. Aside from this boundary condition, the path integral over the fields inside the ball is completely unaffected by the behavior of the field outside the ball, so the integral over the fields inside the ball may be expressed in terms of the values and derivatives of the fields on the surface of the ball, which in turn may be expressed in terms of the fields and their derivatives extrapolated from outside the ball to the interior point x. If we express this integral as a series in products* $o(x)$ of the c-number fields and their derivatives extrapolated to x, then the coefficients can only be functions $U_O^{A_1,A_2,\ldots}(x_1 - x, x_2 - x, \ldots)$ of the coordinate differences. Since the points y_1, y_2, etc. are all far outside the ball $B(R)$, the exclusion of this ball in the action for the fields outside the ball has no effect in the limit $R \to 0$, so in this limit Eq. (20.1.6) becomes

$$\begin{aligned}\langle T\{A_1(x_1), A_2(x_2), \ldots B_1(y_1), B_2(y_2), \ldots\}\rangle_0 \to &\int \left[\prod_{\ell, z} d\phi_\ell(z)\right] \\ &\times b_1(y_1)\, b_2(y_2) \ldots \exp\left(i\int d^4z\,\mathscr{L}(z)\right) \\ &\times \sum_O U_O^{A_1,A_2,\ldots}(x_1 - x, x_2 - x, \ldots)\; o(x) \\ = \sum_O U_O^{A_1,A_2,\ldots}(x_1 - x, x_2 - x, \ldots)\,&\langle T\{O(x), B_1(y_1), B_2(y_2), \ldots\}\rangle_0 \end{aligned} \tag{20.1.7}$$

for x_1, x_2, etc. all approaching x, where $O(x)$ is the quantum-mechanical Heisenberg-picture operator corresponding to $o(x)$. In particular, by Fourier transforming with respect to the y variables and multiplying with appropriate coefficient functions, this yields

$$\langle\beta|T\{A_1(x_1), A_2(x_2), \ldots\}|\alpha\rangle \to \sum_O U_O^{A_1,A_2,\ldots}(x_1 - x, x_2 - x, \ldots)\,\langle\beta|O(x)|\alpha\rangle \tag{20.1.8}$$

for arbitrary states $|\alpha\rangle$ and $\langle\beta|$. Because this applies for arbitrary states, it is the operator product expansion in a generalized version:

$$T\{A_1(x_1), A_2(x_2), \ldots\} \to \sum_O U_O^{A_1,A_2,\ldots}(x_1 - x, x_2 - x, \ldots)\, O(x)\,. \tag{20.1.9}$$

20.2 Momentum Flow*

We shall now consider the simplest example of the operator product expansion: the asymptotic behavior of an $(n+2)$-point Feynman amplitude

* In order for the coefficients in this series to be finite, these products must be renormalized by multiplication with suitable infinite factors. This will be made clearer in the derivation presented in the next section.

* This section lies somewhat out of the book's main line of development, and may be omitted in a first reading.

in the theory of a single real scalar field $\phi(x)$ with mass m and interaction Lagrangian density $-\frac{1}{24}g\phi^4$, when a large four-momentum k flows in one line and out another, with all other external lines held at fixed momentum. This will lead us to a perturbative proof of the leading term in the operator product expansion in this case, but our real purpose here will be to gain some insight into the way that the flow of large momenta through Feynman diagrams leads to this expansion. An appendix to this chapter will discuss the extension of these results to the general case.

Let us define $\Gamma(k;p_1\cdots p_n)$ as the sum of all connected graphs for the n-particle scattering amplitude, whose external lines carry incoming momenta k, $p-k$, and outgoing momenta $p_1\cdots p_n$, where $p\equiv p_1+p_2+\cdots+p_n$. (It will be convenient to specify further that $\Gamma(k;p_1\cdots p_n)$ includes propagators for the external lines with momenta k and $p-k$, but not for those with the fixed momenta $p_1\cdots p_n$.) We wish to show that in any finite order of perturbation theory, for $k\to\infty$,

$$\Gamma(k;p_1\cdots p_n)\to U_{\phi^2}(k)F_{\phi^2}(p_1\cdots p_n)+O(k^{-5})\,, \tag{20.2.1}$$

where $U_{\phi^2}(k)$ is a sum of terms of order** k^{-4}, that is independent of $p_1\cdots p_n$ and of n, and $F_{\phi^2}(p_1\cdots p_n)$ is the amplitude for n ϕ lines with insertion of a single ϕ^2 vertex, times a suitable renormalization constant Z_{ϕ^2} to make it finite. Because $F_{\phi^2}(p_1\cdots p_n)$ is a matrix element of the renormalized operator $(\phi^2)_R\equiv Z_{\phi^2}\phi^2(0)$, Eq. (20.2.1) corresponds to the statement that the leading term in the operator product expansion for $k\to\infty$ is†

$$\int d^4x\, e^{-ik\cdot x}\,T\{\phi_R(x)\phi_R(0)\}_C\to U_{\phi^2}(k)\,(\phi^2(0))_R\,. \tag{20.2.2}$$

In assessing the asymptotic behavior of Feynman amplitudes, we must take account of the fact that there are parts of the range of integration in momentum space where some of the internal lines carry momenta of the same order as the external line momenta that are going to infinity, while other internal lines do not. The contribution to $\Gamma(k;p_1\cdots p_n)$ from the part of the region of integration where the lines in some subgraph $\mathscr{S}$ have momenta of order k has an asymptotic behavior of order $k^{D_{\mathscr{S}}}$, where $D_{\mathscr{S}}$

** Throughout this chapter, whenever it is said that an amplitude is of order k^A, it should be understood that for $k^\rho=\kappa n^\rho$ where $\kappa\to\infty$ with n^ρ a fixed generic four-vector, the amplitude approaches κ^A times a sum of powers of $\ln\kappa$.

† The subscript C on the time-ordered product here indicates that we are including only connected graphs. We have dropped the superscript $\phi\phi$ on V_{ϕ^2}, because in this section we are concerned only with the operator product expansion for two ϕ fields. These fields are themselves renormalized, because we implicitly include counterterms along with radiative corrections for the propagators of the lines carrying momenta k and $p-k$.

is the dimensionality of the subgraph $\mathscr{S}$, calculated according to the rules of Section 12.1. If $\mathscr{S}$ has m external lines that connect it to the rest of the graph as well as the two external lines with momenta k and $p-k$, then from Eq. (12.1.8) we have $D_{\mathscr{S}} = 4-2-m-4 = -2-m$. (The term -4 arises from the two propagators of the lines with momenta k and $p-k$, which we have specified are to be included in $\Gamma(k;p_1\cdots p_n)$.) *Thus the asymptotic behavior of* Γ *is dominated by the part of the momentum-space integral where the large momentum k flows either through the whole graph or through some subgraph, whichever has the smallest number of external lines.*[3]

For $n=0$ this is always the whole graph; that is, the dominant part of the integral comes from the part of the region of integration where every line carries a momentum of order k, giving an asymptotic behavior of order k^{-2}. In this case the only operator in the operator product expansion contributing to this matrix element is the unit operator, $C=1$. This term is excluded here for $n>0$ because for the present we are limiting ourselves to connected graphs.

For $n=2$ the dominant contribution comes both from the whole graph, *and* from subgraphs in which the two external lines with momenta k and $p-k$ are connected to the other two external lines by a bridge consisting of two internal lines,[††] giving an asymptotic behavior of order k^{-4}. For $n \geq 4$ the dominant contribution comes *only* from subgraphs in which the two external lines with momenta k and $p-k$ are connected to the $n>2$ other external lines by a bridge consisting of two internal lines, again giving an asymptotic behavior of order k^{-4}.

The analysis of the cases $n=2$ and $n \geq 4$ is complicated by the fact that a general graph may contain several of these two-particle bridges. Let us first consider the case $n=2$. We define $I(k,k',p)$ as the sum of all graphs contributing to $\Gamma(k;p_1,p_2)$ (with $p_1 = k'$, $p_2 = p-k'$) that are two-particle-irreducible, in the sense that the two external lines with incoming momenta k and $p-k$ cannot be disconnected from the two external lines with momenta k' and $p-k'$ by cutting through any pair of internal lines. Then $\Gamma(k;k',p-k') - I(k,k',p)$ consists of graphs that can be disconnected in this way, and may therefore be written (see Figure 20.1):

$$\Gamma(k;k',p-k') - I(k,k',p) = \int d^4k''\, I(k,k'',p)\Gamma(k'';k',p-k') \,. \quad (20.2.3)$$

(Like $\Gamma(k;k',p-k')$, the kernel $I(k,k',p)$ includes propagators for the lines

---

[††] The possibility $m=1$ is excluded because the symmetry of this theory under $\phi \to -\phi$ rules out graphs or subgraphs with odd numbers of external lines. The possibility $m=0$ is excluded because Γ is defined to arise only from connected graphs.

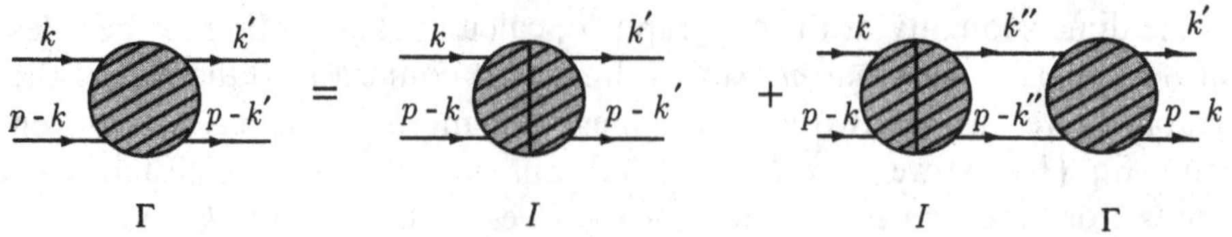

Figure 20.1. Diagrammatic representation of the integral equation (20.2.3). The cross-hatched disks marked Γ represent the sum of all connected Feynman diagrams with the indicated external lines, while the cross-hatched disks marked I, which are divided by a vertical line, represent the sum of all connected diagrams for which the lines on the left cannot be separated from those on the right by cutting through any pair of external lines.

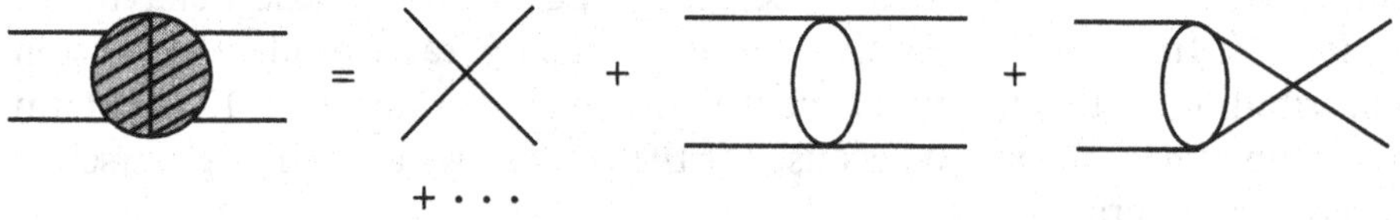

Figure 20.2. Tree and one-loop graphs for the kernel $I(k,\ k',\ p)$ in the theory of a scalar field with interaction Φ^4.

with momenta k and $p - k$, but, to avoid double counting, not for the lines with momenta k' and $p - k'$.) The Feynman diagrams for $I(k,k',p)$ to order g^2 are shown in Figure 20.2.

To evaluate the behavior of the right-hand side of Eq. (20.2.3), let's first consider the asymptotic behavior of the kernel $I(k,k',p)$ when $k \to \infty$ with k' and p fixed. This is dominated by the region of momentum space in which all internal lines carry momenta of order k, which makes a contribution of order k^{-4}, because in any other region the subgraph consisting of internal lines that carry momenta of order k would have more than four external lines, and hence would make a contribution that decreases faster than k^{-4}. It follows that differentiating $I(k,k',p)$ with respect to k' or p would reduce the asymptotic behavior of this kernel by a factor of k^{-1}. Hence for $k \to \infty$ with k' and p fixed, we have

$$I(k,k',p) \to I_\infty(k)\,, \tag{20.2.4}$$

where $I_\infty(k)$ is a function only of k, of order k^{-4}.

Unfortunately, we cannot simply replace $I(k,k',p)$ in Eq. (20.2.3) with this asymptotic limit, because however large k may become, the integral will receive a large contribution from values of k' of order k. To deal with this complication, we shall now employ a trick, based on mathematical induction. In lowest order $\Gamma(k;k',p-k')$ is given by a single vertex with two attached bare propagators

$$\Gamma(k;k',p-k') = \frac{ig}{(2\pi)^4(k^2+m^2)((p-k)^2+m^2)} \qquad \text{(lowest order)}\,,$$

for which it is easy to verify an asymptotic behavior of form Eq. (20.2.1). Let us therefore assume that Eq. (20.2.1) for $n=2$ holds up to some given order N in g — that is, that up to this order, the asymptotic behavior for $k \to \infty$ takes the form

$$\Gamma(k;k',p-k) \to U_{\phi^2}(k)F_{\phi^2}(k',p-k') + O(k^{-5})\,, \tag{20.2.5}$$

and try to verify this behavior to the next order. In order to eliminate the contribution to the integral in Eq. (20.2.3) from values of k' of order k, we rewrite Eq. (20.2.3) as

$$\begin{aligned}\Gamma(k;k',p-k) &= I(k,k',p)\\ &+\int d^4k''\, I(k,k'',p)\Big[\Gamma(k'';k',p-k') - U_{\phi^2}(k'')F_{\phi^2}(k',p-k')\Big]\\ &+F_{\phi^2}(k',p-k')\int d^4k''\, I(k,k'',p)U_{\phi^2}(k'')\,.\end{aligned} \tag{20.2.6}$$

Since $I(k,k',p)$ is at least of first order, we may use Eq. (20.2.5) in the right-hand side of Eq. (20.2.6). Hence in the second term on the right-hand side, the part of the region of integration where k'' is of order k gives a contribution that vanishes like k^{-4+4-5}, and may therefore be neglected compared with the part where k'' remains finite, which yields the convergent integral

$$I_\infty(k)\int d^4k''\,\Big[\Gamma(k'';k',p-k') - U_{\phi^2}(k'')F_{\phi^2}(k',p-k')\Big]\,.$$

Further, since the dominant part of $\int d^4k''\, I(k,k'',p)U_{\phi^2}(k'')$ comes from the region of integration where every internal line of the graphs for $I(k,k'',p)$ carries a momentum of order k, differentiating this integral with respect to p would lower its asymptotic behavior by a factor of order k^{-1}, so asymptotically $I(k,k',p)$ may be replaced in this integral with $I(k,k') \equiv I(k,k',0)$. Therefore for $k \to \infty$, Eq. (20.2.6) becomes

$$\begin{aligned}\Gamma(k;k',p-k') &\to F_{\phi^2}(k',p-k')\int d^4k''\, I(k,k'')U_{\phi^2}(k'')\\ &+I_\infty(k)\left\{1+\int d^4k''\,\Big[\Gamma(k'';k',p-k') - U_{\phi^2}(k'')F_{\phi^2}(k',p-k')\Big]\right\}.\end{aligned} \tag{20.2.7}$$

Let us therefore define $U_{\phi^2}(k)$ and $F_{\phi^2}(k',p-k')$ to order $N+1$ in g in terms of these functions in lower orders of perturbation theory, by

$$U_{\phi^2}(k) = CI_\infty(k) + \int d^4k'\, I(k,k')U_{\phi^2}(k')\,, \tag{20.2.8}$$

$$F_{\phi^2}(k',p-k') = C^{-1}$$
$$\times\left\{1+\int d^4k''\left[\Gamma(k'';k',p-k') - U_{\phi^2}(k'')F_{\phi^2}(k',p-k')\right]\right\}, \tag{20.2.9}$$

where C is a constant that can be chosen as we like. With these definitions, Eq. (20.2.5) follows from Eq. (20.2.7).

It will be convenient to choose the constant C so that $F^2_\phi(k',p-k')$ has the value unity at some renormalization point $k' = k(\mu)$ and $p = p(\mu)$, where $k(\mu)$ and $p(\mu)$ are standard four-momenta of order μ. Then

$$C = 1 + \int d^4k''\Gamma(k'';k(\mu),p(\mu)-k(\mu)) - \int d^4k''\, U_{\phi^2}(k'')\,. \tag{20.2.10}$$

Using Eq. (20.2.5), we see that the divergences in the two integrals in Eq. (20.2.10) cancel.

Eq. (20.2.9) may now be written

$$F_{\phi^2}(k',p-k') = Z_{\phi^2}\left\{1+\int d^4k''\,\Gamma(k'';k',p-k')\right\}, \tag{20.2.11}$$

where

$$Z_{\phi^2} = \left[1+\int d^4k''\,\Gamma(k'';k(\mu),p(\mu)-k(\mu))\right]^{-1}. \tag{20.2.12}$$

We can think of Z_{ϕ^2} as the renormalization constant for the composite operator ϕ^2, defined in such a way that the operator $Z_{\phi^2}\phi^2$ has the finite two-particle matrix element $F_{\phi^2}(k,p-k)$, with the value unity for $k = k(\mu)$ and $p = p(\mu)$.

It is not particularly convenient to calculate $U_{\phi^2}(k)$ or $F_{\phi^2}(k,p-k)$ by using Eqs. (20.2.8) and (20.2.11). Rather, it is simpler to calculate $\Gamma(k;k',p-k')$, and read off $U_{\phi^2}(k)$ or $F_{\phi^2}(k,p-k)$ by comparison with Eq. (20.2.5). By multiplying the function (12.2.26) with the product of propagators for the lines with momenta k and $p-k$, we see that, to one-loop order

$$\Gamma(k;k',p-k') = \left[\frac{-i}{(2\pi)^4(k^2+m^2)}\right]\left[\frac{-i}{(2\pi)^4((p-k)^2+m^2)}\right]\left[-i(2\pi)^4 g\right]$$
$$\times\left\{1-\frac{g}{32\pi^2}\int_0^1 dx\left\{\ln\left(\frac{m^2+4x(1-x)\mu^2/3}{m^2-sx(1-x)}\right)\right.\right.$$
$$\left.\left.+\ln\left(\frac{m^2+4x(1-x)\mu^2/3}{m^2-tx(1-x)}\right)+\ln\left(\frac{m^2+4x(1-x)\mu^2/3}{m^2-ux(1-x)}\right)+\cdots\right\}\right\}, \tag{20.2.13}$$

where s, t, and u are the Mandelstam variables

$$s = -p^2, \qquad t = -(k - k')^2, \qquad u = -(p - k - k')^2,$$

μ is the renormalization scale and g the corresponding renormalized coupling, defined as the Feynman amplitude at $s = t = u = -4\mu^2/3$. This has the asymptotic behavior

$$\Gamma(k; k', p - k') \to \frac{ig}{(2\pi)^4(k^2)^2}$$
$$\times \left\{1 - \frac{g}{32\pi^2} \int_0^1 dx \left\{ \ln\left(\frac{m^2 + 4x(1-x)\mu^2/3}{m^2 + p^2x(1-x)}\right)\right.\right.$$
$$\left.\left. +2\ln\left(\frac{m^2 + 4x(1-x)\mu^2/3}{m^2 + k^2x(1-x)}\right) + \cdots\right\}\right. . \tag{20.2.14}$$

To order g^2, this agrees with Eq. (20.2.5) if we take

$$U_{\phi^2}(k) = \frac{ig}{(2\pi)^4(k^2)^2}$$
$$\times \left\{1 - \frac{g}{16\pi^2} \int_0^1 dx \ln\left(\frac{m^2 + 4x(1-x)\mu^2/3}{m^2 + k^2x(1-x)}\right) + \cdots\right\}, \tag{20.2.15}$$

and

$$F_{\phi^2}(k, p-k) = 1 - \frac{g}{32\pi^2} \int_0^1 dx \ln\left(\frac{m^2 + 4x(1-x)\mu^2/3}{m^2 + p^2x(1-x)}\right) + \cdots . \tag{20.2.16}$$

We have here chosen the renormalization point $k(\mu), p(\mu)$ for the operator ϕ^2 to be related to that for the coupling constant g in such a way that $p(\mu)^2 = 4\mu^2/3$, so that $F_{\phi^2}(k, p-k) = 1$ when $p^2 = 4\mu^2/3$.

Now consider the case where the number n of external lines with fixed momenta is greater than two. In accordance with our earlier discussion, in the limit $k \to \infty$, the leading graphs for $\Gamma(k; p_1 \cdots p_n)$ are those in which the two external lines carrying momenta k and $p - k$ can be disconnected from the n external lines with fixed momenta by cutting through a pair of internal lines:

$$\Gamma(k; p_1 \cdots p_n) \to \int d^4k'\, I(k, k', p)\, \Gamma(k'; p_1 \cdots p_n) . \tag{20.2.17}$$

As before, we cannot simply use the asymptotic limit of the kernel $I(k, k', p)$ for $k \to \infty$ on the right-hand side, because this integral receives important contributions from k' of order k. We deal with this complication by rewriting Eq. (20.2.17) in the form

$$\Gamma(k;p_1\cdots p_n) \to \int d^4k'\, I(k,k',p)\, \Big[\Gamma(k';p_1\cdots p_n) - U_{\phi^2}(k')F_{\phi^2}(p_1\cdots p_n)\Big]$$
$$+ F_{\phi^2}(p_1\cdots p_n)\int d^4k'\, I(k',k,p)\, U_{\phi^2}(k')\,, \qquad (20.2.18)$$

where by mathematical induction, we suppose that up to some given order N

$$\Gamma(k;p_1\cdots p_n) \to U_{\phi^2}(k)F_{\phi^2}(p_1\cdots p_n)\,, \qquad (20.2.19)$$

with correction terms of order $1/k^5$. We can now use the limit (20.2.4) together with Eq. (20.2.8) to rewrite Eq. (20.2.18) as

$$\Gamma(k;p_1\cdots p_n) \to I_\infty(k)\int d^4k'\, \Big[\Gamma(k';p_s\cdots p_n) - U_{\phi^2}(k')F_{\phi^2}(p_1\cdots p_n)\Big]$$
$$+ F_{\phi^2}(p_1\cdots p_n)\Big[U_{\phi^2}(k) - CI_\infty(k)\Big]\,.$$

This agrees with Eq. (20.2.19) to order $N+1$, provided we take

$$CF_{\phi^2}(p_1\cdots p_n) = \int d^4k'\, \Big[\Gamma(k';p_s\cdots p_n) - U_{\phi^2}(k')F_{\phi^2}(p_1\cdots p_n)\Big]$$

or, using Eqs. (20.2.10) and (20.2.12),

$$F_{\phi^2}(p_1\cdots p_n) = Z_{\phi^2}\int d^4k'\, \Gamma(k';p_1\cdots p_n)\,. \qquad (20.2.20)$$

This just says that $F_{\phi^2}(p_1\cdots p_n)$ is the matrix element of the renormalized operator $Z_{\phi^2}\phi^2$, so that Eq. (20.2.1) corresponds to the operator product formula (20.2.2). Note in particular that $U_{\phi^2}(k)$ is the same coefficient function whatever value n or the momenta $p_1\cdots p_n$ may take, as was to be shown.

Strictly speaking, the $(\phi^2)_R$ operator is not the leading term in the expansion of the product of two ϕs. There is also the operator unity, which has lower dimensionality than $(\phi^2)_R$, but was excluded from Eq. (20.2.2) because (as indicated by the subscript C) we are excluding disconnected graphs. As mentioned earlier, the graphs for $\Gamma(k)$ with $n=0$ (and $p^\mu=0$) are dominated for $k\to\infty$ by the region of integration where all internal lines carry momenta of order k, and $U_1(k)$ is the contribution of this region.

Higher-dimensional operators and more general theories will be considered in an appendix to this chapter.

20.3 Renormalization Group Equations for Coefficient Functions

As mentioned earlier, one of the aspects of the operator product expansion that makes it so useful is that the momentum dependence of the coefficient functions is governed by renormalization group equations. This is because they arise from limiting values of sums of Feynman graphs (such as $I_\infty(k)$ in Section 20.2) in which all relevant momenta are going to infinity together, so that masses may be set equal to zero without introducing singularities. Nevertheless the operator product coefficient functions do not obey simple scaling laws, because of renormalization effects: the functions are multiplied with scale-dependent renormalization constants, and depend on scale-dependent renormalized couplings.

Consider the operator product expansion for a Greens function $\Gamma_{\ell\ell'}(k,k',p)$ in which the incoming momenta (collectively labelled k, with sum p) of a set ℓ of lines all go to infinity together, with the set ℓ' of remaining lines having fixed outgoing momenta (collectively labelled k', with sum p):

$$\Gamma_{\ell\ell'}(k,k',p) \to \sum_{\mathcal{O}} U^\ell_{\mathcal{O}}(k) F_{\mathcal{O},\ell'}(k',p)\,. \tag{20.3.1}$$

The function $F_{\mathcal{O},\ell'}(k',p)$ is the matrix element of a renormalized operator $\mathcal{O}_R = \sum_{\mathcal{O}'} Z_{\mathcal{O},\mathcal{O}'}\mathcal{O}'$, so its coefficient $U^\ell_{\mathcal{O}}(k)$ in the product of fields corresponding to the lines ℓ is proportional to $Z^{-1}_{\mathcal{O},\mathcal{O}'}$, as well as to $Z_{\ell',\ell}$, the direct product of all the renormalization matrix factors for the field (or composite) operators in the set ℓ. Hence

$$\mu\frac{d}{d\mu}U^\ell_{\mathcal{O}} = \sum_{\ell'}\gamma_{\ell\ell'}U^{\ell'}_{\mathcal{O}} - \sum_{\mathcal{O}'}U^\ell_{\mathcal{O}'}\gamma_{\mathcal{O}',\mathcal{O}} + \beta(g)\frac{\partial}{\partial g}U^\ell_{\mathcal{O}}\,, \tag{20.3.2}$$

where

$$\mu\frac{\partial}{\partial\mu}Z_{\ell\ell'} = \sum_{\ell''}\gamma_{\ell\ell''}Z_{\ell''\ell'}\,, \qquad \mu\frac{\partial}{\partial\mu}Z_{\mathcal{O}\mathcal{O}'} = \sum_{\mathcal{O}''}\gamma_{\mathcal{O}\mathcal{O}''}Z_{\mathcal{O}''\mathcal{O}'}\,, \tag{20.3.3}$$

and for simplicity we assume a single renormalizable coupling g_μ, defined as the value of some Feynman amplitude at a renormalization point with momenta of order μ, with $\mu\, dg_\mu/d\mu = \beta(g_\mu)$. In order to be able to use dimensional analysis, we multiply all operators by powers of μ in such a way that they become dimensionless. This has the effect that the components of the Z and γ matrices are also dimensionless, with the values in the limit of zero coupling given by

$$\gamma_{\ell\ell'} \to \delta_{\ell\ell'}N(\ell)\,, \qquad \gamma_{\mathcal{O}\mathcal{O}'} \to \delta_{\mathcal{O}\mathcal{O}'}N(\mathcal{O})\,, \tag{20.3.4}$$

where $N(\mathcal{O})$ is the dimensionality of the operator $\mathcal{O}$, and $N(\ell)$ is the total dimensionality of the set ℓ of fields (the sum of $s+1$ for each field, where

$s = 0$ for scalar and massless gauge fields, $s = 1/2$ for Dirac fields, etc.) Also, dimensional analysis tells us that for $k^\mu = \kappa n^\mu$ with n^μ fixed, the amplitude can depend on κ only through the ratio κ/μ, aside from a factor $\kappa^{4-4n(\ell)}$ arising from the integrals used in defining the Fourier transform. The solution of Eq. (20.3.2) is then of the form

$$U^\ell_\mathscr{O}(\kappa n) = \kappa^{4-4n(\ell)} \sum_{\ell',\mathscr{O}'} \left[M\left\{ \exp\left(\int^\kappa \frac{d\mu}{\mu}\gamma(g_\mu)\right)\right\}\right]_{\ell,\ell'} \mathscr{U}^{\ell'}_{\mathscr{O}'}(g_\kappa, n)$$
$$\times \left[M\left\{ \exp\left(\int^\kappa \frac{d\mu}{\mu}\gamma(g_\mu)\right)\right\}\right]^{-1}_{\mathscr{O}',\mathscr{O}} , \tag{20.3.5}$$

where M denotes the 'μ-ordered' product, that is, each term in the expansion of the exponentials rearranged so that the factors are in the order of decreasing μ from left to right.

This gives especially simple results where g_μ approaches a fixed point g_* for $\mu \to \infty$. The contribution of large μs to $M\{\exp(\int^\kappa \gamma(g_\mu)d\mu/\mu)\}$ is then a matrix factor $\kappa^{\gamma(g_*)}$ that because of the μ-ordering appears on the *left*. Thus Eq. (20.3.5) becomes

$$U^\ell_\mathscr{O}(\kappa n) = \kappa^{4-4n(\ell)} \sum_{\ell',\mathscr{O}'} \left[\kappa^{\gamma(g_*)}\right]_{\ell,\ell'} \mathscr{C}_{\ell',\mathscr{O}'} \left[\kappa^{-\gamma(g_*)}\right]_{\mathscr{O}',\mathscr{O}} , \tag{20.3.6}$$

where $\mathscr{C}$ is either a constant or a sum of powers of $\ln\kappa$, depending on the rate at which g_κ approaches g_*.

Asymptotically free theories like quantum chromodynamics are a special case of particular physical interest. Here the fixed point is at $g_* = 0$, and, according to Eq. (20.3.4), near this fixed point the γ matrices go as

$$\gamma(g)_{\ell\ell'} \to N(\ell)\delta_{\ell\ell'} + g^2 c_{\ell\ell'} , \quad \gamma(g)_{\mathscr{O}\mathscr{O}'} \to N(\mathscr{O})\delta_{\mathscr{O}\mathscr{O}'} + g^2 c_{\mathscr{O}\mathscr{O}'} . \tag{20.3.7}$$

Also, if we write the renormalization group equation for the coupling in the form

$$\mu\frac{d}{d\mu}g_\mu^2 = -\frac{b}{8\pi^2}g_\mu^4 , \tag{20.3.8}$$

then

$$\int^\kappa \frac{d\mu}{\mu}g_\mu^2 \to -\frac{8\pi^2}{b}\ln g_\kappa^2 + \text{constant} .$$

Using this in Eq. (20.3.5) yields the asymptotic behavior

$$U^\ell_\mathscr{O}(\kappa n) \to \kappa^{4-4n(\ell)+N(\ell)-N(\mathscr{O})} \sum_{\ell',\mathscr{O}'} \left[\left(g_\kappa^2\right)^{-8\pi^2 c/b}\right]_{\ell,\ell'} \mathscr{C}^{\ell'}_{\mathscr{O}'} \left[\left(g_\kappa^2\right)^{8\pi^2 c/b}\right]_{\mathscr{O}',\mathscr{O}} , \tag{20.3.9}$$

where $\mathscr{C}^{\ell'}_{\mathscr{O}'}$ is a constant matrix, equal to $\mathscr{U}^{\ell'}_{\mathscr{O}'}(0, n)$ times constant factors that are not calculable in perturbation theory because they come from the parts of the integrals in Eq. (20.3.5) where g_μ is not small. For $\kappa \to \infty$, the behavior of the coupling constant is $g_\kappa^2 \to 8\pi^2/b \ln \kappa$, so Eq. (20.3.9) may also be written

$$U_{\mathscr{O}}^{\ell}(\kappa n) \to \kappa^{4-4n(\ell)+N(\ell)-N(\mathscr{O})} \sum_{\ell', \mathscr{O}'} \left[\left(\ln \kappa \right)^{8\pi^2 c/b} \right]_{\ell, \ell'} \mathscr{B}_{\mathscr{O}'}^{\ell'} \left[\left(\ln \kappa \right)^{-8\pi^2 c/b} \right]_{\mathscr{O}', \mathscr{O}} , \tag{20.3.10}$$

where $\mathscr{B}_{\mathscr{O}'}^{\ell'}$ is another constant matrix. The condition for Eqs. (20.3.9) and (20.3.10) to be valid is that $g_\kappa^2/8\pi^2 \ll 1$, but it is not necessary for $\ln \kappa$ to be so large that only one eigenvector of the c matrix contributes to the asymptotic behavior. Eq. (20.3.10) will be used to study deep inelastic scattering in Section 20.6.

20.4 Symmetry Properties of Coefficient Functions

The usefulness of the operator product expansion is greatly enhanced by the fact that the coefficient functions exhibit the full symmetry of the underlying theory, even where part or all of that symmetry is spontaneously broken.[4] To prove this, we consider the operator product expansion for a product of renormalized operators $\mathscr{O}_i(x)$ that transform linearly under some symmetry with conserved current $J^\mu(x)$, in the sense that

$$[J^0(\mathbf{x}, t), \mathscr{O}_i(\mathbf{y}, t)] = -\delta^3(\mathbf{x} - \mathbf{y}) \sum_j t_{ij} \mathscr{O}_j(\mathbf{y}, t) , \tag{20.4.1}$$

where t_{ij} is a constant matrix. We can write the operator product expansion as the statement that, as $x_1, \cdots x_n$ approach x together (with $x_1 - x, \cdots x_n - x$ all having fixed ratios)

$$\langle \beta | T\{\mathscr{O}_{i_1}(x_1) \cdots \mathscr{O}_{i_n}(x_n)\} | \alpha \rangle \to \sum_i U_i^{i_1 \cdots i_n}(x_1 - x, \cdots x_n - x) \langle \beta | \mathscr{O}_i(x) | \alpha \rangle . \tag{20.4.2}$$

Now suppose that the symmetry with current J^μ is spontaneously broken, with the corresponding Goldstone boson π satisfying

$$\langle \text{VAC} | J^\mu(0) | \pi \rangle = \frac{F p_\pi^\mu}{(2\pi)^{3/2} \sqrt{2 p_\pi^0}} . \tag{20.4.3}$$

Then, as we saw in Section 19.2, the matrix element of the operator product between states with an additional low energy Goldstone boson is

$$\langle \beta | T\{\mathscr{O}_{i_1}(x_1) \cdots \mathscr{O}_{i_n}(x_n)\} | \pi \alpha \rangle = \frac{1}{(2\pi)^{3/2} \sqrt{2 p_\pi^0} F} \times \int d^4x \frac{\partial}{\partial x^\mu} \langle \beta | T\{\mathscr{O}_{i_1}(x_1) \cdots \mathscr{O}_{i_n}(x_n) J^\mu(x)\} | \alpha \rangle , \tag{20.4.4}$$

because only the Goldstone pole term survives in the integral on the right-hand side of Eq. (20.4.4). Using Eq. (20.4.1) and the conservation of the current, this can be put in the form

$$\langle\beta|T\{\mathcal{O}_{i_1}(x_1)\cdots\mathcal{O}_{i_n}(x_n)\}|\pi\alpha\rangle = -\frac{1}{(2\pi)^{3/2}\sqrt{2p_\pi^0}F}$$

$$\times\sum_{r=1}^{n}\sum_{j_r} t_{i_r j_r}\langle\beta|T\{\mathcal{O}_{i_1}(x_1)\cdots\mathcal{O}_{j_r}(x_r)\cdots\mathcal{O}_{i_n}(x_n)\}|\alpha\rangle\,. \quad (20.4.5)$$

Now apply the operator product expansion to both sides of this formula. In the limit as $x_1\cdots x_n$ all approach x together, we find

$$\sum_i U_i^{i_1\cdots i_n}(x_1-x,\cdots x_n-x)\langle\beta|\mathcal{O}_i(x)|\pi\alpha\rangle = -\frac{1}{(2\pi)^{3/2}\sqrt{2p_\pi^0}F}$$

$$\times\sum_{r=1}^{n}\sum_{j_r}\sum_i t_{i_r j_r}\,U_i^{i_1\cdots j_r\cdots i_n}(x_1-x,\cdots x_n-x)\langle\beta|\mathcal{O}_i(x)|\alpha\rangle\,. \quad (20.4.6)$$

But as a special case of Eq. (20.4.5), we have

$$\langle\beta|\mathcal{O}_i(x)|\pi\alpha\rangle = -\frac{1}{(2\pi)^{3/2}\sqrt{2p_\pi^0}F}\sum_j t_{ij}\langle\beta|\mathcal{O}_j(x)|\alpha\rangle\,. \quad (20.4.7)$$

Since all this holds for arbitrary states $\langle\beta|$ and $|\alpha\rangle$, the coefficients of $\langle\beta|\mathcal{O}_j(x)|\alpha\rangle$ on both sides of Eq. (20.4.6) must be equal, so

$$0 = -\sum_i t_{ij}U_i^{i_1\cdots i_n}(x_1-x,\cdots x_n-x)+\sum_{r=1}^{n}\sum_{j_r} t_{i_r j_r}\,U_j^{i_1\cdots j_r\cdots i_n}(x_1-x,\cdots x_n-x)\,. \quad (20.4.8)$$

This can be restated as the condition that $U_i^{i_1\cdots i_n}(x_1-x,\cdots x_n-x)$ is invariant under the symmetry generated by t, with the action of this symmetry on the lower index contragredient to its action on the upper indices, in the sense that the matrix t is replaced with $-t^{\mathrm{T}}$. This is the same relation that would be expected if the symmetry generated by J^μ were not spontaneously broken.

20.5 Spectral Function Sum Rules

Spectral function sum rules are constraints on the spectral functions of various currents.[5] We will start here with a set of currents J_a^μ that are arbitrary, except for being Lorentz four-vectors, and then later consider more special examples. To define their spectral functions, we use Lorentz invariance to write

$$\sum_N \delta^4(p-p_N)\langle \text{VAC}|J^\mu_\alpha(0)|N\rangle\langle \text{VAC}|J^\nu_\beta(0)|N\rangle^* = (2\pi)^{-3}\theta(p^0)$$
$$\times \left[\left(\eta^{\mu\nu} - p^\mu p^\nu/p^2\right)\rho^{(1)}_{\alpha\beta}(-p^2) + p^\mu p^\nu \rho^{(0)}_{\alpha\beta}(-p^2)\right] , \tag{20.5.1}$$

in analogy with Eqs. (19.2.19) and (19.2.20) or Eq. (10.7.4). Taking the Fourier transform and using the completeness of the states $|N\rangle$, this can be written

$$\langle J^\mu_\alpha(x)\, J^\nu_\beta(0)\rangle_{\text{VAC}} = \int d\mu^2$$
$$\times \left[\eta^{\mu\nu}\rho^{(1)}_{\alpha\beta}(\mu^2) - \left(\rho^{(0)}_{\alpha\beta}(\mu^2) + \rho^{(1)}_{\alpha\beta}(\mu^2)/\mu^2\right)\partial^\mu\partial^\nu\right]\Delta_+(x;\mu^2)\, , \tag{20.5.2}$$

where $\Delta_+(x;\mu^2)$ is the function defined in Eq. (5.2.7):

$$\Delta_+(x;\mu^2) = \frac{1}{(2\pi)^3}\int d^4p\, \theta(p^0)\,\delta(p^2+\mu^2)e^{ip\cdot x}\, . \tag{20.5.3}$$

Assuming the currents to have been chosen as Hermitian operators, it is immediately apparent from Eq. (20.5.1) that $\rho^{(1)}_{\alpha\beta}(\mu^2)$ and $\rho^{(0)}_{\alpha\beta}(\mu^2)$ are positive Hermitian matrices. Also, taking x^μ to be spacelike (for which $\Delta_+(x)$ is even) and using translation invariance and causality in Eq. (20.5.2), we see that $\rho^{(1)}_{\alpha\beta}(\mu^2)$ and $\rho^{(0)}_{\alpha\beta}(\mu^2)$ are also symmetric.

Now, for $x \to 0$ with $x^2 > 0$, the function $\Delta_+(x;\mu^2)$ goes as

$$\Delta_+(x;\mu^2) \to \frac{1}{4\pi^2 x^2} + \frac{\mu^2}{8\pi^2}\left[\ln\left(\frac{\gamma\mu\sqrt{x^2}}{2}\right) - \frac{1}{2}\right] + O(x^2)\, , \tag{20.5.4}$$

where γ is the Euler constant. The first few terms in the vacuum expectation value of the expansion of $J^\mu_\alpha(x)\, J^\nu_\beta(0)$ are thus

$$\langle J^\mu_\alpha(x)\, J^\nu_\beta(0)\rangle_{\text{VAC}} \to -\frac{1}{2\pi^2}\left[\frac{\eta_{\mu\nu}}{(x^2)^2} - \frac{4x_\mu x_\nu}{(x^2)^3}\right]\int d\mu^2\left(\rho^{(0)}_{\alpha\beta}(\mu^2) + \rho^{(1)}_{\alpha\beta}(\mu^2)/\mu^2\right)$$
$$-\frac{\eta_{\mu\nu}}{4\pi^2 x^2}\int d\mu^2\, \rho^{(0)}_{\alpha\beta}(\mu^2)\mu^2 + \frac{x_\mu x_\nu}{2\pi^2 (x^2)^2}\int d\mu^2\left(\rho^{(0)}_{\alpha\beta}(\mu^2)\mu^2 + \rho^{(1)}_{\alpha\beta}(\mu^2)\right)$$
$$+O(\ln x^2)\, . \tag{20.5.5}$$

Hence if some linear combination $\sum_{\alpha\beta} c_{\alpha\beta}\langle J^\mu_\alpha(x)\, J^\nu_\beta(0)\rangle_{\text{VAC}}$ of the two-point functions has a singularity as $x \to 0$ which can be shown to be weaker than of order $1/x^4$, we have

$$\sum_{\alpha\beta} c_{\alpha\beta} \int d\mu^2 \left(\rho^{(0)}_{\alpha\beta}(\mu^2) + \rho^{(1)}_{\alpha\beta}(\mu^2)/\mu^2\right) = 0\,, \tag{20.5.6}$$

while if its singularity is also weaker than $1/x^2$, then

$$\sum_{\alpha\beta} c_{\alpha\beta} \int d\mu^2 \; \rho^{(1)}_{\alpha\beta}(\mu^2) = 0 \tag{20.5.7}$$

and

$$\sum_{\alpha\beta} c_{\alpha\beta} \int d\mu^2 \; \rho^{(0)}_{\alpha\beta}(\mu^2)\mu^2 = 0\,. \tag{20.5.8}$$

Eqs. (20.5.6), (20.5.7), and (20.5.8) are known respectively as the first, second, and third spectral function sum rules.

Let us see how this works out in the case of greatest interest, where the $J^\mu_\alpha(x)$ are conserved currents in a theory like quantum chromodynamics. The conservation of current tells us that (20.5.1) vanishes when contracted with p_μ, so $\rho^{(0)}_{\alpha\beta}(-p^2)$ must be proportional to $\delta(-p^2)$, and therefore automatically satisfies the third spectral function sum rule (20.5.8) for any $c_{\alpha\beta}$. With $\rho^{(0)}_{\alpha\beta}(-p^2) \propto \delta(-p^2)$, it can receive contributions only from the terms $|B_a\rangle$ in the sum over states in (20.5.1) consisting of a single massless particle B_a of zero spin, which in practice means a Goldstone boson. For such one-particle states, Lorentz invariance gives

$$\langle \text{VAC}|J^\mu_\alpha(0)|B_a\rangle = \frac{iF_{\alpha a}\, p^\mu_B}{(2\pi)^{3/2}\sqrt{2p^0_B}}\,. \tag{20.5.9}$$

Using the relation $\delta(p^0 - |\mathbf{p}|)/2p^0 = \theta(p^0)\delta(-p^2)$, we see that

$$\rho^{(0)}_{\alpha\beta}(-p^2) = \delta(-p^2)\sum_a F_{\alpha a}F^*_{\beta a}\,. \tag{20.5.10}$$

In contrast, $\rho^{(1)}_{\alpha\beta}(-p^2)$ is non-vanishing only for $-p^2 > 0$.

To be more specific, consider a renormalizable asymptotically free gauge theory with a number N of massless (or nearly massless) spin 1/2 fermions belonging to the same representation of the gauge group. Quantum chromodynamics fits this description, with $N = 3$ if we neglect the masses of the u, d, and s quarks, and with $N = 2$ if only the u and d are taken to be massless. As we have seen in Chapter 19, such a theory has an $SU(N) \times SU(N)$ global symmetry* under which the left- and right-handed

* There is also a $U(1)$ symmetry which is vectorial, in the sense that it acts the same way on the left- and right-handed parts of the light fermion fields. This is just the conservation of light quark number, and will not concern us here. The axial $U(1)$

parts of the light fermion fields transform under the representations $(N, 1)$ and $(1, N)$, respectively, where 'N' and '1' denote the defining and identity representations of $SU(N)$, respectively. The currents of the left- and right-handed $SU(N)$ symmetries are

$$J^{\mu}_{La}(x) = -i\bar{\psi}(x)\gamma^{\mu}(1+\gamma_5)\lambda_a\psi(x)\,, \quad J^{\mu}_{Ra}(x) = -i\bar{\psi}(x)\gamma^{\mu}(1-\gamma_5)\lambda_a\psi(x)\,, \tag{20.5.11}$$

where the λ_a form a complete set of Hermitian traceless matrices acting on the 'flavor' index that distinguishes the N light quarks, as in Eq. (19.7.2) for $N = 3$. These currents have dimensionality (in powers of mass) $+3$, so the coefficient of an operator $\mathscr{O}$ of dimensionality $d(\mathscr{O})$ in the expansion (20.3.6) of the product of two currents is expected (apart from logarithms) to have a singularity of order $x^{-6+d(\mathscr{O})}$ as the separation x of their arguments goes to zero. Thus, if the expansion of a linear combination of products of currents contains the operator unity, then the corresponding linear combination of spectral functions goes as x^{-6}, and therefore in general satisfies neither the first nor the second spectral function sum rule; if the lowest-dimensional operator in the vacuum expectation value of this expansion is a fermion bilinear with zero or one derivatives, then the corresponding linear combination of spectral functions goes as x^{-3} or x^{-2}, and therefore satisfies the first spectral function sum rule but generally not the second; and if the lowest-dimensional operator in the vacuum expectation value of this expansion is a fermion bilinear with two or more derivatives or a fermion quadrilinear then the corresponding linear combination of spectral functions is less singular than x^{-2}, and therefore satisfies both the first and second spectral function sum rules.

In order to tell which operators appear in the expansion of the product of two currents, we need to classify the $SU(N) \times SU(N)$ representations contained in the product, and to tell whether these operators have non-vanishing vacuum expectation values, we have to ask which of them are invariant under the subgroup of $SU(N) \times SU(N)$ that is not spontaneously broken. To answer these questions, we note that the currents $J^{\mu}_{La}(x)$ and $J^{\mu}_{Ra}(x)$ transform under $SU(N) \times SU(N)$ respectively according to the $(A, 1)$ and $(1, A)$ representations of $SU(N) \times SU(N)$, where A and 1 are the adjoint and identity representations of $SU(N)$.

Also, we shall assume that the subgroup of $SU(N) \times SU(N)$ that is not spontaneously broken is the vectorial $SU(N)_V$ whose currents are $J^{\mu}_{La}(x) + J^{\mu}_{Ra}(x)$, as is the case in quantum chromodynamics and (as we saw in Section 19.9) a wide range of other theories. We also assume that parity conservation is not spontaneously broken. These unbroken

symmetry of the Lagrangian, which acts differently on the left- and right-handed parts of the light fermion fields, is broken by the quantum effects discussed in Section 23.5.

symmetries have the consequence that

$$\rho^{(1)}_{La,Lb}(\mu^2) = \rho^{(1)}_{Ra,Rb}(\mu^2) = \delta_{ab}\left[\rho^{(1)}_V(\mu^2) + \rho^{(1)}_A(\mu^2)\right] \tag{20.5.12}$$

and

$$\rho^{(1)}_{La,Rb}(\mu^2) = \rho^{(1)}_{Ra,Lb}(\mu^2) = \delta_{ab}\left[\rho^{(1)}_V(\mu^2) - \rho^{(1)}_A(\mu^2)\right] , \tag{20.5.13}$$

where $\delta_{ab}\rho^{(1)}_V(\mu^2)$ and $\delta_{ab}\rho^{(1)}_A(\mu^2)$ are the spectral functions defined by (20.5.2) for the currents

$$J^\mu_{Va} = -i\bar{\psi}\gamma^\mu\lambda_a\psi , \qquad J^\mu_{Aa} = -i\bar{\psi}\gamma^\mu\gamma_5\lambda_a\psi , \tag{20.5.14}$$

with generators λ_a and $\lambda_a\gamma_5$, respectively. Also

$$F_{Lab} = -F_{Rab} = \delta_{ab}F , \tag{20.5.15}$$

where Eq. (20.5.9) here reads

$$\langle \text{VAC}|J^\mu_{Aa}(0)|B_b\rangle = \frac{iF\delta_{ab}\,p^\mu_B}{(2\pi)^{3/2}\sqrt{2p^0_B}} \tag{20.5.16}$$

so that (20.5.10) reads

$$\rho^{(0)}_{La\,Lb}(\mu^2) = \rho^{(0)}_{Ra\,Rb}(\mu^2) = -\rho^{(0)}_{La\,Rb}(\mu^2) = -\rho^{(0)}_{Ra\,Lb}(\mu^2) = F^2\delta(\mu^2)\delta_{ab} . \tag{20.5.17}$$

It will be convenient to consider separately the products of currents of like and unlike chirality.

Like Chirality: The products $J^\mu_{La}(x)J^\nu_{Lb}(0)$ and $J^\mu_{Ra}(x)J^\nu_{Rb}(0)$ transform respectively as the $(A\times A, 1)$ and $(1, A\times A)$ representations of $SU(N)\times SU(N)$, where A and 1 are the adjoint and identity representations, respectively. For any group, $A \times A$ contains the identity representation, so the unit operator appears in the expansions of these products, with equal coefficients proportional to δ_{ab}. Hence only the traceless parts of the operator product can satisfy spectral function sum rules. But as we have seen, the spectral functions have no traceless part, so the like-chirality spectral functions cannot satisfy any sum rules.

Unlike Chirality: The products $J^\mu_{La}(x)J^\nu_{Rb}(0)$ and $J^\mu_{Ra}(x)J^\nu_{Lb}(0)$ both transform as the (A, A) representation of $SU(N) \times SU(N)$. The unit operator (and operators like $F_{\alpha\mu\nu}F^{\mu\nu}_\alpha$) are of course $SU(N) \times SU(N)$ singlets, and therefore cannot appear in the expansion of these products. Non-derivative fermion bilinears like $\bar{\psi}\psi$ transform according to the $(\bar{N}, N)$ and $(N, \bar{N})$ representations of $SU(N)\times SU(N)$, so they also cannot appear in the expansions of the products of currents with unlike chirality. The only gauge- and Lorentz-invariant fermion bilinears with a single derivative involve the gauge-covariant derivative operator $\gamma^\mu D_\mu$ acting on ψ, which the field equations tell us vanishes. Thus these spectral functions satisfy both the first and second spectral function sum rules, which here read

$$\int d\mu^2 \left[\rho_V^{(1)}(\mu^2) - \rho_A^{(1)}(\mu^2)\right] / \mu^2 = F^2 \tag{20.5.18}$$

and

$$\int d\mu^2 \left[\rho_V^{(1)}(\mu^2) - \rho_A^{(1)}(\mu^2)\right] = 0 \, . \tag{20.5.19}$$

In the original work on spectral function sum rules,[5] the $SU(2) \times SU(2)$ spectral functions were assumed to be sharply peaked at values of μ that for $\rho_V^{(1)}(\mu^2)$ could be assumed to be $m_\rho = 770$ MeV, and for $\rho_A^{(1)}(\mu^2)$ was taken to be at an unknown mass m_A. That is,

$$\rho_V^{(1)}(\mu^2) \simeq g_\rho^2 \delta(\mu^2 - m_\rho^2) \, , \qquad \rho_A^{(1)}(\mu^2) \simeq g_A^2 \delta(\mu^2 - m_A^2) \, .$$

Eqs. (20.5.18) and (20.5.19) then read**

$$\frac{g_\rho^2}{m_\rho^2} - \frac{g_A^2}{m_A^2} = F_\pi^2$$

and

$$g_\rho^2 = g_A^2 \, .$$

Eliminating the unknown g_A, this gives a formula

$$g_\rho^2 = F_\pi^2 \left(\frac{1}{m_\rho^2} - \frac{1}{m_A^2} \right)^{-1} \, .$$

Originally in 1967 this result was used together with a formula[6] $g_\rho^2 = 2F_\pi^2 m_\rho^2$ (whose justification was unclear but which agreed with experimental measurements of the rate of the decay $\rho \to e^+ + e^-$) to derive the result

$$m_A = \sqrt{2} m_\rho \, .$$

The status of a possible a_1 resonance with the right quantum numbers to couple to the axial-vector current (that is, 1^+ with $T = 1$ and $\mathrm{C}(a_1^0) = +1$) at around a mass $\sqrt{2} m_\rho$ remained unclear for many years, but a resonance with these quantum numbers is now reasonably well established at a mass 1230 MeV $= 1.6 \, m_\rho$. It seems preferable today to take the ratio m_A/m_ρ as an input, using either the value $\sqrt{2}$ suggested by certain models[7] or the experimental value 1.6, and use it to predict g_ρ.

** Taking the $SU(2)$ generators λ_a to be the Pauli matrices, as in Eq. (19.7.2), we have $F = F_\pi = 184$ MeV.

Since 1967 not only g_ρ but the whole vector spectral function $\rho_V^{(1)}(\mu^2)$ of the $SU(3) \times SU(3)$ current in quantum chromodynamics has been calculated with precision for a wide range of energies from the measured cross section for the process $e^+ + e^- \to \gamma \to$ hadrons, using the fact that the electromagnetic current is a linear combination of $SU(3)$ currents. The axial-vector spectral function $\rho_A^{(1)}(\mu^2)$ of the $SU(3) \times SU(3)$ currents can in principle be measured in the process $\bar{\nu} + e \to$hadrons, since the charged components of the currents (20.5.14) are the same as the hadron currents to which the leptons are coupled. However, although antineutrino–electron scattering has been studied experimentally, the low rate of these reactions precludes the use of colliding beams, so that the electron target is essentially at rest. In order to reach typical hadronic energies of, say, 3 GeV in the center-of-mass frame, it would be necessary to have neutrino energies of $(3\,\text{GeV})^2/2m_e \simeq 10$ TeV in the laboratory frame. Intense beams of neutrinos with energies this high will not be available for many years, if ever. Fortunately, it has become possible to study the spectral functions in the process $\tau \to \nu +$ hadrons, but the hadron energies here are strictly bounded by $m_\tau = 1.7$ GeV. It is also possible to use effective chiral Lagrangians to calculate the spectral functions at small μ^2, and to use quantum chromodynamics to calculate $\rho_V^{(1)} - \rho_A^{(1)}$ at large μ^2, where it is quite small. A careful 1993 analysis by Donoghue and Golowich[8] of all these inputs shows that the spectral function integrals are indeed dominated by the ρ and a_1 resonances, and gives results that are consistent with the first and second spectral function sum rules.

20.6 Deep Inelastic Scattering

The renormalization group coupled with the operator product expansion found one of its most important applications in the analysis of deep inelastic lepton–nucleon scattering. We will first review the early phenomenological models for these reactions, and then see how the operator product expansion justifies these models and provides corrections to them.

Consider a process in which an electron of four-momentum k collides with a nucleon N of four-momentum p, yielding a electron of four-momentum k' and a general unobserved hadron state H, over which we sum. To calculate the spin-averaged inclusive cross section, we shall need to know the quantity

$$(m_N/p_N^0)W^{\mu\nu}(q,p) \equiv \frac{1}{2}\sum_{\sigma_N}\sum_H \delta^4(p_H - p - q)\langle H|J^\mu(0)|N\rangle\langle H|J^\nu(0)|N\rangle^* \,, \tag{20.6.1}$$

where J^μ is the electromagnetic current (divided here by a factor e) and

$q = k - k'$ is the momentum transfer from the electrons to the hadrons. Lorentz invariance tells us that $W^{\mu\nu}(q, p)$ must be a linear combination of $p^\mu p^\nu$, $p^\mu q^\nu$, $q^\mu p^\nu$, $q^\mu q^\nu$, and $\eta^{\mu\nu}$, with coefficients that can only depend on the two independent scalar functions of q and p: q^2 and $\nu \equiv -q \cdot p/m_N$. Current conservation requires that $q_\mu W^{\mu\nu} = q_\nu W^{\mu\nu} = 0$, so W must take the form*

$$W^{\mu\nu}(q,p) = -\left(\frac{q^\mu q^\nu}{q^2} - \eta^{\mu\nu}\right) W_1(\nu, q^2)$$
$$+\frac{1}{m_N^2}\left(p^\mu - \frac{p\cdot q}{q^2} q^\mu\right)\left(p^\nu - \frac{p\cdot q}{q^2} q^\nu\right) W_2(\nu, q^2) . \qquad (20.6.2)$$

Also, Eq. (20.6.1) shows that $W^{\mu\nu *} = W^{\nu\mu}$, so W_1 and W_2 are real, and $W^{\mu\nu}$ is a positive matrix, so W_1 and W_2 are positive. The differential cross section in the nucleon rest frame is

$$\frac{d^2\sigma}{d\Omega d\nu} = \left(\frac{d\sigma}{d\Omega}\right)_{\text{MOTT}} \left(W_2 + 2W_1 \tan^2(\theta/2)\right) , \qquad (20.6.3)$$

where $d\Omega = \sin\theta \, d\theta \, d\phi$ is the solid angle into which the electron is scattered, and $(d\sigma/d\Omega)_{\text{MOTT}}$ is the differential cross section for relativistic elastic scattering by a point spinless particle

$$\left(\frac{d\sigma}{d\Omega}\right)_{\text{MOTT}} = \frac{e^4}{4E_e^2} \frac{\cos^2(\theta/2)}{\sin^4(\theta/2)} , \qquad (20.6.4)$$

with $E_e = -k \cdot p/m_N$ the incident electron energy in the nucleon rest frame.

One might expect that for fixed values of $-p_H^2 = -q^2 + 2m_N\nu + m_N^2$, the differential cross section should fall off very rapidly as $q^2 \to \infty$, because it should be proportional to the square of the form-factor for the transition from the nucleon to whatever particles or resonance have mass near $-p_H^2$. It was therefore somewhat of a surprise that, two years after the 1966 opening of the Stanford Linear Accelerator Center, a SLAC–MIT collaboration headed by Friedman, Kendall, and Taylor[9] discovered that in fact $\nu W_2(q^2, \nu)$ is roughly constant in q^2 for fixed values of $\omega \equiv 2m_N\nu/q^2 > 1$. (To be specific, $W_2(q^2, \nu)$ for the proton was fitted to the curve $\nu W_2(q^2, \nu) \simeq 0.35 - 0.004\omega$ for $E_e = 10$, 13.5, and 16 GeV, and $\theta = 6°$ and $10°$. These experiments were insensitive to W_1, because $\tan^2(10°/2) = 7.6 \times 10^{-3}$.) Note that in this limit $-p_H^2 \to (\omega - 1)q^2 \to \infty$, which is why this is called 'deep-inelastic' scattering.

At about the same time Bjorken[10] was using current algebra to argue that $W_2(q^2, \nu)$ and $W_1(q^2, \nu)$ satisfy scaling laws: for q^2 and ν going to

* The same formalism can be used for other deep-inelastic lepton scattering processes, such as $\nu_\mu + p \to \mu^- + H$, except that since parity is not conserved in these processes there is an additional term in $W^{\mu\nu}(q, p)$ proportional to $\epsilon^{\mu\nu\rho\sigma} p_\rho q_\sigma$. For simplicity, we will limit ourselves here to deep-inelastic electron scattering.

infinity together,

$$\nu W_2(\nu, q^2) \to F_2(\omega)\,, \qquad W_1(\nu, q^2) \to F_1(\omega)\,, \tag{20.6.5}$$

where again $\omega \equiv 2m_N\nu/q^2$. A more intuitive explanation was given a little later by Feynman.[11] He supposed that in deep-inelastic scattering on a highly relativistic nucleon of momentum $\mathbf{p}$, the nucleon behaves as if it consists of 'partons' of various types labelled i, each with a probability $\mathscr{F}_i(x)dx$ of having momentum between $x\mathbf{p}$ and $(x+dx)\mathbf{p}$. Then for each i,

$$\int dx\, \mathscr{F}_i(x) = 1\,. \tag{20.6.6}$$

The condition that the nucleon has total momentum $\mathbf{p}$ yields the additional sum rule

$$\int_0^1 \sum_i \mathscr{F}_i(x)\, x\, dx = 1\,. \tag{20.6.7}$$

For elastic scattering of an electron (with m_e neglected) from a parton of four-momentum xp, we have $x^2m_N^2 = -(q+xp)^2 = -q^2 - 2\nu m_N x + x^2 m_N^2$, so $\nu = q^2/2m_N x$. The inelastic cross section in this model is thus**

$$\frac{d^2\sigma}{d\Omega d\nu} = \left(\frac{d\sigma}{d\Omega}\right)_{\text{MOTT}} \sum_i Q_i^2 \int_0^1 dx \left(1 + \frac{q^2}{2m_N^2 x^2}\tan^2\left(\frac{\theta}{2}\right)\right)\delta\left(\nu - \frac{q^2}{2m_N x}\right) \tag{20.6.8}$$

where Q_i is the charge of the ith type of parton in units of e. (The $\tan^2(\theta/2)$ term here is appropriate for a 'Dirac' parton, with magnetic moment $eQ_i/2m_N x$.) Comparing this with Eq. (20.6.3) gives

$$W_2(\nu, q^2) = \sum_i Q_i^2 \int_0^1 dx\, \mathscr{F}_i(x)\delta\left(\nu - \frac{q^2}{2m_N x}\right) = \frac{1}{\nu\omega}\sum_i Q_i^2 \mathscr{F}_i\left(\frac{1}{\omega}\right)\,, \tag{20.6.9}$$

$$W_1(\nu, q^2) = \frac{1}{2}\sum_i Q_i^2 \int_0^1 dx\, \mathscr{F}_i(x)\frac{q^2}{2m_N^2 x^2}\delta\left(\nu - \frac{q^2}{2m_N x}\right) = \frac{\omega\nu}{2m_N} W_2(\nu, q^2)\,. \tag{20.6.10}$$

** This may be derived from the formulas (8.7.7) and (8.7.38) for Compton scattering on a spin 1/2 particle of four-momentum p and mass m in a general Lorentz frame. For this purpose, it is useful to note that in the gauge used to derive these formulas, for which the initial and final real polarization vectors e_μ and e'_μ satisfy $e^2 = e'^2 = 1$, $e \cdot p = e' \cdot p = e \cdot k = e' \cdot k' = 0$, the polarization sum gives

$$\sum_{e,e'} \left(e \cdot e'\right)^2 = 2 + \frac{m^4 (k \cdot k')^2}{(k \cdot p)^2 (k' \cdot p)^2} + \frac{2m^2 (k \cdot k')}{(k \cdot p)(k' \cdot p)}\,.$$

This agrees with the Bjorken scaling rules (20.6.5), with

$$F_2(\omega) = \frac{1}{\omega}\sum_i Q_i^2 \mathscr{F}_i\left(\frac{1}{\omega}\right) \tag{20.6.11}$$

and

$$F_1(\omega) = (\omega/2m_N)F_2(\omega)\,. \tag{20.6.12}$$

Eq. (20.6.12) was originally derived by Callan and Gross.[12] It agrees with experiment within about 10–15%.

On the assumption that the proton and neutron respectively consist of two u and one d or one u and two d quarks, plus any number of neutral partons, Eqs. (20.6.11) and (20.6.6) yield the sum rule for W_2

$$\int_1^\infty F_2(\omega)\frac{d\omega}{\omega} = \sum_i Q_i^2 = \begin{cases} 1 & p \\ 2/3 & n \end{cases}. \tag{20.6.13}$$

The integral receives a large contribution from large ω, where F_2 is difficult to measure. If we also assume that the total momentum of the nucleon is equally shared among three quarks (and no neutral partons) then in place of Eq. (20.6.7) we have the stronger relation $\int \mathscr{F}_i(x)x\,dx = 1/3$ for each quark, which with Eq. (20.6.11) yields

$$\int_1^\infty F_2(\omega)\frac{d\omega}{\omega^2} = \tfrac{1}{3}\sum_i Q_i^2 = \begin{cases} 1/3 & p \\ 2/9 & n \end{cases}.$$

This integral is easier to measure, and turns out to disagree badly with the sum rule, showing that a large fraction of the nucleon momentum is carried by the neutral partons.

None of the above phenomenology depends on any specific field theory. It was the operator product expansion that finally provided a way of applying an underlying field theory to deep-inelastic scattering. In particular, the operator product expansion made it clear that one needed an asymptotically free field theory to explain the scaling assumptions (20.6.5) (and to calculate the corrections to scaling). This asymptotically free theory was ultimately provided by quantum chromodynamics.

To apply the operator product expansion to deep-inelastic scattering, first Fourier transform Eq. (20.6.1). Using translation invariance and the completeness of the hadron states $|H\rangle$, this gives

$$(m_N/p_N^0)W^{\mu\nu}(q,p) \equiv \frac{1}{2(2\pi)^4}\sum_{\sigma_N}\int d^4z\, e^{-iq\cdot z}\langle N|J^\nu(z)\,J^\mu(0)|N\rangle\,. \tag{20.6.14}$$

The asymptotic behavior of $W^{\mu\nu}(q,p)$ as $q \to \infty$ is therefore related to the singularity of the operator product at $z \to 0$.

Feynman diagram calculations of the coefficient functions in operator product expansions refer directly not to the expansion for matrix elements like those appearing in Eq. (20.6.14) for $W^{\mu\nu}$, but rather to matrix elements of the two-point Greens function

$$(m_N/p_N^0)T^{\mu\nu}(q,p) \equiv \frac{1}{2(2\pi)^4}\sum_{\sigma_N}\int d^4z\, e^{-iq\cdot z}\langle N|T\{J^\nu(z),\, J^\mu(0)\}\,|N\rangle\,. \tag{20.6.15}$$

We can express this in terms of the structure functions for $T^{\mu\nu}$, defined by the analog to Eq. (20.6.2):

$$T^{\mu\nu}(q,p) = -\Big(\frac{q^\mu q^\nu}{q^2} - \eta^{\mu\nu}\Big)T_1(\nu,q^2) + \frac{1}{m_N^2}\Big(p^\mu - \frac{p\cdot q}{q^2}q^\mu\Big)\Big(p^\nu - \frac{p\cdot q}{q^2}q^\nu\Big)T_2(\nu,q^2)\,. \tag{20.6.16}$$

The connection between the $T_r(\nu,q^2)$ and $W_r(\nu,q^2)$ (with $r = 1,2$) is provided[13] by the dispersion relations† for fixed q^2:

$$T_r(\nu,q^2) = \frac{1}{2}W_r(-\nu,q^2) + \frac{1}{2}W_r(\nu,q^2) + \frac{1}{2\pi i}\int_{-\infty}^{\infty} d\nu'\,\frac{W_r(-\nu',q^2) - W_r(\nu',q^2)}{\nu' - \nu}\,, \tag{20.6.17}$$

with the denominator in the integrand interpreted as a principal value function. The functions $W_r(\nu,q^2)$ vanish except for $\nu > q^2/2m_N$, so the dispersion relations may be rewritten

$$T_r(\nu,q^2) = \frac{1}{2}W_r(-\nu,q^2) + \frac{1}{2}W_r(\nu,q^2) - \frac{1}{2\pi i}\int_{q^2/2m_N}^{\infty} d\nu'\,W_r(\nu',q^2)\Big(\frac{1}{\nu'+\nu} + \frac{1}{\nu'-\nu}\Big)\,. \tag{20.6.18}$$

We can categorize the operators that contribute to the operator product expansion for T_r according to the irreducible representation of the Lorentz group to which they belong. The only Lorentz-covariant functions of a single four-vector p^μ with $p^2 = -m_N^2$ fixed are proportional to the symmetric tensors $p^{\mu_1}\cdots p^{\mu_s}$, so the only operators that can contribute to the spin-averaged nucleon expectation value are symmetric traceless tensors $\mathcal{O}_{si}^{\mu_1\cdots\mu_s}$, the subscript i distinguishing any different operators with this tensor structure. (For reasons that will become clear, we are using the same label i to distinguish operators that we used in the parton model to

† For $q^2 = 0$, these may be derived exactly as in the derivation of the dispersion relation (10.8.16) for forward photon scattering. The derivation for fixed $q^2 \neq 0$ is more difficult.

label types of partons.) These operators have matrix elements of the form

$$\frac{1}{2}\sum_{\sigma_N}\langle N|\mathscr{O}_{si}^{\mu_1\cdots\mu_s}|N\rangle = (m_N/p_N^0)\Big[p^{\mu_1}\cdots p^{\mu_s} - \text{traces}\Big]\,\langle\mathscr{O}_{si}\rangle\,, \qquad (20.6.19)$$

where the $\langle\mathscr{O}_{si}\rangle$ are constant coefficients. Such an operator makes a contribution to $T^{\mu\nu}(q,p)$ proportional to s factors of the four-vector p, and hence makes a contribution to $T_1(\nu,q^2)$ and $T_2(\nu,q^2)$ proportional to ν^s and ν^{s-2}, respectively. (We are dropping terms involving p^2, because such terms will be suppressed by factors of m_N^2/q^2 or $m_N^2/p\cdot q$.) If we ignore logarithmic corrections, then in an asymptotically free theory the q^2 dependence of the coefficients is of the forms $(q^2)^{(-4+6-d(s,i)-s)/2}$ and $(q^2)^{(-4+6-d(s,i)-s+2)/2}$, respectively, where $d(s,i)$ is the dimensionality of the operator $\mathscr{O}_{si}$.†† Using $\nu \propto q^2\omega$, we see that the contributions of an operator $\mathscr{O}_{si}$ to the structure functions are asymptotically of the forms

$$T_{1,si} \propto \nu^s (q^2)^{(2-d(s,i)-s)/2} \propto \omega^s (q^2)^{(2-\tau(s,i))/2} \qquad (20.6.20)$$

and

$$\nu T_{2,si} \propto \nu^{s-1}(q^2)^{(4-d(s,i)-s)/2} \propto \omega^{s-1}(q^2)^{(2-\tau(s,i))/2}\,, \qquad (20.6.21)$$

where $\tau(s,i)$ is the 'twist' of the operator $\mathscr{O}_{si}$, defined as[14]

$$\tau(s,i) \equiv d(s,i) - s\,. \qquad (20.6.22)$$

We see that the dominant terms in T_1 and νT_2 for $q^2 \to \infty$ with fixed ω are contributed by operators of minimum twist. Also, Eq. (20.6.18) shows there are no terms in $T_r(\nu,q^2)$ of odd order in ν, so the only operators $\mathscr{O}_{si}$ that can contribute here are those with even s.

The symmetric traceless tensors of rank s with minimum dimensionality are the operators

$$(\mathscr{O}_{sf})_{\mu_1\cdots\mu_s} \equiv (i^{s-2}/s!)\,\bar\psi_f\gamma_{\{\mu_1}\overleftrightarrow{D}_{\mu_2}\cdots\overleftrightarrow{D}_{\mu_s\}}\psi_f \qquad (20.6.23)$$

and

$$(\mathscr{O}_{s0})_{\mu_1\cdots\mu_s} \equiv (i^{s-2}/2s!)\,F_{\alpha\,\nu\{\mu_1}\overleftrightarrow{D}_{\mu_3}\cdots\overleftrightarrow{D}_{\mu_s}F^{\nu}_{\alpha\,\mu_2\}}\,, \qquad (20.6.24)$$

where f labels quark flavors; D_μ is the gauge-covariant derivative; and the brackets indicate a sum over permutations and subtraction of trace terms for the enclosed spacetime indices. (The symbol $\leftrightarrow$ indicates half the difference of the derivative acting to the right and left. We take this difference of derivatives, because their sum would vanish in any

†† The −4 in the exponents arises from the integral over z in Eq. (20.6.14), and the +6 in these exponents is the dimensionality of the two electric current operators. The terms $-s/2$ and $-(s-2)/2$ serve to compensate for the powers of q^μ already in ν^s and ν^{s-2}, respectively.

matrix element between states of equal four-momenta.) These operators have dimensionality $3+(s-1)=4+(s-2)=2+s$, so they all have twist $\tau=2$. Thus an infinite number of operators contribute for $q^2\to\infty$, all of which give contributions to T_1 and νT_2 that depend on ω but only logarithmically on q^2. Asymptotic freedom thus confirms Bjorken scaling, but only up to logarithmic corrections. Keeping only operators of twist two, we see that (as anticipated in our notation) for each s there is indeed one operator for each parton type, with i running over quark types f (lumping quarks and antiquarks together) and also a value $i=0$ for the gluon.

Now let us consider the logarithmic corrections. The asymptotic behavior of the coefficient functions in an asymptotically free theory is governed by Eq. (20.3.9). As discussed in Section 10.4, when we ignore electromagnetic radiative corrections no renormalization factors are needed for the electromagnetic current, because these currents are conserved. Thus the matrix $c_{\ell\ell'}$ in Eq. (20.3.9) vanishes. The matrix $c_{\mathcal{O},\mathcal{O}'}$ has no elements connecting operators of different Lorentz transformation type, so that in the notation we are now using, $c_{si,s'i'}=\delta_{ss'}c_{ii'}(s)$. Thus Eq. (20.3.9) takes the form

$$T_1(\nu,q^2)\to\sum_{sij}\omega^s\mathscr{A}_{si}\left[\left(g_q^2/g_\ell^2\right)^{8\pi^2c(s)/b}\right]_{ij}\langle\mathcal{O}_{sj}\rangle\,, \tag{20.6.25}$$

$$\nu T_2(\nu,q^2)\to\sum_{sij}\omega^{s-1}\mathscr{B}_{si}\left[\left(g_q^2/g_\ell^2\right)^{8\pi^2c(s)/b}\right]_{ij}\langle\mathcal{O}_{sj}\rangle\,, \tag{20.6.26}$$

where $\mathscr{A}$ and $\mathscr{B}$ are constants appearing in the coefficient function for the operator $\mathcal{O}_{si}$ in the operator product expansion of the product of two electric currents, ℓ^2 is a particular value of q^2 at which we choose to define the coefficient functions, b is the constant in the one-loop renormalization group equation (20.3.8) for the strong coupling constant g_q, and $\langle\mathcal{O}_{si}\rangle$ is the constant coefficient in the matrix elements (20.6.19).

The operator product expansion coefficients are independent of the particular process in question and unaffected by quark trapping, so we can calculate the coefficients $\mathscr{A}_{si}$ and $\mathscr{B}_{si}$ by considering a fictitious simpler process, the scattering of an electron by a free quark of flavor f. The renormalized operators $\mathcal{O}_{si}$ may conveniently be defined so that the one-quark matrix elements of the operators (20.6.23) and (20.6.24) are given by the tree approximation:

$$\langle f',\sigma'|\mathcal{O}_{sf}|f'',\sigma''\rangle=\frac{(-1)^s i}{s!}\left(\bar{u}'\gamma^{\{\mu_1}u\right)p^{\mu_2}\cdots p^{\mu_s\}}\delta_{ff'}\delta_{ff''}\,, \tag{20.6.27}$$

$$\langle f',\sigma'|\mathcal{O}_{s0}|f'',\sigma''\rangle=0\,. \tag{20.6.28}$$

Averaging Eq. (20.6.27) over $\sigma'=\sigma''$ and comparing the result with

(20.6.19) yields

$$\langle \mathcal{O}_{sf} \rangle_{f'} \frac{m_f}{p^0} p^{\mu_1} p^{\mu_2} \cdots p^{\mu_s} = \delta_{ff'} \frac{(-1)^s i}{2s!} \mathrm{Tr} \left\{ \left(\frac{-i \not{p} + m_f}{2p^0} \right) \gamma^{\{\mu_1} p^{\mu_2} \cdots p^{\mu_s\}} \right\} ,$$

so that (for even s)

$$\langle \mathcal{O}_{sf} \rangle_{f'} = \delta_{ff'} / m_f \tag{20.6.29}$$

and

$$\langle \mathcal{O}_{s0} \rangle_{f'} = 0 \,. \tag{20.6.30}$$

With this definition of the operators $\mathcal{O}_{si}$, at a renormalization momentum scale ℓ^2 which is large enough for the operator product expansion functions to be calculated using the tree approximation, these functions can be derived from the tree-approximation formulas for the electron-quark scattering cross section. Retracing the parton-model derivations of Eqs. (20.6.9) and (20.6.10), we see that W_r for an electron scattering on a quark of flavor f is given in the tree approximation by‡

$$\nu W_{2,f} = (m_N / m_f) Q_f^2 \delta(\omega - 1) \,, \tag{20.6.31}$$

$$W_{1,f} = Q_f^2 \delta(\omega - 1) / 2m_f \,. \tag{20.6.32}$$

Inserting these results in the dispersion relations (20.6.18) yields, for $\omega \neq 1$,

$$T_{1,f} = \frac{Q_f^2}{2\pi i m_f} \frac{m_N}{m_f} \frac{1}{\omega^2 - 1} \,, \tag{20.6.33}$$

$$\nu T_{2,f} = \frac{2Q_f^2}{2\pi i} \frac{\omega^2}{\omega^2 - 1} \,, \tag{20.6.34}$$

results that may also be obtained directly from the Feynman graphs for electron–quark scattering. Comparing the coefficients of ω^s in Eqs. (20.6.33) and (20.6.34) with those in Eqs. (20.6.25) and (20.6.26) at the renormalization point $q^2 = \ell^2$ and using Eqs. (20.6.29) and (20.6.30), we find that

$$\mathscr{A}_{si} = \frac{iQ_i^2}{2\pi} \,, \qquad \mathscr{B}_{si} = \frac{i m_N Q_i^2}{\pi} \,, \tag{20.6.35}$$

with the gluon charge Q_0 of course taken as zero. To repeat, even though these values have been derived for electron–quark scattering, $\mathscr{A}_{si}$ and $\mathscr{B}_{si}$ are factors in the operator product expansion of two electric currents, and therefore do not depend on the state in which we calculate the expectation

‡ The factor (m_N/m_f) is inserted in Eq. (20.6.31) because ν is defined as $-q \cdot p/m_N$ instead of $-q \cdot p/m_f$, while it is W_2/m_N^2 rather than W_2/m_f^2 that appears in Eq. (20.6.2). Eq. (20.6.32) then follows from Eq. (20.6.10). The quantity ω may be written in a mass-independent way, as $\omega = -2q \cdot p/q^2$.

value of this operator product. So we can use Eq. (20.6.35) in Eqs. (20.6.25) and (20.6.26), and find

$$T_1(\nu,q^2) \to \frac{i}{2\pi}\sum_{sij}\omega^s Q_i^2\left[\left(g_q^2/g_\ell^2\right)^{8\pi^2 c(s)/b}\right]_{ij}\langle\mathcal{O}_{sj}\rangle\,, \tag{20.6.36}$$

$$\nu T_2(\nu,q^2) \to \frac{i m_N}{\pi}\sum_{sij}\omega^{s-1} Q_i^2\left[\left(g_q^2/g_\ell^2\right)^{8\pi^2 c(s)/b}\right]_{ij}\langle\mathcal{O}_{sj}\rangle\,. \tag{20.6.37}$$

Now let us at last return to the structure functions $W_r(\nu,q^2)$. We note that the coefficient of $\omega^s = (2m_N\nu/q^2)^s$ in $T_r(\nu,q^2)$ is given by Eq. (20.6.18) as

$$\frac{-2}{2\pi i}\left(\frac{q^2}{2m_N}\right)^s\int_{q^2/2m_N}^{\infty} d\nu'\,\nu'^{-1-s}W_r(\nu',q^2) = \frac{i}{\pi}\int_1^\infty d\omega\,\omega^{-1-s}W_r\left(\frac{\omega q^2}{2m_N},q^2\right)\,.$$

Comparing with Eqs. (20.6.36) and (20.6.37), we find

$$\int_1^\infty d\omega\,\omega^{-1-s}W_1\left(\frac{\omega q^2}{2m_N},q^2\right) \to \frac{1}{2}\sum_{ij}Q_i^2\left[\left(g_q^2/g_\ell^2\right)^{8\pi^2 c(s)/b}\right]_{ij}\langle\mathcal{O}_{sj}\rangle\,, \tag{20.6.38}$$

$$\int_1^\infty d\omega\,\omega^{-s}\nu W_2\left(\frac{\omega q^2}{2m_N},q^2\right) \to m_N\sum_{ij}Q_i^2\left[\left(g_q^2/g_\ell^2\right)^{8\pi^2 c(s)/b}\right]_{ij}\langle\mathcal{O}_{sj}\rangle\,. \tag{20.6.39}$$

The W_r satisfying these equations may evidently be expressed in a manner similar to the equations (20.6.9) and (20.6.10) of the parton model

$$W_1\left(\frac{\omega q^2}{2m_N},q^2\right) \to \sum_i Q_i^2\mathcal{F}_i(1/\omega,q^2)\,, \tag{20.6.40}$$

$$\nu W_2\left(\frac{\omega q^2}{2m_N},q^2\right) \to \frac{2m_N}{\omega}\sum_i Q_i^2\mathcal{F}_i(1/\omega,q^2)\,, \tag{20.6.41}$$

where $\mathcal{F}_i$ is a parton distribution function, now defined by the moment equations

$$\int_0^1 dx\,x^{s-1}\mathcal{F}_i(x,q^2) = \frac{1}{2}\sum_j\left[\left(g_q^2/g_\ell^2\right)^{8\pi^2 c(s)/b}\right]_{ij}\langle\mathcal{O}_{sj}\rangle\,. \tag{20.6.42}$$

In particular, we see that asymptotic freedom implies not only a corrected version of Bjorken scaling, but also the Callan–Gross relation (20.6.12) between W_1 and W_2.

There is an elegant reformulation of the moment equations (20.6.42) due to Altarelli and Parisi[15] that has become widely used in studies of deep-inelastic scattering. Note that Eqs. (20.6.42) together with the renormalization group equation (20.3.8) imply the differential equations

$$q^2 \frac{d}{dq^2} \int_0^1 dx\, x^{s-1} \mathscr{F}_i(x,q^2) = -g_q^2 \sum_j c_{ij}(s) \int_0^1 dx\, x^{s-1} \mathscr{F}_j(x,q^2)\,. \tag{20.6.43}$$

These equations, together with the 'initial' condition at the renormalization point ℓ^2

$$\int_0^1 dx\, x^{s-1} \mathscr{F}_i(x,\ell^2) = \tfrac{1}{2}\langle \mathscr{O}_{si}\rangle\,, \tag{20.6.44}$$

have a unique solution, so they can be used instead of the moment equations. Now, Eq. (20.6.43) is satisfied by the solution of the differential equation for $\mathscr{F}_i$:

$$q^2 \frac{d}{dq^2} \mathscr{F}_i(x,q^2) = \frac{g_q^2}{4\pi^2} \sum_j \int_x^1 \frac{dy}{y} P_{ij}\left(\frac{x}{y}\right) \mathscr{F}_j(y,q^2)\,, \tag{20.6.45}$$

where the matrix function $P_{ij}(z)$ is subject to the conditions

$$\int_0^1 z^{s-1} P_{ij}(z)\, dz = -4\pi^2 c_{ij}(s)\,. \tag{20.6.46}$$

(The factor $4\pi^2$ is conventional.)

The matrix $c_{ij}(s)$ was calculated in quantum chromodynamics by Georgi and Politzer[16] and Gross and Wilczek.[17] They assumed that there are N flavors of quarks that are sufficiently light to be treated as massless, while quarks of all other flavors are treated as very heavy and integrated out, so that they can be ignored except for their effect on the strong coupling constant. Their results in the basis provided by the operators (20.6.23) and (20.6.24) are

$$c_{00}(s) = \frac{1}{2\pi^2}\left\{ C_1 \left[\frac{1}{12} - \frac{1}{s(s-1)} - \frac{1}{(s+1)(s+2)} + \sum_{t=2}^{s} \frac{1}{t} \right] + \frac{N}{3} C_2 \right\}\,, \tag{20.6.47}$$

$$c_{f0}(s) = \frac{1}{\pi^2} C_2 \left[\frac{1}{s+2} + \frac{2}{s(s+1)(s+2)} \right]\,, \tag{20.6.48}$$

$$c_{0f}(s) = \frac{1}{8\pi^2} C_3 \left[\frac{1}{s+1} + \frac{2}{s(s-1)} \right]\,, \tag{20.6.49}$$

$$c_{ff'}(s) = \frac{1}{8\pi^2} C_3 \left[1 - \frac{2}{s(s+1)} + 4 \sum_{t=2}^{s} \frac{1}{t} \right] \delta_{ff'}\,, \tag{20.6.50}$$

where 0 and f denote the operators (20.6.24) and (20.6.23), respectively; the constants C_1 and C_2 are defined by Eqs. (17.5.33) and (17.5.34) (with only a single quark flavor included in the trace in Eq. (17.5.34)); N is the number of quark flavors; and C_3 is defined (using the notation of Section 17.4) by

$$t_\alpha t_\alpha = C_3 g^2 1 . \tag{20.6.51}$$

In the realistic case of an $SU(3)$ gauge group with quarks in its defining representation **3**, these numbers are

$$C_1 = 3 , \qquad C_2 = \tfrac{1}{2} , \qquad C_3 = \tfrac{4}{3} . \tag{20.6.52}$$

These rather complicated results become simpler when expressed in term of the Altarelli–Parisi functions. It is straightforward to check that Eq. (20.6.46) is satisfied if

$$P_{ff'} = \delta_{ff'} \left[\frac{4}{3} \left(\frac{1+x^2}{(1-x)_+} \right) + 2\delta(1-x) \right] , \tag{20.6.53}$$

$$P_{f0} = x^2 - x + \frac{1}{2} , \tag{20.6.54}$$

$$P_{0f} = \frac{4}{3} \left[\frac{2}{x} - 2 + x \right] , \tag{20.6.55}$$

$$P_{00} = 6 \left[\frac{1-x}{x} + x(1-x) + \frac{x}{(1-x)_+} + \frac{11}{12}\delta(1-x) \right] - \frac{N}{3}\delta(1-x) , \tag{20.6.56}$$

where in an integral over x up to $x = 1$, $1/(1-x)_+$ is defined by

$$\frac{f(x)}{(1-x)_+} \equiv \frac{f(x) - f(1)}{1-x} . \tag{20.6.57}$$

For each s there is an obvious $(N-1)$-fold degenerate eigenvalue of the matrix $c(s)$, equal to the coefficient of $\delta_{ff'}$ in Eq. (20.6.50):

$$c(s, \text{adjoint}) = \frac{1}{8\pi^2} C_3 \left[1 - \frac{2}{s(s+1)} + 4 \sum_{t=2}^{s} \frac{1}{t} \right] , \tag{20.6.58}$$

with eigenoperators given by all independent linear combinations of the operators (20.6.23) with coefficients a_f satisfying $\sum_f a_f = 0$, which belong to the adjoint representation of the unbroken global $SU(N)$ symmetry group of quantum chromodynamics with N quark flavors. In addition for each s there are two eigenoperators belonging to the singlet representation of $SU(N)$, given by linear combinations of the operator (20.6.24) and the sum over f of the operators (20.6.23). These eigenoperators and the corresponding eigenvalues can be found by diagonalizing the 2×2 matrix:

$$c(s)_{\text{singlet}} = \begin{bmatrix} c_{00}(s) & c_{0f}(s)N \\ c_{f0}(s) & c(s, \text{adjoint}) \end{bmatrix} . \tag{20.6.59}$$

For $s = 2$ this matrix takes the form

$$c(2)_{\rm singlet} = \begin{bmatrix} NC_2/6\pi^2 & NC_3/6\pi^2 \\ C_2/3\pi^2 & C_3/3\pi^2 \end{bmatrix} . \tag{20.6.60}$$

This has one zero eigenvalue, corresponding to the linear combination of $\mathcal{O}^{\mu\nu}_{20}$ and $\sum_f \mathcal{O}^{\mu\nu}_{2f}$ that is equal to the energy-momentum tensor, which like J^μ is not renormalized. The other eigenvalue of the matrix (20.6.60) is given by its trace, $NC_2/6\pi^2 + C_3/3\pi^2$. Now, the minimum of the eigenvalues of $c_{ij}(s)$ for a given s must be at least as large as the minimum of the eigenvalues of $c_{ij}(s')$ for any $s' < s$, because otherwise for sufficiently large q^2 the integral (20.6.42) would eventually become larger for s than for s', in contradiction with the fact that this integral is a strictly decreasing function of s. Because the minimum eigenvalue for $s = 2$ is zero, we can conclude that all of the other eigenvalues for $s > 2$ are positive. In fact, they are all positive-definite, because there are no unrenormalized operators for $s > 2$. Hence strict Bjorken scaling occurs only in the extreme limit where $g_q^2 \to 0$, where only the contribution of the energy-momentum tensor survives. The prediction of violations of strict Bjorken scaling have been confirmed experimentally by exhaustive studies of deep-inelastic electron–nucleon and muon–nucleon scattering.

20.7 Renormalons*

Since the beginning of quantum field theory theorists have wondered whether the perturbation series for physical matrix elements converges, and if it does not, then what can be done about it? Very early in the modern period Dyson[18] observed that the number of diagrams of nth order typically grows as $n!$, which suggested that the perturbation series has zero radius of convergence.

There is a well-known technique known as a Borel transformation[19] for improving the convergence of a power series whose nth order term grows as $n!$, either to make the series converge, or at least to improve the behavior of the series so that it can be used as an asymptotic expansion for a larger range of coupling constants. For a given series

$$F(g) = \sum_n f_n g^n \tag{20.7.1}$$

we consider the related series

$$B(z) \equiv \sum_n f_n z^n / n! \,. \tag{20.7.2}$$

* This section lies somewhat out of the book's main line of development, and may be omitted in a first reading.

If f_n grows no faster than $n!$ then $B(z)$ will generally have at least a finite radius of convergence. The question is, can we recover the original series (20.7.1) from the resummed series (20.7.2)? Using the familiar formula

$$\int_0^\infty \exp(-z/g)\, z^n\, dz = n!\, g^{n+1}$$

we see that, at least formally,

$$gF(g) = \int_0^\infty \exp(-z/g)\, B(z)\, dz\,. \tag{20.7.3}$$

Singularities of $B(z)$ anywhere in the complex plane limit the radius of convergence of the series (20.7.2), but this is not an insuperable problem if these singularities are off the positive real axis. To calculate $F(g)$ using Eq. (20.7.3) we need $B(z)$ only for real positive values of z less than or of order g, which can be obtained from the power series (20.7.1) if the singularities of $B(z)$ in the complex plane are all at distances from the origin much greater than g. Even if a few poles z_1, z_2, etc. have moduli of order g or less, we can calculate $B(z)$ out to values of z of order g by using the power series for $(z-z_1)(z-z_2)\cdots B(z)$, but for this purpose we have to know where the poles are.

Singularities of $B(z)$ on the positive real axis are much worse, for they invalidate Eq. (20.7.3). The contour in this integral may be distorted to avoid singularities on the positive real axis, but then we have an ambiguity: do we distort the contour above or below the singularity?

This section will show that some of the singularities of the Borel transform $B(z)$ are associated with solutions of the classical field equations known as *instantons*, while other singularities, known as *renormalons*, are associated with terms in the operator product expansion. In quantum chromodynamics it is the renormalons that obstruct the use of the Borel transformation to sum the perturbation series.

It was Lipatov[20] who in 1976 showed that some of the singularities of the Borel transform $B(z)$ are associated with the existence of classical solutions of the field equations. Consider a function $F(g)$ defined by a Euclidean path integral:

$$F(g) \equiv \int [d\phi] \exp\left(I[\phi, g]\right)\,. \tag{20.7.4}$$

(The use of Euclidean path integration is discussed in Appendix A of Chapter 23.) The coefficients in the power series (20.7.1) are given by

$$\begin{aligned} f_n &= \frac{1}{2\pi i}\int [d\phi] \oint dg\; g^{-n-1} \exp\left(I[\phi, g]\right) \\ &= \frac{1}{2\pi i}\int [d\phi] \oint dg\; \exp\Big(I[\phi, g] - (n+1)\ln g\Big)\,, \end{aligned} \tag{20.7.5}$$

where $\oint$ denotes the integral counterclockwise over a closed curve in the complex g plane surrounding the point $g = 0$. For very large n, it is reasonable to suppose that the integral is dominated by the point ϕ_n, g_n where the argument of the exponential in the last line of Eq. (20.7.5) is stationary in both ϕ and g:

$$\left.\frac{\delta I[\phi, g_n]}{\delta\phi(x)}\right|_{\phi=\phi_n} = 0\,, \tag{20.7.6}$$

$$\left.\frac{\partial I[\phi_n, g]}{\partial g}\right|_{g=g_n} = \frac{n+1}{g_n}\,. \tag{20.7.7}$$

For instance, suppose that $I[\phi, g]$ is the action for a massless scalar field

$$I[\phi, g] = -\frac{1}{2}\int \partial_i\phi\partial_i\phi\, d^4x - \frac{g}{24}\int \phi^4\, d^4x\,, \tag{20.7.8}$$

the sum running over the Euclidean coordinate directions 1, 2, 3, 4. Then the field equation (20.7.6) reads

$$\Box\phi_n = \tfrac{1}{6}g_n\phi_n^3\,. \tag{20.7.9}$$

We will see that g_n is negative, so the solution has

$$\phi_n(x) = (-g_n)^{-1/2}\chi(x)\,, \tag{20.7.10}$$

with $\chi(x)$ the g-independent solution of the equation

$$\Box\chi = -\tfrac{1}{6}\chi^3\,. \tag{20.7.11}$$

The condition (20.7.7) tells us that

$$-\frac{1}{24}\int d^4x\, \phi_n^4 = \frac{n+1}{g_n}$$

or in terms of the rescaled field (20.7.10)

$$g_n = -\frac{1}{24(n+1)}\int d^4x\, \chi^4\,. \tag{20.7.12}$$

At this stationary point the action (20.7.8) becomes

$$\begin{aligned} I[\phi_n, g_n] &= -\frac{1}{2}\int \partial_i\phi_n\partial_i\phi_n\, d^4x - \frac{g_n}{24}\int \phi_n^4\, d^4x = \frac{g_n}{24}\int \phi_n^4\, d^4x \\ &= -n-1\,. \end{aligned} \tag{20.7.13}$$

Evaluating Eq. (20.7.5) at the stationary point then gives for $n \to \infty$**

$$f_n \approx g_n^{-n-1} \exp\Big(I(\phi_n, g_n)\Big) = (n+1)^{n+1} \left(-\frac{e}{24}\int \chi^4\, d^4x\right)^{-n-1}$$
$$\approx n!\left(-\frac{1}{24}\int \chi^4\, d^4x\right)^{-n} . \tag{20.7.14}$$

The leading singularity in $B(z)$ is therefore a pole at $z = z_1$, where

$$z_1 = -\frac{1}{24}\int \chi^4\, d^4x . \tag{20.7.15}$$

Because this is negative, it does not prevent us from carrying out the integration in Eq. (20.7.3). To calculate the pole position (20.7.15), we note that the field equation (20.7.11) has the solution

$$\chi = \frac{4\sqrt{3}a}{r^2 + a^2} , \tag{20.7.16}$$

where $r = (x_i x_i)^{1/2}$, and a is an arbitrary parameter. This solution of the field equation is an elementary example of the 'instanton' solutions to be discussed in Section 23.5. (These are called instantons because instead of being concentrated along a worldline they are concentrated near a point in spacetime — in this case, the origin.) Fortunately the pole position does not depend on a:

$$z_1 = -96\pi^2 a^4 \int_0^\infty \frac{r^2\, dr^2}{(r^2 + a^2)^4} = -16\pi^2 . \tag{20.7.17}$$

We see that the perturbation series for $B(z)$ can be used in Eq. (20.7.3) if $g \ll 16\pi^2$. If $g/16\pi^2$ is of the order or greater than unity, we may still be able to calculate $B(z)$ from the perturbation series for $(z + 16\pi^2)B(z)$.

We will see in Section 23.5 that there are instanton solutions in non-Abelian gauge theories like quantum chromodynamics, but these also yield relatively harmless singularities of $B(z)$ on the negative real axis. The real problem in quantum chromodynamics is with a different class of singularities, known as renormalons.[21] These were first discovered through the realization[22] that single $2n$-th order diagrams, like that shown in Figure 20.3, can make individual contributions that grow like $n!$, and hence according to Eq. (20.7.2) may lead to additional singularities in $B(z)$. In this particular case the singularity is known as an *infrared* renormalon,

** Here the symbol '$\approx$' should be interpreted to mean 'asymptotically equal up to factors of constants and powers of n.' These factors arise from the factor $\sqrt{12\pi n}$ in Stirling's formula for $(n+1)!$, from the ratio of $(n+1)!$ and $n!$, and from the integral over the fluctuations of g and $\phi(x)$ around the stationary point. Since we are not going to calculate the factors from the last source, there is no point in keeping track of the factors from the first two sources either.

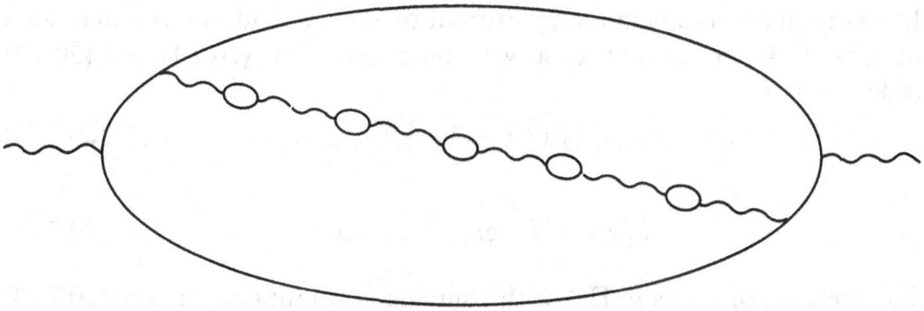

Figure 20.3. One of a class of N-loop Feynman diagrams that grow as $N!$. Solid lines are fermions; wavy lines are gauge bosons.

as it arises from virtual momenta that are much smaller than those used to define the running quantum chromodynamics coupling g_μ). Fortunately, it is possible to use the operator product expansion to locate the infrared renormalons, without having to look at individual Feynman diagrams.

For a simple but important example, consider the sum $\Pi^{\mu\nu}_{\alpha\beta}(q)$ of all vacuum diagrams in quantum chromodynamics with insertions of four-vector currents J^μ_α and J^ν_β carrying a four-momentum q into and out of the diagram. ($\Pi^{\mu\nu}(q)$ gives the hadronic contribution to the electroweak vacuum polarization, and its imaginary part yields the cross section for e^+–e^- and electron–antineutrino annihilation into hadrons.) As we saw in Section 20.5, this receives contributions that go as q^2 (the Fourier transform of x^{-6}) from the operator 1, contributions that go as q^{-2} from the operator $F^{\mu\nu}_\alpha F_{\alpha\mu\nu}$, contributions that go as q^{-4} from four-fermion operators, and so on. (For instance, the contribution of the operator $F^{\mu\nu}_\alpha F_{\alpha\mu\nu}$ arises from diagrams like Figure 20.3, with a momentum much less than q flowing through the chain of bubbles.) Dimensional analysis tells us that these momentum-dependent factors must be accompanied with vacuum matrix elements proportional to Λ^0, Λ^4, Λ^6, etc. But if we calculate Feynman diagrams using a running coupling defined at a scale $\mu \gg \Lambda$, where the coupling constants are small, then according to Eq. (18.7.7)

$$\Lambda^2 = \mu^2 \exp\left(\frac{12\pi}{(33-2n_f)\alpha_s(\mu)}\right) , \tag{20.7.18}$$

where n_f is the number of quark flavors with mass much less than μ. Operators of dimension $d > 0$ would make a contribution in the operator product expansion with the coupling-constant dependence

$$\Lambda^d \propto \exp\left(\frac{6\pi d}{(33-2n_f)\alpha_s(\mu)}\right) . \tag{20.7.19}$$

In quantum chromodynamics, perturbation theory yields a series in powers of $\alpha_s \equiv g^2/4\pi$ rather than g, so we would write Eqs. (20.7.1) and (20.7.3) in terms of α_s:

$$F(\alpha_s) = \sum_n f_n \alpha_s^n \,, \tag{20.7.20}$$

$$\alpha_s F(\alpha_s) = \int_0^\infty \exp(-z/\alpha_s)\, B(z)\, dz \,. \tag{20.7.21}$$

The presence of terms in $\Pi^{\mu\nu}$ with coupling-constant dependence (20.7.19) indicates that $B(z)$ must have singularities (not necessarily poles) at

$$z_1 = \frac{6\pi d}{(33 - 2n_f)} \,. \tag{20.7.22}$$

These are on the positive real axis, and make the integral (20.7.21) ambiguous. Thus Borel summation cannot be used to deal with low-energy effects in quantum chromodynamics.

The fact that diagrams with small virtual momenta impede the use of perturbation theory is of course nothing new. As we saw in Section 20.2, the whole point of the operator product expansion is to separate the parts of Feynman diagrams where every line carries a large momentum, which in asymptotically free theories can be calculated using perturbation theory, from the contribution of the parts of Feynman diagrams through which small momenta flow, which cannot be calculated perturbatively.

Appendix Momentum Flow: The General Case

In this appendix we shall consider the asymptotic behavior of an amplitude in a general renormalizable quantum field theory when the momenta of any set of two or more external lines become large, taking into account operators containing arbitrary numbers of field factors and derivatives, with dimensionality up to some limit N. To deal with this problem we shall have to introduce a more compact notation than that of Sections 20.1 or 20.2. A letter ℓ, ℓ', etc. will denote a set of external lines i of specified types entering or leaving a Feynman diagram or part of a Feynman diagram. A letter k, k', etc. will denote the set of four-momenta k_i of such lines, subject to the condition that their sum has some fixed value p. The amplitude $\Gamma_{\ell\ell'}(k, k', p)$ is the sum of all graphs with a set ℓ of incoming lines carrying momenta k and a set ℓ' of outgoing lines carrying momenta k', including the final bare propagators for the set ℓ but not ℓ'.

As shown by the power-counting theorem[3] quoted in Section 12.1, the part of the region of integration where the momenta of order k flow only through a subgraph with external lines ℓ and ℓ'' makes a contribution to

$\Gamma_{\ell\ell'}(k,k',p)$ of order $k^{d(\ell,\ell'')}$, where $d(\ell,\ell'')$ is the dimensionality (in powers of mass or momentum) of this subgraph:

$$d(\ell,\ell'') = 4 - \sum_{i\in\ell,\ell''}(1+s_i) - \sum_{i\in\ell}(2-2s_i)\,. \tag{20.A.1}$$

(The last term in Eq. (20.A.1) arises from the propagators for the lines in the set ℓ.) It will be convenient to rewrite Eq. (20.A.1) as

$$d(\ell,\ell') = 4 - 4n(\ell) - N(\ell') + N(\ell)\,, \tag{20.A.2}$$

where $n(\ell)$ is the number of lines in the set ℓ, and $N(\ell)$ is the total dimensionality of the fields for these lines:

$$n(\ell) \equiv \sum_{i\in\ell} 1\,, \qquad N(\ell) \equiv \sum_{i\in\ell}(s_i+1)\,. \tag{20.A.3}$$

Here s_i is the 'spin' of a line of type i, in the sense used in Section 12.3: the dimensionality of the field of type i is $1+s_i$, and the bare propagator of such a field goes as k^{-2+2s_i}. (For scalars and gauge bosons, $s_i = 0$; for spin 1/2, $s_i = 1/2$.).

By taking account of the asymptotic behavior associated with these subgraphs, we wish to show that as $k \to \infty$ (all components going to infinity together, in generic directions), with k' and p fixed, the asymptotic behavior of $\Gamma_{\ell\ell'}(k,k',p)$ in each order of perturbation theory is of the form

$$\Gamma_{\ell\ell'}(k,k',p) \to \sum_{\mathcal{O}}^{(N)} U^{\ell}_{\mathcal{O}}(k)\,F_{\mathcal{O},\ell'}(k',p) + o\left(k^{4-4n(\ell)+N(\ell)-N}\right)\,, \tag{20.A.4}$$

where the sum runs over operators $\mathcal{O}$ with dimensionality $N(\mathcal{O}) \leq N$; the function $U^{\ell}_{\mathcal{O}}(k)$ is of order $k^{4+N(\ell)-4n(\ell)-N(\mathcal{O})}$; and $o(k^A)$ denotes terms that vanish faster (by at least one factor $1/k$) than k^A.

To isolate the contribution of operators of dimensionality $\leq N$, we shall define an 'N-irreducible' amplitude $I^N_{\ell\ell'}(k,k',p)$ as the sum of all graphs for $\Gamma_{\ell\ell'}(k,k',p)$ in which the lines in the set ℓ cannot be disconnected from the lines in the set ℓ' by cutting through any set ℓ'' of internal lines with $N(\ell'') \leq N$. Since the difference $\Gamma - I^N$ consists of graphs that *can* be disconnected in this way, it may be written as

$$\Gamma_{\ell\ell'}(k,k',p) - I^N_{\ell\ell'}(k,k',p) = \sum_{\ell''}^{(N)} \int dk''\, I^N_{\ell\ell''}(k,k'',p)\,\Gamma_{\ell''\ell'}(k'',k',p)\,, \tag{20.A.5}$$

where $\sum^{(N)}_{\ell''}$ denotes the sum over sets ℓ'' of particle lines with $N(\ell'') \leq N$, and $\int dk''$ is the integral over the components of the four-momenta in the set ℓ'', subject to the constraint that the sum of these momenta is p.

The asymptotic behavior of the kernel $I^N_{\ell,\ell''}(k,k'',p)$ is much simpler than that of $\Gamma_{\ell,\ell''}(k,k'',p)$. For $k \to \infty$ with k'' and p fixed it is dominated by the part of the region of integration where every internal line carries a momentum of order k, which gives an asymptotic behavior $k^{d(\ell,\ell'')}$, because the terms where only some subregion carries such large momenta would have to be connected to the rest of the graph by a bridge ℓ''' of lines with $N(\ell''') > N \geq N(\ell'')$, and this would be smaller as $k \to \infty$ by at least a factor $k^{N(\ell'')-N-1}$. Thus differentiating $I^N_{\ell,\ell''}(k,k'',p)$ with respect to any component of k'' or p reduces its asymptotic behavior to order $k^{d(\ell,\ell'')-1}$. But differentiating d times lowers the asymptotic behavior to $k^{d(\ell,\ell'')-d}$ only if $d \leq N - N(\ell'') + 1$, because for higher derivatives we get a larger contribution from the region of integration where the subgraph carrying momenta of order k is connected to the rest of the graph by a bridge of lines ℓ''' with total dimensionality greater than N, with the derivatives all acting on lines carrying momenta of order k'' or p. Therefore in order to take into account the contribution of operators containing derivatives of fields, we may write the asymptotic behavior of I^N as

$$I^N_{\ell\ell''}(k,k'',p) = \sum_{\nu:d_\nu+N(\ell'')\leq N} I^N_{\ell\ell''\nu}(k)\, P_{\ell''\nu}(k'',p) + o\left(k^{d(\ell,\ell'')-N+N(\ell'')}\right), \quad (20.A.6)$$

where $P_{\ell''\nu}(k'',p)$ are a complete set of homogeneous polynomials of order d_ν in the $n(\ell'')$ momenta k'' and p, and $I^N_{\ell\ell''\nu}(k)$ are functions only of k, of order $k^{d(\ell,\ell'')-d}$ as $k \to \infty$.

We cannot immediately use Eq. (20.A.6) in Eq. (20.A.5), because the integral over k'' receives important contributions from the region where some of the k'' are of order k. To deal with this, we use mathematical induction: we assume that Eq. (20.A.4) holds up to some given order in perturbation theory, and use it on the right-hand side of Eq. (20.A.5) to calculate the asymptotic behavior of Γ to the next order of perturbation theory. We rewrite Eq. (20.A.5) in the form

$$\begin{aligned}\Gamma_{\ell\ell'}(k,k',p) &= I^N_{\ell\ell'}(k,k',p)\\ &+ \sum_{\ell''}^{(N)} \int dk''\, I^N_{\ell\ell''}(k,k'',p) \left[\Gamma_{\ell''\ell'}(k'',k',p) - \sum_{\mathcal{O}}^{(N)} U^{\ell''}_{\mathcal{O}}(k'')\, F_{\mathcal{O},\ell'}(k',p)\right]\\ &+ \sum_{\mathcal{O}}^{(N)} F_{\mathcal{O},\ell'}(k',p) \sum_{\ell'':\, N(\ell'')\leq N} \int dk''\, I^N_{\ell\ell''}(k,k'',p) U^{\ell''}_{\mathcal{O}}(k'')\,. \end{aligned} \quad (20.A.7)$$

According to Eq. (20.A.4), the quantity in square brackets in the second term on the right-hand side of Eq. (20.A.7) vanishes for $k'' \to \infty$ faster than $(k'')^{4-4n(\ell'')+N(\ell'')-N}$, so the product of this factor with a polynomial $P_{\ell''\nu}(k'')$ of order $\nu \leq N - N(\ell'')$ vanishes faster than $(k'')^{4-4n(\ell'')}$, and therefore has a finite integral over the $4(n(\ell'')-1)$ independent components of k''.* Hence we may use Eq. (20.A.6) in this term, and find

$$\Gamma_{\ell\ell'}(k,k',p) \to \sum_{\nu:\, d_\nu+N(\ell')\leq N} I^N_{\ell\ell'\nu}(k)\, P_{\ell'\nu}(k',p)$$
$$+\sum_{\ell''}^{(N)} \sum_{\nu:\, d_\nu+N(\ell'')\leq N} I^N_{\ell\ell''\nu}(k) \int dk''\, P_{\ell''\nu}(k'',p)$$
$$\times \left[\Gamma_{\ell''\ell'}(k'',k',p) - \sum_{\mathcal{O}}^{(N)} U^{\ell''}_{\mathcal{O}}(k'')\, F_{\mathcal{O},\ell'}(k',p)\right]$$
$$+\sum_{\mathcal{O}}^{(N)} F_{\mathcal{O},\ell'}(k',p) \sum_{\ell''}^{(N)} \int dk''\, I^N_{\ell\ell''}(k,k'',p) U^{\ell''}_{\mathcal{O}}(k'')\,, \quad (20.A.8)$$

the correction being smaller than the terms shown by a factor $1/k$. (Of course, the first term on the right-hand side of Eq. (20.A.8) is present only if $N(\ell') \leq N$.)

Now, for each value of ℓ and ν with $d_\nu + N(\ell) \leq N$, there is an operator $\mathcal{O}$ with field factors corresponding to the lines in ℓ and with d_ν derivatives, such that in zeroth-order perturbation theory the vertex function, with incoming momentum p carried by the operator $\mathcal{O}$ and outgoing momenta k carried by external lines ℓ, is the polynomial $P_{\ell\nu}(k,p)$. Then the corresponding complete vertex function for a renormalized operator $\mathcal{O}_R = \sum_{\mathcal{O}'} Z_{\mathcal{O},\mathcal{O}'}\mathcal{O}'$ is

$$F_{\mathcal{O},\ell}(k,p) = \sum_{\mathcal{O}'} Z_{\mathcal{O},\mathcal{O}'}\left\{\delta_{\ell_{\mathcal{O}'},\ell}\, P_{\ell\nu_{\mathcal{O}'}}(k,p) + \int dk'\, P_{\ell_{\mathcal{O}'}\nu_{\mathcal{O}'}}(k',p)\,\Gamma_{\ell_{\mathcal{O}'}\ell}(k',k,p)\right\}, \quad (20.A.9)$$

where $\ell_{\mathcal{O}}$ and $\nu_{\mathcal{O}}$ label the types of fields and spacetime derivatives in the operator $\mathcal{O}$. We see then that Eq. (20.A.4) is satisfied, with

$$U^\ell_{\mathcal{O}}(k) = \sum_{\mathcal{O}'}^{(N)} I^N_{\ell\mathcal{O}'}(k)\left[Z^{-1}_{\mathcal{O}',\mathcal{O}} - \int dk''\, U^{\ell_{\mathcal{O}'}}_{\mathcal{O}}(k'')\, P_{\mathcal{O}'}(k'')\right]$$
$$+\sum_{\ell''}^{(N)} \int dk''\, I^N_{\ell\ell''}(k,k'')\, U^{\ell}_{\mathcal{O}''}(k'')\,, \quad (20.A.10)$$

* This overlooks the possibility that even though power-counting indicates the convergence of the integral in the second term on the right-hand side of Eq. (20.A.7) over the region where all k'' go to infinity together, for $n(\ell'') \geq 3$ subintegrations may diverge. For this reason the arguments given in this appendix do not constitute a proof of the operator product expansion except for the simple case treated in Section 20.2, where we consider the momenta of just two external lines to go to infinity, and we look for the terms in the power series expansion associated with operators that are quadratic in the fields.

in which we now use the abbreviations

$$I_{\ell\mathcal{O}}(k) \equiv I_{\ell\ell_{\mathcal{O}}\nu_{\mathcal{O}}}(k)\,, \qquad P_{\mathcal{O}}(k) \equiv P_{\ell_{\mathcal{O}}\nu_{\mathcal{O}}}(k)\,. \tag{20.A.11}$$

(The p dependence of $\Gamma_{\ell_{\mathcal{O}''}\ell'}$ is dropped for the same reason as in Section 20.2: whether or not k'' is comparable with k, p is negligible here compared with k.) We will define the normalization constant $Z_{\mathcal{O}',\mathcal{O}}$ so that at a renormalization point $k(\mu)$, $p(\mu)$ the function $F_{\mathcal{O},\ell}(k(\mu),p(\mu))$ has the same value $\delta_{\ell\ell_{\mathcal{O}}}P_{\mathcal{O}}(k(\mu),p(\mu))$ that it would have in the absence of interactions:

$$\delta_{\ell\ell_{\mathcal{O}}}P_{\mathcal{O}}(k(\mu),p(\mu)) = F_{\mathcal{O},\ell}(k(\mu),p(\mu)) = \sum_{\mathcal{O}'} Z_{\mathcal{O},\mathcal{O}'}$$
$$\times\left\{\delta_{\ell_{\mathcal{O}'},\ell}\,P_{\ell\nu_{\mathcal{O}'}}(k(\mu),p(\mu)) + \int dk'\,P_{\ell_{\mathcal{O}'}\nu_{\mathcal{O}'}}(k',p(\mu))\,\Gamma_{\ell_{\mathcal{O}'}\ell}(k',k(\mu),p(\mu))\right\}. \tag{20.A.12}$$

For $\Gamma = 0$ this has a solution $Z = 1$ which is unique (because the polynomials for a given set of lines are supposed to be linearly independent), so by continuity Eq. (20.A.12) will have a unique solution for coupling constants in some finite range. Eq. (20.A.10) therefore provides a recursive definition of the coefficient functions $U^{\ell}_{\mathcal{O}}(k)$ appearing in the general operator product expansion (20.A.4).

Problems

1. Consider a theory of a fermion field ψ interacting with a scalar field ϕ with interactions of the form $\bar{\psi}\psi$ and ϕ^4. List the operators that appear in the operator product expansion of $\bar{\psi}(x)\psi(0)$ with coefficient functions that (judging from perturbation theory) are singular and non-vanishing for $x \to 0$. Describe how you would calculate these coefficient functions to one-loop order.

2. Consider quantum chromodynamics with N massless quarks, and define the spectral functions of the scalar and pseudoscalar quark bilinears by

$$\sum_N \langle \text{VAC}|\bar{\psi}(0)\lambda_\alpha\psi(0)|\text{VAC}\rangle\langle\text{VAC}|\bar{\psi}(0)\lambda_\beta\psi(0)|\text{VAC}\rangle^*$$
$$= (2\pi)^{-3}\theta(p^0)\rho^S_{\alpha\beta}(-p^2)\,,$$
$$\sum_N \langle \text{VAC}|\bar{\psi}(0)\gamma_5\lambda_\alpha\psi(0)|\text{VAC}\rangle\langle\text{VAC}|\bar{\psi}(0)\gamma_5\lambda_\beta\psi(0)|\text{VAC}\rangle^*$$
$$= (2\pi)^{-3}\theta(p^0)\rho^P_{\alpha\beta}(-p^2)\,,$$

where λ_α are a complete set of traceless Hermitian $N \times N$ matrices, normalized so that $\text{Tr}(\lambda_\alpha \lambda_\beta) = 2\delta_{\alpha\beta}$. What spectral function sum rules are satisfied by linear combinations of $\rho^S_{\alpha\beta}(\mu^2)$ and $\rho^P_{\alpha\beta}(\mu^2)$?

3. Derive Eq. (20.6.8) in the parton model from the formulas (8.7.7) and (8.7.38) for Compton scattering.
4. List the gauge-invariant symmetric traceless tensors of twist four in quantum chromodynamics.
5. In the massless scalar field theory with interaction $-g\phi^4/24$ (with $g > 0$), where in the complex plane would you expect the function (20.7.3) to have renormalon singularities?

References

1. K. Wilson, *Phys. Rev.* **179**, 1499 (1969)
2. W. Zimmerman, in *Lectures on Elementary Particles and Quantum Field Theory — 1970 Brandeis University Summer Institute in Theoretical Physics*, eds. S. Deser, H. Pendleton, and M. Grisaru, (MIT Press, Cambridge, 1970).
3. S. Weinberg, *Phys. Rev.* **118**, 838 (1960).
4. C. Bernard, A. Duncan, J. LoSecco, and S. Weinberg, *Phys. Rev.* **D12**, 792 (1975).
5. S. Weinberg, *Phys. Rev. Lett.* **18**, 507 (1967). For the general case, see Ref. 4.
6. K. Kawarabayashi and M. Suzuki, *Phys. Rev. Lett.* **16**, 255 (1966); Riazuddin and Fayyazuddin, *Phys. Rev.* **147**, 1071 (1966).
7. M. Ademollo, G. Veneziano, and S. Weinberg, *Phys. Rev. Lett.* **22**, 83 (1969).
8. J. F. Donoghue and E. Golowich, *Phys. Rev.* **D49**, 1513 (1994).
9. This result was reported at the 1968 'Rochester' conference at Vienna, and published in E. D. Bloom *et al.*, *Phys. Rev. Lett.* **23**, 930 (1969); M. Breidenbach *et al.*, *Phys. Rev. Lett.* **23**, 935 (1969).
10. J. D. Bjorken, *Phys. Rev.* **179**, 1547 (1969).
11. R. P. Feynman, *Phys. Rev. Lett.* **23**, 1415 (1969); *Photon-Hadron Interactions* (Benjamin, New York, 1972).

12. C. G. Callan and D. J. Gross, *Phys. Rev. Lett.* **22**, 156 (1969). The reader should be warned that the symbol ω as used by Callan and Gross is $2/\omega$ in the notation used here.

13. N. Christ, B. Hasslacher, and A. H. Mueller, *Phys. Rev.*, **D6**, 3543 (1972).

14. D. Gross and S. Treiman, *Phys. Rev.* **D4**, 1059 (1971).

15. G. Altarelli and G. Parisi, *Nucl. Phys.* **B126**, 298 (1972).

16. H. Georgi and H. D. Politzer, *Phys. Rev.* **D9**, 416 (1974).

17. D. J. Gross and F. Wilczek, *Phys. Rev.* **D9**, 980 (1974).

18. F. J. Dyson, *Phys. Rev.* **85**, 631 (1952).

19. See, e.g., G. N. Hardy, *Divergent Series* (Oxford University Press, Oxford, 1949).

20. L. N. Lipatov, Leningrad Nuclear Physics Institute report, 1976 (unpublished); *Proceedings of the XVIII International Conference on High Energy Physics at Tbilisi*, 1976.

21. G. 't Hooft, in *The Whys of Subnuclear Physics – Proceedings of the 1977 Erice Summer School*, ed. A. Zichichi (Plenum, New York, 1978). For a collection of articles on renormalons and high orders of perturbation theory, see *Large Order Behavior of Perturbation Theory*, eds. J. C. Le Guillou and J. Zinn-Justin (North-Holland, Amsterdam, 1990). For further work on infrared renormalons, see A. H. Mueller, *Phys. Lett. B* **308**, 355 (1993); *Nucl. Phys.* **B250**, 327 (1995); A. Duncan and S. Pernice, *Phys. Rev. D* **51**, 1956 (1995), and articles quoted therein.

22. B. Lautrup, *Phys. Lett.* **76B**, 109 (1977).

21

Spontaneously Broken Gauge Symmetries

The 1961 theorem[1] that broken symmetry implies massless spin zero Goldstone bosons was seen at first as a serious obstacle to the search for broken symmetries in nature. A few years later several authors noted an exception to this theorem, already mentioned in Section 19.2: the Goldstone bosons are absent where the broken symmetry is local, rather than global.[2] Instead, these degrees of freedom show up as the helicity zero states of the vector particles associated with the broken local symmetries, which thereby acquire a mass. This phenomenon, now generally known as the Higgs mechanism, was not at first applied in any sort of realistic theory, perhaps because by the mid-1960s it had become clear that the pion is a Goldstone boson of a spontaneously broken approximate symmetry, and attention therefore shifted away from the effort to avoid Goldstone bosons. But soon after, spontaneously broken local symmetries turned out to provide the natural framework for understanding the weak and electromagnetic interactions of the elementary particles.[3]

21.1 Unitarity Gauge

We saw in Chapter 19 that in a theory with a global symmetry group G that is spontaneously broken to a subgroup H, there is a massless 'Goldstone' boson for every independent broken symmetry, in the sense that the mass matrix M^2_{nm} of real spinless fields $\phi_n(x)$ has a zero eigenvalue with eigenvector $\sum_m (t_\alpha)_{nm} v_m$ for each independent broken symmetry generator t_α of G. (We are here considering the case where these Goldstone bosons are included among the elementary spinless particles represented by scalar or pseudoscalar fields ϕ_n appearing in the Lagrangian; the more general case will be taken up in Section 21.4.) We also saw that we could rotate away these Goldstone modes, by subjecting the fields to a G transformation $\gamma^{-1}(x)$

$$\tilde{\phi}_n(x) = \sum_m \gamma^{-1}_{nm}(x)\phi_m(x) , \tag{21.1.1}$$

such that the new fields are orthogonal to the Goldstone directions*

$$0 = \sum_{nm} \tilde{\phi}_n(x)(t_\alpha)_{nm} v_m \,, \tag{21.1.2}$$

where v_n is the vacuum expectation value, $v_n \equiv \langle \phi_n(0) \rangle_{\rm VAC}$. After rotating the fields so that they satisfy Eq. (21.1.2), the Goldstone boson fields then reemerged as the spacetime-dependent parameters in $\gamma(x)$. The point of this procedure was that since the Lagrangian was invariant under the transformation (21.1.1) with constant $\gamma(x)$, all dependence on $\gamma(x)$ dropped out except where $\gamma(x)$ is acted on by derivatives.

On the other hand, if the Lagrangian is invariant not only under constant G transformations but also under G transformations that depend on spacetime position, then the transformation (21.1.1) is a true symmetry of the theory, and all dependence on $\gamma(x)$ drops out of the Lagrangian, so that we can simply replace $\phi_n(x)$ everywhere with $\tilde{\phi}_n(x)$. This is a choice of *gauge*, fixed by imposing the condition (21.1.2) on $\tilde{\phi}(x)$ rather than by imposing conditions on the gauge fields themselves. (For instance, in the electrodynamics of a charged scalar field $\phi \equiv \phi_1 + i\phi_2$, we can choose a gauge like Lorentz or Coulomb gauge by imposing conditions $\partial_\mu A^\mu = 0$ or $\nabla \cdot \mathbf{A} = 0$, but we can also choose a gauge by imposing a condition on ϕ, as for instance that ϕ be real, or in other words by rotating the two-vector $\{{\rm Re}\phi, {\rm Im}\phi\}$ into the 1-direction.) The gauge defined by Eq. (21.1.2) is called *unitarity gauge*,[3a] because in this gauge it will be obvious that the theory does not have any degrees of freedom with negative probability, like timelike gauge bosons. More generally, the unitarity gauge makes manifest the menu of physical particles of the theory.

Eq. (21.1.2) shows that there are no Goldstone boson fields in unitarity gauge. Since the theory is gauge-invariant this means that there are no physical Goldstone bosons, whatever gauge we choose. What about the vector bosons? If the ϕ_n are elementary canonically normalized scalar fields, the Lagrangian will contain a term

$$\mathscr{L}_\phi = -\frac{1}{2} \sum_n \left(\partial_\mu \tilde{\phi}_n - i \sum_{m,\alpha} t^\alpha_{nm} A_{\alpha\,\mu} \tilde{\phi}_m \right)^2 \,, \tag{21.1.3}$$

where t^α runs over all the generators of the gauge group G. (From now on we shall drop tildes, it being understood henceforth in this section that we are already in unitarity gauge.) We are assuming that the symmetry G is broken by the vacuum expectation value v_n of ϕ_n, so in order to see the

* Here we are working with a real reducible or irreducible representation of the symmetry algebra, for which the matrices it_α are real. The transition to a complex representation is described below.

nature of the particle spectrum, we define a shifted field ϕ':

$$\phi_n \equiv v_n + \phi'_n \,. \tag{21.1.4}$$

(It is sometimes convenient to take v_n in Eqs. (21.1.2) and (21.1.4) as the vacuum expectation value in the tree approximation. In order to generate a useful perturbation theory it is only necessary that v_n should agree with the true vacuum expectation value in lowest order.) Expanding Eq. (21.1.3) to second order in ϕ' and A, we have:

$$\mathscr{L}_{\phi,\text{QUAD}} = -\frac{1}{2}\sum_n \left(\partial_\mu \phi'_n - i\sum_{m,\alpha} t^\alpha_{nm} A_{\alpha\mu} v_m\right)^2 \,. \tag{21.1.5}$$

Using Eq. (21.1.2), we see that the ϕ'–A cross term vanishes, yielding

$$\mathscr{L}_{\phi,\text{QUAD}} = -\frac{1}{2}\sum_n \partial_\mu \phi'_n \partial^\mu \phi'_n - \frac{1}{2}\sum_{\alpha\beta} \mu^2_{\alpha\beta} A_{\alpha\mu} A^\mu_\beta \,, \tag{21.1.6}$$

where

$$\mu^2_{\alpha\beta} \equiv -\sum_{nm\ell} t^\alpha_{nm} t^\beta_{n\ell} v_m v_\ell \,. \tag{21.1.7}$$

Combining this with the quadratic terms in the Yang–Mills Lagrangian $-\frac{1}{4} F_{\alpha\mu\nu} F^{\mu\nu}_\alpha$, we see that the vector particles have a mass matrix $\mu^2_{\alpha\beta}$. In our notation the generators t^α_{nm} are proportional to gauge coupling constants, so Eq. (21.1.7) yields vector boson masses that are also proportional to these coupling constants.

Let's take a look at some of the algebraic properties of $\mu^2_{\alpha\beta}$. Since t^α_{nm} is imaginary and antisymmetric (and hence Hermitian) the matrix $\mu^2_{\alpha\beta}$ is real, symmetric, and positive. Also, if a certain real linear combination of generators $\sum_\alpha c_\alpha t_\alpha$ is unbroken, then

$$\sum_{\alpha,m} c_\alpha (t_\alpha)_{nm} v_m = 0 \,, \tag{21.1.8}$$

in which case Eq. (21.1.7) shows that, as noted by Kibble,[2]

$$\sum_\beta \mu^2_{\alpha\beta} c_\beta = 0 \,. \tag{21.1.9}$$

That is, we still have a massless gauge boson for every unbroken gauge symmetry. The converse is also true; from Eq. (21.1.7) we see that for arbitrary real constants c_α

$$\sum_{\alpha\beta} \mu^2_{\alpha\beta} c_\alpha c_\beta = \sum_n \left(\sum_{\alpha,m} c_\alpha \cdot i t^\alpha_{nm} v_m\right)^2 \geq 0 \,, \tag{21.1.10}$$

and this can vanish only if c_α satisfies Eq. (21.1.8).

In particular, if there is just one unbroken symmetry $\sum_\alpha c_\alpha t_\alpha$, then the general gauge field can be written

$$A^\mu_\alpha = c_\alpha A^\mu + \cdots \,, \tag{21.1.11}$$

where '$\cdots$' denotes a linear combination of the gauge fields of definite non-zero mass, and c_α is the coefficient of t_α in the single unbroken generator q,

$$q = \sum_\alpha c_\alpha \, t_\alpha \,, \tag{21.1.12}$$

so that the field A^μ will appear with zero coefficient in the mass term $-\frac{1}{4}\sum_{\alpha\beta}\mu^2_{\alpha\beta}A^\mu_\alpha A_{\mu\beta}$. In order for the coefficient of $-\frac{1}{4}(\partial_\mu A_\nu - \partial_\nu A_\mu)^2$ in the kinetic gauge Lagrangian $-\frac{1}{4}\sum_\alpha(\partial_\mu A_{\alpha\nu} - \partial_\nu A_{\alpha\mu})^2$ to have the canonical value 1, the c_α must also be normalized

$$\sum_\alpha c_\alpha^2 = 1 \,. \tag{21.1.13}$$

Note that q *is* the charge to which A^μ is coupled, in the sense that

$$\sum_\alpha t_\alpha \, A^\mu_\alpha = q \, A^\mu + \cdots \,, \tag{21.1.14}$$

where '$\cdots$' again denotes terms involving massive gauge fields. We shall use these general results in studying the electroweak theory in Section 21.3.

These results have been derived here for scalar fields that form a real representation of the gauge group, but they can be straightforwardly converted to a form appropriate for complex representations. We saw in Section 19.6 that a complex scalar field $\chi(x)$ that transforms according to a representation of the gauge group with Hermitian generators T_α may be written as a set of real fields

$$\phi(x) = \begin{pmatrix} \mathrm{Re}\,\chi(x) \\ \mathrm{Im}\,\chi(x) \end{pmatrix} \,, \tag{21.1.15}$$

which furnish a real representation of the gauge group with generators

$$i\,t^\alpha = \begin{pmatrix} -\mathrm{Im}\,T^\alpha & -\mathrm{Re}\,T^\alpha \\ \mathrm{Re}\,T^\alpha & -\mathrm{Im}\,T^\alpha \end{pmatrix} \,. \tag{21.1.16}$$

Inserting Eqs. (21.1.15) and (21.1.16) in Eq. (21.1.7) gives the vector boson mass matrix (21.1.7) as

$$\mu^2_{\alpha\beta} = \mathrm{Re}\left(\langle\chi\rangle^\dagger_{\rm VAC},\, T_\alpha\, T_\beta\, \langle\chi\rangle_{\rm VAC}\right) = \tfrac{1}{2}\left(\langle\chi\rangle^\dagger_{\rm VAC},\, \{T_\alpha,\, T_\beta\}\langle\chi\rangle_{\rm VAC}\right) \,. \tag{21.1.17}$$

Now let's take a closer look at the vector field propagator. Including the quadratic term in the Yang–Mills Lagrangian, the part of the total

Lagrangian that is quadratic in A is

$$-\tfrac{1}{4}\sum_{\alpha}(\partial_\lambda A_{\alpha\nu}-\partial_\nu A_{\alpha\lambda})^2-\tfrac{1}{2}\sum_{\alpha\beta}\mu^2_{\alpha\beta}\,A_{\alpha\lambda}A^\lambda_\beta$$
$$=\tfrac{1}{2}\sum_{\alpha\beta}A_\alpha{}^\nu\mathscr{D}_{\alpha\nu,\beta\lambda}(\partial)A_\beta{}^\lambda+\text{ total derivatives}\,,\qquad(21.1.18)$$

where

$$\mathscr{D}_{\alpha\nu,\beta\lambda}(\partial)=\delta_{\alpha\beta}\Big[\eta_{\nu\lambda}\Box-\partial_\nu\partial_\lambda\Big]-\mu^2_{\alpha\beta}\,\eta_{\nu\lambda}\,.\qquad(21.1.19)$$

Suppose for simplicity that all gauge symmetries are broken, so that $\mu^2_{\alpha\beta}$ has no zero eigenvalues. According to the general rules described in Section 9.4, the gauge-field propagator in momentum space is

$$\Delta_{\alpha\nu,\beta\lambda}(k)=-(\mathscr{D}^{-1})_{\alpha\nu,\beta\lambda}(ik)=\Big[(k^2+\mu^2)^{-1}(\eta_{\nu\lambda}+\mu^{-2}k_\nu k_\lambda)\Big]_{\alpha\beta}\,.\qquad(21.1.20)$$

Because of the $k_\nu k_\lambda$ term, the propagator has an asymptotic behavior $\Delta(k)\sim O(k^0)$, which does not allow us to use the usual power-counting arguments to prove renormalizability. Fortunately, as we shall see in the next section, there is another gauge in which renormalizability is obvious, at the price of obscuring the particle content of the theory.

* * *

It is important to note that although Goldstone bosons suddenly reappear in the physical spectrum in the limit of zero gauge couplings, physical matrix elements are perfectly continuous in this limit. This is because in unitarity gauge the gauge bosons do not entirely decouple for zero gauge couplings. Consider the matrix element for a scattering process $A+B\to C+D$, where A and C belong to some representation of the gauge algebra, and B and D belong to some different representation of the gauge algebra. The S-matrix element for this process receives a contribution from vector boson exchange

$$S_{CD,AB}=i(2\pi)^4\delta^4(p_A+p_B-p_C-p_D)\langle C|J^\nu_{N\alpha}|A\rangle\,\Delta_{\alpha\nu,\beta\lambda}(k)\langle D|J^\lambda_{N\beta}|B\rangle\,,\qquad(21.1.21)$$

where $k=p_A-p_C=p_D-p_B$, and $J^\nu_{N\alpha}$ is the current to which the gauge bosons are coupled, with the subscript N to remind us that in this gauge we omit the Goldstone boson pole term in this current. This current is proportional to gauge coupling constants, so the only terms in Eq. (21.1.21) that survive in the limit of zero gauge couplings are the ones involving the matrix μ^{-2}, which becomes singular in this limit. Hence for zero gauge couplings, the S-matrix element is

$$S_{CD,AB} \to i(2\pi)^4\delta^4(p_A+p_B-p_C-p_D)k_\nu k_\lambda(\mu^{-2})_{\alpha\beta}\langle C|J_{N\alpha}^{\nu}|A\rangle \frac{1}{k^2}\langle D|J_{N\beta}^{\lambda}|B\rangle\,. \tag{21.1.22}$$

The gauge coupling constants in the currents are cancelled by the gauge coupling constants in the matrix μ^2.

Now let us compare this result with the contribution of Goldstone boson exchange. For vanishing gauge couplings there is a set of Goldstone bosons B_α associated with generator t_α (assuming for simplicity that all gauge symmetries are spontaneously broken) with the kinematic term in the Lagrangian given by $-\frac{1}{2}\partial_\mu\phi_\alpha\partial^\mu\phi_\beta\sum_n Z_{\alpha n}Z_{\beta n}$, where $Z_{\gamma n}$ is the component of the Goldstone boson B_α in the spinless field ϕ_n, defined by Eq. (19.2.39). In order that the Goldstone boson fields be canonically normalized, we must therefore have

$$\sum_n Z_{\alpha n}Z_{\beta n} = \delta_{\alpha\beta}\,. \tag{21.1.23}$$

According to Eq. (19.2.49), the Goldstone boson B_α couples to the currents $J_{N\beta}^{\mu}$ with a coupling constant $F^{-1}_{\alpha\beta}$, where $F_{\alpha\beta}$ is the coupling of the Goldstone boson associated with generator t_α to the current $J_{N\beta}^{\nu}$ defined by Eq. (19.2.38). Thus the exchange of Goldstone bosons would give a scattering matrix element

$$S_{CD,AB} = i(2\pi)^4\delta^4(p_A+p_B-p_C-p_D)k_\nu k_\lambda F^{-1}_{\alpha\gamma}F^{-1}_{\beta\gamma}\langle C|J_{N\alpha}^{\nu}|A\rangle \frac{1}{k^2}\langle D|J_{N\beta}^{\lambda}|B\rangle\,. \tag{21.1.24}$$

But Eqs. (19.2.40), (21.1.7), and (21.1.23) give the vector boson mass matrix as

$$\mu^2_{\alpha\beta} = \sum_n F_{\alpha\gamma}Z_{\gamma n}F_{\beta\delta}Z_{\delta n} = F_{\alpha\gamma}F_{\beta\gamma}\,, \tag{21.1.25}$$

so in the limit of vanishing gauge boson coupling, the gauge boson exchange matrix element (21.1.22) is the same as the Goldstone boson exchange matrix element (21.1.24). This argument can be reversed; the requirement of continuity at zero gauge coupling can be used to derive a formula for the gauge boson masses even in the case where all other couplings are strong.

21.2 Renormalizable ξ-Gauges

In 1971 't Hooft[4] showed that path integrals in spontaneously broken gauge theories can be calculated in a gauge in which the vector boson propagators vanish for momentum $k \to \infty$ as k^{-2}, so that these theories satisfy the power-counting test for renormalizability described in Section 12.2. Here we shall describe a larger class of renormalizable gauges, parameterized by an arbitrary constant ξ, that was introduced a little later by Fujikawa, Lee, and Sanda.[5]

In general gauges the kinematic term (21.1.3) in the Lagrangian for the scalar fields of a theory contains a cross term

$$i \sum_{nm\alpha} \partial_\mu \phi'_n \, t^\alpha_{nm} A_\alpha{}^\mu v_m \,,$$

where v_m is the vacuum expectation value of ϕ_m, and ϕ'_n is the shifted field defined by Eq. (21.1.4). In unitarity gauge this term vanished as a consequence of the gauge condition (21.1.2). We will adopt a different approach here, similar to that of Sections 15.5 and 15.6. A functional $B[f]$ is introduced into the path integral, with

$$B[f] = \exp\left(\frac{-i}{2\xi} \int d^4x \sum_\alpha f_\alpha f_\alpha\right) . \tag{21.2.1}$$

This is equivalent to adding a gauge-fixing term in the Lagrangian

$$\mathscr{L}_{gf} = -\frac{1}{2\xi} \sum_\alpha f_\alpha f_\alpha \,. \tag{21.2.2}$$

Instead of taking $f_\alpha = \partial_\mu A^\mu_\alpha$ as in Section 15.5, we shall now take the gauge-fixing function as

$$f_\alpha = \partial_\mu A^\mu_\alpha - i\xi (t_\alpha)_{nm} \phi'_n v_m \,, \tag{21.2.3}$$

which is designed so that the above cross term in (21.1.3) is cancelled by the cross term in Eq. (21.2.2). Unitarity gauge is now a special case; for $\xi \to \infty$, the gauge-fixing functional (21.2.1) is infinitely sharply peaked at a ϕ' that satisfies the unitarity gauge condition (21.1.2). Another special case is provided by the limit $\xi \to 0$; here the gauge-fixing functional is peaked at a gauge field that satisfies the Landau gauge condition $\partial_\mu A^\mu_\alpha = 0$.

We also include in the Lagrangian a quartic polynomial $-P(\phi)$, subject to the gauge-invariance condition

$$\frac{\partial P(\phi)}{\partial \phi_n} (t_\alpha)_{nm} \phi_m = 0 \,. \tag{21.2.4}$$

We must of course also include in the Lagrangian a gauge-field term

$$\mathscr{L}_A = -\frac{1}{4} \sum_\alpha F^{\mu\nu}_\alpha F_{\alpha\mu\nu} \,. \tag{21.2.5}$$

The total Lagrangian density of gauge fields and scalars is then

$$\begin{aligned}\mathscr{L}_{A,\phi} &\equiv \mathscr{L}_A + \mathscr{L}_\phi + \mathscr{L}_{gf} \\ &= -\frac{1}{4}\sum_\alpha F_\alpha^{\mu\nu} F_{\alpha\mu\nu} + \frac{1}{2}\sum_{n\alpha\beta}(t^\alpha\phi)_n(t^\beta\phi)_n A_\alpha^\mu A_{\beta\mu} - \frac{1}{2\xi}\sum_\alpha(\partial_\mu A_\alpha^\mu)\,(\partial_\nu A_\alpha^\nu) \\ &\quad -\frac{1}{2}\sum_n \partial_\mu\phi'_n\partial^\mu\phi'_n + \frac{\xi}{2}\sum_{\alpha nm}(t_\alpha v)_n(t_\alpha v)_m\phi'_n\phi'_m \\ &\quad -P(\phi) + i\sum_{\alpha nm}\partial_\mu\phi'_n(t_\alpha)_{nm}\phi'_m A_\alpha^\mu + \text{total derivatives}\,. \end{aligned} \tag{21.2.6}$$

As we saw in Section 15.6, the introduction of a gauge-fixing functional $B[f]$ requires also the introduction of a ghost field $\omega_\alpha(x)$, with Lagrangian depending on the gauge-transformation properties of f_α. Under a general gauge transformation (with an arbitrary function $\epsilon_\alpha(x)$)

$$\delta A_\alpha^\mu = -\sum_{\beta\gamma} C_{\alpha\beta\gamma}\epsilon_\beta A_\gamma{}^\mu + \partial^\mu\epsilon_\alpha\,, \tag{21.2.7}$$

$$\delta\phi_n = i\sum_{\alpha m}\epsilon_\alpha(t_\alpha)_{nm}\phi_m\,, \tag{21.2.8}$$

we have

$$\delta f_\alpha = \Box\,\epsilon_\alpha - \sum_{\beta\gamma} C_{\alpha\beta\gamma}\partial_\mu(\epsilon_\beta A_\gamma{}^\mu) + \xi\sum_{n\beta}(t_\alpha v)_n\,\epsilon_\beta(t_\beta\phi)_n\,. \tag{21.2.9}$$

According to the general results of Section 15.6, this yields a ghost Lagrangian

$$\mathscr{L}_\omega = \omega_\alpha^*\left[\Box\,\omega_\alpha - \sum_{\beta\gamma} C_{\alpha\beta\gamma}\partial_\mu(\omega_\beta A_\gamma^\mu) + \xi\sum_{n\beta}(t_\alpha v)_n\omega_\beta(t_\beta\phi)_n\right]. \tag{21.2.10}$$

Finally, if the theory involves spin $\frac{1}{2}$ fermions there will also be a general renormalizable term

$$\mathscr{L}_\psi = -\bar\psi(\not\partial - i\not A_\alpha t_\alpha^{(\psi)} + m_0 + \Gamma_n\phi_n)\psi\,, \tag{21.2.11}$$

where $t_\alpha^{(\psi)}$ is the matrix representation of the generators of the gauge group for the fermions (including coupling-constant factors), and m_0 and Γ_n are constant matrices (in general, linear combinations of terms proportional to Dirac matrices 1 and γ_5) satisfying the gauge-invariance conditions

$$\left[t_\alpha^{(\psi)},\,\gamma_4 m_0\right] = 0\,, \tag{21.2.12}$$

$$\left[t_\alpha^{(\psi)},\,\gamma_4\Gamma_n\right] + \sum_m (t_\alpha)_{mn}\,\gamma_4\Gamma_m = 0\,. \tag{21.2.13}$$

(The factor $\gamma_4 \equiv i\gamma^0$ arises from the definition $\bar\psi \equiv \psi^\dagger\gamma_4$. It is relevant only if $t_\alpha^{(\psi)}$ involves terms proportional to γ_5.) The general theorem proved in Sections 15.5 and 15.6 guarantees that the S-matrix calculated from a

Lagrangian given by the sum of Eqs. (21.2.6), (21.2.10), and (21.2.11) is independent of the choice of the parameter ξ appearing in the gauge-fixing function (21.2.3), and so for any ξ will give the same results as the choice $\xi = \infty$ corresponding to unitarity gauge.

To derive the propagators for all these fields, we need the part of the Lagrangian that is quadratic in fields:

$$\begin{aligned}\mathscr{L}_{\rm QUAD} = & -\frac{1}{4}\sum_\alpha (\partial^\mu A_\alpha{}^\nu - \partial^\nu A_\alpha{}^\mu)(\partial_\mu A_{\alpha\nu} - \partial_\nu A_{\alpha\mu}) \\ & -\frac{1}{2}\sum_{\alpha\beta} \mu^2_{\alpha\beta} A^\mu_\alpha A_{\beta\mu} - \frac{1}{2\xi}\sum_\alpha (\partial_\mu A^\mu_\alpha)(\partial_\nu A_\alpha{}^\nu) \\ & -\frac{1}{2}\sum_n (\partial_\mu \phi'_n)(\partial^\mu \phi'_n) - \frac{1}{2}\sum_{nm} M^2_{nm}\phi'_n\phi'_m \\ & -\bar{\psi}(\not{\partial} + m)\psi - \partial_\mu \omega^*_\alpha \partial^\mu \omega_\alpha - \xi \sum_{\alpha\beta} \mu^2_{\alpha\beta}\omega^*_\alpha \omega_\beta \\ & + \text{ total derivatives}\,, \end{aligned} \qquad (21.2.14)$$

where $\mu^2_{\alpha\beta}$ is the vector boson mass matrix (21.1.7):

$$\mu^2_{\alpha\beta} = -\sum_n (t^\alpha v)_n (t^\beta v)_n\,, \qquad (21.2.15)$$

and M^2_{nm} and m are new scalar and fermion mass matrices:

$$\begin{aligned} M^2_{nm} &= \left.\frac{\partial^2 P(\phi)}{\partial\phi_n \partial\phi_m}\right|_{\phi=v} - \frac{\xi}{2}\sum_\alpha (t_\alpha v)_n (t_\alpha v)_m\,, \\ m &= m_0 + \sum_n \Gamma_n v_n\,. \end{aligned} \qquad (21.2.16)$$

We see from Eq. (21.2.14) that the ghosts have gauge-dependent masses, equal to $\sqrt{\xi}$ times the corresponding vector boson masses.

These expressions give the particle masses in the zeroth order of perturbation theory. To this order, the vacuum expectation value v_n is just the location of the minimum of the polynomial 'potential' $P(\phi)$:

$$\left.\frac{\partial P(\phi)}{\partial \phi_n}\right|_{\phi=v} = 0\,. \qquad (21.2.17)$$

Also, as we saw in Section 19.2, it follows from Eqs. (21.2.4) and (21.2.17) that

$$\sum_m \left.\frac{\partial^2 P(\phi)}{\partial\phi_n\partial\phi_m}\right|_{\phi=v} (t_\alpha v)_m = 0 \qquad (21.2.18)$$

for all α. It follows then that in place of the Goldstone modes with mass zero, the scalar boson mass-squared matrix in Eq. (21.2.16) has eigenvalues equal to $\sqrt{\xi}$ times the non-zero vector boson masses. That is, if $\mu^2_{\alpha\beta}$ has an eigenvector c_β with eigenvalue μ^2, then $\sum_\beta c_\beta t_\beta v$ is an eigenvector of M^2 with an eigenvalue $\xi\mu^2$:

$$\sum_m M^2_{nm}\left(\sum_\beta c_\beta t_\beta v\right)_m = \xi \sum_{\alpha\beta} \mu^2_{\alpha\beta} c_\beta (t_\alpha v)_n = \xi\mu^2 \left(\sum_\alpha c_\alpha t_\alpha v\right)_n \ . \quad (21.2.19)$$

The other eigenvectors of M^2_{nm} are then orthogonal to all of these, and hence to all $t_\alpha v$, so these eigenvectors and the corresponding eigenvalues are just the same as for the matrix $(\partial^2 P(\phi)/\partial\phi_n\partial\phi_m)_{\phi=v}$. We see that for the unitarity gauge value $\xi \to \infty$, the Goldstone bosons are so heavy that they drop out of the theory, while the other boson masses are as usual.

The propagators are calculated by the usual rules: if the free-particle part of the Lagrangian takes the form (after integration by parts) $-\zeta^\dagger \mathscr{D}(\partial)\zeta$ for a complex field ζ or $-\frac{1}{2}\zeta^T \mathscr{D}(\partial)\zeta$ for a real field ζ, then the propagator of this field is $\mathscr{D}^{-1}(ik)$. This gives the propagators:

$$A: \quad \Delta_{\alpha\mu,\beta\nu}(k) = \left[\frac{1}{k^2+\mu^2}\left(\eta_{\mu\nu} - \frac{(1-\xi)k_\mu k_\nu}{k^2+\mu^2\xi}\right)\right]_{\alpha\beta} , \quad (21.2.20)$$

$$\phi: \quad \Delta_{nm}(k) = (k^2+M^2)^{-1}_{nm} + \xi\sum_{\alpha,\beta}(t_\alpha v)_n (t_\beta v)_m (k^2)^{-1}(k^2+\xi\mu^2)^{-1}_{\alpha\beta} , \quad (21.2.21)$$

$$\psi: \quad \Delta(k) = [-i\not{k} + m]\big/(k^2+m^2) , \quad (21.2.22)$$

$$\omega: \quad \Delta_{\alpha\beta}(k) = (k^2+\xi\mu^2)^{-1}_{\alpha\beta} \ . \quad (21.2.23)$$

The poles in Eq. (21.2.20) at unphysical mass squares proportional to ξ are cancelled by the poles at the same masses in Eq. (21.2.21). Note that now for finite ξ all propagators have the same asymptotic behaviors as in the unbroken symmetry case, as required for renormalizability. In particular, the $k_\mu k_\nu$ term in the vector boson propagator no longer presents any problem for renormalizability, because it is accompanied with an extra factor $(k^2 + \mu^2\xi)^{-1}$. We can even drop this term by choosing Feynman gauge, with $\xi = 1$. It is only in the unitarity gauge case where $\xi \to \infty$ that this factor fails to give the propagator the asymptotic behavior needed for renormalizability.

Even with well-behaved propagators, it is still necessary to show that the ultraviolet divergences in these theories are constrained by the broken gauge symmetries in such a way that every infinity can be cancelled by the renormalization of a field or a parameter in the Lagrangian. This can be done by the same techniques as in Sections 17.2 and 17.3, but treating the vacuum expectation values of the spinless fields as external fields rather than fixed quantities that break the symmetries.[5a]

21.3 The Electroweak Theory

The most important application of spontaneously broken gauge theories has been to the theory of weak and electromagnetic interactions.[3] Weak interactions at low energies are well described by an effective Lagrangian given by a sum of products of vector (including axial-vector) currents, as in Eq. (19.4.22). This suggests that these interactions may like electromagnetism be described by some sort of gauge theory. In order to insure the separate conservation of electronic-type and muonic-type leptons and baryons (or quarks) we may guess that the known electronic-type and muonic-type leptons and the quarks all form separate representations of the gauge group. With this assumption, there are only a few possibilities for the structure of the gauge group.

Let's first consider the electronic-type lepton fields. As far as we know, these consist only of the left- and right-handed parts of the electron field e:

$$e_L = \tfrac{1}{2}(1+\gamma_5)e\,, \qquad e_R = \tfrac{1}{2}(1-\gamma_5)e\,, \tag{21.3.1}$$

and a purely left-handed electron-neutrino field ν_{eL}:

$$\gamma_5 \nu_{eL} = \nu_{eL}\,. \tag{21.3.2}$$

The fields in any representation of the gauge group must all have the same Lorentz-transformation properties, so the representations of the gauge group here divide* into a left-handed doublet (ν_{eL}, e_L) and a right-handed singlet e_R. The largest possible gauge group is then

$$SU(2)_L \times U(1)_L \times U(1)_R\,,$$

under which the fields transform as

$$\delta \begin{pmatrix} \nu_e \\ e \end{pmatrix} = i\left[\vec{\epsilon}\cdot\vec{t} + \epsilon_L t_L + \epsilon_R t_R\right] \begin{pmatrix} \nu_e \\ e \end{pmatrix}, \tag{21.3.3}$$

where the generators are

$$\vec{t} = \frac{g}{4}(1+\gamma_5)\left\{\begin{pmatrix} 0 & 1 \\ 1 & 0 \end{pmatrix}, \begin{pmatrix} 0 & -i \\ i & 0 \end{pmatrix}, \begin{pmatrix} 1 & 0 \\ 0 & -1 \end{pmatrix}\right\}, \tag{21.3.4}$$

$$t_L \propto (1+\gamma_5)\begin{pmatrix} 1 & 0 \\ 0 & 1 \end{pmatrix}, \tag{21.3.5}$$

$$t_R \propto (1-\gamma_5)\,, \tag{21.3.6}$$

---

* If we allow gauge couplings that change electron-type lepton number, then it is possible to include the left-handed field $\overline{e_R}$ along with ν_{eL} and e_L in a representation of the gauge group. This was the basis for an early $SO(3)$ variant[6] of the electroweak theory, which has since been ruled out by experiment.

with g a constant to be chosen later. It will be convenient instead of t_L and t_R to consider the generators

$$y \equiv g' \left[\left(\frac{1+\gamma_5}{4} \right) \begin{pmatrix} 1 & 0 \\ 0 & 1 \end{pmatrix} + \left(\frac{1-\gamma_5}{2} \right) \right] \tag{21.3.7}$$

and

$$n_e \equiv g'' \left[\left(\frac{1+\gamma_5}{2} \right) \begin{pmatrix} 1 & 0 \\ 0 & 1 \end{pmatrix} + \left(\frac{1-\gamma_5}{2} \right) \right] , \tag{21.3.8}$$

where g' and g'' are constants like g to be chosen later. The generator y appears along with t_3 in a linear combination that plays a special role in physics; it is the electric charge

$$q = e \begin{pmatrix} 0 & 0 \\ 0 & -1 \end{pmatrix} = \frac{e}{g} t_3 - \frac{e}{g'} y . \tag{21.3.9}$$

Also, n_e is the electron-type lepton number. We want to include both charge-changing weak interactions (like beta decay) and electromagnetism in our theory, so we will assume that there are gauge fields $\vec{A}^\mu$ and B^μ coupled to $\vec{t}$ and y. In addition, we may or may not want to include a gauge field coupled to the one remaining independent linear combination of t_L and t_R, which can be taken as the electron-type lepton number (21.3.8). There are very stringent limits[7] on the long-range forces that would be produced by a massless gauge field coupled to n_e, so in order to include in our theory a gauge field coupled to n_e with a strength g'' comparable to the weak and electromagnetic interactions, we would have to assume that this gauge symmetry is spontaneously broken.** However, there is no experimental evidence for the weak interaction that would be produced by such a gauge coupling (and plenty of evidence by now against it) so we shall simply exclude n_e from the generators of the gauge group. The gauge group is then[8]

$$G = SU(2)_L \times U(1) \tag{21.3.10}$$

with generators $\vec{t}, y$ given by Eqs. (21.3.4) and (21.3.7) respectively. The coupling constants g and g' are to be adjusted so that the gauge fields $\vec{A}^\mu$ and B^μ coupled to these generators are canonically normalized. The most general gauge-invariant and renormalizable Lagrangian that involves just

** Note that this is possible without violating the global conservation law of electronic lepton conservation. We would have to assume that the Lagrangian is invariant under both a global phase transformation acting only on electron-type lepton fields, and also a local phase transformation that acts on electron-type lepton fields as well as on some scalar field that does not interact with leptons. The vacuum expectation value of this scalar would break the local symmetry, giving the gauge boson coupled to n_e a mass, without breaking the global symmetry.

these gauge fields and electronic leptons is then

$$\mathscr{L}_{YM} + \mathscr{L}_e = -\tfrac{1}{4}\left(\partial_\mu \vec{A}_\nu - \partial_\nu \vec{A}_\mu + g\vec{A}_\mu \times \vec{A}_\nu\right)^2 - \tfrac{1}{4}(\partial_\mu B_\nu - \partial_\nu B_\mu)^2$$
$$-\bar{\ell}(\not\partial - i\vec{\not A} \cdot \vec{t}_L - i \not B\, y)\ell \,. \qquad (21.3.11)$$

(We here use the fact that the structure constants of $SU(2)_L$ and $U(1)$ are $C_{ijk} = -i\, g\, \epsilon_{ijk}$ and zero, respectively.)

Of course, of the four gauge fields coupled to $\vec{t}$ and y, only one linear combination, the electromagnetic field A^μ, is actually massless. We therefore must assume that $SU(2)_L \times U(1)$ is spontaneously broken to a subgroup $U(1)_{em}$, with generator given by the charge (21.3.9). The details of the symmetry-breaking mechanism will be considered a little later. However, whatever this mechanism may be, we know that the canonically normalized vector fields corresponding to particles of spin one and definite mass consist of one field of charge $+e$ with mass m_W:

$$W^\mu = \frac{1}{\sqrt{2}}(A_1^\mu + i\,A_2^\mu)\,, \qquad (21.3.12)$$

another of charge $-e$ and the same mass:

$$W^{\mu *} = \frac{1}{\sqrt{2}}(A_1^\mu - i\,A_2^\mu)\,, \qquad (21.3.13)$$

and two electrically neutral fields of mass m_Z and zero respectively, given by orthonormal linear combinations of A_3^μ and B^μ:

$$Z^\mu = \cos\theta\, A_3^\mu + \sin\theta\, B^\mu\,, \qquad (21.3.14)$$

$$A^\mu = -\sin\theta\, A_3^\mu + \cos\theta\, B^\mu\,, \qquad (21.3.15)$$

or equivalently

$$A_3^\mu = \cos\theta\, Z^\mu - \sin\theta\, A^\mu\,, \qquad (21.3.16)$$

$$B^\mu = \sin\theta\, Z^\mu + \cos\theta\, A^\mu\,. \qquad (21.3.17)$$

According to the general result (21.1.11)–(21.1.12), the generator of the unbroken symmetıy, which is here electromagnetic gauge invariance, is given by a linear combination of generators in which the coefficients are the same as the coefficients of the corresponding massless field in the expansion of the canonically normalized gauge fields coupled to these generators. Inspecting Eqs. (21.3.16) and (21.3.17) shows that

$$q = -\sin\theta\, t_3 + \cos\theta\, y\,. \qquad (21.3.18)$$

Comparing this with Eq. (21.3.9) gives then

$$g = -e/\sin\theta\,, \qquad g' = -e/\cos\theta\,. \qquad (21.3.19)$$

The complete lepton–gauge boson coupling can be expressed in terms of the couplings g and g':

$$\begin{aligned}
i\mathscr{L}'_e &= -\overline{\begin{pmatrix} \nu_e \\ e \end{pmatrix}} \left[\sum_\alpha \not{A}_\alpha t_\alpha \right] \begin{pmatrix} \nu_e \\ e \end{pmatrix} \\
&= -\overline{\begin{pmatrix} \nu_e \\ e \end{pmatrix}} \left[\frac{1}{\sqrt{2}} \not{W}(t_{1L} - it_{2L}) + \frac{1}{\sqrt{2}} \not{W}^*(t_{1L} + it_{2L}) \right. \\
&\quad \left. + \not{Z}\,(t_{3L} \cos\theta + y \sin\theta) + \not{A}(-t_{3L}\sin\theta + y\cos\theta) \right] \begin{pmatrix} \nu_e \\ e \end{pmatrix} \\
&= \frac{g}{\sqrt{2}} \left(\bar{e}\, \not{W} \left(\frac{1+\gamma_5}{2} \right) \nu_e \right) + \frac{g}{\sqrt{2}} \left(\bar{\nu}_e\, \not{W}^* \left(\frac{1+\gamma_5}{2} \right) e \right) \\
&\quad - \frac{1}{2} \sqrt{g^2 + g'^2}\, \bar{\nu}_e\, \not{Z} \left(\frac{1+\gamma_5}{2} \right) \nu_e + \frac{(g^2 - g'^2)}{2\sqrt{g^2 + g'^2}}\, \bar{e}\, \not{Z} \left(\frac{1+\gamma_5}{2} \right) e \\
&\quad + g'\bar{e}\, \not{Z} \left(\frac{1-\gamma_5}{2} \right) e - e(\bar{e}\not{A} e)\,.
\end{aligned} \tag{21.3.20}$$

To complete the theory, we must now make some assumption about the mechanism of symmetry breaking. We want this mechanism to give masses not only to the $W^\pm$ and Z^0, but to the electron as well. Now, the only way that this is possible in a renormalizable weakly-coupled theory is to have a scalar field coupled without derivatives to $\bar{\ell}_R$ and ℓ_L (and also $\bar{\ell}_L$ and ℓ_R). Then $SU(2)_L \times U(1)$ invariance requires that the scalar be an $SU(2)_L$ doublet like ℓ_L, but with a shifted value of y and hence of q. We thus assume a 'Yukawa' coupling

$$\mathscr{L}_{\phi e} = -G_e \overline{\begin{pmatrix} \nu_e \\ e \end{pmatrix}}_L \begin{pmatrix} \phi^+ \\ \phi^0 \end{pmatrix} e_R \; + \text{H.c.}\,, \tag{21.3.21}$$

where (ϕ^+, ϕ^0) is a doublet, on which the $SU(2) \times U(1)$ generators are represented by the matrices:

$$\vec{t}^{(\phi)} = \frac{g}{2} \left\{ \begin{pmatrix} 0 & 1 \\ 1 & 0 \end{pmatrix}, \begin{pmatrix} 0 & -i \\ i & 0 \end{pmatrix}, \begin{pmatrix} 1 & 0 \\ 0 & -1 \end{pmatrix} \right\}, \tag{21.3.22}$$

$$y^{(\phi)} = -\,g'/2 \begin{pmatrix} 1 & 0 \\ 0 & 1 \end{pmatrix}, \tag{21.3.23}$$

so that the charge matrix is

$$q^{(\phi)} = \frac{e}{g}\, t_3^{(\phi)} - \frac{e}{g'}\, y^{(\phi)} = e \begin{pmatrix} 1 & 0 \\ 0 & 0 \end{pmatrix}. \tag{21.3.24}$$

It is possible that there are other scalar multiplets in the theory, but for the moment let's suppose that this is the only one.

We must add a gauge-invariant term involving scalar and gauge fields to the Lagrangian. The most general form consistent with $SU(2) \times U(1)$

gauge invariance, Lorentz invariance, and renormalizability is:

$$\mathscr{L}_\phi = -\frac{1}{2}\left|(\partial_\mu - i\vec{A}_\mu \cdot \vec{t}^{(\phi)} - i\,B_\mu\, y^{(\phi)})\phi\right|^2 - \frac{\mu^2}{2}\,\phi^\dagger\phi - \frac{\lambda}{4}\,(\phi^\dagger\phi)^2\,, \qquad (21.3.25)$$

where $\lambda > 0$, and

$$\phi \equiv \begin{pmatrix} \phi^+ \\ \phi^0 \end{pmatrix}\,. \qquad (21.3.26)$$

For $\mu^2 < 0$, there is a tree-approximation vacuum expectation value at the stationary point of the Lagrangian

$$\langle\phi\rangle^\dagger\langle\phi\rangle \equiv v^2 = |\mu^2|/\lambda\,. \qquad (21.3.27)$$

We can always perform an $SU(2) \times U(1)$ gauge transformation to a unitarity gauge, in which $\phi^+ = 0$ and ϕ^0 is Hermitian, with a positive vacuum expectation value. (This is why we normalized the complex doublet ϕ so that an unconventional factor $\frac{1}{2}$ appears in the kinetic term in Eq. (21.3.25); $\mathrm{Re}\,\phi^0$ is the only physical scalar field, and Eq. (21.3.25) makes this a canonically normalized field.) In unitarity gauge the vacuum expectation values of the components of ϕ are

$$\langle\phi^+\rangle = 0\,, \qquad \langle\phi^0\rangle = v > 0\,. \qquad (21.3.28)$$

The scalar Lagrangian (21.3.25) then yields a vector meson mass term

$$-\frac{1}{2}\left|\left(\vec{A}_\mu \cdot \vec{t}^{(\phi)} + B_\mu y^{(\phi)}\right)\langle\phi\rangle\right|^2 = -\frac{1}{2}\left|\left(\frac{g}{2}\vec{A}_\mu \cdot \vec{\tau} - \frac{g'}{2}B_\mu\right)\begin{pmatrix} 0 \\ v \end{pmatrix}\right|^2$$

$$= -\frac{v^2 g^2}{4}\,W_\mu^\dagger W^\mu - \frac{v^2}{8}\,(g^2 + g'^2)\,Z_\mu Z^\mu\,. \qquad (21.3.29)$$

We see that as expected, the photon mass is zero, while the $W^\pm$ and Z^0 have the masses

$$m_W = \frac{v|g|}{2}\,, \qquad m_Z = \frac{v\sqrt{g^2 + g'^2}}{2}\,. \qquad (21.3.30)$$

Also, from Eqs. (21.3.21) and (21.3.28) we see that the electron is given a lowest-order mass

$$m_e = G_e v\,. \qquad (21.3.31)$$

It is difficult to study reactions among electron-type leptons alone, though by now there are data on scattering processes like $\bar{\nu}_e + e^- \to \bar{\nu}_e + e^-$. For high precision data we have to consider reactions that also at least involve muonic-type leptons, such as the well-studied process of muon decay, $\mu^+ \to e^+ + \nu_e + \bar{\nu}_\mu$. It is trivial to extend the above model to include muon-type leptons — just add to the Lagrangian terms $\mathscr{L}_\mu$

and $\mathscr{L}_{\phi\mu}$, like the last terms in Eqs. (21.3.11) and (21.3.21), with the fields e and ν_e replaced with the muon and muon-neutrino fields μ^- and ν_μ, and with G_e replaced with $G_\mu = G_e(m_\mu/m_e)$. Inspection of (21.3.20) and the corresponding term with e and ν_e replaced with μ and ν_μ shows that W exchange between low energy, e-type and μ-type leptons produces the effective interaction

$$\left(\frac{g}{\sqrt{2}}\right)^2 \frac{1}{m_W^2} \left(\bar{e}\gamma^\lambda \left(\frac{1+\gamma_5}{2}\right) \nu_e\right) \left(\bar{\nu}_\mu \gamma_\lambda \left(\frac{1+\gamma_5}{2}\right) \mu\right) + \text{H.c.} \tag{21.3.32}$$

This may be compared with the interaction of the effective '$V - A$' theory which is known to give a good description of muon decay

$$\frac{G_F}{\sqrt{2}} \left(\bar{e}\gamma^\lambda (1+\gamma_5)\, \nu_e\right) \left(\bar{\nu}_\mu \gamma_\lambda (1+\gamma_5)\, \mu\right) + \text{H.c.} \tag{21.3.33}$$

Here G_F is the conventional Fermi coupling constant, known from the muon decay rate to have the value $G_F = 1.16639(2) \times 10^{-5}\,\text{GeV}^{-2}$. Comparing these two expressions, we find

$$g^2/m_W^2 = 4\sqrt{2}\, G_F\,. \tag{21.3.34}$$

This allows an immediate determination of the vacuum expectation value v, given by Eq. (21.3.30) as

$$v = \frac{2m_W}{g} = \frac{1}{2^{1/4} G_F^{1/2}} = 247\,\text{GeV}\,. \tag{21.3.35}$$

Also, Eq. (21.3.31) shows that G_e has the very small value

$$G_e = \frac{0.511\,\text{MeV}}{247\,\text{GeV}} = 2.07 \times 10^{-6}\,. \tag{21.3.36}$$

From Eq. (21.3.30) we see that $m_Z > m_W$. We cannot use Eq. (21.3.30) to determine the actual values of m_Z and m_W without knowing something about g and g′. Using Eqs. (21.3.30) and (21.3.19), we can express m_Z and m_W in terms of the electroweak mixing angle θ:

$$m_W = \frac{ev}{2|\sin\theta|} = \frac{37.3\,\text{GeV}}{|\sin\theta|}\,,$$

$$m_Z = \frac{ev}{2|\sin\theta||\cos\theta|} = \frac{74.6\,\text{GeV}}{|\sin 2\theta|}\,.$$

These are the original results obtained in Ref. 3. Of course, there are radiative corrections of all sorts, most of which depend on details of the theory that have not yet been specified in this section. But there is one particularly large radiative correction that can be readily calculated without further information. The above values for m_W and m_Z were calculated using the conventionally defined electronic charge for e. However, as explained in Section 18.2, this is not precisely the appropriate value to use in calculations of processes at energies $E \gg m_e$; we should instead use the electric charge e_μ defined at a sliding scale μ comparable to the energies of interest. For μ of the order of 90 GeV the effective fine structure constant $e_\mu^2/4\pi$ is about $1/129$ (and quite insensitive to the precise value of μ), so the above values for m_W and m_Z should be multiplied with $\sqrt{137/129}$, giving

$$m_W = \frac{38.4 \text{ GeV}}{|\sin\theta|}\,, \tag{21.3.37}$$

$$m_Z = \frac{76.9 \text{ GeV}}{|\sin 2\theta|}\,. \tag{21.3.38}$$

Whatever the value of θ, these masses are too large for there to have been any hope of detecting the W or Z in the 1960s or early 1970s. Experimental evidence for the electroweak theory had to come instead from the discovery of the new class of weak interactions predicted by the theory, the neutral current processes produced by Z^0 exchange.[9] The first observation of a neutral current process was the 1973 bubble chamber detection of the purely leptonic process of ν_μ–e^- elastic scattering.[10] Although these processes are easy to deal with theoretically, the frequency of events is relatively low, because the cross section is proportional to the square of the center-of-mass energy.[†] It was years before the purely leptonic neutral current reactions could be used to give a reasonably precise value for the parameter $\sin^2\theta$. By 1994, the study of purely leptonic neutral current processes like $\nu_\mu + e^- \to \nu_\mu + e^-$ and $\bar{\nu}_\mu + e^- \to \bar{\nu}_\mu + e^-$ had yielded the value 0.222 ± 0.011, which would give $m_W = 81.5$ GeV and $m_Z = 92.5$ GeV.

Even before the discovery of neutral currents, the electroweak theory had been extended to the weak and electromagnetic interactions of hadrons with each other and with leptons. By the mid-1960s, it had become understood that weak interaction processes in which charge is exchanged between leptons and hadrons are well described at low energy by the effective Lagrangian

$$\frac{G_F}{\sqrt{2}}\left[\bar{e}\gamma_\lambda(1+\gamma_5)\nu_e + \bar{\mu}\gamma_\lambda(1+\gamma_5)\nu_\mu\right] J^\lambda \;+\; \text{H.c.}\,, \tag{21.3.39}$$

where J^λ is an hadronic current. Within the quark model, the commutation and conservation properties of J^λ allowed it to be identified with the quark

[†] The cross section is proportional to G_F^2, so in order to have the dimensions of energy^{-2}, it must also be proportional to some energy squared. Where the center-of-mass energy is much larger than the electron mass, it is the only energy that can appear in this formula.

current

$$J^\lambda = \bar{u}\gamma^\lambda(1+\gamma_5)d\cos\theta_c + \bar{u}\gamma^\lambda(1+\gamma_5)s\sin\theta_c\,. \qquad (21.3.40)$$

Here u, d, s are the fields of the up, down, and strange quarks, and θ_c is another angle, known as the Cabibbo angle.[11] Experiments on processes like $O^{14} \to N^{14*} + e^+ + \nu_e$ and $K^+ \to \pi^0 + e^+ + \nu_e$ confirm that G_F has very nearly the same value as that measured in the purely leptonic process $\mu^+ \to \bar{\nu} + e^+ + \nu_e$, and give for θ_c the value[12] $\sin\theta_c = 0.220 \pm 0.003$. We naturally conclude that the quarks provide another $SU(2) \times U(1)$ doublet

$$\mathscr{Q} = \left(\frac{1+\gamma_5}{2}\right)\begin{bmatrix} u \\ d\cos\theta_c + s\sin\theta_c \end{bmatrix}, \qquad (21.3.41)$$

as well as right-handed singlets, with y values adjusted to give the quark charges $2e/3$ and $-e/3$. By itself this would lead to a serious difficulty. The Z^0 boson interacts with the quark neutral current

$$\sum_{\mathscr{Q}} \bar{\mathscr{Q}}\gamma^\mu(t_{3L}\cos\theta + y\sin\theta)\mathscr{Q} = \sum_{\mathscr{Q}} \bar{\mathscr{Q}}\gamma^\mu(t_{3L}\sec\theta + q\tan\theta)\mathscr{Q}\,, \qquad (21.3.42)$$

with the sum running over all quark doublets $\mathscr{Q}$ like (21.3.41). The charge matrix q is diagonal in quark flavors, but if (21.3.41) were the only quark doublet then the term involving the matrix t_{3L} would contain cross terms proportional to $\bar{s}\gamma^\mu(1+\gamma_5)d$ and $\bar{d}\gamma^\mu(1+\gamma_5)s$, leading to effective Z exchange interactions like $s + \bar{d} \leftrightarrow d + \bar{s}$ and $s + \bar{d} \leftrightarrow \mu^+ + \mu^-$ with the strength of ordinary first-order weak interactions. Such effects would lead to rates for processes like K^0–$\overline{K^0}$ oscillations and $K^0 \to \mu^+ + \mu^-$ many orders of magnitude greater than observed. Also, even without neutral current terms in the Lagrangian, the one-loop diagrams involving the interaction (21.3.39) with the charged current (21.3.40) would lead to an effective interaction $s + \bar{d} \to d + \bar{s}$ which is smaller than an ordinary first-order weak interaction only by a factor of order $\alpha/2\pi$, leading to a rate for K^0–$\overline{K^0}$ oscillations that is still much too large. In order to avoid this last difficulty, it was proposed[13] that there is another term in J^λ; in modern notation,

$$\bar{c}\gamma^\lambda(1+\gamma_5)\left[-d\sin\theta_c + s\cos\theta_c\right], \qquad (21.3.43)$$

where c is a fourth quark, like u with charge $2e/3$. Adding (21.3.43) to (21.3.40), the charged current may be written

$$J^\lambda = (\bar{u}\cos\theta_c - \bar{c}\sin\theta_c)\gamma^\lambda(1+\gamma_5)d + (\bar{u}\sin\theta_c + \bar{c}\cos\theta_c)\gamma^\lambda(1+\gamma_5)s\,.$$

The only reason that the interactions of the W with this current do not conserve strangeness is that the c and u have different masses, leading to transitions between $u\cos\theta_c - c\sin\theta_c$ and $u\sin\theta_c + c\cos\theta_c$. But this means that the loop diagrams for the effective interaction $s + \bar{d} \to d + \bar{s}$

are suppressed by additional factors (since $m_u \ll m_c$) of m_c^2/m_W^2, bringing the rate for K^0–$\overline{K^0}$ oscillations into agreement with experiment.

It was subsequently noted[14] that this also solves the problem of the strangeness changing Z^0 interactions. In the context of the $SU(2) \times U(1)$ gauge theory the combination $-d_L \sin\theta_c + s_L \cos\theta_c$ cannot be a singlet, but must be part of another doublet

$$\left(\frac{1+\gamma_5}{2}\right) \begin{bmatrix} c \\ -d \sin\theta_c + s \cos\theta_c \end{bmatrix} . \tag{21.3.44}$$

Including this doublet in the weak neutral current (21.3.42), the strangeness non-conserving terms proportional to $\bar{s}\gamma^\mu(1+\gamma_5)d$ and $\bar{d}\gamma^\mu(1+\gamma_5)s$ cancel, removing the problem of excessive Z exchange contributions to processes like K^0–$\overline{K^0}$ oscillations and $K^0 \to \mu^+ + \mu^-$. Particles containing the c quark in a c–$\bar{c}$ bound state were discovered[15] in 1974, and indicated a mass $m_c \approx 1.5$ GeV.[††] This completed two generations of quarks and leptons: a (u, d) quark doublet mixed with a (c, s) quark doublet, together with two lepton doublets (ν_e, e) and (ν_μ, μ).

The first sign of a third generation was the discovery of a third charged lepton,[16] the τ. Later a fifth quark type, the b, was discovered,[17] with charge $-e/3$ and a mass of about 4.5 GeV. A sixth, the t with charge $2e/3$, then became theoretically necessary, and after a long interval it too was discovered,[18] with a mass quoted in 1995 as 181 ± 12 GeV.[19] Today the hadronic current J^λ in (21.3.39) is expressed as

$$J^\lambda = \overline{\begin{bmatrix} u \\ c \\ t \end{bmatrix}} \gamma^\lambda(1+\gamma_5)\, V \begin{bmatrix} d \\ s \\ b \end{bmatrix} , \tag{21.3.45}$$

where V is an incompletely known 3×3 unitary matrix, known as the Kobayashi–Maskawa matrix.[20] In the $SU(2) \times U(1)$ gauge theory, this means that there are three quark doublets:

$$\left(\frac{1+\gamma_5}{2}\right) \begin{bmatrix} u \\ V_{ud}d + V_{us}s + V_{ub}b \end{bmatrix} , \tag{21.3.46}$$

$$\left(\frac{1+\gamma_5}{2}\right) \begin{bmatrix} c \\ V_{cd}d + V_{cs}s + V_{cb}b \end{bmatrix} , \tag{21.3.47}$$

$$\left(\frac{1+\gamma_5}{2}\right) \begin{bmatrix} t \\ V_{td}d + V_{ts}s + V_{tb}b \end{bmatrix} . \tag{21.3.48}$$

†† We do not observe quarks in isolation, so their masses are not precisely defined. The mass of the c quark quoted here is roughly half the mass of the J–ψ particle, interpreted as a c–$\bar{c}$ bound state. The b and t quarks are so heavy that their masses can be taken from the masses of the hadrons containing them with little ambiguity.

It is important to recognize that this is just what we should naturally expect on general grounds for three quark doublets. The most general renormalizable $(SU(3)\times SU(2)\times U(1))$-invariant interactions of the scalar doublets ϕ_n with the quarks must in general take the form

$$\mathscr{L}_\phi = -\sum_{ijn} G^n_{ij}\overline{\begin{pmatrix} U_{iL} \\ D_{iL}\end{pmatrix}}\cdot\begin{pmatrix}\phi^0_n \\ \phi^-_n\end{pmatrix} U_{jR} - \sum_{ijn} H^n_{ij}\overline{\begin{pmatrix} U_{iL} \\ D_{iL}\end{pmatrix}}\cdot\begin{pmatrix}-\phi^{-\dagger}_n \\ \phi^{0\dagger}_n\end{pmatrix} D_{jR} + \text{H.c.}\,, \tag{21.3.49}$$

where U_i and D_i with $i = 1, 2, 3$ are three independent quark fields of charge $2e/3$ and $-e/3$, respectively, L and R denote the left- and right-handed parts of the quark fields, and G^n_{ij} and H^n_{ij} are unknown constants. The vacuum expectation values of the neutral scalars then produce a quark mass term

$$\mathscr{L}_m = -\sum_{ij}\overline{U_{iL}}\, m^U_{ij}\, U_{jR} - \sum_{ij}\overline{D_{iL}}\, m^D_{ij}\, D_{jR} + \text{H.c.}\,, \tag{21.3.50}$$

where

$$m^U_{ij} = \sum_n G^n_{ij}\langle\phi^0_n\rangle_{\text{VAC}}\,, \qquad m^D_{ij} = \sum_n H^n_{ij}\langle\phi^0_n\rangle^*_{\text{VAC}}\,. \tag{21.3.51}$$

The matrices m^U_{ij} and m^D_{ij} are not constrained in any way, and in particular may be complex and non-diagonal, in which case parity- and flavor-non-conserving terms appear in $\mathscr{L}_m$. But we can introduce new quark fields $U'_R = A^U_R U_R$, $U'_L = A^U_L U_L$, $D'_R = A^D_R D_R$, $D'_L = A^D_L D_L$, where the As are 3×3 matrices constrained only by the condition that they must be unitary in order to preserve the form of the kinematic term (19.4.1). Then the mass term (21.3.50) takes the same form when rewritten in terms of the primed quark fields, but with the matrices m^U and m^D replaced with

$$m^{U\prime} = A^U_L m^U A^{U\dagger}_R\,, \qquad m^{D\prime} = A^D_L m^D A^{D\dagger}_R\,. \tag{21.3.52}$$

Now it is a general theorem that for any matrix m, it is always possible to choose unitary matrices A and B such that AmB is real and diagonal. (Use the polar decomposition theorem to write $m = HU$, where H is Hermitian and U is unitary, and choose $A = S^\dagger$ and $B = U^\dagger S$, where S is the unitary matrix that diagonalizes H.) We can therefore choose the As so that $m^{U\prime}$ and $m^{D\prime}$ are real and diagonal, in which case the quark fields u, c, t, d, s, and b are to be identified with the components of $U'_L + U'_R$ and $D'_L + D'_R$. The weak doublets are now written as

$$Q_{iL} = \begin{pmatrix} (A^{U\,-1}_L U'_L)_i \\ (A^{D\,-1}_L D'_L)_i \end{pmatrix}\,,$$

but we can just as well take the doublets as linear combinations $A_L^U Q_L$ that have charge $2e/3$ quarks of definite mass u, c, t as their top component, in which case these doublets take the form (21.3.46)–(21.3.48), with

$$V = A_L^U A_L^{D-1} . \tag{21.3.53}$$

Within 90% confidence limits, the best present (1995) values for the absolute values of the elements of the Kobayashi–Maskawa matrix are[20a]

$$\begin{pmatrix} 0.9745 \text{ to } 0.9757 & 0.219 \text{ to } 0.224 & 0.002 \text{ to } 0.005 \\ 0.218 \text{ to } 0.224 & 0.9736 \text{ to } 0.9750 & 0.036 \text{ to } 0.046 \\ 0.004 \text{ to } 0.014 & 0.034 \text{ to } 0.046 & 0.9989 \text{ to } 0.9993 \end{pmatrix} ,$$

with rows labelled u, c, and t, and columns labelled d, s, and b.

If there were only two quark doublets formed from u, d, c, and s quarks, it would be possible to choose the phases of the quark fields so that all V_{ij} are real,‡ so that the V matrix is orthogonal, and the doublets (21.3.46) and (21.3.47) (with b omitted) take the form (21.3.41) and (21.3.44), respectively. In this case the gauge interactions would automatically conserve T and CP. The great importance of the third generation is that it is no longer always possible to choose quark phases so that the V matrix is real, and therefore the gauge interactions can violate T and CP conservation. But for unknown reasons the elements V_{ub}, V_{cb}, V_{td}, and V_{ts} that connect the third generation with the first two are all quite small, so the physics of the first two generations is hardly affected by the presence of the third, which explains in a more-or-less natural way why the Cabibbo assumption (21.3.40) works so well and why the violation of T and CP conservation is so weak. T and CP conservation can also be violated by scalar boson interactions if there are two or more scalar doublets;[20b] here the violation of T and CP conservation is expected to be weak because the scalar doublets couple weakly to light quarks. It is still unknown which of these mechanisms is responsible for the observed violation of T and CP conservation in K_2^0 decay, discussed in Section 3.3.

Neutral current processes involving hadrons, such as neutrino–nucleon deep-inelastic scattering, were discovered in 1973,[21] shortly after the detection of the purely leptonic process $\nu_\mu + e \to \nu_\mu + e$. Because of the much greater mass of the target particle here, it became possible before long to observe large numbers of events, and use them to confirm the electroweak theory and measure its parameters. Additional information on lepton–hadron neutral current interactions came from the observation of parity violation in atomic physics. By 1983 all direct measurements of

‡ Adjust the phases of d and s so that V_{ud} and V_{us} are real. Unitarity then requires that V_{cd} and V_{cs} have the same phase, which can be eliminating by adjusting the phase of c.

$\sin^2\theta$ had become consistent, and gave a combined value $\sin^2\theta = 0.23$, yielding the predictions $m_W = 80.1\,\text{GeV}$ and $m_Z = 91.4\,\text{GeV}$. Then in 1983 the W was discovered, with the Z following soon after.[22] Their measured masses are now (in 1995)

$$m_W = 80.410 \pm 0.180\ \text{GeV}^{23}\,, \qquad m_Z = 91.1887 \pm 0.0022\ \text{GeV}^{24}\,,$$

in satisfactory agreement with the predictions of the electroweak theory.

The very great accuracy of the measurement of the Z mass, which has been achieved by tuning the energy of e^+–e^- collisions to the Z resonance at LEP (CERN's Large Electron Positron collider) and the SLC (Stanford Linear Collider), has changed the way that electroweak data is analyzed. Instead of comparing predictions of W and Z masses with observed values, the Z mass is taken as an experimental input, along with the Fermi coupling constant $G_F = 1.16639(2) \times 10^{-5}\ \text{GeV}^{-2}$ taken from the rate of muon decay (including radiative corrections to order α), and the fine structure constant $\alpha(m_Z) = (128.87 \pm 0.12)^{-1}$, extrapolated from low energy measurements as described in Section 18.2. In this way $\sin^2\theta$ becomes a derived quantity; if defined by Eq. (21.3.38), it takes the value $\sin^2\theta = 0.2312 \pm 0.003$. With these inputs, the electroweak theory can be used to make predictions of other quantities like m_W with sufficient precision that it becomes necessary to take electroweak radiative corrections into account.[25] In one-loop order these radiative corrections involve the masses of the t quark and scalar ('Higgs') boson, and thus can be used to estimate these masses. For instance, before the top quark was discovered the agreement between theory and experiment set bounds on these radiative corrections which implied a top quark mass in the range 130–200 GeV,[26] in agreement with the value subsequently found experimentally. The W mass is predicted (in 1994) to be 80.29 GeV, with an uncertainty of ± 0.02 GeV from uncertainties in the inputs m_Z, G_F, and $\alpha(m_Z)$, and an uncertainty of ± 0.11 GeV from the range of possible values of m_t and m_{Higgs}. One 1995 study[27] concludes that $m_{\text{Higgs}} < 225$ GeV. The precise measurement of m_W expected at the LEP 2 electron–positron collider at CERN will allow a useful estimate of m_{Higgs}.

* * *

The most general renormalizable Lagrangian with the field content and $SU(3) \times SU(2) \times U(1)$ gauge symmetries of the electroweak theory automatically respects baryon and lepton conservation. This is obviously true for the gauge interactions and bare mass terms, because the quarks, antiquarks, leptons, and antileptons all belong to distinct representations of $SU(3) \times SU(2) \times U(1)$. With scalars all belonging to $SU(3)$ neutral $SU(2)$ doublets with $U(1)$ quantum number $\pm 1/2$, the only renormalizable interactions of scalars with fermions and/or antifermions are with quark–

antiquark and lepton–antilepton pairs, which of course conserve baryon and lepton number. (In much the same way, one can see that the charged hadronic currents with which leptons interact are necessarily linear combinations of the currents associated with the spontaneously broken $SU(3) \times SU(3)$ symmetry described in Section 19.7, as assumed without knowing the explanation in the original work on this broken symmetry.)

These results depend critically on the assumption that the standard model is renormalizable. But as we have repeatedly emphasized, the renormalizable Lagrangian of the standard model is expected to be accompanied with non-renormalizable terms of dimensionality $d > 4$, suppressed by $4 - d$ powers of some very large mass M. The leading corrections to the predictions of the renormalizable standard model come from terms with the smallest possible dimensionality greater than four.

The only Lorentz-invariant terms of dimensionality five that can be constructed out of the fermion and other fields of the standard model are at most bilinear in fermion fields and also contain either two scalars, or one scalar and one gauge-invariant derivative, or no scalars and two gauge-invariant derivatives (including their commutator, a field strength tensor). Color $SU(3)$ invariance requires that the fermion fields in such an interaction appear in either a quark–antiquark bilinear or a pair of lepton and/or antilepton fields, all of which operators conserve baryon number. There are a great number of such terms, but to violate lepton number conservation they must involve a product of two lepton fields or of their conjugates. The left-handed lepton doublets (ℓ^-_{Li}, ν_i) and right-handed charged lepton singlets ℓ^-_{Ri} (with $i = e, \mu$, or τ) have $U(1)$ quantum numbers $1/2$ and $+1$, respectively, while the scalar doublet (or doublets) (ϕ^+, ϕ^0) have $U(1)$ quantum number $-1/2$, so we can construct $U(1)$-invariant interactions of dimensionality five out of two left-handed lepton doublets and two scalar doublets. With only a single type of scalar doublet, there is just one such term that satisfies $SU(2)$ and Lorentz invariance:[27a]

$$\sum_{ij} f_{ij}(\overline{\ell^c_{Li}}\phi^+ - \overline{\nu^c_i}\phi^0)(\ell_{Lj}\phi^+ - \nu_j\phi^0)\,, \tag{21.3.54}$$

where i and j are lepton flavor indices, and c denotes the charge conjugate field. At energies below the electroweak breaking scale, this yields an effective interaction

$$\sum_{ij} f_{ij}\, \overline{\nu^c_i}\nu_j\, \langle\phi^0\rangle^2\,. \tag{21.3.55}$$

We expect f_{ij} to be of order $1/M$, perhaps multiplied with small coupling constants, so this gives lepton number non-conserving neutrino masses at most of order[27b] $(300\ \text{GeV})^2/M$. We shall see in Section 21.5 that M

is expected to be of order 10^{15}–10^{18} GeV, so we would expect neutrino masses in the range 10^{-4}–10^{-1} eV, or less if suppressed by small coupling constants. Such masses are too small for direct measurement, but there is no reason for the neutrino mass matrix to be diagonal, so neutrino masses might be detected in oscillations of one neutrino type into another over long flight paths.

A similar analysis shows that there are interactions of dimensionality six that violate both baryon and lepton number conservation, involving three quark fields and one lepton field.[27c] Such interactions would have coupling constants of order M^{-2}, and would lead to processes like proton decay, with rates proportional to M^{-4}.

21.4 Dynamically Broken Local Symmetries*

Our discussion of spontaneously broken local symmetries has so far been entirely within the context of perturbation theory. To some extent, this limitation is inevitable. Whereas for spontaneously broken global symmetries it is possible to prove exact theorems about the existence and interactions of massless Goldstone bosons, the spontaneous breakdown of a local symmetry does not lead to any such precise consequences. Even the existence of massive vector bosons is not really a general theorem; for sufficiently strong gauge coupling these particles decay so rapidly that they lose their identity as distinct resonances of definite spin $j = 1$.

On the other hand, if the gauge couplings like e or g or g' are sufficiently small then the theory with a spontaneously broken local symmetry must be very close to one with a spontaneously broken global symmetry, about which exact theorems can be proved. It is therefore possible to derive useful approximate results for such gauge theories, *even if the other non-gauge couplings are very strong.* One example is provided by the standard $SU(2) \times U(1)$ electroweak theory with a large scalar self-coupling λ (and hence a large scalar mass; see Eq. (21.3.27)). A more intriguing possibility is that the breakdown of electroweak symmetry is due to strong forces associated with some new gauge group acting on a set of new fermions. We will here consider the results that can be obtained for all such theories, without reference to the specific mechanism for spontaneous symmetry breaking.[28]

We assume that in the limit of zero gauge couplings, our theory is invariant under some group G of *global* symmetries, spontaneously broken

* This section lies somewhat out of the book's main line of development, and may be omitted in a first reading.

to a subgroup H. As discussed in Section 19.5, in this case the theory can be written in terms of a set of Goldstone boson fields ξ_a, plus other matter fields $\tilde{\psi}$, whose G-transformation properties are such that the Lagrangian is G-invariant if it is H-invariant and constructed solely from $\tilde{\psi}$ and covariant derivatives $D_{a\mu}$, $D_\mu\tilde{\psi}$ etc., given by Eqs. (19.6.14), (19.6.30), etc.

We now 'turn on' the gauge couplings. The gauge group $\mathcal{G}$ is of course required to be a subgroup $\mathcal{G} \subset G$ of the group G of all symmetries of the theory, and when G is spontaneously broken to H, $\mathcal{G}$ must be spontaneously broken to a subgroup $\mathcal{H}$, equal to the intersection of $\mathcal{G}$ with H. The generators $\mathcal{T}_\alpha$ of the gauge group $\mathcal{G}$ may be expressed as linear combinations of the generators T_A of the full group G:

$$\mathcal{T}_\alpha = \sum_A e_{\alpha A} T_A \,, \tag{21.4.1}$$

with coefficients $e_{\alpha A}$, the gauge couplings, that are taken very small. The index A runs over the labels i, a of the unbroken symmetry generators t_i and the broken symmetry generators x_a. (We are here taking the generators T_A to be conventionally normalized; that is, they are represented by matrices with elements of order unity. In particular, in contrast to the $\mathcal{T}_\alpha$, the structure constants of the x_a and t_i do not include factors of coupling constants.)

In the underlying theory in which G invariance is linearly realized, we introduce the coupling of gauge fields $\mathscr{A}_{\alpha\mu}$ to other fields ψ by replacing ordinary derivatives with gauge-covariant derivatives

$$\left(\partial_\mu - i\sum_\alpha \mathcal{T}_\alpha \mathscr{A}_{\alpha\mu}\right)\psi = \left(\partial_\mu - i\sum_A T_A A_{A\mu}\right)\psi \,, \tag{21.4.2}$$

where

$$A_{A\mu} \equiv \sum_\alpha e_{\alpha A}\mathscr{A}_{\alpha\mu} \,. \tag{21.4.3}$$

The resulting theory is then invariant under *formal* local transformations, under which the fields transform according to

$$\psi \to g\psi \,, \tag{21.4.4}$$

$$\sum_A T_A A_{A\mu} \to g\left(\sum_A T_A A_{A\mu}\right) g^{-1} - i(\partial_\mu g)g^{-1} \,, \tag{21.4.5}$$

where $g(x)$ is an arbitrary spacetime-dependent element of the group G. This is a purely formal invariance, because the gauge couplings in general actually break G, as shown by the fact that the transformation (21.4.5) does not in general preserve the form of the linear combination (21.4.3). Nevertheless, we can temporarily forget about Eq. (21.4.3), treating A^μ_A as an unconstrained classical external field, and analyze the structure of the Lagrangian for the matter fields and their interaction with the gauge fields by requiring that it be invariant under local transformations (21.4.4), (21.4.5). In this way we will insure not only that the Lagrangian is invariant under the true local symmetry subgroup $\mathscr{G}$ (and for $e_{\alpha A} \to 0$ under the larger global symmetry group G), but also that the currents, the variational derivatives of the matter action with respect to $\mathscr{A}^{\mu}_{\alpha}$, will have the correct transformation under the broken global symmetry group G. Later we will restrict A^{μ}_{A} to the form (21.4.3), and treat the field $\mathscr{A}^{\mu}_{\alpha}$ as a quantum field, supplying a suitable kinematic term in the Lagrangian for this field.

In order to explore the implications of the spontaneous breakdown of the invariance group G to its subgroup H, we will proceed as we did in Section 19.6. First, replace ψ and A with new fields $\tilde{\psi}$, $\tilde{A}$:

$$\tilde{\psi} = \gamma^{-1}(\xi)\psi \,, \tag{21.4.6}$$

$$\tilde{A}^{\mu}_{A} = \sum_{B} D_{AB}\left(\gamma^{-1}(\xi)\right) A^{\mu}_{B} \,, \tag{21.4.7}$$

where $\gamma(\xi)$ is the standard G transformation that eliminates the Goldstone boson degrees of freedom in ψ, and $D(g)$ is the representation of G furnished by the gauge fields:

$$g\, T_A\, g^{-1} = \sum_{B} D_{BA}(g)\, T_B \,. \tag{21.4.8}$$

These Goldstone degrees of freedom reappear in the spacetime-dependent parameters ξ_a on which $\gamma(\xi)$ depends. By the same calculations as in Section 19.6, for local as well as global transformations the transformation rule (21.4.4) translates into the transformations

$$\xi^a \to \xi^{a'} = f^a(\xi, g) \,, \tag{21.4.9}$$

$$\tilde{\psi} \to \tilde{\psi}' = h(\xi, g)\tilde{\psi} \,, \tag{21.4.10}$$

where h and f are defined by

$$g\,\gamma(\xi) = \gamma(f(\xi, g))\, h(\xi, g) \,, \tag{21.4.11}$$

with h in the unbroken subgroup H. We also need to work out the transformation rule for $\tilde{A}^{\mu}_{A}$. Recall that, according to Eq. (21.4.5), under these local transformations the linear combinations of gauge fields in Eq. (21.4.2) transform as

$$\sum_{A} T_A A_{A\mu} \to \sum_{A} T_A A'_{A\mu} = g \left[\sum_{A} T_A A_{A\mu} - i g^{-1}\partial_{\mu} g\right] g^{-1} \,.$$

Multiplying on the left and right by $\gamma^{-1}(\xi')$ and $\gamma(\xi')$ respectively, and using Eqs. (21.4.7), (21.4.8), and (21.4.11), we can write this as

$$i\sum_A T_A\tilde{A}'_{A\mu} = h(\xi,g)\left[i\sum_A T_A\tilde{A}_{A\mu} + \gamma^{-1}(\xi)[g^{-1}\partial_\mu g]\gamma(\xi)\right]h^{-1}(\xi,g)\,. \tag{21.4.12}$$

To see how to cancel the inhomogeneous $g^{-1}\partial_\mu g$ term, we note that by differentiating Eq. (21.4.11) and multiplying on the left with its inverse, we have

$$\begin{aligned}\gamma^{-1}(\xi')\partial_\mu\gamma(\xi') &= h(\xi,g)\gamma^{-1}(\xi)[g^{-1}\partial_\mu g]\gamma(\xi)h^{-1}(\xi,g)\\ &+ h(\xi,g)[\gamma^{-1}(\xi)\partial_\mu\gamma(\xi)]h^{-1}(\xi,g) + h(\xi,g)\partial_\mu h^{-1}(\xi,g)\,. \end{aligned}\tag{21.4.13}$$

So to cancel the inhomogeneous term, we must subtract Eq. (21.4.12) from Eq. (21.4.13):

$$\begin{aligned}&\gamma^{-1}(\xi')\partial_\mu\gamma(\xi') - i\sum_A T_A\tilde{A}'_{A\mu}\\ &\quad= h(\xi,g)\left[\gamma^{-1}(\xi)\partial_\mu\gamma(\xi) - i\sum_A T_A\tilde{A}_{A\mu}\right]h^{-1}(\xi,g)\\ &\quad - \left[\partial_\mu h(\xi,g)\right]h^{-1}(\xi,g)\,.\end{aligned}\tag{21.4.14}$$

We therefore define new gauge-covariant quantities $\mathscr{D}$ and $\mathscr{E}$ by

$$\sum_a i\mathscr{D}_{a\mu}x_a + \sum_i i\mathscr{E}_{i\mu}t_i \equiv \gamma^{-1}(\xi)\partial_\mu\gamma(\xi) - i\sum_A T_A\tilde{A}_{A\mu} \tag{21.4.15}$$

with transformation properties

$$\mathscr{D}_{a\mu} \to \mathscr{D}'_{a\mu}\,, \qquad \mathscr{E}_{i\mu} \to \mathscr{E}'_{i\mu}\,, \tag{21.4.16}$$

where

$$\sum_a \mathscr{D}'_{a\mu}x_a = h(\xi,g)\left(\sum_a \mathscr{D}_{a\mu}x_a\right)h^{-1}(\xi,g)\,, \tag{21.4.17}$$

$$\sum_a \mathscr{E}'_{i\mu}t_i = h(\xi,g)\left(\sum_i \mathscr{E}_{i\mu}t_i\right)h^{-1}(\xi,g) + i\Big(\partial_\mu h(\xi,g)\Big)h^{-1}(\xi,g)\,, \tag{21.4.18}$$

just as in Eqs. (19.6.26) and (19.6.27). We can use $\mathscr{E}$ to construct fully covariant derivatives of matter fields

$$\mathscr{D}_\mu\tilde{\psi} = \partial_\mu\tilde{\psi} + i\sum_i t_i\mathscr{E}_{i\mu}\tilde{\psi} \tag{21.4.19}$$

as well as higher derivatives like $\mathscr{D}_\nu\mathscr{D}_\mu\tilde{\psi}$, etc. Because of the inhomogeneous term in Eq. (21.4.12), we cannot freely introduce $\tilde{A}_{A\mu}$ or covariant derivatives like (21.4.19) of $\tilde{A}_{\alpha\mu}$ into the Lagrangian. However, it is easy to construct a 'curl' that transforms covariantly under both local $\mathscr{G}$ transformations and global G transformations. It is

$$\tilde{F}_{A\mu\nu} \equiv \sum_B D_{AB}\Big(\gamma^{-1}(\xi)\Big)\Big(\partial_\mu A_{B\nu} - \partial_\nu A_{B\mu} - \sum_{CD} C_{BCD} A_{C\mu} A_{D\nu}\Big) . \tag{21.4.20}$$

This transforms under formal local G transformations as

$$\tilde{F}_{A\mu\nu} \to \tilde{F}'_{A\mu\nu} = \sum_B \mathscr{D}_{AB}\Big(h(\xi, g)\Big)\tilde{F}_{B\mu\nu} . \tag{21.4.21}$$

The Lagrangian is therefore invariant under formal local G transformations if it is constructed as an arbitrary function of $\tilde{\psi}$, $\mathscr{D}_{a\mu}$, $\mathscr{D}_\mu\tilde{\psi}$, $\tilde{F}_{A\mu\nu}$, *and higher covariant derivatives, that satisfies global H invariance.*

Now let us return to reality, and treat A^μ_A as a quantum field of the restricted form (21.4.3). Eqs. (21.4.15) and (21.4.20) now become

$$\begin{aligned} &\sum_a i\mathscr{D}_{a\mu} x_a + \sum_i i\mathscr{E}_{i\mu} t_i \\ &= \gamma^{-1}\partial_\mu \gamma(\xi) - i\sum_{AB\alpha} T_A D_{AB}(\gamma^{-1}(\xi)) e_{\alpha B} \mathscr{A}_{\alpha\mu} \end{aligned} \tag{21.4.22}$$

and

$$\tilde{F}_{A\mu\nu} = \sum_{B\alpha} D_{AB}\Big(\gamma^{-1}(\xi)\Big) e_{\alpha B}\, \mathscr{F}_{\alpha\mu\nu} , \tag{21.4.23}$$

where

$$\mathscr{F}_{\beta\mu\nu} = \partial_\mu \mathscr{A}_{\beta\nu} - \partial_\nu \mathscr{A}_{\beta\mu} - \sum_{\gamma\delta} \mathscr{C}_{\beta\gamma\delta} \mathscr{A}_{\gamma\mu} \mathscr{A}_{\delta\nu} , \tag{21.4.24}$$

with $\mathscr{C}_{\beta\gamma\delta}$ the structure constant of the gauge group, related to the structure constant of G by

$$\sum_{CD} C_{BCD}\, e_{\gamma C}\, e_{\delta D} = \sum_\beta \mathscr{C}_{\beta\gamma\delta}\, e_{\beta B} . \tag{21.4.25}$$

We include in the Lagrangian as the kinematic term for this field the usual Yang–Mills term

$$\mathscr{L}_{\mathscr{A}} = -\frac{1}{4}\sum_\alpha \mathscr{F}_{\alpha\mu\nu}\mathscr{F}_\alpha{}^{\mu\nu} , \tag{21.4.26}$$

in which, by linear transformations of the $\mathscr{A}_{\alpha\mu}$ and correspondingly of the $e_{\alpha A}$, we have adjusted the coefficient of $\mathscr{F}_{\alpha\mu\nu}\mathscr{F}_\beta{}^{\mu\nu}$ to be just $\delta_{\alpha\beta}$. The linear term in $\mathscr{F}_{\alpha\mu\nu}$ is just $\partial_\mu\mathscr{A}_{\alpha\nu} - \partial_\nu\mathscr{A}_{\alpha\mu}$, so Eq. (21.4.26) has the effect of making $\mathscr{A}_{\alpha\mu}$ a canonically normalized vector field. The effective Lagrangian density is therefore to be taken as a function of $\tilde{\psi}$, $\mathscr{D}_\mu\tilde{\psi}$, and $\mathscr{D}_{a\mu}$ that is invariant under global H transformations, plus possible terms that conserve $\mathscr{G}$ but not G, with coefficients proportional to two or more factors of the $e_{\alpha A}$.

Now let us see what sort of perturbation theory we can construct out of these ingredients. We know that the gauge bosons become massless in the limit $e_{\alpha A} \to 0$, where they decouple from the matter fields that experience the spontaneous symmetry breakdown. Let us therefore tentatively anticipate that their mass for small $e_{\alpha A}$ is of order eM, where e is a typical value of $e_{\alpha A}$ (the generators T_A being normalized to have structure constants of order unity) and M is an energy scale typical of the dynamics that leads to spontaneous symmetry breaking. We shall consider here a general Feynman diagram involving gauge and Goldstone bosons of energy or momentum $Q \lesssim eM$, and with all particles of higher energy or momentum and all heavier matter particles buried in corrections to the coupling constants in an effective field theory. Our perturbation theory will be an expansion in powers of e and Q/M. Following the same analysis as in Sections 19.4–6, the total number of powers of e and/or Q/M in any such diagram is

$$\nu = \sum_i V_i(d_i + e_i - 2) + 2L + 2 \,, \tag{21.4.27}$$

where V_i is the number of vertices of type i; d_i and e_i are the number of derivatives and factors of $e_{\alpha A}$, respectively, in an interaction of type i; and L is the number of loops. With the constraint (21.4.3), a field $A_{A\mu}$ or $\tilde{A}_{A\mu}$ contributes one factor of e. Inspection of Eq. (21.4.15) shows that each Goldstone boson covariant derivative $\mathscr{D}_{a\mu}$ contributes $+1$ to $d_i + e_i$, and inspection of Eqs. (21.4.19) and (21.4.15) shows that each additional covariant derivative of $\mathscr{D}_{a\mu}$ contributes another $+1$ to $d_i + e_i$. All allowed terms in the Lagrangian have $d_i + e_i \geq 2$, so the dominant contributions are those from tree graphs ($L = 0$) constructed entirely from interactions with $d_i + e_i = 2$. The only such interactions are the Goldstone boson kinematic term

$$\mathscr{L} = -\tfrac{1}{2} \sum_{ab} F^2_{ab} \mathscr{D}_{a\mu} \mathscr{D}_b{}^{\mu} \,, \tag{21.4.28}$$

the Yang–Mills term (21.4.26), and possible symmetry-breaking non-derivative terms of second order in the $e_{\alpha A}$.

To see the physical significance of the field ξ_a, note that the linear term in $\mathscr{D}_{\alpha\mu}$ is

$$(\mathscr{D}_{a\mu})_{\rm LIN} = \partial_\mu \xi_a - \sum_\alpha e_{\alpha a} \mathscr{A}_{\alpha\mu} \,. \tag{21.4.29}$$

As shown in the appendix to this chapter, we may always choose a 'unitarity gauge' in which, for all α,

$$\sum_{ab} F^2_{ab}\,\xi_a e_{\alpha b} = 0\,, \tag{21.4.30}$$

which makes the cross term in Eq. (21.4.28) vanish. To clarify the significance of this condition, note that in the special case where all broken symmetries are gauge symmetries, any x_a can be written as a linear combination of gauge generators and unbroken generators

$$\begin{aligned} x_a &= \sum_\alpha c_{a\alpha}\mathscr{T}_\alpha + \sum_i c_{ai}t_i\,, \\ &= \sum_\alpha c_{a\alpha}\left(\sum_a e_{\alpha b}x_b + \sum_i e_{\alpha i}t_i\right) + \sum_i c_{ai}t_i\,, \end{aligned}$$

and therefore

$$\sum_\alpha c_{a\alpha}e_{\alpha b} = \delta_{ab}\,.$$

Contracting Eq. (21.4.30) with $c_{\alpha b}$, we see then that $\xi_a = 0$; there are no Goldstone bosons at all in this gauge. More generally, Eq. (21.4.30) leaves us with just those Goldstone bosons that do *not* correspond to gauge symmetries. Some of these are associated with elements of G that are broken by the gauge interactions, and therefore have masses of second order in gauge couplings; these are called pseudo-Goldstone bosons.

With ξ chosen to satisfy the unitarity gauge condition (21.4.30), the quadratic part of the Lagrangian (21.4.28) is simply

$$(\mathscr{L}_\xi)_{\rm QUAD} = -\tfrac{1}{2}\sum_{ab} F^2_{ab}\partial_\mu\xi_a\partial^\mu\xi_b - \tfrac{1}{2}\sum_{\alpha\beta}\mu^2_{\alpha\beta}\mathscr{A}_{\alpha\mu}\mathscr{A}_\beta{}^\mu\,, \tag{21.4.31}$$

where

$$\mu^2_{\alpha\beta} = \sum_{ab} F^2_{ab}\,e_{\alpha a}e_{\beta b}\,. \tag{21.4.32}$$

This has two important implications. First, we note that ξ_a may be expressed in terms of a canonically orthonormalized field π_a, as

$$\xi_a = \sum_b F^{-1}_{ab}\,\pi_b\,, \tag{21.4.33}$$

where F_{ab} is the positive square root of the positive matrix F^2_{ab}. This shows that F^{-1}_{ab} are the factors analogous to F_π^{-1} that accompany the emission and absorption of low energy Goldstone bosons. Second, since $\mathscr{A}_{\alpha\mu}$ has been defined to be a canonically normalized vector field, Eq. (21.4.31) shows that $\mu^2_{\alpha\beta}$ is the square of the vector boson mass matrix. Eq. (21.4.32) is a universal formula for the vector boson mass matrix, valid to second order in the gauge couplings but to all orders in all other interactions. By using Eq. (21.4.1) in Eq. (21.1.7), it is easy to see that our previous result Eq. (21.1.7) is a special case of Eq. (21.4.32), with

$$F^2_{ab} = -\sum_{nml}(x_a)_{nm}(x_b)_{nl}v_m v_l \, .$$

Eq. (21.4.32) may also be understood on the basis of the continuity arguments outlined (in a somewhat different notation) at the end of Section 21.1. It guarantees that the effects of gauge boson exchange that survive in the limit of zero gauge coupling are the same as the effects that would have been produced by Goldstone boson exchange if there were no gauge coupling.

In general, we cannot calculate the F^2_{ab} matrix, but we do know that it must be invariant under the unbroken subgroup H, in the sense that

$$\sum_d \left[C_{ibd}F^2_{dc} + C_{icd}F^2_{bd}\right] = 0 \, .$$

This condition allows us to put useful constraints on the gauge boson masses (21.4.32).

As an example, consider the case of the electroweak gauge group $SU(2) \times U(1)$, spontaneously broken to the $U(1)$ of electromagnetism. The three broken symmetry generators x_a can be taken as the three generators of $SU(2)$ (called t_1, t_2, t_3 in Section 21.3), without the coupling constant factor g, and the one unbroken symmetry generator t can be taken as the charge q, without the factor e. That is, the x_a and t are represented on the lepton doublets by the matrices

$$\vec{x} = \frac{1}{4}(1+\gamma_5)\left\{\begin{pmatrix} 0 & 1 \\ 1 & 0 \end{pmatrix}, \begin{pmatrix} 0 & -i \\ i & 0 \end{pmatrix}, \begin{pmatrix} 1 & 0 \\ 0 & -1 \end{pmatrix}\right\} ,$$

$$t = -\begin{pmatrix} 0 & 0 \\ 0 & 1 \end{pmatrix} .$$

The gauge generators are then given by

$$\vec{\mathscr{T}} = g\vec{x} \, , \qquad \mathscr{T}_y = g'(x_3 - t) \, . \tag{21.4.34}$$

That is, the non-zero coefficients $e_{\alpha a}$ of x_a in the gauge generator $\mathscr{T}_\alpha$ are

$$e_{11} = e_{22} = e_{33} = g \, , \qquad e_{y3} = g' \, .$$

Also, since t subjects the three-vector $\vec{x}$ to a rotation around the three-axis, this unbroken symmetry requires the matrix F^2_{ab} to have the non-zero components

$$F^2_{11} = F^2_{22} \equiv F^2_C \, , \qquad F^2_{33} \equiv F^2_N \, .$$

According to Eq. (21.4.32), the mass squared matrix of the gauge bosons then has the non-vanishing elements

$$\mu^2_{11} = \mu^2_{22} = g^2 F^2_C \, , \qquad \mu^2_{33} = g^2 F^2_N \, ,$$
$$\mu^2_{3y} = g\, g' F^2_N \, , \qquad \mu^2_{yy} = g'^2 F^2_N \, .$$

Its eigenvalues are

$$m_W^2 = g^2 F_C^2\,, \qquad m_Z^2 = (g^2 + g'^2) F_N^2\,, \qquad m_A^2 = 0\,. \qquad (21.4.35)$$

To go further, we need a relation between F_C and F_N. This is provided if the theory is invariant in the limit $g = g' = 0$ under a global symmetry group G larger than $SU(2) \times U(1)$, which breaks spontaneously to a subgroup H, that includes three-dimensional rotations under which $\vec{x}$ rotates as a three-vector. Such an unbroken symmetry would require F^2_{ab} to be proportional to δ_{ab}, and hence

$$F_C = F_N\,.$$

Any such symmetry is known as a 'custodial' symmetry. It has the consequence that m_Z/m_W is given in terms of the gauge coupling constants by the successful formula discussed in Section 21.3

$$m_Z/m_W = \sqrt{1 + g'^2/g^2} = 1/\sin\theta\,. \qquad (21.4.36)$$

For instance, in the absence of the gauge couplings, the Lagrangian (21.3.28) for the scalar doublet ϕ in the simplest version of the $SU(2)\times U(1)$ electroweak theory may be written

$$(\mathscr{L}_\phi)_{g=g'=0} = -\frac{1}{2}\,\partial_\mu \phi_n \partial^\mu \phi_n - \frac{\mu^2}{2}\,\phi_n \phi_n - \frac{\lambda}{4}(\phi_n \phi_n)^2\,,$$

where

$$\phi_1 \equiv \mathrm{Im}\phi^+\,, \quad \phi_2 \equiv \mathrm{Re}\phi^+\,, \quad \phi_3 \equiv \mathrm{Im}\phi^0\,, \quad \phi_4 \equiv \mathrm{Re}\phi^0\,.$$

This is automatically invariant under an 'accidental' $SO(4) \equiv SU(2) \times SU(2)$ global symmetry group, which is spontaneously broken by the vacuum expectation value of $\mathrm{Re}\phi^0$ down to an approximate unbroken $SO(3)$ custodial subgroup.[29] The result (21.4.36) applies even if there is more than one scalar doublet, because even though in this case the mass and interaction terms in the scalar Lagrangian will not in general respect the custodial symmetry, it is only the kinematic term that enters in the derivation, and this always has the full $SO(4)$ symmetry.

Custodial symmetries can be found in other theories. Consider for instance a theory with no scalar fields, but with new extra-strong vector gauge interactions,[30] called technicolor interactions, that act on a new $SU(2) \times U(1)$ doublet (U_r, D_r) of 'techniquarks' U_r and D_r, with r a technicolor index. As long as the left- and right-handed parts of both U_r and D_r all transform in the same way under the technicolor gauge group, the Lagrangian will be invariant in the limit of vanishing electroweak couplings under the group $SU(2) \times SU(2)$ of independent $SU(2)$ transformations on the left- and right-handed techniquark doublets. According to the arguments described in Section 19.9, the subgroup $SU(2)_V$ consisting of simultaneous $SU(2)$ transformations on both the left- and right-handed techniquark doublets will not be spontaneously broken. It is reasonable to suppose that the technicolor interactions will produce a spontaneous breakdown of $SU(2) \times SU(2)$ to $SU(2)_V$, just as color interactions led to a spontaneous breakdown of the chiral $SU(2) \times SU(2)$ symmetry of quantum chromodynamics (for vanishing u and d quark masses) to its isospin subgroup. Under the unbroken $SU(2)_V$ symmetry the electroweak generator $\vec{x}$ or $\vec{\mathcal{T}}$ rotates as a three-vector, leading again to the relation $F_C = F_N$, and the consequent successful prediction of the relation between the W and Z masses.

The technicolor idea is attractive, because it provides a natural mechanism for breaking the electroweak symmetry at a characteristic scale which is very much smaller than what is often supposed to be the fundamental scale of physics, generally assumed (as in string theories) to be of the order of the Planck mass, or about 10^{18} GeV. It is only necessary to suppose that just below the fundamental scale there is an unbroken gauge group consisting of the $SU(3) \times SU(2) \times U(1)$ of the strong and electroweak interactions, plus a technicolor gauge group, all with comparable small coupling constants. If the technicolor gauge interactions are asymptotically free the technicolor coupling like the QCD color coupling will increase slowly with decreasing energy, becoming strong at an energy much less than the fundamental scale. This energy where the technicolor coupling becomes strong would set the scale for the parameters F_{ab} that appear in our formula, Eq. (21.4.32), for the gauge boson masses, and would therefore presumably be of the order of 300 GeV. Since the increase of coupling with decreasing energy is logarithmic, a moderate difference in the beta-functions for technicolor and color can easily give rise to the three orders of magnitude difference between the energy scales where color and technicolor forces become strong.

Unfortunately, although technicolor provides a very attractive picture of the spontaneous breaking of $SU(2) \times U(1)$, it does not by itself offer a mechanism for giving masses to the quarks and leptons. For this reason it has been suggested to add additional 'extended technicolor' gauge interactions with transformations that link quarks and techniquarks.[31] Such theories have potential problems with flavor-changing neutral current weak interactions, and though these problems may be surmounted, the added complications reduce their attractiveness. The question of elementary weakly coupled scalars versus dynamical symmetry breaking remains open.

21.5 Electroweak–Strong Unification

We saw in Section 15.2 that a gauge theory will have an independent coupling constant for each simple or $U(1)$ subgroup of the gauge group.

Thus the electroweak theory which is based on the gauge group $SU(2) \times U(1)$, has two independent couplings, g and g'. In order to reduce the number of free parameters it was suggested[32] that the $SU(2) \times U(1)$ gauge group might be embedded in a simple $SU(3)$ gauge group, which would give $g' = g/\sqrt{3}$, but this has been ruled out experimentally. After the advent of quantum chromodynamics, theorists confronted a gauge group $SU(3) \times SU(2) \times U(1)$, which has three independent gauge coupling constants: the coupling g_s of quantum chromodynamics, and the couplings g and g' of the electroweak interactions. To reduce these to just a single free parameter, it was proposed to embed $SU(3) \times SU(2) \times U(1)$ in various simple Lie groups:* $SU(4) \times SU(4)$,[33] $SU(5)$,[34] or $SO(10)$.[35] Such models are often known as grand-unified theories.

Fortunately, the consequences of these and a large class of other models for the ratios of the $SU(3) \times SU(2) \times U(1)$ coupling constants are independent of the details of the individual models.[36] This class of models is characterized by the fact that the observed generations of quarks and leptons are the only fermions in the models, or at least the only fermions that are not neutral under $SU(3) \times SU(2) \times U(1)$.

As shown in Section 15.2, for any simple compact Lie group there is a conventional choice of generators T_α with totally antisymmetric structure constants, which in each reducible or irreducible representation D satisfy the normalization condition:

$$\mathrm{Tr}\{T_\alpha T_\beta\} = N_D \delta_{\alpha\beta} \,. \tag{21.5.1}$$

We are assuming that all left-handed fermions form n_g generations:

$$\begin{pmatrix} \nu_e \\ e \end{pmatrix}_L \quad \begin{pmatrix} \nu_\mu \\ \mu \end{pmatrix}_L \quad \begin{pmatrix} \nu_\tau \\ \tau \end{pmatrix}_L \quad \cdots,$$

$$\overline{e_R} \qquad \overline{\mu_R} \qquad \overline{\tau_R} \qquad \cdots,$$

$$\begin{pmatrix} u \\ d \end{pmatrix}_L \quad \begin{pmatrix} c \\ s \end{pmatrix}_L \quad \begin{pmatrix} t \\ b \end{pmatrix}_L \quad \cdots,$$

$$\overline{u_R} \qquad \overline{c_R} \qquad \overline{t_R} \qquad \cdots,$$

$$\overline{d_R} \qquad \overline{s_R} \qquad \overline{b_R} \qquad \cdots.$$

The $SU(3)$ generator $\frac{1}{2}g_s\lambda_3$ has eigenvalues: $+\frac{1}{2}g_s$ for the red quark doublets and the white antiquark singlets; $-\frac{1}{2}g_s$ for the white quark doublets and the red antiquark singlets; and zero for all other left-handed

* The group $SU(4) \times SU(4)$ is made simple by inclusion of a discrete symmetry operator that interchanges the two $SU(4)$s.

fermions; so its square has the trace

$$\mathrm{Tr}\left(\tfrac{1}{2}g_s\lambda_3\right)^2 = 4n_g \times \left(\tfrac{1}{2}g_s\right)^2 + 4n_g \times \left(-\tfrac{1}{2}g_s\right)^2 = 2n_g g_s^2\,. \tag{21.5.2}$$

The $SU(2)$ generator t_3 has eigenvalues: $\frac{1}{2}g$ for the red, white, and blue quarks of charge $\frac{2}{3}$ and the neutrinos; $-\frac{1}{2}g$ for the red, white, and blue quarks of charge $-\frac{1}{3}$ and the charged leptons; and zero for all other fermions, so its square has the trace

$$\mathrm{Tr}\,(t_3)^2 = [3n_g + n_g] \times [(\tfrac{1}{2}g)^2 + (-\tfrac{1}{2}g)^2] = 2n_g g^2\,. \tag{21.5.3}$$

Finally, the $U(1)$ generator $y = t_3 - q$ has eigenvalues: $\frac{1}{2}g'$ for neutrinos and charged leptons; $-g'$ for charged antileptons; $-\frac{1}{6}g'$ for quarks; $\frac{2}{3}g'$ for antiquarks of charge $-\frac{2}{3}$; and $-\frac{1}{3}g'$ for antiquarks of charge $+\frac{1}{3}$, so its square has the trace

$$\begin{aligned}\mathrm{Tr}\,y^2 &= 2n_g(\tfrac{1}{2}g')^2 + n_g(-g')^2 + 6n_g(-\tfrac{1}{6}g')^2 + 3n_g(\tfrac{2}{3}g')^2 + 3n_g(-\tfrac{1}{3}g')^2 \\ &= \tfrac{10}{3}n_g g'^2\,. \end{aligned}\tag{21.5.4}$$

Eq. (21.5.1) requires the traces (21.5.2)–(21.5.4) to be all equal, so in this class of models the embedding of $SU(3) \times SU(2) \times U(1)$ in a simple Lie group imposes the coupling constant relations

$$g_s^2 = g^2 = \tfrac{5}{3}g'^2\,. \tag{21.5.5}$$

Now, Eq. (21.5.5) is in gross disagreement with the observed values of the coupling constants. The ratio $g'^2/g^2 = \frac{3}{5}$ implies an electroweak mixing angle with $\sin^2\theta \equiv g'^2/(g^2+g'^2) = \frac{3}{8}$, while the experimental value is $\sin^2\theta = 0.231$. Even worse, the strong coupling g_s^2 is of course much larger than g^2 or g'^2.

The solution[36] to this problem is that coupling-constant relations like Eq. (21.5.5) apply only to the couplings measured at an energy scale comparable to the typical mass[47] M of the gauge bosons that become massive in the spontaneous breakdown of the simple gauge group to $SU(3)\times SU(2)\times U(1)$. If the energy E at which the couplings is measured is very much less than M, then there will be large radiative corrections proportional to $\ln(M/E)$.

As emphasized in Chapter 19, there are no large logarithms in the relation between couplings measured at nearby energies μ and $\mu - d\mu$, so by integrating this relation from M down to E we can calculate the couplings at energies $E \ll M$ without encountering large logarithms. For this to be done it is only necessary that the couplings should stay small over this whole range. For the $SU(3) \times SU(2) \times U(1)$ couplings with n_g

fermion generations, Eq. (18.7.2) gives**

$$\mu \frac{d}{d\mu} g_s(\mu) = -\frac{g_s^3(\mu)}{4\pi^2}\left(\frac{11}{4} - \frac{n_g}{3}\right) , \tag{21.5.6}$$

$$\mu \frac{d}{d\mu} g(\mu) = -\frac{g^3(\mu)}{4\pi^2}\left(\frac{11}{6} - \frac{n_g}{3}\right) , \tag{21.5.7}$$

$$\mu \frac{d}{d\mu} g'(\mu) = -\frac{g'^3(\mu)}{4\pi^2}\left(-\frac{5n_g}{9}\right) . \tag{21.5.8}$$

The solutions of these equations are

$$\frac{1}{g_s^2(\mu)} = \frac{1}{g_s^2(M)} - \frac{1}{8\pi^2}\left(11 - \frac{4n_g}{3}\right)\ln\left(\frac{M}{\mu}\right) , \tag{21.5.9}$$

$$\frac{1}{g^2(\mu)} = \frac{1}{g^2(M)} - \frac{1}{8\pi^2}\left(\frac{22}{3} - \frac{4n_g}{3}\right)\ln\left(\frac{M}{\mu}\right) , \tag{21.5.10}$$

$$\frac{1}{g'^2(\mu)} = \frac{1}{g'^2(M)} - \frac{1}{8\pi^2}\left(-\frac{20n_g}{9}\right)\ln\left(\frac{M}{\mu}\right) . \tag{21.5.11}$$

Also, Eq. (21.5.5) should be interpreted to mean that

$$g_s^2(M) = g^2(M) = \tfrac{5}{3} g'^2(M) . \tag{21.5.12}$$

We can therefore eliminate the couplings (21.5.12) *and* the number of generations by subtracting Eq. (21.5.10) from Eq. (21.5.9):

$$\frac{1}{g_s^2(\mu)} - \frac{1}{g^2(\mu)} = -\frac{11}{24\pi^2}\ln\left(\frac{M}{\mu}\right) \tag{21.5.13}$$

and by subtracting 3/5 of Eq. (21.5.11) from Eq. (21.5.10):

$$\frac{1}{g^2(\mu)} - \frac{3}{5g'^2(\mu)} = -\frac{11}{12\pi^2}\ln\left(\frac{M}{\mu}\right) . \tag{21.5.14}$$

Taking the ratio of these two equations gives a formula for $\sin^2\theta \equiv g'^2/(g^2 + g'^2)$:

$$\sin^2\theta = \frac{1}{6} + \frac{5e^2(m_Z)}{9g_s^2(m_Z)} . \tag{21.5.15}$$

** The second term in the brackets in Eqs. (21.5.6) and (21.5.7) is given by Eqs. (18.7.2) and (18.7.3) as $-n_f/6$, where n_f are the numbers of fermions in the defining representation of the respective $SU(N)$ gauge groups. But this was calculated under the assumption that left- and right-handed fermions are in the same representation of the gauge group. If we count only left-handed fermions (and antifermions), then the second term in the brackets in Eqs. (21.5.6) and (21.5.7) should be $-n_f/12$. For $SU(3)$ there are two left-handed quark triplets and two left-handed antiquark triplets per generation, so $n_f = 4n_g$, while for $SU(2)$ there are three left-handed quark doublets and one left-handed lepton doublet per generation, so again $n_f = 4n_g$. For $U(1)$, the beta-function is $g'/24\pi^2$ times the sum of the squares of the $U(1)$ charges for left-handed fermions and antifermions (compare Eq. (18.2.38)), which according to Eq. (21.5.4) is $(g'/24\pi^2) \times (10n_g g'^2/3)$.

In this formula we have set μ equal to the typical energy of the processes used to measure $\sin^2\theta$, that is, $\mu \approx m_Z$. This has the advantage that we will be using the renormalization group equations (21.5.6)–(21.5.8) only above m_Z, where they are not strongly affected by the spontaneous breakdown of $SU(2)\times U(1)$. Eqs. (21.5.13) and (21.5.14) can also be combined to give a formula for the unification scale M:

$$\ln\left(\frac{M}{m_Z}\right) = \frac{4\pi^2}{11e^2}\left(1-\frac{8e^2(m_Z)}{3g_s^2(m_Z)}\right), \qquad (21.5.16)$$

where again, to avoid the effects of electroweak symmetry breaking on the renormalization group equations, we have taken μ as of order m_Z.

We saw in Section 18.2 that the value of $e(\mu)$ at $\mu \approx m_Z$ is given by $e(m_Z)^2/4\pi = (128.87 \pm 0.12)^{-1}$. This is the charge defined in the conventional (Gell-Mann–Low) manner, in terms of the vacuum polarization. For purposes of comparison with g_s, g', and g, it is better to use the coupling defined (as in Section 18.6) by modified minimum subtraction: $e(m_Z)^2/4\pi = (127.9 \pm 0.1)^{-1}$. The greatest uncertainty in Eqs. (21.5.13) and (21.5.14) is in the value of $g_s^2(m_Z)$. As discussed in Section 18.7, extrapolation of g_s from lower energy data gives $g_s^2(m_Z)/4\pi = 0.118 \pm 0.006$, while direct measurement from the rate of Z^0 decay into hadrons gives $g_s^2(m_Z)/4\pi = 0.120 \pm 0.0025$. For $g_s^2(m_Z)/4\pi = 0.118$ and $e(m_Z)^2/4\pi = 1/128$, Eqs. (21.5.15) and (21.5.16) give $\sin^2\theta = 0.203$ and $M \approx 1.1 \times 10^{15}$ GeV.

As mentioned in Section 21.3, there is no reason to expect that baryon and lepton number should be conserved by the suppressed non-renormalizable terms in the effective Lagrangian that describes physics at ordinary energies, so we may expect the presence of an $(SU(3)\times SU(2)\times U(1))$-conserving four fermion (three quarks and a lepton) interaction, with a coefficient that on dimensional grounds would be of order M^{-2}. The proton lifetime was first estimated on this basis, and found to be of order 10^{32} years.[36] Such baryon- and lepton-non-conserving four-fermion interactions are produced in models like those of Refs. 33–35 from the exchange of gauge bosons with masses of order M. More generally, once the standard model explained why baryon- and lepton-non-conserving processes are naturally suppressed, there ceased to be any reason to believe that either baryon or lepton number are exactly conserved.

We have seen that the prediction (21.5.15) comes quite close to the measured value 0.23 of $\sin^2\theta$, but the accuracy of the measurements and calculations has become good enough to make clear that they are not in precise agreement. The extra particles in supersymmetric theories appear to remove this discrepancy,[36a] and lead to a value of $M \simeq 2\times 10^{16}$ GeV, about an order of magnitude larger.[36b] It is very interesting that this M is not so very different from the energy 10^{18} GeV at which the gravitational interactions become strong. The larger value of M also has the effect of increasing the proton lifetime, which is proportional to M^4.

21.6 Superconductivity*

Superconductivity is quite different from the elementary particle phenomena that chiefly concern us in this book, but it is worth some consideration here, both as the earliest realistic example of a spontaneously broken gauge symmetry, and also as an exceptionally enlightening example of the power of effective field theories and of the use of topological arguments in field theory.

A superconductor is simply a material in which electromagnetic gauge invariance is spontaneously broken.** Detailed dynamical theories are needed to explain why and at what temperatures this symmetry breaking occurs, but they are not needed to derive the most striking aspects of superconductivity: exclusion of magnetic fields, flux quantization, zero resistivity, and alternating currents at a gap between superconductors held at different voltages. As we shall see here, these consequences of broken gauge invariance can be worked out, in a manner somewhat like our treatment of soft pions, solely on the basis of general properties of the Goldstone mode.[39]

The action for any system will be invariant under gauge transformations, which in cgs units take the form

$$A_\mu(x) \to A_\mu(x) + \partial_\mu \Lambda(x)\,, \tag{21.6.1}$$

$$\psi_n(x) \to \exp\Big(iq_n\Lambda(x)/\hbar\Big)\psi_n(x)\,, \tag{21.6.2}$$

where $\Lambda(x)$ is arbitrary, and q_n is the electric charge destroyed by field ψ_n. All charges are assumed to be integral multiples of the electron charge $-e$, so this group is compact: the phases Λ and $\Lambda + 2\pi\hbar/e$ are regarded as identical. This symmetry group is assumed to be broken in a superconductor by non-vanishing expectation values of operators carrying charge $-2e$ (such as products of two electron fields), so there

---

* This section lies somewhat out of the book's main line of development, and may be omitted in a first reading.

** This is not the way that most experts have historically thought about superconductivity. Early phenomenological theories were known to violate electromagnetic gauge invariance, but this was regarded as more annoying than enlightening. Broken symmetry is never mentioned in the seminal paper by Bardeen, Cooper, and Schrieffer[37] that first gave us a microscopic theory of superconductivity. Anderson[38] subsequently stressed the important role of broken symmetry in superconductors, but even today most textbooks explain superconductivity in terms of detailed dynamical models, with broken symmetry rarely mentioned.

is an unbroken Z_2 subgroup, consisting of gauge transformations with $\Lambda = 0$ and $\Lambda = \pi\hbar/e$.

We introduce a Goldstone boson field $\phi(x)$ by writing all charged fields as

$$\psi_n(x) = \exp\left(iq_n\phi(x)/\hbar\right)\tilde{\psi}_n(x)\,. \tag{21.6.3}$$

The field $\phi(x)$ parameterizes the coset space $U(1)/Z_2$, and so is given the gauge transformation property

$$\phi(x) \to \phi(x) + \Lambda(x)\,. \tag{21.6.4}$$

Because $\phi(x)$ parameterizes $U(1)/Z_2$ rather than $U(1)$, we must identify $\phi(x)$ and $\phi(x) + \pi\hbar/e$. All the $\tilde{\psi}_n(x)$ are gauge-invariant, and so when integrated out leave the Lagrangian as a gauge-invariant functional of ϕ and A^μ alone. It follows that the Lagrangian for the Goldstone and electromagnetic fields may be written as

$$L = -\tfrac{1}{4}\int d^3x\, F_{\mu\nu}F^{\mu\nu} + L_s\left[A_\mu - \partial_\mu\phi\right]\,, \tag{21.6.5}$$

where L_s is an imperfectly known functional. The electric current and charge density here are

$$\mathbf{J}(x) = \frac{\delta L_s}{\delta \mathbf{A}(x)}\,, \tag{21.6.6}$$

$$J^0(x) = -\frac{\delta L_s}{\delta A^0(x)} = -\frac{\delta L_s}{\delta\dot{\phi}(x)}\,. \tag{21.6.7}$$

The equations of motion for the Goldstone boson field are then

$$\frac{\partial}{\partial t}\frac{\delta L_s}{\delta\dot{\phi}(x)} = \frac{\delta L_s}{\delta\phi(x)} = \nabla\cdot\frac{\delta L_s}{\delta\mathbf{A}(x)}\,, \tag{21.6.8}$$

which in light of Eqs. (21.6.6) and (21.6.7) is equivalent to the conservation law of electric charge

$$\nabla\cdot\mathbf{J} + \frac{\partial}{\partial t}J^0 = 0\,. \tag{21.6.9}$$

Now let us see how this formalism explains the remarkable properties of superconductors. About L_s we will only need to assume that the system is stable in the absence of Goldstone or external electromagnetic fields, so that the energy is at least at a local minimum at $A_\mu = \partial_\mu\phi$, with non-vanishing second derivatives with respect to $A_\mu - \partial_\mu\phi$.

One immediate consequence is that, deep in a large superconductor where boundary conditions are unimportant, the electromagnetic field is a pure gauge:

$$A_\mu = \partial_\mu \phi \,, \tag{21.6.10}$$

so that in particular the magnetic field must vanish. This is known as the *Meissner effect.* It is possible to be a little more quantitative about what we mean by 'deep in a large superconductor.' Since the energy is a minimum when Eq. (21.6.10) is satisfied, for small values of $|\mathbf{A} - \nabla\phi|$ it must be of order $|\mathbf{A} - \nabla\phi|^2 L^3/\lambda^2$, where λ is some length depending on the nature of the material, and L^3 is the superconductor volume. If a magnetic field of order B penetrated the superconductor, then we would have $|\mathbf{A} - \nabla\phi|$ of order BL, so the energy cost in allowing the magnetic field into the superconductor would be of order B^2L^5/λ^2. On the other hand, the energy cost in expelling a magnetic field B from a volume L^3 is of order B^2L^3. Hence a weak magnetic field will be expelled from a superconductor if $B^2L^5/\lambda^2 \gg B^2L^3$, or in other words if $L \gg \lambda$. For this reason λ is known as the *penetration depth* of the superconductor.

The same energetic considerations tell us that for any superconducting material, there is a critical magnetic field, above which superconductivity is extinguished. The existence of superconductivity at zero magnetic field means that the material in its normal state has an energy per unit volume which is higher than that in the superconducting state, say by an amount Δ. When a superconductor with linear dimensions much larger than λ is placed in a magnetic field B, the magnetic field is expelled from most of the material, at an energy cost per volume of $B^2/2$. Hence it is energetically favorable for the material to be in the superconducting state if and only if the magnetic field is below a critical value

$$B_c = \sqrt{2\Delta}\,. \tag{21.6.11}$$

(This is for uniform superconductors. As we shall see below, for certain kinds of superconductor it is possible to maintain superconductivity throughout most of the material for magnetic fields in a finite range above B_c by the formation of narrow vortex lines with normal metal at their cores.) A magnetic field $B < B_c$ will penetrate a superconductor to a depth λ, but this does not extinguish superconductivity in this layer; indeed, as shown by the field equation $\nabla \times \mathbf{B} = \mathbf{J}$, it is in this surface layer of the superconductor that electric currents can flow.

Now consider a thick superconducting wire, with thickness much larger than λ, bent into a closed ring. We can draw a closed contour $\mathscr{C}$ running deep inside the wire, along which $|\mathbf{A} - \nabla\phi|$ must vanish. This does not mean that either $\mathbf{A}$ or ϕ vanishes on this contour, but we do know that in going around the ring ϕ must return to an equivalent value, and can therefore only change by an amount $n\pi\hbar/e$, with n a positive or negative integer or zero. It follows then by Stokes's theorem that the magnetic flux through the area $\mathscr{A}$ surrounded by $\mathscr{C}$ is subject to the *flux quantization* rule

$$\int_{\mathscr{A}} \mathbf{B}\cdot d\mathbf{S} = \oint_{\mathscr{C}} \mathbf{A}\cdot d\mathbf{x} = \oint_{\mathscr{C}} \nabla\phi\cdot d\mathbf{x} = \frac{n\pi\hbar}{e}\,. \tag{21.6.12}$$

The electric current that maintains the magnetic flux (21.6.12) flows in a layer of thickness λ just below the surface of the superconducting wire. The quantization of flux shows that this current cannot decay smoothly, but only in jumps at which the flux (21.6.12) drops by multiples of $\pi\hbar/e$, so there can be no ordinary electrical resistance in the superconductor.

The absence of resistance in a superconductor can be shown in a more general context than closed rings, by considering time-dependent effects in superconductors. Note that Eq. (21.6.7) can be interpreted as the statement that $-J^0$ is the canonical conjugate to ϕ. The Hamiltonian H_s is thus to be regarded as a functional of ϕ and J^0 rather than ϕ and $\dot{\phi}$, with the time dependence of ϕ given by the Hamiltonian equation

$$\dot{\phi}(x) = \frac{\delta H_s}{\delta(-J^0(x))}\,. \tag{21.6.13}$$

Now, the 'voltage' $V(x)$ at any point is just the change in energy density per change in the charge density at that point, so Eq. (21.6.13) gives the time dependence of the Goldstone boson field as

$$\dot{\phi}(x) = -V(x)\,. \tag{21.6.14}$$

It follows that a piece of superconducting wire that carries a steady current, with time-independent fields, must have zero voltage difference between its ends, because otherwise $\phi(x)$ would have a time-dependent gradient. A zero voltage difference at finite current is what we mean by zero resistance.

Now consider a gap between two pieces of superconducting material. In the absence of any gradients along the surface of the gap, or any vector potential, gauge invariance requires L_s to depend only on the difference $\Delta\phi$ between the Goldstone boson fields in the two superconductors:

$$L_{\text{junction}} = \mathscr{A} F(\Delta\phi)\,, \tag{21.6.15}$$

where $\mathscr{A}$ is the area of the junction. Furthermore, we can shift ϕ in either superconductor by any integer multiple of $\pi\hbar/e$ without physical effect, so the function F must be periodic[†]

$$F(\Delta\phi) = F(\Delta\phi + \pi\hbar n/e)\,. \tag{21.6.16}$$

[†] This function was calculated by Josephson,[40] who found it to be proportional to $\cos(2e\Delta\phi/\hbar)$, but this is an approximate result, while the periodicity is exact.

A current flows through such a gap, which can be calculated by considering the junction in the presence of a vector potential **A**. Gauge invariance then tells us that in place of $\Delta\phi$, the function F must depend on

$$\Delta_A\phi = \int \mathbf{dx} \cdot (\nabla\phi - \mathbf{A}) \,,$$

the integral being taken over a line joining the two superconductors. Eq. (21.6.6) then shows that the current density is

$$\mathbf{J} = \frac{\delta L_{\text{junction}}}{\delta \mathbf{A}} = -\hat{\mathbf{n}} F'(\Delta_A\phi) \,,$$

where $\hat{\mathbf{n}}$ is the unit vector perpendicular to the gap. We can now take away the vector potential, and find the current

$$\mathbf{J} = -\hat{\mathbf{n}} F'(\Delta\phi) \,. \tag{21.6.17}$$

If we now suppose that the two superconductors are maintained at uniform voltages, with a voltage difference ΔV, then, according to Eq. (21.6.14), the difference in the Goldstone boson fields will have the time dependence

$$\Delta\phi = -t\Delta V + \text{constant} \,. \tag{21.6.18}$$

Using this in Eq. (21.6.17) and recalling Eq. (21.6.16), we see that the current oscillates at a frequency

$$\nu = e|\Delta V|/\pi\hbar \,. \tag{21.6.19}$$

This is the *ac Josephson effect.*[40] It is possible to measure frequencies and voltages with great accuracy, so this effect provides a very accurate method of measuring the constant $e/\hbar$.

As pointed out at the end of Section 19.6, the description of a system with broken symmetry in terms of Goldstone modes alone becomes inadequate when a system is brought close to the point where the broken symmetry becomes unbroken. Under these circumstances the Goldstone mode is accompanied with other modes that have nearly zero frequency in the long-wavelength limit, which together with the Goldstone mode forms a *linear* representation, usually irreducible, of the symmetry group, known as the order parameter. It is plausible to assume that a uniform superconductor in slowly varying external fields is described by a *local* order parameter, because any non-locality would be characterized by microscopic distance scales (such as the mean electron–electron spacing) that are much smaller than the scales of distances over which the electromagnetic and Goldstone boson fields are assumed to be varying. For superconductivity there is no doubt about the nature of this order parameter. The only non-trivial irreducible linear representation of the group $U(1)$ is a real two-vector ψ_n, or equivalently a Goldstone mode ϕ

and modulus field ρ, with

$$\psi_1 + i\psi_2 = \rho \exp\left(2ie\phi/\hbar\right) \equiv \psi \,. \tag{21.6.20}$$

The coefficient of $i\phi$ in the exponential must be $2e/\hbar$ in order that a gauge transformation with $\Lambda = \pi\hbar/e$ (and with no smaller Λ) should leave ψ_n invariant. For a nearly uniform time-independent system close to the symmetry-breaking transition, the order parameter is small and slowly varying in space, and so the Lagrangian in an external vector potential $\mathbf{A}$ may be approximated (from now on using natural units with $\hbar = 1$) by

$$L_s \simeq \int d^3x \left[-\tfrac{1}{2}\left(\nabla\psi_n - 2ie\, t_{nm}\mathbf{A}\psi_m\right)^2 + \tfrac{1}{2}m^2\psi_n\psi_n - \tfrac{1}{4}g(\psi_n\psi_n)^2 \right], \tag{21.6.21}$$

where t is the Hermitian $U(1)$ generator

$$t = \begin{pmatrix} 0 & -i \\ i & 0 \end{pmatrix},$$

and g must be taken positive to give a bounded Hamiltonian. This is the *Ginzburg–Landau* theory of superconductivity.[41] It has been derived by Gor'kov[42] from the microscopic theory of superconductivity presented below, in the case of a short-range potential and a temperature close to the critical temperature at which the material loses its superconductivity.

In terms of ρ and ϕ, Eq. (21.6.21) becomes

$$L_s \simeq \int d^3x \left[-2e^2\rho^2\left(\nabla\phi - \mathbf{A}\right)^2 + \tfrac{1}{2}m^2\rho^2 - \tfrac{1}{4}g\rho^4 - \tfrac{1}{2}\left(\nabla\rho\right)^2 \right]. \tag{21.6.22}$$

The field equations are then

$$\nabla \times \mathbf{B} = 4e^2\rho^2(\nabla\phi - \mathbf{A})\,, \tag{21.6.23}$$

$$\nabla^2\rho = -m^2\rho + g\rho^3 + 4e^2\rho(\nabla\phi - \mathbf{A})^2\,. \tag{21.6.24}$$

The $U(1)$ symmetry is broken if these field equations are satisfied for $\rho \neq 0$, which in a field-free homogeneous material will be the case if $m^2 > 0$, in which case ρ takes the value $\langle\rho\rangle = m/\sqrt{g}$. The penetration depth λ was defined earlier as the inverse square root of the coefficient of $-\frac{1}{2}(\nabla\phi - \mathbf{A})^2$, so here

$$\lambda = \frac{1}{\sqrt{4e^2\langle\rho\rangle^2}} = \frac{\sqrt{g}}{2em}\,. \tag{21.6.25}$$

This is the distance that according to Eq. (21.6.23) characterizes variations in the magnetic field. On the other hand, variations in the modulus ρ are characterized by a distance scale known as the *correlation length*, given according to Eq. (21.6.24) by[††]

$$\xi = 1/m\sqrt{2}\,. \tag{21.6.26}$$

Also, the superconducting state with $\rho = \langle\rho\rangle$ has an energy per unit volume that is less than that of the normal state with $\rho = 0$ by an amount

$$\Delta = \tfrac{1}{2}m^2\langle\rho\rangle^2 - \tfrac{1}{4}\langle\rho\rangle^4 = m^4/4g\,. \tag{21.6.27}$$

Eliminating the parameters m and g from Eqs. (21.6.25)–(21.6.27), we find one important approximate relation among the observable quantities λ, ξ, and Δ:

$$\Delta \simeq \frac{1}{8\,e^2\lambda^2\xi^2}\,. \tag{21.6.28}$$

The modular field becomes important in the dynamics of superconducting vortex lines. These arise when a superconductor of a certain type is placed in a magnetic field that is strong enough so that it is energetically favorable for tubes of magnetic flux known as vortex lines to penetrate the material.[43] (The conditions for the appearance of vortex lines are discussed below.) By drawing a closed curve $\mathscr{C}$ around the tube at a distance much larger than the penetration depth, where the magnetic field vanishes, and repeating the argument contained in Eq. (21.6.12), we see that the magnetic flux through the area $\mathscr{A}$ surrounded by $\mathscr{C}$ must be equal to the change of ϕ around the curve, and hence equal to an integer multiple of the flux quantum π/e, just as for the flux through a thick superconducting ring. Where this flux is not zero, there must be a line within each tube along which electromagnetic gauge invariance is *not* broken. To see this, note that if we shrink the curve $\mathscr{C}$ into the region of high magnetic field it becomes no longer true that $\nabla\phi = \mathbf{A}$, but the change of ϕ around the curve must remain an integer multiple of π/e, and so by continuity cannot change. Thus we must eventually encounter a line (conceivably of finite thickness) along which ρ vanishes, so that ϕ becomes ill-defined. (This is an elementary example of the sort of topological reasoning we shall use in Chapter 23.) Near this line we must take both ρ and ϕ into account as dynamical variables.

The quantization of magnetic flux shows that a superconducting vortex line of minimum flux π/e is stable. A vortex line of higher flux cannot simply disappear, but magnetic flux quantization alone does not prevent it from breaking up into vortex lines of smaller flux. Bogomol'nyi[43a] has shown that vortex lines of flux $n\pi/e$ with $n > 1$ are unstable against breakup into n vortex lines of flux π/e if and only if $\lambda > \xi$.

[††] The factor $\sqrt{2}$ is included along with m because at $\rho = \langle\rho\rangle$ the derivative of the function $-m^2\rho + g\rho^3$ in Eq. (21.6.24) is $2m^2$.

For this and other reasons it is convenient to divide superconducting materials into two classes: *type I superconductors* (most pure metals, except niobium) have $\xi > \lambda$, while *type II superconductors* (niobium and most alloys) have $\lambda > \xi$. The corresponding distinction in the electroweak standard model is between theories where the scalar mass (analogous to $1/\xi$) is less than or greater than the W and Z masses (analogous to $1/\lambda$).

From the definitions of the correlation length ξ and penetration depth λ, it follows that the modular parameter will rise from zero at the central line of a vortex to its equilibrium value $\langle\rho\rangle$ in a distance of the order of the correlation length ξ, while the magnetic field will decay with distance from the central line in a distance of the order of the penetration depth λ. Hence a vortex solution in a type I superconductor with $\xi \gg \lambda$ would consist of a thin inner cylinder of nearly normal metal, within which the magnetic field drops to zero, surrounded by a much thicker outer cylinder within which the modular parameter rises to its asymptotic value $\langle\rho\rangle$. In contrast, a vortex in a type II superconductor with $\lambda \gg \xi$ consists of a thin inner cylinder of constant magnetic field, within which the modular parameter rises to its asymptotic value $\langle\rho\rangle$, surrounded by a much thicker outer cylinder of superconducting material within which the magnetic field falls to zero. Vortex solutions exist for both types of superconductor and for any magnetic field, but as we shall now see, it is only in type II superconductors and in a finite range of magnetic fields that vortex lines are energetically favored.

Because each vortex line has a cross-sectional area of order $\pi\xi^2$ within which the material is in its normal state or nearly so, the extra energy per volume required to create these vortex lines is of order $\mathcal{N}\pi\xi^2\Delta$, where $\mathcal{N}$ is the number of vortex lines per area. This vortex density is limited by the condition that $\mathcal{N} < 1/\pi\xi^2$, since otherwise the cylinders of normal metal would overlap, and the material would be considered to be in its normal state. The magnetic field must be expelled from a fraction $1 - \mathcal{N}\pi\lambda^2$ of the material if $\mathcal{N} < 1/\pi\lambda^2$, and from the whole material if $\mathcal{N} > 1/\pi\lambda^2$, so the energy per volume of the vortex state, relative to the superconducting state in the absence of magnetic fields, is

$$W_V \approx \mathcal{N}\pi\xi^2\Delta + \tfrac{1}{2}B^2 \times \begin{cases} 1 - \mathcal{N}\pi\lambda^2 & \mathcal{N} \leq 1/\pi\lambda^2 \\ 0 & \mathcal{N} \geq 1/\pi\lambda^2 \end{cases} . \qquad (21.6.29)$$

(Numerical factors like $\frac{1}{2}$ and π are kept here to remind the reader of the origin of these expressions, but should not be taken literally.) For comparison, the energy per volume of the normal metal exceeds that in the superconducting state by $W_N = +\Delta$, and the energy per volume required to expel all magnetic fields from a superconductor is $W_S = B^2/2$. We can decide which state is present at a given magnetic field by checking which of W_S, W_N, or W_V is smallest.

In type I superconductors we must distinguish between magnetic fields less or greater than the critical field $B_c \equiv \sqrt{2\Delta}$. Recalling that $\mathcal{N} < 1/\pi\xi^2$, and here $\xi > \lambda$, we also have $\mathcal{N} < 1/\pi\lambda^2$. Hence for $B < B_c$ Eq. (21.6.29) gives $W_V > \frac{1}{2}B^2 + \mathcal{N}\pi(\xi^2 - \lambda^2)\Delta > W_S$, so there can be no vortex lines. Also, for such fields $W_N > W_S$, so the material is superconducting. On the other hand, for $B > B_c$ Eq. (21.6.29) implies that $W_V > \Delta[1 + \mathcal{N}\pi(\xi^2 - \lambda^2)] > W_N$, so here again there are no vortex lines. Also for such fields $W_S > W_N$, so the material is in its normal state.

In type II superconductors we need to distinguish between magnetic fields in three ranges: $B < B_{c1}$, $B_{c1} < B < B_{c2}$, and $B > B_{c2}$, where B_{c1} and B_{c2} are a pair of critical fields, of order

$$B_{c1} \approx \sqrt{2\Delta}\,\xi/\lambda\,, \qquad B_{c2} \approx \sqrt{2\Delta}\,\lambda/\xi\,.$$

As we have seen, for type II superconductors the only stable vortex lines are those with the minimum flux‡ π/e, so in a magnetic field B we should put the number of vortices per area $\mathcal{N}$ in Eq. (21.6.9) equal to eB/π. For $B < B_{c1}$ we can use Eq. (21.6.8) to show that $eB/\pi < 1/\pi\lambda^2$. The coefficient of the vortex density in Eq. (21.6.29) is therefore the positive quantity $\pi\xi^2\Delta - \frac{1}{2}B^2\lambda^2$, so here $W_V > W_S$, and there are no vortex lines for $B < B_{c1}$. Also, for such fields $W_N > W_S$, so the material is entirely superconducting. For $B > B_{c1}$, Eq. (21.6.28) shows that the vortex density $\mathcal{N} = eB/\pi$ is greater than $1/\pi\lambda^2$, so in the vortex state the magnetic field completely penetrates the superconductor, and the energy per volume is given by Eqs. (21.6.29) and (21.6.28) as

$$W_V \approx \mathcal{N}\pi\xi^2\Delta = eB\xi^2\Delta \approx (B/B_{c2})W_N \approx (B_{c1}/B)W_S\,.$$

Hence for $B_{c1} < B < B_{c2}$ we have $W_V < W_N$ and $W_V < W_S$, so the material is in its vortex state. For $B > B_{c2}$ we still have $W_V < W_S$ but now $W_N < W_V$, so the vortices disappear and all superconductivity is extinguished. The ability of type II superconductors with $\lambda \gg \xi$ to carry magnetic fields much higher than the critical value $B_c \approx \sqrt{\Delta}$ deduced

‡ The derivation of flux quantization given above for an isolated vortex line is not fully applicable here. As we shall see, the separation of the vortex lines for $B > B_{c1}$ is less than the penetration depth λ, so it is not possible to find a contour $\mathscr{C}$ on which $\mathbf{A} - \nabla\phi = 0$ by simply drawing a circle around the vortex line at a distance from the vortex much greater than λ. Instead, it is necessary to appeal to considerations of continuity. (M. Tinkham, private communication.) Suppose we draw an arbitrary continuous curve between the centers of any two vortex lines. As shown below, on this curve the vector $\mathbf{A} - \nabla\phi$ will be very large close to either vortex, but pointing to opposite sides of the curve. Thus there is at least one point on each such curve where $\mathbf{A} - \nabla\phi = 0$. Because $\mathbf{A} - \nabla\phi$ is gauge-invariant, it must be continuous, so there is a closed contour $\mathscr{C}$ around each vortex line where $\mathbf{A} - \nabla\phi = 0$.

earlier is important in technological applications of superconductivity, including magnets in high energy accelerators.

The Ginzburg–Landau theory holds only where the material is near the transition between its normal and superconducting states, so let us apply it near the center of a vortex line, where ρ drops to zero. Close to the center of a vortex line we can ignore its curvature, and assume cylindrical symmetry. The field $\mathbf{A} - \nabla\varphi$ is taken to have only an azimuthal component:

$$\left(\mathbf{A} - \nabla\phi\right)_\theta = A(r)\,, \tag{21.6.30}$$

so that the magnetic field has only an axial component:

$$B_z = \left(\nabla \times (\mathbf{A} - \nabla\phi)\right)_z = A'(r) + A(r)/r\,, \tag{21.6.31}$$

while ρ is a function only of r. The structure of the vortex line is thus governed by the pair of coupled differential equations:

$$A''(r) + r^{-1}A'(r) - r^{-2}A(r) = \frac{\rho^2(r)A(r)}{\lambda^2\langle\rho\rangle^2}\,, \tag{21.6.32}$$

$$\rho''(r) + r^{-1}\rho'(r) + \frac{1}{2\xi^2}\left(\rho(r) - \frac{\rho^3(r)}{\langle\rho\rangle^2}\right) = 4e^2\rho(r)A^2(r)\,. \tag{21.6.33}$$

When r is small compared with both the correlation length ξ and the penetration depth λ we can neglect the terms in Eqs. (21.6.32) and (21.6.33) proportional to $1/\xi^2$ and $1/\lambda^2$, and find the simplified equations

$$A'' + A'/r - A/r^2 = 0\,, \tag{21.6.34}$$

$$\rho'' + \rho'/r = 4e^2A^2\rho\,. \tag{21.6.35}$$

Eq. (21.6.34) has the general solution

$$A(r) = \frac{Br}{2} + \frac{C}{2er}\,, \tag{21.6.36}$$

where B is a constant that according to Eq. (21.6.31) is the magnetic field along the vortex line, and C is a real constant that is so far arbitrary. Using this in Eq. (21.6.35) shows that for $r \to 0$, the solution for $\rho(r)$ is a linear combination of $r^{|C|}$ and $r^{-|C|}$. The order parameter $\rho\exp(2ie\phi)$ must be a smooth function of position, so we can conclude that $|C|$ is a positive integer ℓ, with $\rho \propto r^\ell$ and $\phi = \pm\ell\varphi/2e+$ constant for $r \to 0$. Note that this solution for ϕ is consistent with Eq. (21.6.36) with $C = \pm\ell$; the azimuthal component of $\nabla\phi$ approaches $\pm\ell/2er$, while analyticity requires the azimuthal component of $\mathbf{A}$ to vanish as $r \to 0$. By a non-singular gauge transformation we can arrange that $\phi = \pm\ell\varphi/2e$ everywhere, so by the same reasoning as in Eq. (21.6.12) the magnetic flux carried by a single vortex line in a superconductor much larger than the penetration depth is $\pm\pi\ell/e$. We see not only that, as expected, the order parameter vanishes at the center of the vortex line, but that it vanishes with a power of r equal to the magnitude of the magnetic flux in units of π/e.

This solution for ϕ obeys the 'quantization' condition that $\phi(2\pi)-\phi(0)$ is an integral multiple of π/e, which implies the quantization of magnetic flux. In the theory of Goldstone boson and electromagnetic fields based on Eq. (21.6.5) this condition on ϕ has to be imposed by hand on the solution of the field equations, while the Ginzburg–Landau equations 'know' about this condition, because these equations are based on an appropriate choice of order parameter.

* * *

Although the most dramatic properties of superconductors can be derived directly from the assumption that electromagnetic gauge invariance is spontaneously broken, a microscopic theory of superconductivity is needed to understand how and when this occurs. The derivation of the microscopic superconductivity theory of Bardeen, Cooper, and Schrieffer[37] has been recast[44,44a] in the language of effective field theory, using power-counting methods similar to those that we have used here in Sections 19.5 and 21.4. For this purpose, suppose we integrate out the degrees of freedom associated with the ions in a superconductor,‡‡ leaving only an effective interaction among electrons. For simplicity, we shall work at zero temperature, and at first assume no external field, so that the Lagrangian is invariant under translations and the time reversal operation T. We shall also assume spin-independent forces, so that the Lagrangian is invariant under $SU(2)$ transformations that act on spin indices alone, but we shall not need to assume invariance under rotations acting on momenta. Electrons are then characterized by a momentum $\mathbf{p}$ and spin index $s=\pm\frac{1}{2}$, and are described by annihilation and creation operators $a(\mathbf{p},s,t)$ and $a^\dagger(\mathbf{p},s,t)$, with a Lagrangian of form

$$
\begin{aligned}
L = & -\sum_s \int d^3p\, a^\dagger(\mathbf{p},s,t)\left[-i\frac{\partial}{\partial t}+E(\mathbf{p})\right]a(\mathbf{p},s,t) \\
& + \sum_{s_1 s_2 s_3 s_4} \int d^3p_1\, d^3p_2\, d^3p_3\, d^3p_4\, V_{s_1 s_2 s_3 s_4}(\mathbf{p}_1,\mathbf{p}_2,\mathbf{p}_3,\mathbf{p}_4) \\
& \quad \times a^\dagger(\mathbf{p}_1,s_1,t)\, a^\dagger(\mathbf{p}_2,s_2,t)\, a(\mathbf{p}_3,s_3,t)\, a(\mathbf{p}_4,s_4,t)\delta^3(\mathbf{p}_1+\mathbf{p}_2-\mathbf{p}_3-\mathbf{p}_4) \\
& + \cdots,
\end{aligned}
\tag{21.6.37}
$$

‡‡ Strictly speaking, it is not possible to integrate out all degrees of freedom except electrons, because the phonon is a Goldstone boson whose frequency vanishes for very large wavelengths, like a massless particle in relativistic theories. However those effects of phonon interactions that cannot be represented as effective electron–electron interactions are suppressed by inverse factors of the ion masses.

where '$\cdots$' represents terms with six or more creation and annihilation operators, and $E(\mathbf{p})$ is the electron energy minus the chemical potential. For free electrons, $E(\mathbf{p}) = \mathbf{p}^2/2m_e - E_F$ where E_F is the energy of electrons at the Fermi surface. Interactions will inevitably change this function, but it is natural to expect that since $E(\mathbf{p})$ vanishes for free electrons at momenta $\mathbf{p}$ on the sphere $|\mathbf{p}| = \sqrt{2m_e E_F}$, in the presence of interactions it will still vanish on some closed Fermi surface $\mathscr{S}$:

$$E(\mathbf{p}) = 0 \text{ for } \mathbf{p} \text{ on } \mathscr{S} . \tag{21.6.38}$$

For reasons that will become apparent, we shall integrate out all electrons except those within a thin shell of thickness κ around the Fermi surface.[♮] (Later we will be able to remove the cut-off κ by introducing a renormalized electron–electron potential.) The remaining degrees of freedom are then electrons with momenta of the form

$$\mathbf{p} = \mathbf{k} + \hat{\mathbf{n}}(\mathbf{k})\ell , \tag{21.6.39}$$

where $\mathbf{k}$ is on the Fermi surface $\mathscr{S}$, $\hat{\mathbf{n}}(\mathbf{k})$ is the unit vector normal to the surface at $\mathbf{k}$, and $0 \leq \ell \leq \kappa$. For such momenta,

$$E(\mathbf{p}) = v_F(\mathbf{k})\ell , \tag{21.6.40}$$

where

$$v_F(\mathbf{k}) \equiv \hat{\mathbf{n}}(\mathbf{k}) \cdot \Big(\nabla_{\mathbf{p}} E(\mathbf{p})\Big)_{\mathbf{p}=\mathbf{k}} . \tag{21.6.41}$$

The electron propagator in wave number–frequency space is then

$$\frac{1}{\omega - v_F(\mathbf{k})\ell + i\epsilon} . \tag{21.6.42}$$

Now let us consider how a general connected matrix element scales with κ as $\kappa \to 0$. We have an integral over frequencies for every loop, and a propagator for every internal line, so the integral of the product of propagators over all frequencies will have an ℓ dependence ℓ^{L-I}, where L is the number of loops and I is the number of internal lines. To count the number of ℓ integrals, it is important to note that for generic momenta the delta-function in the interaction term in Eq. (21.6.37) constrains the $\mathbf{k}$s, not the ℓs. (For instance, if momentum conservation constrains the momenta $\mathbf{p}_1$ and $\mathbf{p}_2$ of two electron lines to have total momentum $\mathbf{P} \neq 0$, then the integral over $\mathbf{p}_1$ runs over the intersection of two closed shells of thickness κ, one centered on $\mathbf{P}$ and the other on zero. The intersection of these shells is a closed ring of thickness κ, so we have to integrate

[♮] For forces that are not spin-independent, there are two Fermi surfaces, one for each eigenvalue of the matrix $E_{s's}(\mathbf{p})$, and we integrate over all electrons in each spin eigenstate except those within a thin shell around the corresponding Fermi surface.

over one **k**-component which gives position around the ring, and two ℓs which give position within the cross section of the ring.) Thus there are I integrals over ℓs, and since the integrand goes as ℓ^{L-I}, the matrix element will vary as

$$M \propto \kappa^L \,. \tag{21.6.43}$$

The number of loops is related to the number of internal lines and the numbers V_i of vertices of type i by the familiar relation

$$L = I - \sum_i V_i + 1 \,. \tag{21.6.44}$$

Also, the number of internal lines is related to the numbers of vertices and the number E of external lines by another familiar relation

$$2I + E = \sum_i n_i V_i \,, \tag{21.6.45}$$

where n_i is the number of electron operators in the interaction of type i. Eliminating I from Eqs. (21.6.44) and (21.6.45) gives

$$L = 1 - \tfrac{1}{2}E + \tfrac{1}{2}\sum_i V_i(n_i - 2) \,. \tag{21.6.46}$$

Terms in the action with $n_i = 2$ just serve to change the function $E(\mathbf{p})$, and hence to shift the Fermi surface. True interactions have $n_i > 2$, and so it appears from Eqs. (21.6.43) and (21.6.46) that they yield terms in the matrix element that make relatively negligible contributions for $\kappa \to 0$. In the language of Section 18.5, this would mean that all interactions are irrelevant operators. This is why electrons near the Fermi surface in normal metals behave pretty much like free particles.

There is, however, an exception to this conclusion. If a pair of electron lines disappears into the vacuum, then translation invariance requires the momenta of these two lines to be equal in magnitude and opposite in direction, so if one momentum is near the Fermi surface, then the other is also. (Time reversal invariance requires that $E(\mathbf{p})$ is even in $\mathbf{p}$, so even though the Fermi surface is generally not spherical, it is still true that if $E(\mathbf{p}) = 0$ then $E(-\mathbf{p}) = 0$.) The integral over the momentum $\mathbf{p}$ of one of these lines is then over a shell of thickness κ, or in other words over two ks and *one* ℓ. For each interaction involving two such lines, we have one rather than two integrals over ℓs, so that instead of reducing the order of magnitude of the matrix element by a factor κ as indicated in Eqs. (21.6.43) and (21.6.46), such an interaction has no effect on the order of magnitude of the matrix element. In other words, interactions involving four electron operators become marginal rather than irrelevant if they act between electron lines that eventually disappear into the vacuum.

To see the consequences of this exception, it is convenient to use a trick, known as the *Hubbard–Stratonovich transformation*,[45] that was mentioned briefly in Section 10.7. We are now going to include slowly varying external electromagnetic fields, so it will be convenient to work in coordinate space. Gauge invariance tells us that the Lagrangian (21.6.37) then becomes

$$L = -\sum_s \int d^3x \, \psi_s^\dagger(\mathbf{x},t) \left[-i\frac{\partial}{\partial t} - A_0(\mathbf{x},t) + E\Big(-i\nabla + \mathbf{A}(\mathbf{x},t) \Big) \right] \psi_s(\mathbf{x},t)$$
$$+ \sum_{s_1\, s_2\, s_3\, s_4} \int d^3x_1\, d^3x_2\, d^3x_3\, d^3x_4\, V_{s_1\, s_2\, s_3\, s_4}(\mathbf{x}_1,\, \mathbf{x}_2,\, \mathbf{x}_3,\, \mathbf{x}_4)$$
$$\times\, \psi_{s_1}^\dagger(\mathbf{x}_1,t)\, \psi_{s_2}^\dagger(\mathbf{x}_2,t)\, \psi_{s_3}(\mathbf{x}_3,t)\, \psi_{s_4}(\mathbf{x}_4,t)\,, \tag{21.6.47}$$

where

$$\psi_s(\mathbf{x},t) \equiv (2\pi)^{-3/2} \int d^3p \, \exp(i\mathbf{p}\cdot\mathbf{x})\, a(\mathbf{p},s,t)\,. \tag{21.6.48}$$

We are now dropping terms with more than four electron operators, because they are all irrelevant. To this Lagrangian we add the term

$$\Delta L = \frac{1}{4} \int d^3x_1\, d^3x_2\, d^3x_3\, d^3x_4 \sum_{s_1 s_2 s_2 s_4} V_{s_1 s_2 s_2 s_4}(\mathbf{x}_1, \mathbf{x}_2, \mathbf{x}_3, \mathbf{x}_4)$$
$$\times \left[\Psi_{s_2 s_1}^\dagger(\mathbf{x}_2,\, \mathbf{x}_1,\, t) - \psi_{s_1}^\dagger(\mathbf{x}_1,\, t)\, \psi_{s_2}^\dagger(\mathbf{x}_2,\, t) \right]$$
$$\times \left[\Psi_{s_3 s_4}(\mathbf{x}_3,\, \mathbf{x}_4,\, t) - \psi_{s_3}(\mathbf{x}_3,\, t)\, \psi_{s_4}(\mathbf{x}_4,\, t) \right] \tag{21.6.49}$$

and do path integrals over the new 'pair' field $\Psi(\mathbf{x},\, s,\, \mathbf{x}',\, s',\, t)$ as well as the electron field ψ. This is allowed because ΔL is quadratic in the pair field, with a field-independent coefficient of the second-order term, so, according to the appendix of Chapter 9, the integration over $\Psi_{ss'}(\mathbf{x},\, \mathbf{x}',\, t)$ in path integrals just has the effect of setting $\Psi_{ss'}(\mathbf{x},\, \mathbf{x}',\, t)$ equal to the stationary point $\Psi_{ss'}(\mathbf{x},\, \mathbf{x}',\, t) = \psi_s(\mathbf{x},\, t)\, \psi_{s'}(\mathbf{x}',\, t)$ of the Lagrangian, at which $\Delta L = 0$. This additional term is chosen so that the terms in $L+\Delta L$ that are quartic in electron fields cancel, leaving only terms quadratic in these fields:

$$L + \Delta L$$
$$= -\sum_s \int d^3x \, \psi_s^\dagger(\mathbf{x},t) \left\{ -i\frac{\partial}{\partial t} - A_0(\mathbf{x},t) + E\Big(-i\nabla + \mathbf{A}(\mathbf{x},t) \Big) \right\} \psi_s(\mathbf{x},t)$$
$$-\frac{1}{4} \sum_{s_1\, s_2\, s_3\, s_4} \int d^3x_1\, d^3x_2\, d^3x_3\, d^3x_4\, V_{s_1\, s_2\, s_3\, s_4}(\mathbf{x}_1,\, \mathbf{x}_2,\, \mathbf{x}_3,\, \mathbf{x}_4)$$
$$\times \Big[\psi_{s_1}^\dagger(\mathbf{x}_1,t)\, \psi_{s_2}^\dagger(\mathbf{x}_2,t)\, \Psi_{s_3 s_4}(\mathbf{x}_3,\, \mathbf{x}_4,\, t) + \Psi_{s_2 s_1}^\dagger(\mathbf{x}_2,\, \mathbf{x}_1,\, t) \psi_{s_3}(\mathbf{x}_3,t)\, \psi_{s_4}(\mathbf{x}_4,t)$$
$$- \Psi_{s_2 s_1}^\dagger(\mathbf{x}_2,\, \mathbf{x}_1,\, t)\, \Psi_{s_3 s_4}(\mathbf{x}_3,\, \mathbf{x}_4,\, t) \Big]\,. \tag{21.6.50}$$

Now to the point. We have shown that the interaction V is irrelevant except where it acts on a pair of electron lines that disappear into the vacuum. When we calculate the quantum effective action $\Gamma[\Psi]$ in the presence of an external, slowly varying pair field $\Psi_{ss'}(\mathbf{x}, \mathbf{x}', t)$, this means that we drop all diagrams except those that become disconnected when we slice through any internal Ψ line. But the general definition of the quantum effective action requires that we drop all diagrams that *do* become disconnected when we slice through any internal line. *Therefore we must not include any internal Ψ lines at all.* Because the electron field enters quadratically in Eq. (21.6.50), the only graphs that survive when we integrate out the electron field are a one-vertex graph arising from the last term in the square brackets in Eq. (21.6.50), plus a one-loop graph given, by the same reasoning as in Section 16.2, by the logarithm of the determinant of the coefficient of the terms quadratic in electron fields:

$$\begin{aligned}\Gamma[\Psi] = \frac{1}{4}\sum_{s_1\, s_2\, s_3\, s_4}\int dt \int d^3x_1\, d^3x_2\, d^3x_3\, d^3x_4\, V_{s_1\, s_2\, s_3\, s_4}(\mathbf{x}_1, \mathbf{x}_2, \mathbf{x}_3, \mathbf{x}_4) \\ \times \Psi^\dagger_{s_2 s_1}(\mathbf{x}_2, \mathbf{x}_1, t)\, \Psi_{s_3 s_4}(\mathbf{x}_3, \mathbf{x}_4, t) \\ -\frac{i}{2}\ln \mathrm{Det}\begin{pmatrix} A & B \\ B^\dagger & -A^T \end{pmatrix} + \text{constant}\,. \qquad (21.6.51)\end{aligned}$$

where A and B are the 'matrices':

$$A_{\mathbf{x}'s't',\mathbf{x}st} \equiv \delta_{s's}\left\{-i\frac{\partial}{\partial t} - A_0(\mathbf{x},t) + E\Big(-i\nabla + \mathbf{A}(\mathbf{x},t)\Big)\right\}\delta^3(\mathbf{x}'-\mathbf{x})\delta(t'-t)\,, \qquad (21.6.52)$$

$$B_{\mathbf{x}'s't',\mathbf{x}st} \equiv -\Delta_{s's}(\mathbf{x}', \mathbf{x}, t)\delta(t'-t)\,, \qquad (21.6.53)$$

and Δ is the *gap function*:

$$\Delta_{s's}(\mathbf{x}', \mathbf{x}, t) \equiv \frac{1}{2}\sum_{\sigma'\sigma}\int d^3y\, d^3y'\; V_{s's\sigma'\sigma}(\mathbf{x}',\mathbf{x},\mathbf{y}',\mathbf{y})\,\Psi_{s's}(\mathbf{x}', \mathbf{x}, t)\,. \qquad (21.6.54)$$

(Here and below, we ignore an additive constant in Γ arising from the modes that have been integrated out.) This is one problem in which the effective action can be calculated without any assumption about the smallness of the interaction in the Lagrangian.

In order to simplify our further discussion, let us now specialize to the case of a spin-singlet pair field

$$\Psi_{+-}(\mathbf{x}', \mathbf{x}, t) = -\Psi_{-+}(\mathbf{x}', \mathbf{x}, t) \equiv \Psi(\mathbf{x}', \mathbf{x}, t)\,, \qquad (21.6.55)$$

where subscripts + and − stand for spin indices +1/2 and −1/2, respectively.[♮♮] Using invariance under rotations of spin indices, the components of the potential needed here are then

$$V_{+-+-}(\mathbf{x}', \mathbf{x}, t) - V_{+--+}(\mathbf{x}', \mathbf{x}, t) = -V_{-++-}(\mathbf{x}', \mathbf{x}, t) + V_{-+-+}(\mathbf{x}', \mathbf{x}, t)$$
$$\equiv 2V(\mathbf{x}', \mathbf{x}, t) . \tag{21.6.56}$$

Then Eq. (21.6.54) gives

$$\Delta_{+-}(\mathbf{x}', \mathbf{x}, t) = -\Delta_{-+}(\mathbf{x}', \mathbf{x}, t) \equiv \Delta(\mathbf{x}', \mathbf{x}, t) , \tag{21.6.57}$$
$$\Delta_{++}(\mathbf{x}', \mathbf{x}, t) = \Delta_{--}(\mathbf{x}', \mathbf{x}, t) = 0 . \tag{21.6.58}$$

The quantum effective action (21.6.51) is now

$$\Gamma[\Psi] = \int dt \int d^3x_1\, d^3x_2\, d^3x_3\, d^3x_4\, V(\mathbf{x}_1, \mathbf{x}_2, \mathbf{x}_3, \mathbf{x}_4)$$
$$\times\, \Psi^\dagger(\mathbf{x}_2, \mathbf{x}_1, t)\, \Psi(\mathbf{x}_3, \mathbf{x}_4, t)$$
$$-i \ln \mathrm{Det} \begin{pmatrix} \mathscr{A} & \mathscr{B} \\ \mathscr{B}^\dagger & -\mathscr{A}^T \end{pmatrix} , \tag{21.6.59}$$

where

$$\mathscr{A}_{\mathbf{x}'t',\mathbf{x}t} \equiv \left\{ -i\frac{\partial}{\partial t} - A_0(\mathbf{x}, t) + E\Big(-i\nabla + \mathbf{A}(\mathbf{x}, t)\Big) \right\} \delta^3(\mathbf{x}' - \mathbf{x})\delta(t' - t), \tag{21.6.60}$$

$$\mathscr{B}_{\mathbf{x}'t',\mathbf{x}t} \equiv -\Delta(\mathbf{x}', \mathbf{x}, t)\, \delta(t' - t), \tag{21.6.61}$$

and

$$\Delta(\mathbf{x}', \mathbf{x}, t) \equiv \int d^3y\, d^3y'\; V(\mathbf{x}', \mathbf{x}, \mathbf{y}', \mathbf{y})\, \Psi(\mathbf{y}', \mathbf{y}, t) . \tag{21.6.62}$$

Let's first use these results to consider the translationally invariant case, with no external electromagnetic fields. Then the pair and gap fields can be expressed as Fourier transforms

$$\Psi(\mathbf{x}', \mathbf{x}) = \int d^3p\; e^{i\mathbf{p}\cdot(\mathbf{x}'-\mathbf{x})}\Psi(\mathbf{p}) , \tag{21.6.63}$$

$$\Delta(\mathbf{x}', \mathbf{x}) = (2\pi)^{-3} \int d^3p\; e^{i\mathbf{p}\cdot(\mathbf{x}'-\mathbf{x})}\Delta(\mathbf{p}) , \tag{21.6.64}$$

and the electron–electron potential appears here in the form

$$\int d^3x_1\, d^3x_2\, d^3x_3\, d^3x_4\; e^{i\mathbf{p}'\cdot(\mathbf{x}_1-\mathbf{x}_2)}\, e^{i\mathbf{p}\cdot(\mathbf{x}_3-\mathbf{x}_4)}\, V(x_1, x_2, x_3, x_4)$$
$$\equiv \mathscr{V}_4\, V(\mathbf{p}', \mathbf{p}) , \tag{21.6.65}$$

[♮♮] In liquid He^3 the non-vanishing components of the pair field form a spin triplet, with components $\Psi_1 = \Psi_{++}$, $\Psi_0 = \sqrt{2}\Psi_{+-} = \sqrt{2}\Psi_{-+}$, $\Psi_{-1} = \Psi_{--}$.

where $\mathscr{V}_4$ is the spacetime volume

$$\mathscr{V}_4 \equiv \int d^3x \int dt\, 1\,. \tag{21.6.66}$$

By going over to wave number–frequency space, the 'matrices' (21.6.60) and (21.6.61) become diagonal, and the calculation of the determinant becomes trivial. The effective potential $V[\Psi]$ (not to be confused with the electron–electron potential) was defined in Section 16.1 as minus the effective action per spacetime volume:

$$\begin{aligned} V[\Psi] &\equiv -\Gamma[\Psi]/\mathscr{V}_4 \\ &= -\int d^3p\, d^3p'\; \Psi^*(\mathbf{p}')V(\mathbf{p}',\mathbf{p})\Psi(\mathbf{p}) \\ &\quad + \frac{i}{(2\pi)^4}\int d\omega\; d^3p\; \ln\left(1 - \frac{|\Delta(\mathbf{p})|^2}{\omega^2 - E^2(\mathbf{p}) + i\epsilon}\right) \end{aligned} \tag{21.6.67}$$

with

$$\Delta(\mathbf{p}) = -\int d^3p'\; V(\mathbf{p},\mathbf{p}')\,\Psi(\mathbf{p}')\,. \tag{21.6.68}$$

(The $i\epsilon$ term can be inferred from the Feynman rules for the electron loop graphs, and as shown in Section 9.2, it ultimately arises from the conditions on the electron field at $t \to \pm\infty$.) Wick rotating, integrating over ω, and expressing Ψ in terms of Δ, this becomes

$$\begin{aligned} V[\Delta] &= -\int d^3p\, d^3p'\; \Delta^*(\mathbf{p}')V^{-1}(\mathbf{p}',\mathbf{p})\Delta(\mathbf{p}) \\ &\quad - \frac{1}{(2\pi)^3}\int d^3p\left[\sqrt{E^2(\mathbf{p}) + |\Delta(\mathbf{p})|^2} - E(\mathbf{p})\right]. \end{aligned} \tag{21.6.69}$$

As we saw in Section 16.1, the field equations are just the condition that the effective action is stationary, which in our case yields the famous *gap equation* for the equilibrium gap function $\Delta_0(\mathbf{p})$:

$$\begin{aligned} 0 &= \left.\frac{\delta V[\Delta]}{\delta\Delta^*(\mathbf{p})}\right|_{\Delta=\Delta_0} \\ &= -\int d^3p'\; V^{-1}(\mathbf{p},\mathbf{p}')\Delta_0(\mathbf{p}') - \frac{1}{2(2\pi)^3}\frac{\Delta_0(\mathbf{p})}{\sqrt{E^2(\mathbf{p}) + |\Delta_0(\mathbf{p})|^2}} \end{aligned}$$

or, in a more familiar form,

$$\Delta_0(\mathbf{p}) = -\frac{1}{2(2\pi)^3}\int d^3p'\frac{V(\mathbf{p},\mathbf{p}')\,\Delta_0(\mathbf{p}')}{\sqrt{E^2(\mathbf{p}') + |\Delta_0(\mathbf{p}')|^2}}\,. \tag{21.6.70}$$

All along, all integrals over momenta have been implicitly understood to be limited to momenta of the form (21.6.39) within a thin shell of thickness κ around the Fermi surface. The effective potential (21.6.69) is thus

$$V[\Delta] = -\kappa^2 \int_{\mathscr{S}} d^2k\, d^2k'\ \Delta^*(\mathbf{k}')V^{-1}(\mathbf{k}',\mathbf{k})\Delta(\mathbf{k})$$
$$-\frac{1}{(2\pi)^3}\int_0^\kappa \int_{\mathscr{S}} d^2k\left[\sqrt{\ell^2 v_F^2(\mathbf{k}) + |\Delta(\mathbf{k})|^2} - \ell\, v_F(\mathbf{k})\right]. \quad (21.6.71)$$

The effective potential is now to be understood as a functional only of the gap function Δ on the Fermi surface.

Since κ is arbitrary, the potential $V(\mathbf{k}',\mathbf{k})$ must be given a κ-dependence such that $V[\Delta]$ is κ-independent. Most applications[44] of renormalization group methods to superconductivity have followed the Wilson approach outlined in Section 12.4, deriving a differential equation for the dependence of $V(\mathbf{k}',\mathbf{k})$ on κ, and studying the behavior of its solutions for $\kappa \to 0$. For the sake of flexibility it is useful to note that one can just as well adopt the approach of Gell-Mann and Low, and introduce a renormalized electron–electron potential[44a]

$$V_\mu^{-1}(\mathbf{k}',\mathbf{k}) \equiv -\left.\frac{\delta^2 V[\Delta]}{\delta\Delta^*(\mathbf{k}')\delta\Delta(\mathbf{k})}\right|_{\Delta(\mathbf{k})=\Delta(\mathbf{k})^*=\mu}, \quad (21.6.72)$$

where μ is a sliding renormalization scale, like that introduced in Section 18.2. By expressing the original electron–electron potential in terms of V_μ, Eq. (21.6.71) becomes

$$V[\Delta] = -\int_{\mathscr{S}} d^2k\, d^2k'\ \Delta^*(\mathbf{k}')V_\mu^{-1}(\mathbf{k}',\mathbf{k})\Delta(\mathbf{k})$$
$$-\frac{1}{(2\pi)^3}\int_0^\kappa d\ell \int_{\mathscr{S}} d^2k\ \left[\sqrt{\ell^2 v_F^2(\mathbf{k}) + |\Delta(\mathbf{k})|^2} - \ell v_F(\mathbf{k})\right.$$
$$\left. - \frac{|\Delta(\mathbf{k})|^2}{2(\ell^2 v_F^2(\mathbf{k}) + \mu^2)^{1/2}} + \frac{\mu^2|\Delta(\mathbf{k})|^2}{4(\ell^2 v_F^2(\mathbf{k}) + \mu^2)^{3/2}}\right]. \quad (21.6.73)$$

The integral over ℓ now converges if we take the cutoff κ to infinity, and yields

$$V[\Delta] = -\int_{\mathscr{S}} d^2k\, d^2k'\ \Delta^*(\mathbf{k}')V_\mu^{-1}(\mathbf{k}',\mathbf{k})\Delta(\mathbf{k})$$
$$+\frac{1}{2(2\pi)^3}\int_{\mathscr{S}} d^2k\ \frac{|\Delta(\mathbf{k})|^2}{v_F(\mathbf{k})}\left[\ln\left(\frac{|\Delta(\mathbf{k})|}{\mu}\right) - 1\right]. \quad (21.6.74)$$

The condition that $V[\Delta]$ be stationary at $\Delta = \Delta_0$ yields the gap equation in the more useful form

$$\Delta_0(\mathbf{k}) = \frac{1}{2(2\pi)^3}\int_{\mathscr{S}} d^2k'\ V_\mu(\mathbf{k},\mathbf{k}')v_F^{-1}(\mathbf{k}')\Delta_0(\mathbf{k}')\ln\left(\frac{|\Delta_0(\mathbf{k}')|}{\mu}\right). \quad (21.6.75)$$

Using this in Eq. (21.6.74) shows that the energy density of the superconducting state is less than that of the normal state by an amount

$$\Delta \equiv V(0) - V(\Delta_0) = \int_{\mathscr{S}} d^2k \, \frac{|\Delta_0(\mathbf{k})|^2}{2(2\pi)^3 v_F(\mathbf{k})} \, . \tag{21.6.76}$$

Of course the μ dependence of the electron–electron potential V_μ must be chosen to satisfy a renormalization group equation that insures that the effective potential (21.6.74) is independent of the arbitrary renormalization scale μ:

$$\mu \frac{d}{d\mu} V_\mu^{-1}(\mathbf{k}', \mathbf{k}) = -\frac{\delta^2(\mathbf{k}' - \mathbf{k})}{2(2\pi)^3 v_F(\mathbf{k})}$$

or, equivalently,

$$\mu \frac{d}{d\mu} V_\mu(\mathbf{k}', \mathbf{k}) = \frac{1}{2(2\pi)^3} \int d^2k'' \; V_\mu(\mathbf{k}', \mathbf{k}'') \, v_F^{-1}(\mathbf{k}'') \, V_\mu(\mathbf{k}'', \mathbf{k}) \, . \tag{21.6.77}$$

This can be usefully rewritten in terms of the Hermitian kernel

$$K_\mu(\mathbf{k}', \mathbf{k}) \equiv \frac{1}{2(2\pi)^3} \, v_F^{-1/2}(\mathbf{k}') \, v_F^{-1/2}(\mathbf{k}) \, V_\mu(\mathbf{k}', \mathbf{k}) \, , \tag{21.6.78}$$

as

$$\mu \frac{d}{d\mu} K_\mu = K_\mu^2 \, . \tag{21.6.79}$$

The eigenvectors $u_n(\mathbf{k})$ of $K_\mu(\mathbf{k}, \mathbf{k}')$ are then independent of μ, while the eigenvalues take the form $1/\ln(\Lambda_n/\mu)$, where the Λ_n are integration constants, like the Λ of quantum chromodynamics. The potential therefore takes the form

$$V_\mu(\mathbf{k}', \mathbf{k}) = 2(2\pi)^3 \, v_F^{1/2}(\mathbf{k}) \, v_F^{1/2}(\mathbf{k}') \sum_n \frac{u_n(\mathbf{k}) \, u_n^*(\mathbf{k}')}{\ln(\Lambda_n/\mu)} \, , \tag{21.6.80}$$

where we have chosen the eigenvectors to be orthonormal, so that the completeness relation takes the form

$$\sum_n u_n(\mathbf{k}) u_n^*(\mathbf{k}') = \delta^2(\mathbf{k} - \mathbf{k}') \, . \tag{21.6.81}$$

The effective potential (21.6.74) may then be written as

$$V[\Delta] = \frac{1}{2(2\pi)^3} \int d^2k \, d^2k' \sum_n \frac{\Delta^*(\mathbf{k}') \, u_n^*(\mathbf{k}') \, \Delta(\mathbf{k}) \, u_n(\mathbf{k})}{\sqrt{v_F(\mathbf{k}') \, v_F(\mathbf{k})}} \left[\ln \left(\frac{|\Delta(\mathbf{k})|}{\Lambda_n} \right) - 1 \right] . \tag{21.6.82}$$

For a material with local rotational invariance the Fermi surface is a sphere and the eigenvectors $u_n(\mathbf{k})$ are spherical harmonic functions of the coordinates on this sphere, but Eq. (21.6.82) holds without any assumption of rotational invariance.

We can now use this formalism to get an idea of when superconductivity occurs. The logarithm in Eq. (21.6.82) is large and negative for very small Δ, and large and positive for very large Δ, so as the overall scale of Δ increases from zero to infinity, $V[\Delta]$ drops from zero to negative values and then rises to infinity. It therefore always has a minimum at a non-vanishing value of Δ for any electron–electron potential. But this result is subject to an important qualification. When we take the cut-off κ to infinity, the integral in Eq. (21.6.73) is effectively cut off at ℓ of order $|\Delta|/v_F$, while the Fermi surface has a radius of order k_F, where $8\pi k_F^3/3(2\pi)^3$ equals the electron number density. Since we have been assuming that the electrons taken into account here are in a *thin* shell around the Fermi surface, this derivation is only valid if $|\Delta| \ll \omega_D$, where ω_D is the Debye frequency $k_F v_F$. In particular, in the case of rotational invariance and a direction-independent gap function, the effective potential reaches a local minimum for a gap function of the order of $\Lambda_{s\text{ wave}}$, so the symmetry is spontaneously broken as long as $\Lambda_{s\text{ wave}} \ll \omega_D$. Another way of stating this result is that for s-wave superconductivity, the s-wave projection of the electron–electron potential (21.6.80) must be attractive if renormalized at a scale $\mu \approx \omega_D$. *But it does not matter how strong this renormalized attractive potential is.*

This feature of superconductivity, that a Goldstone boson forms for an attractive potential however weak the potential may be, is a consequence of the existence of a Fermi surface, which enhances long-range effects. In quantum field theories without elementary spinless fields like those of Section 21.5, we would not normally expect a spontaneous symmetry breakdown in empty space unless the interactions are sufficiently strong.

Now let us return to the case of an external electromagnetic field. As usual, we can introduce the Goldstone boson field $\phi(\mathbf{x},t)$ by writing each charged field of the theory, which here is just the gap field $\Delta(\mathbf{x},\mathbf{x}',t)$ or equivalently the pair field $\Psi(\mathbf{x},\mathbf{x}',t)$, as a gauge transformation with gauge parameter $\phi(\mathbf{x},t)$ acting on the corresponding gauge-invariant field, distinguished with a tilde:

$$\Psi(\mathbf{x},\mathbf{x}',t) = \exp\Big(-i\phi(\mathbf{x},t)\Big)\tilde{\Psi}(\mathbf{x},\mathbf{x}',t)\exp\Big(-i\phi(\mathbf{x}',t)\Big)\,. \qquad (21.6.83)$$

The effective action is then given by using Eq. (21.6.83) in Eq. (21.6.59). By a gauge transformation, we can then remove the ϕ dependence in Eq. (21.6.83), provided we replace $A_\mu(x)$ in Eq. (21.6.60) with $A_\mu(x) - \partial_\mu\phi(x)$. When the material is not only superconducting but far from the transition between normal and superconducting states, we may also integrate out the gauge-invariant degrees of freedom associated with $\tilde{\Psi}(\mathbf{x},\mathbf{x}',t)$, which simply means that we replace it with its equilibrium value $\Psi_0(\mathbf{x},\mathbf{x}',t)$. The part of the effective action that depends on the

Goldstone and external electromagnetic fields is then

$$\Gamma[\phi, A] = \Gamma_{\Delta=0}[A] - i \ln \mathrm{Det}\begin{bmatrix} \mathscr{A} & \mathscr{B} \\ \mathscr{B}^\dagger & -\mathscr{A}^T \end{bmatrix} + i \ln \mathrm{Det}\begin{bmatrix} \mathscr{A} & 0 \\ 0 & -\mathscr{A}^T \end{bmatrix}, \tag{21.6.84}$$

where now

$$\mathscr{A}_{\mathbf{x}'t',\mathbf{x}t} = \left[-i\frac{\partial}{\partial t} + eA_0(\mathbf{x},t) - e\dot{\phi}(\mathbf{x},t) + E\Big(-i\nabla + e\mathbf{A}(\mathbf{x},t) - e\nabla\phi(\mathbf{x},t)\Big)\right] \times \delta^3(\mathbf{x}' - \mathbf{x})\delta(t'-t)\,, \tag{21.6.85}$$

$$\mathscr{B}_{\mathbf{x}'t',\mathbf{x}t} = \Delta_0(\mathbf{x}' - \mathbf{x})\delta(t'-t)\,. \tag{21.6.86}$$

Quantitative properties of the superconductor such as the penetration depth can be read off[46] from the expansion of Eq. (21.6.84) in powers of $A_0(\mathbf{x}, t) - \dot{\phi}(\mathbf{x}, t)$ and $\mathbf{A}(\mathbf{x}, t) - \nabla\phi(\mathbf{x}, t)$.

Appendix General Unitarity Gauge

In this appendix we shall show that in general spontaneously broken gauge theories it is always possible to adopt a 'unitarity' gauge in which the Goldstone boson fields satisfy Eq. (21.4.30):

$$\sum_{ab} F^2_{ab}\,\xi_a e_{\alpha b} = 0\,. \tag{21.A.1}$$

Using the exponential parameterization for all groups, we note first that any element of G, in at least a finite neighborhood of the identity, may be put in the form

$$g = \exp\left(-i\sum_\alpha \theta_\alpha \mathscr{T}_\alpha\right)\exp\left(i\sum_a \phi_a x_a\right)\exp\left(i\sum_i \mu_i t_i\right), \tag{21.A.2}$$

with ϕ_a subject to the linear constraint that for all α

$$\sum_{ab} F^2_{ab}\,\phi_a\, e_{\alpha b} = 0\,. \tag{21.A.3}$$

This is easy to see when g is infinitesimally close to the identity. Any such g may be written

$$g = 1 + i\sum_a \phi^0_a x_a + i\sum_i \mu^0_i t_i\,, \tag{21.A.4}$$

with ϕ^0_a and μ^0_i infinitesimal. Equivalently,

$$g = 1 + i\sum_i \phi_a x_a + i\sum_i \mu_i t_i - i\sum_\alpha \theta_\alpha \mathscr{T}_\alpha\,, \tag{21.A.5}$$

where θ_α is an arbitrary infinitesimal, and

$$\phi_a(\theta) \equiv \phi_a^0 + \sum_\alpha \theta_\alpha e_{\alpha a} , \tag{21.A.6}$$

$$\mu_i(\theta) \equiv \mu_i^0 + \sum_\alpha \theta_\alpha e_{\alpha i} . \tag{21.A.7}$$

For any given ϕ_a^0, we can choose θ_α to minimize the positive quantity

$$\sum_{ab} F_{ab}^2 \phi_a(\theta)\phi_b(\theta) . \tag{21.A.8}$$

At this minimum, the quantity (21.A.8) is stationary with respect to variations of θ_α, so $\phi_a(\theta)$ satisfies Eq. (21.A.3). For ϕ, μ, and θ infinitesimal, Eq. (21.A.5) is the same as Eq. (21.A.2), so we see that the set of all gs of the form (21.A.2) (with ϕ_a satisfying Eq. (21.A.3)) includes all gs infinitesimally close to the identity. It follows then from continuity that the same is true of all gs in at least some finite neighborhood of the identity.

Next, consider the particular group element

$$g = \gamma(\xi) = \exp\left(i \sum_a \xi_a x_a\right) \tag{21.A.9}$$

and write it in the form (21.A.2):

$$\gamma(\xi) = \exp\left(-i \sum_\alpha \theta_\alpha(\xi) \mathscr{T}_\alpha\right) \gamma\left(\phi(\xi)\right) \exp\left(i \sum_i \mu_i(\xi) t_i\right) , \tag{21.A.10}$$

with $\phi_a(\xi)$ subject to Eq. (21.A.3). This just says that the gauge transformation $\exp\left(i \sum_\alpha \theta_\alpha(\xi) \mathscr{T}_\alpha\right)$ transforms ξ_a into

$$\xi'_a = \phi_a(\xi) . \tag{21.A.11}$$

Now dropping the prime, we have thus succeeded in constructing a gauge in which ξ_a satisfies Eq. (21.A.1), as was to be shown.

Problems

1. Calculate the effective ghost Lagrangian in a 'generalized unitarity gauge,' with $B[f]$ given as usual by Eq. (15.5.22), but now where $f_\alpha(x) = i\phi_n(x)(t_\alpha)_{nm}\langle\phi_m(0)\rangle_{VAC}$, with ϕ_n real scalar fields, and t_α imaginary antisymmetric matrices representing the Lie algebra of the gauge group. What is the ghost propagator? Is this part of the Lagrangian renormalizable?

2. What would be the effect in the $SU(2) \times U(1)$ electroweak theory if the gauge symmetry were broken by the vacuum expectation value

of a field ϕ_3 belonging to a real triplet $\vec{\phi} = (\phi^+, \phi^0, \phi^-)$ instead of the usual complex doublet (ϕ^0, ϕ^-)?

3. Consider the standard electroweak theory, with a single scalar doublet. In one-loop order, calculate the effect of Z^0 and neutral scalar exchange on the anomalous magnetic moment of the muon.

4. What is the lowest-order magnetic moment of the W^+ and Z^0 particles in the standard electroweak theory?

5. What would be the effect of the discovery of a fourth generation of quarks and leptons on the predictions of the unified theories of strong and electroweak interactions discussed in Section 21.5?

6. Suppose that several fields with incommensurate values of the electric charge had nonvanishing vacuum expectation values in a superconductor. What effect would this have on the properties of superconductors discussed in Section 21.6?

References

1. J. Goldstone. *Nuovo Cimento* **9**, 154 (1961); J. Goldstone, A. Salam, and S. Weinberg, *Phys. Rev.* **127**, 965 (1962).

2. P. W. Higgs, *Phys. Lett.* **12**, 132 (1964); *Phys. Rev. Lett.* **13**, 508 (1964); *Phys. Rev.* **145**, 1156 (1966); F. Englert and R. Brout, *Phys. Rev. Lett.* **13**, 321 (1964); G. S. Guralnik, C. R. Hagen, and T. W. B. Kibble, *Phys. Rev. Lett.* **13**, 585 (1964); T. W. B. Kibble, *Phys. Rev.* **155**, 1554 (1967); S. Weinberg, *Phys. Rev. Lett.* **18**, 507 (1967): footnote 7.

3. S. Weinberg, *Phys. Rev. Lett.* **19**, 1264 (1967). A. Salam, in *Elementary Particle Physics*, ed. N. Svartholm (Almqvist and Wiksells, Stockholm,1968): p. 367.

3*a*. The Feynman rules for general gauge theories in unitarity gauge were given by S. Weinberg, *Phys. Rev.* **D7**, 1068 (1973).

4. G. 't Hooft, *Nucl. Phys.* **B35**, 167 (1971); B. W. Lee, *Phys. Rev.* **D5**, 823 (1972).

5. K. Fujikawa , B. W. Lee, and A. Sanda, *Phys. Rev.* **D6**, 2923 (1972).

5*a*. B. W. Lee and J. Zinn-Justin, *Phys. Rev.* **D5**, 3121, 3137 (1972); *Phys. Rev.* **D7**, 1049 (1972); G. 't Hooft and M. Veltman, *Nucl. Phys.* **B50**, 318 (1972); B. W. Lee, *Phys. Rev.* **D9**, 933 (1974).

6. H. Georgi and S. L. Glashow, *Phys. Rev. Lett.* **28**, 1494 (1972).

7. T. D. Lee and C. N. Yang, *Phys. Rev.* **98**, 101 (1955).

8. Models with an incomplete $SU(2) \times U(1)$ symmetry were proposed by S. L. Glashow, *Nucl. Phys.* **22**, 579 (1961); A. Salam and J. Ward, Phys. Lett. **13**, 168 (1964). Also see J. Schwinger, *Ann. Phys. (N.Y.)* **2**, 407 (1957).

9. Speculations about possible neutral currents go back to work of G. Gamow and E. Teller, *Phys. Rev.* **51**, 288 (1937); N. Kemmer, *Phys. Rev.* **52**, 906 (1937); G. Wentzel, *Helv. Phys. Acta* **10**, 108 (1937); S. Bludman, *Nuovo Cimento* **9**, 433 (1958); J. Leites-Lopes, *Nucl. Phys.* **8**, 234 (1958).

10. Neutral currents were first observed in a single $\bar{\nu}_\mu$–e^- scattering event at CERN, by F. J. Hasert *et al.* (Aachen–Brussels–CERN–Ecole Polytechnique–Milan–Orsay–London collaboration), *Phys. Lett.* **46**, 121 (1973).

11. N. Cabibbo, *Phys. Rev. Lett.* **10**, 531 (1963).

12. I. S. Towner, E. Hagberg, J. C. Hardy, V. T. Koslowsky, and G. Savard, Chalk River preprint nucl-th/9507005 (1995).

13. S. L. Glashow, J. Iliopoulos, and L. Maiani, *Phys. Rev.* **D2**, 1285 (1970).

14. S. Weinberg, *Phys. Rev. Lett.* **27**, 1688 (1971); *Phys. Rev.* **5**, 1413 (1972).

15. J. J. Aubert *et al.*, *Phys. Rev. Lett.* **33**, 1404 (1974); J. E. Augustin *et al.*, *Phys. Rev. Lett.* **33**, 1406 (1974).

16. M. L. Perl *et al.*, *Phys. Rev. Lett.* **35**, 1489 (1975).

17. S. W. Herb *et al.*, *Phys. Rev. Lett.* **39**, 252 (1977).

18. F. Abe *et al.*, *Phys. Rev. Lett.* **74**, 2626 (1995); S. Abachi *et al.*, *Phys. Rev. Lett.* **74**, 2632 (1995).

19. The values given by the CDF and D0 collaborations at Fermilab are respectively $176 \pm 8 \pm 10$ GeV (F. Abe *et al.*, Ref. 18.) and $199^{+19}_{-21} \pm 22$ GeV (S. Abachi *et al.*, Ref. 18), with the first and second errors statistical and systematic, respectively. These results are interpreted to give a combined value of 181 ± 12 GeV by J. Ellis, G. L. Fogli, and E. Lisi, CERN-BARI preprint hep-ph/9507424.

20. M. Kobayashi and K. Maskawa, *Prog. Theor. Phys.* **49**, 282 (1972).

20*a*. F. J. Gilman, K. Kleinknecht, and B. Renk, Carnegie Mellon–Mainz preprint CMU-HUP95-19–DOE-ER/40682- 107 (1995), to be published in the 1995 *Review of Particle Properties.*

20*b*. S. Weinberg, *Phys. Rev. Lett.* **37**, 657 (1976). For earlier models in which scalar fields are responsible for CP and T non-conservation see T. D. Lee, *Phys. Rev.* **D8**, 1226 (1973); *Phys. Rep.* **9C**, 143 (1974).

21. F. J. Hasert *et al.*, *Phys. Lett.* **46B**, 138 (1973); P. Musset, *Jour. de Physique* **11/12**, T34 (1973). Neutral current events were seen at about the same time by the Harvard–Pennsylvania–Wisconsin–Fermilab group at Fermilab, but publication of their paper was delayed, so they took the opportunity to rebuild their detector, and at first did not find the same signal. Evidence for neutral currents was published by this group in A. Benvenuti *et al.*, *Phys. Rev. Lett.* **32**, 800 (1974).

22. G. Arnison *et al.* *Phys. Lett.* **122B**, 103 (1983); **126B**, 398 (1983); **129B**, 273 (1983); **134B**, 469 (1984); **147B**, 241 (1984).

23. F. Abe et al. (CDF collaboration), *Phys. Rev. Lett.* **75**, 11 (1995). The W mass is measured here in observations of the decays $W \to \mu + \nu$ and $W \to e + \nu$.

24. CERN report LEPEWWG/95-01, unpublished (1995).

25. M. Green and M. Veltman, *Nucl. Phys.* **B169**, 137 1980); V. A. Novikov, L. B. Okun', and M. I. Vysotsky, *Nucl. Phys.* **B397**, 35 (1992). For later studies, see P. Langacker, in *Precision Tests of the Standard Model* (World Scientific, Singapore, 1994); P. Bamert, C. P. Burgess, and I. Maksymyk, McGill–Neuchatal–Texas preprint hep-ph/9505339 (1995); W. Hollik, Karlsruhe preprint hep-ph/9507406 (1995).

26. S. Fanchotti, B. Kniehl, and A. Sirlin, *Phys. Rev.* **D48**, 307 (1973), and references cited therein.

27. J. Ellis, G. L. Fogli, and E. Lisi, Ref. 19.

27*a*. S. Weinberg, *Phys. Rev. Lett.* **43**, 1566 (1979).

27*b*. For instance, in the so-called see-saw mechanism, a neutrino mass of this order would be produced by exchange of a heavy neutral lepton of mass M; see M. Gell-Mann, P. Ramond, and R. Slansky, in *Supergravity*, ed. P. van Nieuwenhuizen and D. Freedman (North Holland, Amsterdam, 1979): p. 315; T. Yanagida, *Prog. Theor. Phys.* **B135**, 66 (1978).

27*c*. S. Weinberg, Ref. 27*a*; F. Wilczek and A. Zee, *Phys. Rev. Lett.* **43**, 1571 (1979). Also see S. Weinberg, *Phys. Rev.* **D22**, 1694 (1980).

28. This general formalism was described by S. Weinberg, *Phys. Rev.* **D13**, 974 (1976). For earlier work on dynamical symmetry breaking in gauge theories, see R. Jackiw and K. Johnson, *Phys. Rev.* **D8**, 2386 (1973); J. M. Cornwall and R. E. Norton, *Phys. Rev.* **D8**, 3338 (1973).

29. This interpretation of the results of the standard model is due to L. Susskind, Ref. 30.

30. S. Weinberg, ref. 28 and *Phys. Rev.* **D19**, 1277 (1979); L. Susskind, *Phys. Rev.* **D19**, 2619 (1979). The term 'technicolor' is due to Susskind.

31. S. Dimopoulos and L. Susskind, *Nucl. Phys.* **B15**, 237 (1979); E. Eichten and K. Lane, *Phys. Lett.* **90B**, 125 (1980); S. Weinberg, unpublished work quoted by Eichten and Lane. For a review, see E. Farhi and L Susskind, *Phys. Rep.* **74**, 277 (1981).

32. S. Weinberg, *Phys. Rev.* **D5**, 1962 (1972).

33. J. Pati and A. Salam, *Phys. Rev. Lett.* **31**, 275 (1973).

34. H. Georgi and S. L. Glashow, *Phys. Rev. Lett.* **32**, 438 (1974).

35. H. Georgi, in *Particles and Fields — 1974*, ed. C. Carlson (Amer. Inst. of Physics, New York, 1975).

36. H. Georgi, H. R. Quinn, and S. Weinberg, *Phys. Rev. Lett.* **33**, 451 (1974).

36*a*. S. Dimopoulos and H. Georgi, *Nucl. Phys.* **B193**, 150 (1981); J. Ellis, S. Kelley, and D. V. Nanopoulos, *Phys. Lett.* **B260**, 131 (1991); U. Amaldi, W. de Boer, and H. Furstmann, *Phys. Lett.* **B260**, 447 (1991). For more recent analyses of the data, see P. Langacker and N. Polonsky, *Phys. Rev.* **D47**, 4028 (1993); **D49**, 1454 (1994); L. J. Hall and U. Sarid, *Phys. Rev. Lett.* **70**, 2673 (1993).

36*b*. S. Dimopoulos, S. Raby, and F. Wilczek, *Phys. Rev.* **D24**, 1681 (1981).

37. J. Bardeen, L. N. Cooper, and J. R. Schrieffer, *Phys. Rev.* **108**, 1175 (1957).

38. P. M. Anderson, *Phys. Rev.* **130**, 439 (1963).

39. S. Weinberg, *Prog. Theor. Phys. Suppl.* No. 86, 43 (1986).

40. B. D. Josephson, *Phys. Lett.* **1**, 25 (1962).

41. V. L. Ginzburg and L. D. Landau, *JETP (USSR)* **20**, 1064 (1950).

42. L. P. Gor'kov, *Soviet Phys. JETP* **9**, 1364 (1959).

43. A. A. Abrikosov, *Soviet Phys. JETP* **5**, 1174 (1957).

43*a*. E. B. Bogomol'nyi, *Sov. J. Nucl. Phys.* **24**, 449 (1976). For numerical calculations, see E. B. Bogomol'nyi and A. I. Vainshtein, *Sov. J. Nucl. Phys.* **23**, 588 (1976).

44. G. Benfatto and G. Gallavotti, *J. Stat. Phys.* **59**, 541 (1990); *Phys. Rev.* **42**, 9967 (1990); J. Feldman and E. Trubowitz, *Helv. Phys. Acta* **63**, 157 (1990); **64**, 213 (1991); **65**, 679 (1992); R. Shankar, *Physica* **A177**, 530 (1991); *Rev. Mod. Phys.* **66**, 129 (1993); J. Polchinski, in *Recent Developments in Particle Theory, Proceedings of the 1992 TASI*, eds. J. Harvey and J. Polchinski (World Scientific, Singapore, 1993).

44*a*. S. Weinberg, *Nucl. Phys.* **B413**, 567 (1994).

45. R. L. Stratonovich, *Sov. Phys. Dokl.* **2**, 416 (1957); J. Hubbard, *Phys. Rev. Lett.* **3**, 77 (1959).

46. A. A. Abrikosov, L. P. Gor'kov, I. Ye. Dzyaloshinskii, *Quantum Field Theoretical Methods in Statistical Physics*, 2nd edn., transl. by D. E. Brown (Pergamon Press, Oxford, 1965).

47. In order to give a precise meaning to the mass M it is necessary to take into account both two-loop corrections to the renormalization group equations (21.5.6)–(21.5.8), and one-loop 'threshold corrections' to condition (21.5.5). See S. Weinberg, *Phys. Lett.* **82B**, 387 (1979).

22

Anomalies

There are subtleties in the implications of symmetries in quantum field theory that have no counterpart in classical theories. Even in renormalizable theories, the infinities in quantum field theory require that some sort of regulator or cut-off be used in actual calculations. The regulator may violate symmetries of the theory, and even when this regulator is removed at the end of the calculation it may leave traces of this symmetry violation. This problem first emerged in trying to understand the decay rate of the neutral pion, in the form of an anomaly that violates a *global* symmetry of the strong interactions. Anomalies can also violate *gauge* symmetries, but in this case the theory becomes inconsistent, so that the condition of anomaly cancellation may be used as a constraint on physical gauge theories. The importance of anomalies will become even more apparent in the next chapter, where we shall study the non-perturbative effects of anomalies in the presence of topologically non-trivial field configurations.

22.1 The π^0 Decay Problem

By the mid-1960s the picture of the pion as a Goldstone boson associated with a spontaneously broken $SU(2) \otimes SU(2)$ symmetry of the strong interactions had scored a number of successes, outlined here in Chapter 19. However, this picture also had a few outstanding failures. The most disturbing had to do with the rate of the dominant decay mode of the neutral pion, $\pi^0 \to 2\gamma$. It was the solution of this problem that led to the discovery of symmetry-breaking anomalies.

After integrating out all heavy and trapped particles, we would expect the effective Lagrangian for $\pi^0 \to 2\gamma$ to be given by the unique gauge- and Lorentz-invariant term with no more than two derivatives:

$$\mathscr{L}_{\pi\gamma\gamma} = g\pi^0 \epsilon^{\mu\nu\rho\lambda} F_{\mu\nu} F_{\rho\lambda}\,, \tag{22.1.1}$$

where g is an unknown constant with the dimensions $[mass]^{-1}$. The methods of Section 3.4 can then be used to calculate the rate for $\pi^0 \to 2\gamma$

$$\Gamma(\pi^0 \to 2\gamma) = \frac{m_\pi^3 g^2}{\pi} . \tag{22.1.2}$$

One might naively expect g to be of order

$$g \approx \frac{e^2}{8\pi^2 F_\pi} , \tag{22.1.3}$$

where $F_\pi \simeq 190$ MeV is used as a typical strong interaction mass scale, and a factor $1/8\pi^2$ is inserted because the graphs responsible for $\pi^0 \to 2\gamma$ would have to include at least one loop. For instance, in 1949, using the pre-QCD theory of pions and nucleons with interaction Lagrangian $iG_{\pi N}\vec{\pi} \cdot \bar{N}2\vec{t}\gamma_5 N$, Steinberger[1] calculated that the contribution to g from triangle graphs with a single proton loop is

$$g_\triangle = \frac{e^2 G_{\pi N}}{32\pi^2 m_N} . \tag{22.1.4}$$

This is numerically not very different from Eq. (22.1.3), because the Goldberger–Treiman relation (see Section 19.4) gives $G_{\pi N} = 2m_N g_A/F_\pi$.

This estimate of the amplitude for $\pi^0 \to 2\gamma$ leaves out the special constraints imposed by the $SU(2) \otimes SU(2)$ symmetry. The electromagnetic interaction violates most of this symmetry, but at least formally it has no effect on the $U(1) \times U(1)$ subgroup generated by the electrically neutral generators of $SU(2) \otimes SU(2)$. Acting on quarks, the infinitesimal pseudoscalar element of this subgroup has the effect:

$$\delta u = i\epsilon\gamma_5 u , \qquad \delta d = -i\epsilon\gamma_5 d , \tag{22.1.5}$$

so that the electric current is invariant:

$$\delta\left[\frac{2}{3}\bar{u}\gamma^\mu u - \frac{1}{3}\bar{d}\gamma^\mu d\right] = 0 . \tag{22.1.6}$$

(This argument was made by Sutherland[2] and Veltman[3] before the advent of quantum chromodynamics, with the proton and neutron in place of the u and d quarks.) Because the π^0 is the Goldstone boson associated with this symmetry, a non-derivative interaction like (22.1.1) can arise only from the breaking of this symmetry by quark masses, and so must be proportional to $m_\pi^2 \propto m_u + m_d$. With this in mind, we would expect the constant g to be suppressed[2,3] by an extra factor m_π^2/m_N^2:

$$g \approx \frac{e^2}{8\pi^2 F_\pi}\left(\frac{m_\pi^2}{m_N^2}\right) . \tag{22.1.7}$$

(In place of m_N we might have used the chiral-symmetry-breaking scale $2\pi F_\pi = 1200$ MeV discussed in Section 19.3, with little change in our results). There are also chiral-invariant effective $\pi^0\gamma\gamma$ interactions involving the derivative of the π^0 field. Lorentz invariance requires that these involve at least two additional derivatives. Using the homogeneous Maxwell equation $\partial_\mu F_{\nu\lambda} - \partial_\nu F_{\mu\lambda} = -\partial_\lambda F_{\mu\nu}$ and integrating by parts, we see that there is only one independent chiral-invariant coupling with just two extra derivatives, given by inserting a d'Alembertian acting on the π^0 field in Eq. (22.1.1). On the pion mass shell this is just the same as an interaction (22.1.1) with an extra factor m_π^2, leading to the same estimate (22.1.7). It was sometimes said that chiral symmetry would forbid the decay $\pi^0 \to 2\gamma$, but it is more accurate to say that chiral symmetry would make the rate for this process go as m_π^7 instead of m_π^3 for $m_\pi \to 0$.

The difficulty is that the observed rate for $\pi^0 \to 2\gamma$ is much larger than would be expected from Eq. (22.1.7), and is in fact much closer to what would be inferred from the naive result Eq. (22.1.3). To be specific, from Eqs. (22.1.7) and (22.1.2) we would expect a decay rate

$$\Gamma(\pi^0 \to 2\gamma) \approx \frac{m_\pi^7 \alpha^2}{4\pi^3 F_\pi^2 m_N^4} = 1.9 \times 10^{13} \text{ s}^{-1} , \tag{22.1.8}$$

while using Eq. (22.1.3) in place of Eq. (22.1.7) would give

$$\Gamma(\pi^0 \to 2\gamma) \approx \frac{m_\pi^3 \alpha^2}{4\pi^3 F_\pi^2} = 4.4 \times 10^{16} \text{ s}^{-1} . \tag{22.1.9}$$

The observed rate is $\Gamma(\pi^0 \to 2\gamma) = (1.19 \pm 0.08) \times 10^{16}$ s^{-1}, which is in satisfactory agreement with the naive rough estimate (22.1.9), and almost three orders of magnitude larger than the 'improved' result (22.1.8)! One is forced to the conclusion that something anomalous here is invalidating the chiral symmetry that led us to introduce the additional factor m_π^2/m_N^2 in g. Similar problems arise in trying to understand the rates of some other processes, such as $\eta^0 \to \gamma + \gamma$.

In 1969 the source of this anomaly was traced by Bell and Jackiw[4] to the violation of chiral symmetry by the regulator that is needed in order to derive consequences of the conservation of the neutral axial vector current for one-loop Feynman diagrams. Their result was confirmed, generalized, and extended to higher orders by Adler,[5] who independently discovered chiral anomalies in a study of axial vector Ward identities in quantum electrodynamics. It was subsequently realized in 1979 by Fujikawa[6] that in the path-integral formulation of field theory the chiral-symmetry-breaking anomaly enters only in the measure used to define the path integral over fermion fields. As we shall see in the next section, this approach makes it simple to evaluate the amplitude for $\pi^0 \to 2\gamma$ produced by this anomaly to all orders of perturbation theory. After that, we shall return to the direct calculation of anomalies in more general theories, and discuss various applications.

22.2 Transformation of the Measure: The Abelian Anomaly

We now turn to a calculation of anomalies of the sort relevant to π^0 decay. For this purpose we adopt Fujikawa's interpretation[6] of the anomaly as a symptom of the impossibility of defining a suitably invariant measure for integrations over fermionic field variables. Fujikawa's analysis of this problem was based on the use of path integrals in Euclidean space, and on an expansion of the fermionic variables of integration in eigenfunctions of the gauge-invariant Dirac operator, which is Hermitian in four-dimensional Euclidean space. Here we shall first present a less rigorous derivation, based on the familiar path integrals in Minkowski space, which will allow us to obtain the correct answer with a minimum of work. The Euclidean approach will be taken up briefly at the end of this section, and used to derive a famous index theorem.

We will start by evaluating the anomaly in the transformation of the measure under an arbitrary local matrix transformation $\psi(x) \to U(x)\psi(x)$ of a column $\psi_n(x)$ of massless complex spin 1/2 fermion fields that interact non-chirally with a set of gauge fields $A^\mu_\alpha(x)$ (such as the u and d quark fields interacting with the electromagnetic vector potential $A^\mu(x)$ in the problem of calculating the rate for $\pi^0 \to 2\gamma$). These are fermionic variables, so the measure is transformed not with the determinant of the transformation matrix, but with its inverse:

$$[d\psi][d\bar{\psi}] \to (\mathrm{Det}\,\mathscr{U}\;\mathrm{Det}\,\bar{\mathscr{U}})^{-1}[d\psi][d\bar{\psi}]\,, \tag{22.2.1}$$

where

$$\mathscr{U}_{xn,ym} \equiv U(x)_{nm}\delta^4(x-y)\,, \tag{22.2.2}$$

$$\bar{\mathscr{U}}_{xn,ym} \equiv [\gamma_4 U(x)^\dagger \gamma_4]_{nm}\delta^4(x-y) \tag{22.2.3}$$

and $\gamma_4 \equiv i\gamma^0$ is the matrix used to define $\bar{\psi} = \psi^\dagger\gamma_4$. The indices n, m run over flavor labels and Dirac spin indices.

The reader may wonder at this point why we bother to include factors of γ_4 in Eq. (22.2.3), since they just amount to a unitary transformation that should not affect the determinants. The answer is that in order for calculations to be meaningful we will find it necessary to regulate the sum over fermion modes in calculating propagators and determinants, and we shall find that the γ_4 factors do affect the regulated determinants. Whether or not we include the γ_4 factors thus depends on the method of regularization that is used to make these hand-waving manipulations meaningful. We include the factors of γ_4 because we wish to regulate in such a way as to preserve Lorentz invariance, and in the cases of interest here it is $\gamma_4\, U(x)^\dagger\, \gamma_4$ rather than $U(x)^\dagger$ that transforms as a scalar.

First, let us consider the case where $U(x)$ is a unitary non-chiral transformation

$$U(x) = \exp[i\alpha(x)t] \,, \tag{22.2.4}$$

with t an ordinary Hermitian matrix (not involving γ_5, but not necessarily traceless) and $\alpha(x)$ an arbitrary real function of x. In this case $\mathscr{U}$ is pseudounitary:

$$\bar{\mathscr{U}}\mathscr{U} = 1 \,, \tag{22.2.5}$$

so the measure is invariant under this sort of transformation. In particular, the symmetry under the gauge group itself, where t is one of the non-chiral generators t_α, is not spoiled by any anomalies.

Next, consider a local chiral transformation, with

$$U(x) = \exp[i\gamma_5\alpha(x)t] \,, \tag{22.2.6}$$

with t again an ordinary Hermitian matrix and $\alpha(x)$ again an arbitrary real function of x. In this case $\mathscr{U}$ is pseudo-Hermitian:

$$\bar{\mathscr{U}} = \mathscr{U} \,. \tag{22.2.7}$$

The measure is not invariant under the chiral transformation; rather, we have

$$[d\psi][d\bar{\psi}] \to (\text{Det}\,\mathscr{U})^{-2}[d\psi][d\bar{\psi}] \,. \tag{22.2.8}$$

Now let us specialize to the case of an *infinitesimal* local chiral transformation. Taking $\alpha(x)$ to be infinitesimal in Eq. (22.2.6), we have here

$$[\mathscr{U} - 1]_{nx,my} = i\alpha(x)[\gamma_5\, t]_{nm}\, \delta^4(x-y) \,. \tag{22.2.9}$$

Using the identity $\text{Det}M = \exp \text{Tr} \ln M$ and the limiting formula $\ln(1+x) \to x$ for $x \to 0$, the measure now has the transformation property

$$[d\psi][d\bar{\psi}] \to \exp\left\{i\int d^4x\; \alpha(x)\mathscr{A}(x)\right\} [d\psi][d\bar{\psi}] \,, \tag{22.2.10}$$

where $\mathscr{A}$ is the anomaly function:

$$\mathscr{A}(x) = -2\,\text{Tr}\left\{\gamma_5 t\right\} \delta^4(x-x) \,, \tag{22.2.11}$$

with 'Tr' here denoting a trace over both Dirac and species indices. The measure $[d\psi][d\bar{\psi}]$ appears in the path integral weighted with a factor $\exp\{i\int d^4x\; \mathscr{L}(x)\}$, so the factor $\exp\left\{i\int d^4x\; \alpha(x)\mathscr{A}(x)\right\}$ in the transformation rule (22.2.10) for the measure has the same effect as if the Lagrangian density $\mathscr{L}(x)$ were not invariant under these transformations, but instead $\mathscr{L}(x) \to \mathscr{L}(x) + \alpha(x)\mathscr{A}(x)$. Hence when we use an effective Lagrangian with the fermions integrated out, to take account of the anomaly we must be sure to include a non-invariant term so that

$$\mathscr{L}_{\text{eff}}(x) \to \mathscr{L}_{\text{eff}}(x) + \alpha(x)\mathscr{A}(x)\,. \tag{22.2.12}$$

It now remains to calculate the anomaly function $\mathscr{A}(x)$.

At first sight it does not look like we can get any definite result for the anomaly. The delta-function is infinite, but the trace vanishes. To do better, we must introduce a regulator to make $\delta^4(x-x)$ meaningful. We can do this in a gauge-invariant manner by inserting a differential operator $f(-\not{D}_x^2/M^2)$ acting on the delta-function before taking its argument to zero:

$$\mathscr{A}(x) = -2\left[\operatorname{Tr}\left\{\gamma_5 t\, f(-\not{D}_x^2/M^2)\right\}\delta^4(x-y)\right]_{y\to x}\,. \tag{22.2.13}$$

Here D_x is the Dirac differential operator in the presence of the gauge field $A_\alpha^\mu(x)$:

$$(D_x)_\mu \equiv \frac{\partial}{\partial x^\mu} - i t_\alpha A_{\alpha\mu}(x)\,. \tag{22.2.14}$$

Also, M is some large mass, eventually to be taken to infinity, and $f(s)$ is a smooth function, subject only to the condition* that as s goes from 0 to ∞, $f(s)$ must drop smoothly from 1 to 0:

$$f(0) = 1\,, \qquad f(\infty) = 0\,, \tag{22.2.15}$$

$$sf'(s) = 0 \quad \text{at} \quad s = 0 \text{ and } s = \infty\,. \tag{22.2.16}$$

Note that we do not take the regulator function to be a function of $\not{\partial}$, because we want to preserve gauge invariance, and we do not take it to be a function of $D^\mu D_\mu$, because we need it to regulate not only the determinant but also the fermion propagator $\not{D}^{-1}$.

To evaluate the expression (22.2.13), we use the Fourier representation of the delta-function, and write the anomaly function as

$$\begin{aligned}\mathscr{A}(x) &= -2\int \frac{d^4k}{(2\pi)^4}\left[\operatorname{Tr}\left\{\gamma_5\, t\, f(-\not{D}_x^2/M^2)\right\} e^{ik\cdot(x-y)}\right]_{y=x} \\ &= -2\int \frac{d^4k}{(2\pi)^4}\operatorname{Tr}\left\{\gamma_5\, t\, f\Big(-[i\not{k}+\not{D}_x]^2/M^2\Big)\right\}\,.\end{aligned}$$

(A derivative with respect to x gives zero when acting on the extreme right in the second expression, but not when acting on $A_{\alpha\mu}(x)$.) Rescaling the momentum k^μ by a factor M, this is

$$\mathscr{A}(x) = -2M^4\int \frac{d^4k}{(2\pi)^4}\operatorname{Tr}\left\{\gamma_5\, t\, f\Big(-[i\not{k}+\not{D}_x/M]^2\Big)\right\}\,. \tag{22.2.17}$$

* For instance, we might take $f(s) = e^{-s^2}$, as done originally by Fujikawa, or $f(s) = 1/(1+s)$.

The argument of the cut-off function may be written

$$-\left[i\,\not{k}+\frac{\not{D}_x}{M}\right]^2 = k^2 - \frac{ik\cdot D_x}{M} - \left(\frac{\not{D}_x}{M}\right)^2 . \tag{22.2.18}$$

Eq. (22.2.17) receives contributions in the limit $M \to \infty$ only from terms in the expansion of $f(-[i\ \not{k}+\ \not{D}_x/M]^2)$ that have no more than four factors of $1/M$, and also only from terms that contain at least four Dirac gamma matrices, because otherwise the trace over Dirac indices vanishes. This leaves us with only the terms of second order in $\not{D}_x^2$:

$$\mathscr{A}(x) = -\int \frac{d^4k}{(2\pi)^4} f''(k^2)\,\mathrm{Tr}\left\{\gamma_5\, t\ \not{D}_x^4\right\} , \tag{22.2.19}$$

which are now independent of the regulator mass M.

To evaluate the integral over k, we perform the same sort of rotation of the k^0 contour of integration as in evaluating Feynman diagrams, so that in effect k^0 is replaced with ik^4, with k^4 running from $-\infty$ to $+\infty$. (This is a step that can really only be justified by working with Euclidean path integrals from the beginning.) The integral is then in effect

$$\int d^4k\, f''(k^2) = i\int_0^\infty 2\pi^2\kappa^3\, d\kappa\, f''(\kappa^2) . \tag{22.2.20}$$

By repeated integration by parts using Eq. (22.2.16) and then Eq. (22.2.15), this is

$$\int d^4k\, f''(k^2) = i\pi^2\int_0^\infty ds\, s\, f''(s) = -i\pi^2\int_0^\infty ds\, f'(s) = i\pi^2 . \tag{22.2.21}$$

To calculate the trace, we write

$$\begin{aligned}\not{D}_x^2 &= \tfrac{1}{4}\{(D_x)^\mu,(D_x)^\nu\}\ \{\gamma_\mu,\gamma_\nu\} + \tfrac{1}{4}\,[(D_x)^\mu,(D_x)^\nu]\ [\gamma_\mu,\gamma_\nu] \\ &= D_x^2 - \tfrac{1}{4}it_\alpha F_\alpha^{\mu\nu}\ [\gamma_\mu,\gamma_\nu] .\end{aligned} \tag{22.2.22}$$

The only term in $\not{D}_x^4$ that contributes to the trace over Dirac indices is the one involving a product of four Dirac matrices, which gives

$$\mathrm{tr_D}\ \{\gamma_5\ [\gamma_\mu,\gamma_\nu]\ [\gamma_\rho,\gamma_\sigma]\} = 16\,i\,\epsilon_{\mu\nu\rho\sigma} , \tag{22.2.23}$$

where '$\mathrm{tr_D}$' denotes a trace over Dirac indices only, and as usual $\epsilon_{\mu\nu\rho\sigma}$ is the totally antisymmetric tensor with $\epsilon^{0123} = +1$. Using Eqs. (22.2.21)–(22.2.23) in Eq. (22.2.19) then gives the anomaly function as

$$\mathscr{A}(x) = -\frac{1}{16\pi^2}\,\epsilon_{\mu\nu\rho\sigma}\,F_\alpha^{\mu\nu}(x)\,F_\beta^{\rho\sigma}(x)\,\mathrm{tr}\,\{t_\alpha\, t_\beta\, t\} , \tag{22.2.24}$$

with 'tr' here denoting a trace only over indices labelling the various fermion species. In the special case in which t is the unit matrix, the quantity (22.2.24) is known as the *Chern–Pontryagin density*.

This result can be expressed in terms of the current associated with the anomalous symmetry. If we assume for simplicity that the action itself is invariant under the symmetry transformation $\psi(x) \to \psi(x) + it\gamma_5\alpha\psi(x)$ with constant infinitesimal parameter α, then as discussed in Section 7.3, the change in the action when we make such a transformation with a spacetime-*dependent* parameter $\alpha(x)$ may be written as $\delta I = \int d^4x\, J_5^\mu(x)\partial_\mu\alpha(x)$, where $J_5^\mu(x)$ is the current that becomes conserved when the field operators satisfy the dynamical equations that make the action stationary with respect to arbitrary variations in the fields. When we make the change of variables $\delta\psi(x) = it\gamma_5\alpha(x)\psi(x)$, the change in the path integral over the fermionic fields is

$$\delta \int [d\psi][d\bar{\psi}]\, e^{iI} = i \int d^4x \int [d\psi][d\bar{\psi}][\mathscr{A}(x)\alpha(x) + J_5^\mu(x)\partial_\mu\alpha(x)]e^{iI} \,. \tag{22.2.25}$$

But this is a mere change of variables, and so for arbitrary $\alpha(x)$ it cannot affect the path integral. Therefore for arbitrary gauge fields

$$\langle \partial_\mu J_5^\mu \rangle_A = \mathscr{A} = -\frac{1}{16\pi^2}\epsilon_{\mu\nu\rho\sigma}\, F_\alpha^{\mu\nu}\, F_\beta^{\rho\sigma}\, \mathrm{tr}\, \{t_\alpha\, t_\beta\, t\} \,, \tag{22.2.26}$$

where for any operator $\mathscr{O}$, $\langle\mathscr{O}\rangle_A$ is the quantum average of $\mathscr{O}$ in a fixed background field $A^\mu(x)$:

$$\langle\mathscr{O}\rangle_A \equiv \frac{\int [d\psi][d\bar{\psi}]\, e^{iI}\,\mathscr{O}}{\int [d\psi][d\bar{\psi}]\, e^{iI}} \,. \tag{22.2.27}$$

Incidentally, it is possible to rewrite Eq. (22.2.26) as a conservation condition. Consider the special case where $\mathrm{tr}\,\{t_\alpha\, t_\beta\, t\}$ is proportional to $\delta_{\alpha\beta}$:

$$\mathrm{tr}\,\{t_\alpha\, t_\beta\, t\} = N\delta_{\alpha\beta} \,. \tag{22.2.28}$$

Define a current known as the *Chern–Simons class*:

$$\begin{aligned} G^\mu &\equiv 2\epsilon^{\mu\nu\lambda\rho}\left[A_{\gamma\nu}\partial_\lambda A_{\gamma\rho} + \frac{1}{3}C_{\alpha\beta\gamma}A_{\alpha\nu}A_{\beta\lambda}A_{\gamma\rho}\right] \\ &= \epsilon^{\mu\nu\lambda\rho}\left[A_{\gamma\nu}\, F_{\gamma\lambda\rho} - \frac{1}{3}C_{\alpha\beta\gamma}A_{\alpha\nu}A_{\beta\lambda}A_{\gamma\rho}\right] \,, \end{aligned} \tag{22.2.29}$$

which satisfies the identity

$$\partial_\mu G^\mu = \frac{1}{2}\epsilon^{\mu\nu\lambda\rho}\, F_{\gamma\mu\nu}\, F_{\gamma\lambda\rho} \,. \tag{22.2.30}$$

Eq. (22.2.30) allows us to rewrite Eq. (22.2.26) as the condition

$$\partial_\mu K^\mu = 0 \,, \tag{22.2.31}$$

where

$$K^\mu \equiv \left\langle J_5^\mu \right\rangle_A + \frac{N}{8\pi^2}\, G^\mu \,. \tag{22.2.32}$$

However, we cannot use the conservation of the current K^μ to argue, as we did in the previous section, that the process $\pi^0 \to 2\gamma$ is suppressed, because in making that argument we assumed not only the chiral symmetry associated with axial-vector current conservation but also electromagnetic gauge invariance, and as we see from Eq. (22.2.29), the current K^μ though conserved is not gauge-invariant.

Our derivation of the formula (22.2.24) for the anomaly shows that if we had evaluated the anomaly function using a differential operator $f(-\not{\partial}_x^2/M^2)$ in place of $f(-\not{D}_x^2/M^2)$ in Eq. (22.2.13), we would have obtained a vanishing anomaly function. In fact, with this regulator procedure the axial-vector current is K^μ, not J_5^μ. The trouble with this procedure, as mentioned above, is that the regulator operator is no longer gauge-invariant, resulting in the presence of a non-gauge-invariant term in K^μ. There is no regulator procedure for fermion propagators as well as determinants that is both gauge- and chiral-invariant.

We may now return to the problem that gave anomalies their start, and use our results to calculate the actual rate of the process $\pi^0 \to 2\gamma$. The symmetry of interest here is generated by charge-neutral chiral transformations of the light quark fields:

$$\delta u = i\alpha\gamma_5 u\,, \qquad \delta d = -i\alpha\gamma_5 d\,. \tag{22.2.33}$$

This symmetry is anomaly-free in pure quantum chromodynamics, because the u and d belong to the same representation of the color gauge group, so that their contributions to the gluon–gluon terms in the anomaly of the symmetry (22.2.33) cancel. On the other hand, in the presence of an electromagnetic field $A^\mu(x)$, this symmetry has an anomaly

$$\mathscr{A}(x) = -\frac{1}{16\pi^2}\epsilon_{\mu\nu\rho\sigma}F^{\mu\nu}(x)F^{\rho\sigma}(x)\,\mathrm{tr}\left\{q^2\tau_3\right\},$$

where q is the quark charge matrix and τ_3 is the diagonal 2×2 matrix with elements $+1$ for u and -1 for d. If we assume as usual that there are N_c u quarks of charge $2e/3$ and an equal number of d quarks of charge $-e/3$, the trace is

$$\mathrm{tr}\left\{q^2\tau_3\right\} = N_c\left(\frac{2e}{3}\right)^2(+1) + N_c\left(\frac{-e}{3}\right)^2(-1) = \frac{N_c e^2}{3}\,,$$

so the anomaly function here is

$$\mathscr{A}(x) = -\frac{N_c e^2}{48\pi^2}\epsilon_{\mu\nu\rho\sigma}F^{\mu\nu}(x)F^{\rho\sigma}(x)\,. \tag{22.2.34}$$

We must then include in the effective Lagrangian terms that under the chiral transformation (22.2.33) give it the transformation (22.2.12), that is,

$$\delta\mathscr{L}_{\rm eff}(x) = \alpha\mathscr{A}(x) = -\frac{N_c e^2}{48\pi^2}\epsilon_{\mu\nu\rho\sigma}F^{\mu\nu}(x)F^{\rho\sigma}(x)\alpha\,. \tag{22.2.35}$$

Under the transformation (22.2.33) the pion field transforms as

$$\delta\pi^0 = \alpha F_\pi \,, \tag{22.2.36}$$

where $F_\pi = 184$ MeV is the pion decay amplitude introduced in Chapter 19. (The normalization convention for this constant is fixed by our definition of the symmetry generator as $\gamma_5\tau_3 = 2\gamma_5 t_3$.) It follows that we must include in the effective Lagrangian a term

$$\frac{\pi^0(x)\mathscr{A}(x)}{F_\pi} = -\frac{N_c e^2}{48\pi^2 F_\pi}\epsilon_{\mu\nu\rho\sigma}\, F^{\mu\nu}(x)\, F^{\rho\sigma}(x)\, \pi^0(x)\,. \tag{22.2.37}$$

Comparing this with the general formula (22.1.1) for the effective Lagrangian for the decay $\pi^0 \to 2\gamma$, we see that the constant g in Eq. (22.1.1) must have the value[5]

$$g = \frac{N_c e^2}{48\pi^2 F_\pi}\,. \tag{22.2.38}$$

(This shows that our previous crude order-of-magnitude estimate (22.1.3) was too large by just a factor $6/N_c$.) The rate (22.1.2) for pion decay is thus predicted to be

$$\Gamma(\pi^0 \to 2\gamma) = \frac{N_c^2\alpha^2 m_\pi^3}{144\pi^3 F_\pi^2} = \left(\frac{N_c}{3}\right)^2 \times 1.11 \times 10^{16}\ \mathrm{s}^{-1}\,. \tag{22.2.39}$$

The observed rate is $\Gamma(\pi^0 \to 2\gamma) = (1.19 \pm 0.08) \times 10^{16}\ \mathrm{s}^{-1}$, in good agreement with the theoretical result (22.2.39) if and only if $N_c = 3$. The success of this calculation was one of the first pieces of firm evidence that there are three colors of quarks.

Remarkably, as we have seen in the previous section, shortly after the discovery of the π^0 Steinberger calculated g from a single proton-loop diagram, and obtained the result $g = e^2G/32\pi^2 m_N$, where G is the pseudoscalar pion–nucleon coupling constant. This result would precisely agree with Eq. (22.2.38) for $N_c = 3$ if we used the Goldberger–Treiman relation with $g_A = 1$ to set $G = 2m_N/F_\pi$. The correct result is larger than the Steinberger result by a factor $g_A^2 = 1.56$. The reason that Steinberger got nearly the right result is because the answer is determined by the triangle anomaly, which is proportional to $\mathrm{tr}\,\{q^2\tau_3\}$. For one proton, this trace has the value e^2, which is the same as the quark value found above for three colors of quarks.

* * *

As mentioned earlier, a more rigorous derivation of the anomaly can be given by using path integrals in Euclidean spacetime. (The use of Euclidean path integrals is briefly discussed in Appendix A of Chapter 23.) We introduce a Euclidean fourth coordinate $x_4 = ix^0 = -ix_0$, and correspondingly $\partial_4 = -i\partial_0$, $\gamma_4 \equiv i\gamma^0$, and $A_{4\alpha} = iA^0_\alpha$. The spacetime volume is then expressed as $d^4x = -i(d^4x)_E$, where $(d^4x)_E$ is the Euclidean volume element $(d^4x)_E = dx_1dx_2dx_3dx_4$. In Euclidean spacetime the fields $\psi(x)$ and $\bar\psi(x)$ must be regarded as entirely independent, with local chiral transformations defined by $\delta\psi(x) = i\alpha(x)t\gamma_5\psi(x)$, $\delta\bar\psi(x) = -i\alpha(x)\bar\psi(x)t\gamma_5$. The transformation of the measure is then again given by Eq. (22.2.10), with the anomaly function $\mathscr{A}(x)$ given by Eq. (22.2.11). Introducing a regulator function as before, this again gives the formula (22.12.13) for $\mathscr{A}(x)$. One great advantage of the Euclidean approach is that with x_4 and $A_{4\alpha}$ real, the Dirac operator $i\not{D}$ in Eq. (22.2.13) is Hermitian:

$$i\not{D} = \Big[i\partial_i + t_\alpha A_{i\alpha}\Big]\gamma_i\,, \tag{22.2.40}$$

with i, j, etc. here summed over the values 1, 2, 3, 4. It therefore has orthonormal spinor eigenfunctions $\varphi_\kappa(x)$:

$$i\not{D}\varphi_\kappa = \lambda_\kappa\varphi_\kappa\,, \tag{22.2.41}$$

$$\int (d^4x)_E\ \varphi_\kappa(x)^\dagger\varphi_{\kappa'}(x) = \delta_{\kappa\kappa'}\,, \tag{22.2.42}$$

with real eigenvalues λ_κ. We also are assuming, as throughout this section, that t commutes with $i\not{D}$, so that we can choose the φ_κ so that $t\varphi_\kappa = t_\kappa\varphi_\kappa$. These eigenfunctions satisfy a completeness relation

$$\sum_\kappa \varphi_\kappa(x)\varphi_\kappa^\dagger(y) = \delta^4(x-y)1\,, \tag{22.2.43}$$

where '1' is the 4×4 unit matrix. Therefore the anomaly function may now be written as the limit of a manifestly convergent sum:

$$\begin{aligned}\mathscr{A}(x) &= -2\lim\nolimits_{M\to\infty}\mathrm{Tr}\Big\{\gamma_5\, t\, f(-\not{D}^2/M^2)\sum_\kappa \varphi_\kappa(x)\varphi_\kappa^\dagger(x)\Big\}\\ &= -2\lim\nolimits_{M\to\infty}\sum_\kappa t_\kappa\, f(\lambda_\kappa^2/M^2)\,\Big(\varphi_\kappa^\dagger(x)\gamma_5\varphi_\kappa(x)\Big)\,. \end{aligned}\tag{22.2.44}$$

In just the same way that we derived formula (22.2.24) for the anomaly function, we can show here that

$$\mathscr{A}(x) = \frac{1}{16\pi^2}\,\epsilon^E_{ijkl}\,F_{ij\alpha}\,F_{kl\beta}\ \mathrm{tr}\,\{t_\alpha t_\beta t\}\,, \tag{22.2.45}$$

where ϵ^E_{ijkl} is the totally antisymmetric tensor with $\epsilon^E_{1234} = +1$. (The difference of sign in Eqs. (22.2.24) and (22.2.45) arises because Eq. (22.2.45) is missing two factors of i as compared with Eq. (22.2.24): one factor i from Eq. (22.2.23), because

$$\mathrm{tr}_D\{\gamma_5[\gamma_i,\gamma_j][\gamma_k,\gamma_l]\} = 16\,\epsilon^E_{ijkl}\,,$$

together with another factor i from the replacement of d^4k with $(d^4k)_E$ in Eq. (22.2.20).)

Now, given any eigenfunction $\varphi_\kappa(x)$ of $i\not{D}$ and t with eigenvalue $\lambda_\kappa \neq 0$, there is another normalized eigenfunction $\varphi_{\kappa^-}(x)$ with eigenvalues $\lambda_{\kappa^-} = -\lambda_\kappa$ and t_κ, given by $\varphi_{\kappa^-}(x) = \gamma_5\varphi_\kappa(x)$. (Recall that in the notation used throughout this book, γ_5 is the Hermitian matrix $\gamma_5 = -i\gamma_1\gamma_2\gamma_3\gamma_0 = \gamma_1\gamma_2\gamma_3\gamma_4$.) Because $\phi_\kappa(x)$ and $\phi_{\kappa^-}(x)$ are eigenvectors of a Hermitian operator with different eigenvalues they are orthogonal, so $\int d^4x \left(\varphi_\kappa^\dagger(x)\gamma_5\varphi_\kappa(x)\right) = 0$. This leaves just a sum over the eigenfunctions with $\lambda_\kappa = 0$. These are not generally paired; rather, since γ_5 anticommutes with $i\not{D}$, they can be chosen as simultaneous orthornormal eigenfunctions φ_u, φ_v of $i\not{D}$ with eigenvalue zero *and* of γ_5 with eigenvalues $+1$ and -1, respectively.

$$i\not{D}\varphi_u = 0\,, \qquad \gamma_5\varphi_u = \varphi_u\,,$$
$$i\not{D}\varphi_v = 0\,, \qquad \gamma_5\varphi_v = -\varphi_v\,. \tag{22.2.46}$$

Using the fact that $f(0) = 1$, Eq. (2.2.44) then becomes

$$\mathscr{A}(x) = -2\left[\sum_u t_u\left(\varphi_u^\dagger(x)\varphi_u(x)\right) - \sum_v t_v\left(\varphi_v^\dagger(x)\varphi_v(x)\right)\right]. \tag{22.2.47}$$

Now, since the φ_u and φ_v are normalized as in Eq. (22.2.42), the integral of Eq. (22.2.47) gives

$$\int (d^4x)_E\mathscr{A}(x) = -2\Big[\sum_u t_u - \sum_v t_v\Big]\,, \tag{22.2.48}$$

with the sums over u and v running over left- and right-handed zero modes of the operator $i\not{D}$, respectively. In particular, for the case where t is the unit matrix, by using Eq. (22.2.45) this may be expressed as a relation between a functional of the gauge field and the numbers of the zero modes of the Dirac operator with definite chiralities:

$$-\frac{1}{32\pi^2}\int (d^4x)_E\ \epsilon^E_{ijkl}\,F_{\alpha ij}\,F_{\beta kl}\ \mathrm{tr}\left[t_\alpha t_\beta\right] = n_+ - n_-\,, \tag{22.2.49}$$

where here $n_\pm$ are the numbers of zero modes of $\not{D}$ that have eigenvalues ± 1 for γ_5. This is the celebrated *Atiyah–Singer index theorem.*[6a] Among other things, it shows that under variations in the gauge field the integral on the left-hand side of Eq. (22.2.49) cannot change smoothly, but only by integers, so this integral can only depend on the topology of the gauge field. Its dependence on this topology will be described in Section 23.5.

22.3 Direct Calculation of Anomalies: The General Case

We saw in Section 22.2 how to use the elegant Fujikawa approach to calculate anomalies of chiral symmetries in gauge theories like quantum chromodynamics, where the gauge interactions are non-chiral and fermion number is conserved. This method can also be used to deal with more general problems, but there it becomes less straightforward.[7] In this section we will treat the anomaly by direct calculation, as it was originally done. This will provide a useful additional perspective on the anomalies, and will incidentally allow us to discuss anomalies in general theories with very little extra trouble.

To deal with the general case, we will unite *all* left-handed fermion fields (including antifermions, where that distinction is meaningful) in a single column χ. For instance, if ψ is a column containing all quark and lepton (as opposed to antiquark and antilepton) fields, then

$$\chi \equiv \begin{bmatrix} \frac{1}{2}(1+\gamma_5)\psi \\ \frac{1}{2}\,[\beta\mathscr{C}\,(1-\gamma_5)\,\psi]^* \end{bmatrix} = \begin{bmatrix} \frac{1}{2}(1+\gamma_5)\psi \\ \frac{1}{2}(1+\gamma_5)\beta\mathscr{C}\psi^* \end{bmatrix}, \tag{22.3.1}$$

where $\mathscr{C}$ is the matrix defined in Section 5.4 by

$$\mathscr{C}\gamma_\mu^{\rm T}\mathscr{C}^{-1} = -\gamma_\mu\,,$$

which is needed in order that all components of χ should belong to the same $(1/2, 0)$ representation of the Lorentz group. Under an infinitesimal fermion number (e. g., baryon number or baryon number minus lepton number) conserving gauge transformation:

$$\delta\psi = i\theta_\alpha \left[\, \tfrac{1}{2}(1+\gamma_5)t_\alpha^L + \tfrac{1}{2}(1-\gamma_5)t_\alpha^R \right] \psi \tag{22.3.2}$$

this column undergoes the transformation

$$\delta\chi = i\epsilon_\alpha T_\alpha \chi\,, \tag{22.3.3}$$

where

$$T_\alpha = \begin{bmatrix} t_\alpha^L & 0 \\ 0 & -t_\alpha^{R*} \end{bmatrix} = \begin{bmatrix} t_\alpha^L & 0 \\ 0 & -(t_\alpha^R)^{\rm T} \end{bmatrix}. \tag{22.3.4}$$

We shall not limit ourselves here to theories that conserve fermion number, so the T_α will now be any Hermitian representations of the gauge algebra, not necessarily of the block-diagonal form (22.3.4). For the moment we will consider only massless fermions, returning to the effect of fermion masses a little later.

We shall be concerned here with the one-loop three-point function

$$\Gamma^{\mu\nu\rho}_{\alpha\beta\gamma}(x,y,z) \equiv \langle T\{j_\alpha^\mu(x),\, j_\beta^\nu(y),\, j_\gamma^\rho(z)\}\rangle_{\rm VAC}\,, \tag{22.3.5}$$

where j_α^μ is the fermionic current, calculated in terms of free fields:

$$j_\alpha^\mu = -i\bar{\chi}T_\alpha\gamma^\mu\chi\,. \tag{22.3.6}$$

The two Feynman diagrams of Figure 22.1 give

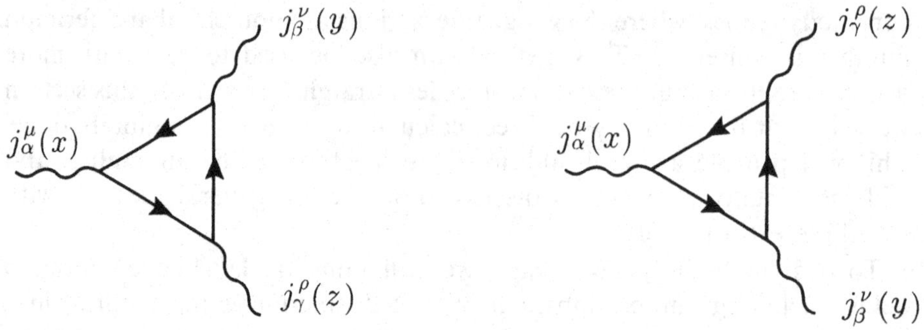

Figure 22.1. Two triangle diagrams for the anomaly in the current $j_\alpha^\mu(x)$. Solid lines are fermions; wavy lines indicate fictitious gauge fields, coupled to the currents.

$$\Gamma^{\mu\nu\rho}_{\alpha\beta\gamma}(x,y,z) = -i\mathrm{Tr}\left[S(x-y)T_\beta\gamma^\nu P_L S(y-z)T_\gamma\gamma^\rho P_L S(z-x)T_\alpha\gamma^\mu P_L\right]$$
$$-i\mathrm{Tr}\left[S(x-z)T_\gamma\gamma^\rho P_L S(z-y)T_\beta\gamma^\nu P_L S(y-x)T_\alpha\gamma^\mu P_L\right]\,, \tag{22.3.7}$$

where P_L is the projection operator on the left-handed fermion fields

$$P_L = \left(\frac{1+\gamma_5}{2}\right) \tag{22.3.8}$$

and $S(x)$ is the propagator of a massless fermion field:

$$S(x) = \frac{-i}{(2\pi)^4}\int d^4p\,\left(\frac{-i\,\not{p}}{p^2-i\epsilon}\right)\,e^{ip\cdot x}\,. \tag{22.3.9}$$

(For further comments on this use of the Feynman rules, see the end of this section.) Collecting factors, Eq. (22.3.7) now reads

$$\Gamma^{\mu\nu\rho}_{\alpha\beta\gamma}(x,y,z) = \frac{i}{(2\pi)^{12}}\int d^4k_1\,d^4k_2\,e^{-i(k_1+k_2)\cdot x}\,e^{ik_1\cdot y}\,e^{ik_2\cdot z}\int d^4p$$
$$\times\left\{\mathrm{tr}\left[\frac{\not{p}-\not{k}_1+\not{a}}{(p-k_1+a)^2-i\epsilon}\gamma^\nu\frac{\not{p}+\not{a}}{(p+a)^2-i\epsilon}\gamma^\rho\frac{\not{p}+\not{k}_2+\not{a}}{(p+k_2+a)^2-i\epsilon}\gamma^\mu\frac{1+\gamma_5}{2}\right]\right.$$
$$\times\,\mathrm{tr}\left[T_\beta T_\gamma T_\alpha\right]$$
$$+\mathrm{tr}\left[\frac{\not{p}-\not{k}_2+\not{b}}{(p-k_2+b)^2-i\epsilon}\gamma^\rho\frac{\not{p}+\not{b}}{(p+b)^2-i\epsilon}\gamma^\nu\frac{\not{p}+\not{k}_1+\not{b}}{(p+k_1+b)^2-i\epsilon}\gamma^\mu\frac{1+\gamma_5}{2}\right]$$
$$\left.\times\,\mathrm{tr}\left[T_\gamma T_\beta T_\alpha\right]\right\}\,, \tag{22.3.10}$$

where 'tr' here denotes a trace over either Dirac or group indices, depending on its argument. We have introduced the arbitrary constant four-vectors a and b because, although the expression (22.3.10) is convergent and hence independent of the labelling of the momenta carried by internal lines, the evaluation of $\partial_\mu \Gamma^{\mu\nu\rho}_{\alpha\beta\gamma}$ involves the manipulation of divergent integrals that do depend on this labelling. We will see that the freedom to choose a^μ and b^μ corresponds to a freedom to move the anomaly in these integrals from one current to another, but does not allow us to eliminate the anomaly altogether.

In taking the divergence of Eq. (22.3.10), we use the identities

$$\not{k}_1 + \not{k}_2 = (\not{p} + \not{k}_2 + \not{a}) - (\not{p} - \not{k}_1 + \not{a}) = (\not{p} + \not{k}_1 + \not{b}) - (\not{p} - \not{k}_2 + \not{b})$$

and find

$$\frac{\partial}{\partial x^\mu}\Gamma^{\mu\nu\rho}_{\alpha\beta\gamma}(x,y,z) = \frac{1}{(2\pi)^{12}}\int d^4k_1\, d^4k_2\, e^{-i(k_1+k_2)\cdot x}\, e^{ik_1\cdot y}\, e^{ik_2\cdot z}\int d^4p$$
$$\times\Bigg\{\mathrm{tr}\left[T_\beta\, T_\gamma\, T_\alpha\right]\ \mathrm{tr}\left[\frac{\not{p} - \not{k}_1 + \not{a}}{(p-k_1+a)^2 - i\epsilon}\gamma^\nu \frac{\not{p} + \not{a}}{(p+a)^2 - i\epsilon}\gamma^\rho \frac{1+\gamma_5}{2}\right]$$
$$-\mathrm{tr}\left[T_\beta\, T_\gamma\, T_\alpha\right]\ \mathrm{tr}\left[\frac{\not{p} + \not{a}}{(p+a)^2 - i\epsilon}\gamma^\rho \frac{\not{p} + \not{k}_2 + \not{a}}{(p+k_2+a)^2 - i\epsilon}\gamma^\nu \frac{1+\gamma_5}{2}\right]$$
$$+\mathrm{tr}\left[T_\gamma\, T_\beta\, T_\alpha\right]\ \mathrm{tr}\left[\frac{\not{p} - \not{k}_2 + \not{b}}{(p-k_2+b)^2 - i\epsilon}\gamma^\rho \frac{\not{p} + \not{b}}{(p+b)^2 - i\epsilon}\gamma^\nu \frac{1+\gamma_5}{2}\right]$$
$$-\mathrm{tr}\left[T_\gamma\, T_\beta\, T_\alpha\right]\ \mathrm{tr}\left[\frac{\not{p} + \not{b}}{(p+b)^2 - i\epsilon}\gamma^\nu \frac{\not{p} + \not{k}_1 + \not{b}}{(p+k_1+b)^2 - i\epsilon}\gamma^\rho \frac{1+\gamma_5}{2}\right]\Bigg\}. \tag{22.3.11}$$

At this point it is convenient to separate the three-point function into terms symmetric and antisymmetric in the group indices, by writing

$$\mathrm{tr}\left[T_\beta\, T_\gamma\, T_\alpha\right] = D_{\alpha\beta\gamma} + \tfrac{1}{2} iN\, C_{\alpha\beta\gamma}$$

and

$$\mathrm{tr}\left[T_\gamma\, T_\beta\, T_\alpha\right] = D_{\alpha\beta\gamma} - \tfrac{1}{2} iN\, C_{\alpha\beta\gamma}\,,$$

where $D_{\alpha\beta\gamma}$ is the totally symmetric quantity

$$D_{\alpha\beta\gamma} = \tfrac{1}{2}\mathrm{tr}\left[\{T_\alpha,\, T_\beta\}\, T_\gamma\right] \tag{22.3.12}$$

and the coefficient of the structure constant $C_{\alpha\beta\gamma}$ is defined by

$$\mathrm{tr}[T_\alpha T_\beta] = N\,\delta_{\alpha\beta}\,.$$

The terms that are antisymmetric in group indices are in general non-zero, but they do not represent any breakdown of the symmetry. Just as in the derivation of the Ward identity in Section 10.4, in a formal calculation of the divergence of the matrix element (22.3.5) we encounter contributions from the time-derivatives of the theta-functions in the time-ordered product, equal to

$$\left[\frac{\partial}{\partial x^\mu}\Gamma^{\mu\nu\rho}_{\alpha\beta\gamma}(x,y,z)\right]_{\text{formal}} = -iC_{\alpha\beta\delta}\delta^4(x-y)\langle j^\nu_\delta(y)\, j^\rho_\gamma(z)\rangle_{\text{VAC}}$$
$$-iC_{\alpha\gamma\delta}\delta^4(x-z)\langle j^\nu_\beta(y) j^\rho_\delta(z)\rangle_{\text{VAC}}\,. \qquad (22.3.13)$$

It is not hard to see that the antisymmetric terms in Eq. (22.3.11) just reproduce Eq. (22.3.13). The anomaly is contained in the symmetric part of Eq. (22.3.11):

$$\left[\frac{\partial}{\partial x^\mu}\Gamma^{\mu\nu\rho}_{\alpha\beta\gamma}(x,y,z)\right]_{\text{anom}} = \frac{1}{(2\pi)^{12}}D_{\alpha\beta\gamma}\int d^4k_1\, d^4k_2\, e^{-i(k_1+k_2)\cdot x}\, e^{ik_1\cdot y}\, e^{ik_2\cdot z}$$
$$\times \int d^4p \left\{ \text{tr}\left[\frac{\not p - \not k_1 + \not a}{(p-k_1+a)^2 - i\epsilon}\gamma^\nu \frac{\not p + \not a}{(p+a)^2 - i\epsilon}\gamma^\rho \frac{1+\gamma_5}{2}\right]\right.$$
$$-\text{tr}\left[\frac{\not p + \not a}{(p+a)^2 - i\epsilon}\gamma^\rho \frac{\not p + \not k_2 + \not a}{(p+k_2+a)^2 - i\epsilon}\gamma^\nu \frac{1+\gamma_5}{2}\right]$$
$$+\text{tr}\left[\frac{\not p - \not k_2 + \not b}{(p-k_2+b)^2 - i\epsilon}\gamma^\rho \frac{\not p + \not b}{(p+b)^2 - i\epsilon}\gamma^\nu \frac{1+\gamma_5}{2}\right]$$
$$\left.-\text{tr}\left[\frac{\not p + \not b}{(p+b)^2 - i\epsilon}\gamma^\nu \frac{\not p + \not k_1 + \not b}{(p+k_1+b)^2 - i\epsilon}\gamma^\rho \frac{1+\gamma_5}{2}\right]\right\}. \qquad (22.3.14)$$

Grouping together the first and fourth traces, and the third and second traces, this may be put in the form

$$\left[\frac{\partial}{\partial x^\mu}\Gamma^{\mu\nu\rho}_{\alpha\beta\gamma}(x,y,z)\right]_{\text{anom}} = \frac{1}{(2\pi)^{12}}D_{\alpha\beta\gamma}\int d^4k_1\, d^4k_2\, e^{-i(k_1+k_2)\cdot x}\, e^{ik_1\cdot y}\, e^{ik_2\cdot z}$$
$$\times \left\{ \text{tr}\left[\gamma^\kappa\gamma^\nu\gamma^\lambda\gamma^\rho\frac{1+\gamma_5}{2}\right] I_{\kappa\lambda}(a-b-k_1, b, b+k_1)\right.$$
$$\left.+\text{tr}\left[\gamma^\kappa\gamma^\rho\gamma^\lambda\gamma^\nu\frac{1+\gamma_5}{2}\right] I_{\kappa\lambda}(b-a-k_2, a, a+k_2)\right\}, \qquad (22.3.15)$$

where

$$I_{\kappa,\lambda}(k,c,d) \equiv \int d^4p\Big[f_{\kappa\lambda}(p+k,c,d) - f_{\kappa\lambda}(p,c,d)\Big]\,, \qquad (22.3.16)$$

$$f_{\kappa\lambda}(p,c,d) \equiv \frac{(p+c)_\kappa(p+d)_\lambda}{[(p+c)^2 - i\epsilon][(p+d)^2 - i\epsilon]}\,. \qquad (22.3.17)$$

To evaluate these integrals consider the expansion of the function $f_{\kappa\lambda}(p+k,c,d)$ in powers of k:

$$f_{\kappa\lambda}(p+k,c,d) = \sum_{n=0}^{\infty}\frac{1}{n!}k^{\mu_1}\cdots k^{\mu_n}\frac{\partial^n f_{\kappa\lambda}(p,c,d)}{\partial p^{\mu_1}\cdots\partial p^{\mu_n}}\,.$$

The zeroth-order term $f_{\kappa\lambda}(p,c,d)$ clearly cancels in Eq. (22.3.16). All other terms in Eq. (22.3.16) are integrals of derivatives with respect to p, and hence after Wick rotation may be written as surface integrals over a large three-sphere, say of radius P. The nth derivative of f then yields the surface integral of a function that behaves as $P^{-2-(n-1)}$, while the area of the three-sphere of radius P goes as P^3, so the only terms that can contribute for $P \to \infty$ are those with $n = 1$ and $n = 2$:

$$I_{\kappa\lambda}(k, c, d) = k^\mu \int d^4p \frac{\partial f_{\kappa\lambda}(p, c, d)}{\partial p^\mu} + \tfrac{1}{2} k^\mu k^\nu \int d^4p \, \frac{\partial^2 f_{\kappa\lambda}(p, c, d)}{\partial p^\mu \partial p^\nu} \, . \quad (22.3.18)$$

A straightforward calculation then gives

$$I_{\kappa\lambda}(k, c, d) = \tfrac{1}{6} i\pi^2 \Big[2k_\lambda c_\kappa + 2k_\kappa d_\lambda - k_\lambda d_\kappa - k_\kappa c_\lambda - \eta_{\kappa\lambda} \, k \cdot (k + c + d) \Big] \, . \quad (22.3.19)$$

We must now separately consider the terms in the traces in Eq. (22.3.15) arising from the 1 and the γ_5 in the projection matrix $\frac{1}{2}(1 + \gamma_5)$. The terms arising from the 1 involve $\text{tr}[\gamma^\kappa \gamma^\nu \gamma^\lambda \gamma^\rho]$, which is symmetric in κ and λ, and also in ν and ρ, so the integrals here appear in the combination

$$\begin{aligned} &I_{\kappa\lambda}(a - b - k_1, b, b + k_1) + I_{\lambda\kappa}(a - b - k_1, b, b + k_1) \\ &+ I_{\kappa\lambda}(b - a - k_2, a, a + k_2) + I_{\lambda\kappa}(b - a - k_2, a, a + k_2) \, . \end{aligned}$$

Using Eq. (22.3.19), it is not difficult to see that this vanishes if and only if we choose the arbitrary constant vectors so that

$$a = -b \, . \quad (22.3.20)$$

Furthermore, this is a choice that avoids the non-chiral anomaly for all three currents, because in $(\partial/\partial y^\nu)\Gamma^{\mu\nu\rho}_{\alpha\beta\gamma}(x, y, z)$, a and b would be replaced with $a' = k_2 + a$ and $b' = -k_2 + b$, while in $(\partial/\partial z^\rho)\Gamma^{\mu\nu\rho}_{\alpha\beta\gamma}(x, y, z)$, a and b would be replaced with $a'' = k_1 + a$ and $b'' = -k_1 + b$, so that taking $a = -b$ also insures that $a' = -b'$ and $a'' = -b''$.

We are left with the term in the trace involving γ_5. This is totally antisymmetric

$$\text{tr} \left[\gamma^\kappa \gamma^\nu \gamma^\lambda \gamma^\rho \gamma_5 \right] = -4i\epsilon^{\kappa\nu\lambda\rho} \, , \quad (22.3.21)$$

where $\epsilon^{\kappa\nu\lambda\rho}$ is the totally antisymmetric tensor with $\epsilon^{0123} = 1$. Using this in Eq. (22.3.15) with $b = -a$ gives

$$\begin{aligned} \left[\frac{\partial}{\partial x^\mu} \Gamma^{\mu\nu\rho}_{\alpha\beta\gamma}(x, y, z) \right]_{\text{anom}} &= \frac{2}{(2\pi)^{12}} D_{\alpha\beta\gamma} \int d^4k_1 \, d^4k_2 \, e^{-i(k_1 + k_2)\cdot x} \\ &\quad \times e^{ik_1 \cdot y} \, e^{ik_2 \cdot z} \, \pi^2 \epsilon^{\kappa\nu\lambda\rho} a_\kappa \, (k_1 + k_2)_\lambda \, . \end{aligned} \quad (22.3.22)$$

We could eliminate the anomaly (22.3.22) in the current $J^\mu_\alpha(x)$ by taking $a \propto k_1 + k_2$. Although this is possible, it would not eliminate the anomaly altogether; it would appear in $(\partial/\partial y^\nu)\Gamma^{\mu\nu\rho}_{\alpha\beta\gamma}(x, y, z)$ or $(\partial/\partial z^\rho)\Gamma^{\mu\nu\rho}_{\alpha\beta\gamma}(x, y, z)$. The symmetry of the problem indicates that the anomaly will be absent in $(\partial/\partial y^\nu)\Gamma^{\mu\nu\rho}_{\alpha\beta\gamma}(x, y, z)$ if and only if $(k_2 + a) - (-k_2 + b) \propto k_1$, or in other words $a+k_2 \propto k_1$, and will be absent in $(\partial/\partial z^\rho)\Gamma^{\mu\nu\rho}_{\alpha\beta\gamma}(x,y,z)$ if and only if $(-k_1+a)-(k_1+b) \propto k_2$, or in other words $a-k_1 \propto k_2$. It is possible to choose a^μ to satisfy any two of these conditions, so that the anomaly can be removed from any two of the currents, but for non-parallel k_1 and k_2 it is evidently impossible to simultaneously satisfy all three conditions — $a \propto k_1+k_2$ and $a+k_2 \propto k_1$ and $a-k_1 \propto k_2$. (For instance, the first two conditions would imply that $a=-k_1-k_2$, in contradiction with the third condition.) We therefore conclude that although we have some freedom in deciding which current exhibits the anomaly, *the non-vanishing of $D_{\alpha\beta\gamma}$ shows definitely that there is an anomaly in at least one of the currents $J^\mu_\alpha(x)$, $J^\nu_\beta(y)$, or $J^\rho_\gamma(z)$*. This is one of the chief results of these calculations, and will be exploited in the next section as a source of constraints on the matter content of gauge theories.

We have seen that the evaluation of the anomalies depends on the choice we make of the shift vector a^μ. Unfortunately, there is no one choice that is uniformly satisfactory, so we have to choose a^μ in accordance with the special features of the problem at hand.

In one class of problems of great importance, $J^\mu_\alpha(x)$ is the current of a global symmetry, while $J^\nu_\beta(y)$ and $J^\rho_\gamma(z)$ are currents of gauge symmetries, that is, currents to which gauge fields are coupled. (We dealt with such a problem in the previous section.) In such cases, we *must* choose a^μ so that the anomaly is solely in $J^\mu_\alpha(x)$, not in $J^\nu_\beta(y)$ or $J^\rho_\gamma(z)$. As we have seen, this requires that $a+k_2 \propto k_1$ and $a-k_1 \propto k_2$, which leads to the unique result that

$$a = k_1 - k_2\,. \tag{22.3.23}$$

Adopting this value of a^μ, the anomaly (22.3.22) is

$$\begin{aligned}\left[\frac{\partial}{\partial x^\mu}\Gamma^{\mu\nu\rho}_{\alpha\beta\gamma}(x,y,z)\right]_{\text{anom}} &= \frac{1}{(2\pi)^{12}}D_{\alpha\beta\gamma}\int d^4k_1\, d^4k_2\, e^{-i(k_1+k_2)\cdot x}\, e^{ik_1\cdot y}\, e^{ik_2\cdot z} \\ &\quad\times 4\pi^2\epsilon^{\kappa\nu\lambda\rho}k_{1\kappa}k_{2\lambda} \\ &= -\frac{1}{4\pi^2}D_{\alpha\beta\gamma}\epsilon^{\kappa\nu\lambda\rho}\frac{\partial\delta^4(y-x)}{\partial y^\kappa}\frac{\partial\delta^4(z-x)}{\partial z^\lambda}\,. \end{aligned}\tag{22.3.24}$$

We note in passing that a result like this can only arise from a theory that involves massless particles.[7a] Otherwise, we would expect the Fourier transform of $\Gamma^{\mu\nu\rho}_{\alpha\beta\gamma}(x,y,z)$ to have a power series expansion around zero momentum. The only terms in such an expansion that could possibly lead to a current divergence of the form (22.3.24) are pseudotensors of first order in momenta

$$\begin{aligned}\left[\Gamma^{\mu\nu\rho}_{\alpha\beta\gamma}(k_1,k_2)\right]_{\text{anom}} &\equiv \int d^4y\, d^4z\, e^{-ik_1\cdot y}e^{-ik_2\cdot z}\left[\Gamma^{\mu\nu\rho}_{\alpha\beta\gamma}(0,y,z)\right]_{\text{anom}} \\ &= \epsilon^{\mu\nu\rho\sigma}\left[A_{\alpha\beta\gamma}k_{1\sigma}+B_{\alpha\beta\gamma}k_{2\sigma}\right],\end{aligned}$$

where $A_{\alpha\beta\gamma}$ and $B_{\alpha\beta\gamma}$ are constants. According to Eq. (22.3.24),

$$(k_1+k_2)_\mu \left[\Gamma^{\mu\nu\rho}_{\alpha\beta\gamma}(k_1,k_2)\right]_{\rm anom} = \frac{i}{4\pi^2}D_{\alpha\beta\gamma}\epsilon^{\kappa\nu\lambda\rho}k_{1\kappa}k_{2\lambda}$$

so

$$A_{\alpha\beta\gamma} - B_{\alpha\beta\gamma} = \frac{i}{4\pi^2}D_{\alpha\beta\gamma}\,.$$

But the symmetry of the amplitude among the three currents requires that

$$A_{\alpha\beta\gamma}k_{1\sigma} + B_{\alpha\beta\gamma}k_{2\sigma} = -A_{\alpha\gamma\beta}k_{2\sigma} - B_{\alpha\gamma\beta}k_{1\sigma} = -A_{\gamma\alpha\beta}(-k_{1\sigma}-k_{2\sigma}) - B_{\gamma\alpha\beta}k_{2\sigma}$$

so, equating the coefficients of $k_{1\sigma}$ and $k_{2\sigma}$,

$$A_{\alpha\beta\gamma} = -B_{\alpha\gamma\beta} = A_{\gamma\alpha\beta}\,, \qquad B_{\alpha\beta\gamma} = -A_{\alpha\gamma\beta} = A_{\gamma\alpha\beta} - B_{\gamma\alpha\beta}\,.$$

Taking the difference of these equations gives

$$A_{\alpha\beta\gamma} - B_{\alpha\beta\gamma} = A_{\alpha\gamma\beta} - B_{\alpha\gamma\beta} = B_{\gamma\alpha\beta}\,.$$

Since this is proportional to $D_{\alpha\beta\gamma}$ it must be totally symmetric, so $B_{\alpha\beta\gamma}$ and hence also $A_{\alpha\beta\gamma}$ are totally symmetric. But then the conditions for the symmetry of the three-point function read $A = -B = B - A$, which implies that $A_{\alpha\beta\gamma} = B_{\alpha\beta\gamma} = 0$, in contradiction to Eq. (22.3.24). By the same reasoning, it is not possible to cancel the anomaly by adding a local term to the interaction whose contribution to the divergence of the current J^μ_α cancels that given by Eq. (22.3.24).

Returning now to Eq. (22.3.24), we can express this result in terms of the vacuum expectation value of the current J^μ_α in the presence of gauge fields coupled to the currents J^ν_β and J^ρ_γ. The triangle graphs make a contribution to the current in the presence of the gauge fields

$$\langle J^\mu_\alpha(x)\rangle_\triangle = -\tfrac{1}{2}\int d^4y\, d^4z\, \Gamma^{\mu\nu\rho}_{\alpha\beta\gamma}(x,y,z)\, A^\beta_\nu(y)\, A^\gamma_\rho(z)\,. \qquad (22.3.25)$$

Using Eq. (22.3.24), this has the anomalous divergence

$$[\langle\partial_\mu J^\mu_\alpha(x)\rangle_\triangle]_{\rm anom} = -\frac{1}{8\pi^2}D_{\alpha\beta\gamma}\,\epsilon^{\kappa\nu\lambda\rho}\,\partial_\kappa A^\beta_\nu(x)\,\partial_\lambda A^\gamma_\rho(x)\,. \qquad (22.3.26)$$

There are additional diagrams, shown in Figure 22.2, that also exhibit anomalies. Gauge invariance requires that the diagrams of Figure 22.2 should add up to give the gauge-invariant result

$$[\langle\partial_\mu J^\mu_\alpha(x)\rangle]_{\rm anom} = -\frac{1}{32\pi^2}D_{\alpha\beta\gamma}\,\epsilon^{\kappa\nu\lambda\rho}\,F^\beta_{\kappa\nu}(x)\,F^\gamma_{\lambda\rho}(x)\,. \qquad (22.3.27)$$

As a check, consider a fermion-conserving theory, with generators T_α of the form (22.3.4). The constant $D_{\alpha\beta\gamma}$ in the anomaly (22.3.26) is here given by

$$D_{\alpha\beta\gamma} = \tfrac{1}{2}{\rm tr}[\{t^L_\alpha, t^L_\beta\}t^L_\gamma] - \tfrac{1}{2}{\rm tr}[\{t^R_\alpha, t^R_\beta\}t^R_\gamma]\,. \qquad (22.3.28)$$

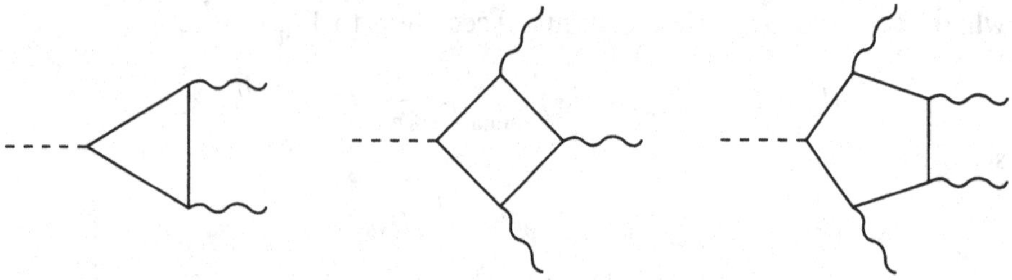

Figure 22.2. One-loop diagrams for the anomaly in a current indicated by the dashed line. Solid lines are fermions; wavy lines are gauge fields with which they interact.

More specifically, in Section 22.2 we calculated the divergence of an axial-vector current J_5^μ, with $t^L = -t^R \equiv t$, due to gauge-field interactions with vector currents J_β^ν and J_γ^ρ (called J_α^ν and J_β^ρ in Section 22.2) with $t_\beta^L = t_\beta^R \equiv t_\beta$, and likewise for t_γ. Hence in this case $D_{\alpha\beta\gamma}$ is replaced with $\text{tr}[\{t, t_\beta\}t_\gamma] = \text{tr}[\{t_\beta, t_\gamma\}t]$, and Eq. (22.3.27) becomes

$$[\langle\partial_\mu J_5^\mu(x)\rangle]_{\text{anom}} = -\frac{1}{32\pi^2}\text{tr}[\{t_\beta, t_\gamma\}t]\,\epsilon^{\kappa\nu\lambda\rho}\,F_{\kappa\nu}^\beta(x)\,F_{\lambda\rho}^\gamma(x)\,, \tag{22.3.29}$$

in agreement with Eq. (22.2.26).

Where no gauge fields are coupled to any of the currents $J_\alpha^\mu(x)$, $J_\beta^\nu(y)$, or $J_\gamma^\rho(z)$, the choice of the shift vector a^μ is a matter of convenience. When some but not all currents are associated with symmetries that are spontaneously broken, we keep the unbroken symmetries manifest if we choose a^μ so that there are no anomalies in the currents corresponding to the unbroken symmetries. (This will be important in Section 22.7.) For instance, in quantum chromodynamics and similar theories, where the generators T_α of global symmetries like chiral $SU(3) \times SU(3)$ take the form (22.3.4), all currents are either vector, with $t_\alpha^R = t_\alpha^L$, corresponding to unbroken symmetries, or axial-vector, with $t_\alpha^R = -t_\alpha^L$, corresponding to broken symmetries. We see from Eq. (22.3.28) that in this case the only triangle graphs with anomalies are those with one axial-vector and two vector currents, or three axial-vector currents. In the case of one axial-vector and two vector currents, we choose a^μ so that there is no anomaly that interferes with the conservation of the vector currents. Thus, if $J_\alpha^\mu(x)$ is the axial-vector current and $J_\beta^\nu(y)$ and $J_\gamma^\rho(z)$ the two vector currents, then as in Eq. (22.3.23) we must take $a^\mu = k_1^\mu - k_2^\mu$, so that the anomaly is given by Eq. (22.3.24). On the other hand, for three axial-vector currents, there is no reason to require that any one of them should be free of anomalies. Instead, it is natural to give a^μ a value that respects the symmetry among the three currents. Guided by Lorentz invariance, suppose we try $a = \alpha k_1 + \beta k_2$, where α and β are constants.

Then symmetry would require that the momentum of *each* internal line should be p plus α times the momentum flowing out at the end of the line plus β times the momentum flowing out at the beginning of the line. That is, if we require that $a = \alpha k_1 + \beta k_2$, then symmetry requires also that $a - k_1 = -\alpha(k_1+k_2)+\beta k_1$ and $a+k_2 = \alpha k_2 - \beta(k_1+k_2)$. These three relations are satisfied for non-parallel k_1 and k_2 if and only if $\alpha = -\beta = 1/3$, so that

$$a = \tfrac{1}{3}(k_1 - k_2) \,. \tag{22.3.30}$$

Using this in Eq. (22.3.22) and comparing with Eq. (22.3.23), we see that the anomaly in the axial-vector current in a Feynman amplitude for three axial-vector currents is one-third what it would be for one axial-vector and two vector currents.

The divergence of the current contains additional anomalies from the graphs of Figure 22.2. Gauge invariance is no guide here, because even the triangle graph does not yield conserved currents. The total anomaly has been calculated by Bardeen[8] for the chiral $SU(3) \times SU(3)$ symmetry of the strong interactions, with momentum shift vectors chosen so that vector currents are conserved and graphs with all axial currents are symmetric among these currents. Although this $SU(3) \times SU(3)$ symmetry (which acts on quark flavors rather than colors) is not gauged in quantum chromodynamics, it is convenient to express the anomaly as a failure of gauge invariance in a functional $\Gamma[V, A]$ of fictitious weakly coupled gauge fields: an octet of vector fields $V_a^\mu(x)$ and an octet of axial-vector fields $A_a^\mu(x)$. We also introduce infinitesimal gauge-transformation operators*

$$i\mathscr{Y}_a(x) = -\frac{\partial}{\partial x^\mu}\frac{\delta}{\delta V_{a\mu}(x)} - f_{abc}V_{b\mu}(x)\frac{\delta}{\delta V_{c\mu}(x)} - f_{abc}A_{b\mu}(x)\frac{\delta}{\delta A_{c\mu}(x)} \tag{22.3.31}$$

and

$$i\mathscr{X}_a(x) = -\frac{\partial}{\partial x^\mu}\frac{\delta}{\delta A_{a\mu}(x)} - f_{abc}V_{b\mu}(x)\frac{\delta}{\delta A_{c\mu}(x)} - f_{abc}A_{b\mu}(x)\frac{\delta}{\delta V_{c\mu}(x)}\,, \tag{22.3.32}$$

where f_{abc} are the $SU(3)$ structure constants. As mentioned above, the labelling of internal line momenta is chosen so that the vector current is not anomalous:

$$\mathscr{Y}_a\Gamma[V, A] = 0\,, \tag{22.3.33}$$

* The operators $\mathscr{Y}_a$ and $\mathscr{X}_a$ here are what in Ref. 8 were called X_a and Y_a. This switch in notation is made to maintain consistency with the notation of Chapter 19, where broken symmetry generators are consistently labelled X_a.

but then an anomaly appears in the axial-vector currents

$$\mathscr{X}_a\Gamma[V,A] = \frac{i}{32\pi^2}\epsilon^{\mu\nu\rho\sigma}\text{Tr}\left\{\lambda_a\Big[V_{\mu\nu}V_{\rho\sigma} + \tfrac{1}{3}A_{\mu\nu}A_{\rho\sigma} - \tfrac{32}{3}A_\mu A_\nu A_\rho A_\sigma\right.$$
$$\left. + \tfrac{8}{3}i(A_\mu A_\nu V_{\rho\sigma} + A_\mu V_{\rho\sigma}A_\nu + V_{\rho\sigma}A_\mu A_\nu)\Big]\right\}, \qquad (22.3.34)$$

where λ_a are the $SU(3)$ matrices given by Eq. (19.7.2), and

$$V_\mu \equiv \tfrac{1}{2}\lambda_a V_{\mu a}\,, \qquad A_\mu \equiv \tfrac{1}{2}\lambda_a A_{\mu a}\,, \qquad (22.3.35)$$

$$V_{\mu\nu} = \partial_\mu V_\nu - \partial_\nu V_\mu - i[V_\mu, V_\nu] - i[A_\mu, A_\nu]\,, \qquad (22.3.36)$$

$$A_{\mu\nu} = \partial_\mu A_\nu - \partial_\nu A_\mu - i[V_\mu, A_\nu] - i[A_\mu, V_\nu]\,. \qquad (22.3.37)$$

The factor 1/3 accompanying the second term on the right-hand side in Eq. (22.3.34) has already been explained as a consequence of the different choice of a^μ in the AVV and AAA graphs. In Section 22.6 we will describe consistency conditions that allow the cubic and quartic terms in Eq. (22.3.34) to be calculated from the quadratic terms.

For anomalies involving symmetries that are *all* spontaneously broken, there is no reason to choose a^μ in a way that distinguishes among the different currents, and instead it is natural to label internal fermion lines in a way that is symmetric among the attached gauge boson lines. As we have seen, this means that the triangle graph is to be calculated with the choice $a = \frac{1}{3}(k_1 - k_2)$. This gives a triangle anomaly one-third of the value given by (22.3.26). When square and pentagon graphs are included, this result becomes[8a]

$$[\langle D_\mu J^\mu_\alpha\rangle]_{\text{anom}} = -\frac{1}{24\pi^2}\epsilon^{\kappa\nu\lambda\rho}\,\text{Tr}\left\{T_\alpha\Big[\partial_\kappa A_\nu\,\partial_\lambda A_\rho - \tfrac{1}{2}i\partial_\kappa A_\nu\, A_\lambda A_\rho\right.$$
$$\left. + \tfrac{1}{2}iA_\kappa\,\partial_\nu A_\lambda\, A_\rho - \tfrac{1}{2}iA_\kappa A_\nu \partial_\lambda A_\rho\Big]\right\}, \qquad (22.3.38)$$

where $A^\mu \equiv A^\mu_\alpha T_\alpha$. We will not derive this result here, because we have already seen that in such cases the terms quadratic in A_μ are given by one-third the corresponding terms in Eq. (22.3.26), and in Section 22.6 we shall be able to use consistency conditions to derive the remaining terms in Eq. (22.3.38) from the quadratic terms.

Now we must consider possible corrections to these results. A careful derivation of the anomaly to all orders of perturbation theory was given by Adler and Bardeen;[9] what follows is intended to give only the gist of their analysis.

The argument that led to Eq. (22.3.22) may be repeated in any order of perturbation theory, and shows that the anomaly in general arises from momentum-space integrals that can be reexpressed as surface terms. In consequence, as we saw here in Eq. (22.3.18), the only graphs for the current divergence that contribute to the anomaly are those for which the integral over the momentum circulating in the fermion loop has dimensionality (in powers of momentum) zero or greater. Interactions of the fermions in the loop with virtual gauge bosons would reduce the dimensionality of the integral over the fermion loop momentum enough to eliminate the anomaly, so the anomaly receives no contribution from such radiative corrections. (It is true that the integral over the momenta of the virtual gauge bosons as well as the fermion loop would have non-negative dimensionality, but the gauge boson propagators can be regulated, as the fermion propagator cannot, without interfering with the chiral symmetry in question.) The anomaly is affected by interactions of the gauge bosons attached to the fermion loop with other gauge bosons and fermion loops, but these just serve to renormalize operators like $\epsilon^{\kappa\nu\lambda\rho}\, F^{\beta}_{\kappa\nu}(x)\, F^{\gamma}_{\lambda\rho}(x)$. By the same argument, any fermion mass that respects the symmetries in question (if this were possible) would not change the anomaly, because extracting factors of this mass would lower the dimensionality of the momentum-space integral.

The last remark of the previous paragraph raises the question of whether we can calculate the anomaly without knowing all the fermions in a theory, heavy fermions as well as light or massless ones. Yes, we can; we shall now show that *no fermion that is allowed by a given symmetry to have a mass can contribute to the anomaly for that symmetry*. In the general class of theories considered here, a mass term in the Lagrangian density would take the form

$$\mathscr{L}_{\text{mass}} = -\sum_{nn'\sigma\sigma'} \chi_{\sigma n}\epsilon_{\sigma\sigma'}M_{nn'}\chi_{\sigma' n'} + \text{H.c.}\,, \tag{22.3.39}$$

where σ is the two-component spinor index of the $(\frac{1}{2},0)$ representation of the Lorentz group, $\epsilon_{\sigma\sigma'}$ is the antisymmetric matrix with $\epsilon_{\frac{1}{2},-\frac{1}{2}} = +1$, needed for Lorentz invariance, and M is a symmetric mass matrix.** Now, in order for $\mathscr{L}_{\text{mass}}$ to respect gauge invariance, the mass matrix must

** In fermion-number-conserving theories where χ has the form (22.3.1), M is related to the usual mass matrix m by

$$M = \frac{1}{2}\begin{pmatrix} 0 & m \\ m^{\mathrm{T}} & 0 \end{pmatrix}.$$

With all inversion phases equal to unity, parity, charge conjugation, and time-reversal invariance respectively have the further consequences that m is Hermitian, symmetric, or real.

satisfy

$$- T_\alpha^{\rm T} M = M T_\alpha . \tag{22.3.40}$$

The index n may be replaced with an index r labelling different irreducible representations of the gauge group, and an index s labelling components within each irreducible representation, so that

$$(T_\alpha)_{rs,r's'} = \delta_{rr'}(T_\alpha^{(r)})_{s,s'} , \tag{22.3.41}$$

and we write

$$M_{rs,r's'} = (M^{(rr')})_{ss'} . \tag{22.3.42}$$

Eq. (22.3.40) then becomes

$$- T_\alpha^{(r){\rm T}} M^{(r,r')} = M^{(r,r')} T_\alpha^{(r')} . \tag{22.3.43}$$

(This is not summed over r or r'.) Schur's lemma[10] tells us that whenever the matrices of a pair of irreducible representations are related by such an equation, the matrix that relates them is either zero or non-singular. (See Section 5.5.) Thus, either (i) $M^{(r,r')} = 0$, or else (ii) $-T_\alpha^{(r){\rm T}}$ and $T_\alpha^{(r')}$ are related by a similarity transformation, and likewise for T_β and T_γ. In the latter case the contributions to the anomaly constant (22.3.12) from fermions belonging to the individual irreducible representations r and r' are related by

$$D^{(r)}_{\alpha\beta\gamma} = -D^{(r')}_{\alpha\beta\gamma} , \tag{22.3.44}$$

so the anomaly either vanishes (for $r = r'$) or cancels between the two representations (for $r \neq r'$).The anomalies in a given set of symmetries therefore are unaffected by the possible presence of fermions with a mass allowed by these symmetries.

* * *

There is a fine point in the use of the Feynman rules to calculate the three-point function (22.3.7), to which we now return. In using the conventional fermion propagator (22.3.9), we have in effect doubled the number of fermion fields; in addition to the purely left-handed fields $\chi(x)$ given by Eq. (22.3.1), the propagator (22.3.9) includes right-handed modes (unrelated to the fields $(1 - \gamma_5)\psi$ of fermion-number-conserving theories) that do not interact with the gauge fields. Combining these non-interacting right-handed fields with the interacting left-handed fields in a single spinor Ψ, the fermion Lagrangian density is $-\bar{\Psi} \not{D} \Psi$, where $\not{D}$ is now

$$\not{D} = \not{\partial} - i \not{A}_\alpha T_\alpha \left(\frac{1 + \gamma_5}{2} \right) . \tag{22.3.45}$$

The one-loop vacuum functional for spin $\frac{1}{2}$ fields interacting with real or fictitious gauge fields is just Det $\not{D}$. The failure of gauge invariance in this determinant can be blamed on the fact that the operator (22.3.45) is the sum of two terms

$$\not{D} = \not{\partial}\left(\frac{1-\gamma_5}{2}\right) + \not{D}\left(\frac{1+\gamma_5}{2}\right) \tag{22.3.46}$$

of which only the second is gauge-invariant.

There is another way of looking at anomalies,[10a] which provides a framework for calculations of a large class of anomalies, of coordinate as well as gauge symmetries in spaces of any dimensionality. Instead of working with the operator (22.3.45), which has a well-defined determinant, one deals instead with just the second term in Eq. (22.3.46)

$$\not{D}_L \equiv \not{D}\left(\frac{1+\gamma_5}{2}\right) . \tag{22.3.47}$$

This is perfectly gauge-invariant, but does not have a well-defined determinant, because this operator does not map the space of fermion fields of one handedness into itself, but rather into the space of fermion fields of the other handedness. One can try to define a gauge-invariant vacuum functional Det $\not{D}_L$ by writing differential equations for Det $\not{D}_L$ in the space of gauge fields, modulo gauge transformations, but then one may encounter obstructions. There are local obstructions due to violations of the necessary integrability conditions, which correspond to the anomalies we have been discussing. Even where these local obstructions are absent, in cases where the infinite-dimensional space of gauge-field configurations, modulo gauge transformations, is not simply connected, topological obstructions can prevent the definition of single-valued functionals on this space. Such a global anomaly was found by Witten[10b] for the gauge group $SU(2)$ (which is free of local anomalies) with an odd number of massless left-handed fermions in $SU(2)$ doublets.

22.4 Anomaly-Free Gauge Theories

We have calculated the effect of anomalies on the conservation of a general current J^μ_α. Where this current is itself coupled to a gauge field, gauge invariance requires that the anomaly be absent. We saw in the previous section that the anomaly is proportional to the completely symmetric constant factor $D_{\alpha\beta\gamma}$ defined by Eq. (22.3.12), so for gauge currents we must have[11]

$$D_{\alpha\beta\gamma} \equiv \tfrac{1}{2}\text{tr}\left[\{T_\alpha, T_\beta\}\, T_\gamma\right] = 0\,, \tag{22.4.1}$$

where T_α is the representation of the gauge algebra on the set of all left-handed fermion and antifermion fields, and 'tr' again denotes a sum over these fermion and antifermion species. This condition may be satisfied for any gauge group if the fermion fields furnish a suitable reducible or irreducible representation of the group. In addition, there are some gauge groups for which Eq. (22.4.1) is satisfied for fermions in *any* representation of the group.[12] (The Batalin–Vilkovisky formalism is used in Section 22.6 to give a purely algebraic proof that for such gauge groups the anomaly is absent to all orders of perturbation theory.)

The condition (22.4.1) is obviously satisfied if the left-handed fermion (and antifermion) fields furnish a representation of the gauge algebra that is equivalent to its complex conjugate, in the sense that

$$(iT_\alpha)^* = S\,(iT_\alpha)\,S^{-1}$$

or equivalently (since we always take T_α Hermitian)

$$T_\alpha^T = -S\,T_\alpha S^{-1}\,. \tag{22.4.2}$$

(Inserting Eq. (22.4.2) in Eq. (22.3.12) gives $D_{\alpha\beta\gamma} = -D_{\alpha\beta\gamma}$.) Such a representation may be either *real*, in which case it is possible by a similarity transformation $T'_\alpha = R\,T_\alpha\,R^{-1}$ to convert the representation to a form in which T'_α is imaginary and antisymmetric, or *pseudoreal*, in which case this is impossible. (For instance, the three-dimensional irreducible representation of $SU(2)$ is real, while the two-dimensional representation is pseudoreal.) There is no anomaly for gauge algebras that have only real or pseudoreal representations, namely[13] $SO(2n+1)$ (including $SU(2) \equiv SO(3)$), $SO(4n)$ for $n \geq 2$, $USp(2n)$ for $n \geq 3$, G_2, F_4, E_7, and E_8, and all of their direct sums. A few other algebras also have only representations for which $D_{\alpha\beta\gamma}$ vanishes, even though some representations are neither real nor pseudoreal.[12] These are $SO(4n+2)$ (except for $SO(2) \equiv U(1)$ and $SO(6) \equiv SU(4)$) and E_6, and their direct sums with each other and the above algebras. Anomalies are thus only possible for gauge algebras that include $SU(n)$ (for $n \geq 3$) or $U(1)$ factors. As it happens, these are among the most important gauge algebras in today's physics. The standard model is based on the gauge group $SU(3) \times SU(2) \times U(1)$, so we must rely on cancellations among the quarks and leptons to make the theory free of anomalies.

Table 22.1 gives a classification of the left-handed spinor fields of the first generation of the standard model according to the representations they furnish of the color $SU(3)$ group and the electroweak $SU(2)$ group, and the value of the $U(1)$ quantum number $y/g' = t_3/g - q/e$.

We can now check whether $D_{\alpha\beta\gamma}$ vanishes when T_α, T_β, and T_γ run over all of the generators of $SU(3) \times SU(2) \times U(1)$. We need only consider those combinations of generators for which the product of T_α, T_β, and T_γ

Table 22.1. First-generation left-handed fermion and antifermion fields of the standard model.

Fermions	$SU(3)$	$SU(2)$	$U(1)$ $[y/g']$
$\begin{pmatrix} u \\ d \end{pmatrix}_L$	3	2	−1/6
u_R^*	$\bar{3}$	1	+2/3
d_R^*	$\bar{3}$	1	−1/3
$\begin{pmatrix} \nu_e \\ e \end{pmatrix}_L$	1	2	1/2
e_R^*	1	1	−1

is neutral under $SU(3) \times SU(2) \times U(1)$, since $D_{\alpha\beta\gamma}$ obviously vanishes for all the others. We can make invariants out of either zero, two, or three $SU(3)$ generators (because 8×8 and $8 \times 8 \times 8$ both contain singlets), zero, two, or three $SU(2)$ generators, and any number of $U(1)$ generators, so we only need to check the following cases:

***SU*(3)–*SU*(3)–*SU*(3)**: Here $D_{\alpha\beta\gamma}$ vanishes because the left-handed fermions furnish a representation $3 + 3 + \bar{3} + \bar{3} + 1 + 1 + 1$ of $SU(3)$, which is real.

***SU*(3)–*SU*(3)–*U*(1)**: Here the anomaly is proportional to

$$\sum_{3,\bar{3}} y/g' = -\frac{1}{6} - \frac{1}{6} + \frac{2}{3} - \frac{1}{3} = 0\,.$$

***SU*(2)–*SU*(2)–*SU*(2)**: There is no anomaly here because $SU(2)$ only has real or pseudoreal representations.

***SU*(2)–*SU*(2)–*U*(1)**: Here the anomaly is proportional to

$$\sum_{\text{doublets}} y/g' = 3\left(-\frac{1}{6}\right) + \frac{1}{2} = 0\,.$$

$U(1)$–$U(1)$–$U(1)$: Here the anomaly is proportional to

$$\sum (y/g')^3 = 6\left(-\frac{1}{6}\right)^3 + 3\left(\frac{2}{3}\right)^3 + 3\left(-\frac{1}{3}\right)^3 + 2\left(\frac{1}{2}\right)^3 + (-1)^3 = 0\,.$$

We see that all anomalies cancel for the gauge symmetries of the standard model.[13a] This result can be neatly understood by noting that $SU(3) \times SU(2) \times U(1)$ may be embedded in $SO(10)$.[14] All of the representations of $SO(10)$ are anomaly-free, so the same property is inherited by any reducible representation of $SU(3) \times SU(2) \times U(1)$ that furnishes a complete representation of $SO(10)$. As it happens, the left-handed fields of a single generation of quarks, leptons, antiquarks, and antileptons plus one additional $(SU(3) \times SU(2) \times U(1))$-singlet forms a complete 16-dimensional representation of $SO(10)$ (the fundamental spinor representation), so for this set of left-handed fermions there are no $SU(3) \times SU(2) \times U(1)$ anomalies. The singlet would not contribute to such anomalies anyway, so there are no anomalies in the gauge symmetries of the standard model.

There is one more anomaly that needs to be evaluated. All species of fermions interact with gravitation in the same way. Calculation of the fermion loop graph for the expectation value of the current $\bar{\chi} T \gamma^\mu \chi$ (where χ is a column of all left-handed fermion and antifermion fields) in the presence of an external gravitational field yields an anomaly[15] $\partial_\mu(\bar{\chi} T \gamma^\mu \chi)$ proportional to

$$\text{tr}\{T\} \epsilon^{\mu\nu\rho\sigma} R_{\mu\nu\kappa\lambda} R_{\rho\sigma}{}^{\kappa\lambda}\,.$$

In particular, to avoid a gravitational violation of a gauge symmetry like (22.3.3), the generators must satisfy

$$\text{tr}\{T_\alpha\} = 0\,. \tag{22.4.3}$$

Like the pure gauge anomaly, this vanishes for gauge generators satisfying Eq. (22.4.2), so this anomaly receives no contribution from fermions that form real or pseudoreal representations of the gauge group, and it can therefore be calculated taking into account only those fermions whose masses arise from a breaking of the gauge symmetry. Also, this condition is automatically satisfied by the generators of any *simple* subalgebras like $SU(2)$ or $SU(3)$ of the gauge algebra. ($\text{tr}\{T_\alpha\}$ is just a number which commutes with all the T_β, so if it is non-zero then the algebra is not simple.) Hence we only need to check that Eq. (22.4.3) is satisfied by the $U(1)$ generators of the gauge algebra. In the standard model, the sum of the values of the weak hypercharge y for all left-handed fermions is

$$\sum y/g' = 6\left(-\frac{1}{6}\right) + 3\left(\frac{2}{3}\right) + 3\left(-\frac{1}{3}\right) + 2\left(\frac{1}{2}\right) + (-1) = 0\,,$$

so there are no gravitational anomalies in the standard model currents.

The requirement of vanishing of anomalies may be used as a guide in formulating realistic theories. For instance, the values of the weak hypercharges y for various $SU(3) \times SU(2)$ multiplets were originally taken from experiment, but one may wonder why these weak hypercharges (and the corresponding average electric charges in each multiplet) take the observed values. To answer this question, suppose we assign arbitrary weak hypercharges a, b, c, d, and e to the multiplets (u_L, d_L), u_R^*, d_R^*, (ν_L, e_L), and e_R^*, respectively. The conditions for anomaly cancellation tell us that:

$SU(3)$–$SU(3)$–$U(1)$:

$$\sum_{3,\bar{3}} y = 2a + b + c = 0\ ;$$

$SU(2)$–$SU(2)$–$U(1)$:

$$\sum_{\text{doublets}} y = 3a + d = 0\ ;$$

$U(1)$–$U(1)$–$U(1)$:

$$\sum y^3 = 6a^3 + 3b^3 + 3c^3 + 2d^3 + e^3 = 0\ ;$$

graviton–graviton–$U(1)$:

$$\sum y = 6a + 3b + 3c + 2d + e = 0\ .$$

Aside from the possibility of interchanging u_R and d_R, these equations have only two solutions, which we may call $U(1)$ and $U(1)'$:

$$U(1): \quad b/a = -4\ , \quad c/a = 2\ , \quad d/a = -3\ , \quad e/a = 6\ ,$$
$$U(1)': \quad b = -c\ , \quad a = c = d = e = 0\ .$$

Furthermore, these solutions are exclusive; we cannot suppose that *both* $U(1)$ and $U(1)'$ are local symmetries, because then we would encounter a $U(1)'$–$U(1)'$–$U(1)$ anomaly proportional to $(-4) + (+2) \neq 0$ and a $U(1)'$–$U(1)$–$U(1)$ anomaly proportional to $(-4)^2 - (+2)^2 \neq 0$. The $U(1)$ generator is just the weak hypercharge of the standard electroweak theory (with the overall constant factor a absorbed into the definition of g'), while the $U(1)'$ symmetry resembles nothing observed in nature. This little calculation provides a rational explanation of the assignment of y values, or equivalently of electric charges, in the standard model, and it shows that if all gauge anomalies must cancel within a single generation of quarks and leptons, then it is not possible to couple a gauge boson to any other $U(1)$ quantum number, in addition to weak hypercharge.

On the other hand, although it is reasonable to guess that the $SU(3) \times SU(2) \times U(1)$ gauge bosons of the standard model may couple only to the known quarks and leptons (both to explain why no other fermions have been discovered and to preserve the beautiful cancellation of anomalies in the standard model), there may be other $U(1)'$ gauge bosons that couple to other undetected $(SU(3)\times SU(2)\times U(1))$-neutral fermions as well as to the known quarks and leptons. Suppose we denote the $U(1)'$ quantum numbers y' of the multiplets (u_L, d_L), u_R^*, d_R^*, (ν_L, e_L), and e_R^* as a', b', c', d', and e', respectively. Since we don't know anything about possible $(SU(3)\times SU(2)\times U(1))$-neutral fermions, the requirement of cancellation of the $U(1)'$–$U(1)'$–$U(1)'$ and graviton–graviton–$U(1)'$ anomalies does not help us to constrain a', b', c', d', or e'. The remaining conditions of anomaly cancellation tell us that:

***SU*(3)–*SU*(3)–*U*(1)′:**

$$\sum_{3,\bar{3}} y' = 2a' + b' + c' = 0\ ;$$

***SU*(2)–*SU*(2)–*U*(1)′:**

$$\sum_{\text{doublets}} y' = 3a' + d' = 0\ ;$$

***U*(1)–*U*(1)–*U*(1)′:**

$$\sum y^2 y' = 6a' + 3(-4)^2 b' + 3(2)^2 c' + 2(-3)^2 d' + (6)^2 e' = 0\ ;$$

***U*(1)–*U*(1)′–*U*(1)′:**

$$\sum y y'^2 = 6a'^2 + 3(-4)b'^2 + 3(2)c'^2 + 2(-3)d'^2 + (6)e'^2 = 0\ .$$

The general solution has y' a linear combination of y and a quantum number $B-L$ (where B and L are the conventional baryon and lepton numbers) that takes values $1/3$, $-1/3$, $-1/3$, -1, and $+1$ for the multiplets (u_L, d_L), u_R^*, d_R^*, (ν_L, e_L), and e_R^*, respectively. If $B-L$ *is* a local symmetry, with a coupling that is not many orders of magnitude smaller than e, then it must be spontaneously broken, since ordinary bodies have macroscopic values of $B-L$. To avoid conflict with observations of neutral currents, the characteristic scale F of the symmetry breaking would have to be larger than that of the electroweak interactions, but not necessarily many orders of magnitude larger. Thus a neutral vector boson somewhat heavier than the Z^0 and coupled to $B-L$ seems like the most plausible addition to the standard model.

This has all been for a single generation of the standard model. With three generations there are many more anomaly-free symmetries. One class of symmetries that are not broken by anomalies or (as far as we know) by anything else consists of the differences in the numbers of leptons of the various flavors. Together with $B-L$, this will be important in Section 23.5 in classifying the baryon- and lepton-non-conserving processes produced by anomalies.

22.5 Massless Bound States*

It is sometimes speculated that quarks and leptons may be bound states of more fundamental particles. If these hypothetical fundamental particles had asymptotically free gauge interactions, like those in quantum chromodynamics, then we would expect them to be trapped, which would explain why they are not observed. However, there is a difficulty with this picture. No internal structure of the quarks and leptons has ever been observed, so the characteristic energy scale Λ' (analogous to the $\Lambda \approx 200$ MeV of quantum chromodynamics) of these gauge interactions must be very high. For instance, as remarked in Section 12.3, the agreement between theory and experiment for the magnetic moment of the muon indicates that $\Lambda' > 3$ TeV. But, aside from Goldstone bosons, we would normally expect the bound states in such a theory to have masses of order Λ', or perhaps $2\pi\Lambda'$, just as in quantum chromodynamics the proton mass is of order $2\pi\Lambda_{\text{QCD}}$. This expectation is of course in sharp contradiction with the fact the observed masses of the quarks and leptons are much less than Λ'. To put this problem another way, if the leptons and quarks are bound states, then why is their size (as measured by anomalous magnetic moments, etc.) so much smaller than their Compton wavelengths?

One way to answer this question is to suppose that, unlike quantum chromodynamics, this theory has unbroken chiral symmetries that keep the quarks and leptons massless, aside from small corrections from other interactions. In general, a chiral symmetry is any symmetry for which the massless elementary fields of some given helicity (including the complex conjugates of the fields of opposite helicity) furnish a complex representation. By operating on the vacuum with products of the elementary fields we can construct other states of definite helicity that also furnish complex representations of these symmetries. If any of these states are actual composite particles then they must be massless, because all helicity components of the state of a massive particle must furnish the same representation of any symmetry that commutes with rotations, so a given helicity component of a particle together with the antiparticle of the opposite helicity component would together furnish a real representation. Of course, it may not be so easy to tell which of the states constructed in this way correspond to actual composite massless particles, but if they do, their masslessness is natural, because in order for some massless particles of a given helicity that belong to a complex representation of the symmetry

* This section lies somewhat out of the book's main line of development, and may be omitted in a first reading.

group to become massive as we change some parameter of the theory, their symmetry properties would have to change discontinuously from a complex to a real representation.

Although this reasoning shows that there are theories in which it is natural to have massless or very light composite particles, it does not give any indication of when this actually happens. The question is an interesting one, quite apart from the problem of understanding quarks and leptons as possible composite particles. 't Hooft[16] has suggested a powerful way of answering this question, based on considerations of anomalies. Briefly, if the underlying theory has global chiral symmetries (not broken by gauge anomalies, and not spontaneously broken) consisting of transformations on the elementary left-handed spin $\frac{1}{2}$ fermions (and antifermions) χ with symmetry generators T_α, T_β, etc., and if the anomaly constant $\text{tr}[\{T_\alpha, T_\beta\}T_\gamma]$ of these global symmetries is non-zero, then the spectrum of bound states must include spin $\frac{1}{2}$ massless particles on whose left-handed states the same symmetries induce transformations with generators $\mathscr{T}_\alpha$, $\mathscr{T}_\beta$, etc., with the same anomaly constant

$$\text{tr}[\{\mathscr{T}_\alpha, \mathscr{T}_\beta\}\mathscr{T}_\gamma] = \text{tr}[\{T_\alpha, T_\beta\}T_\gamma] \,. \tag{22.5.1}$$

't Hooft's argument was as follows. Imagine that some weakly-interacting gauge bosons are coupled to generators T_α, T_β, etc. of the global symmetries of the underlying theory. Suppose also that although some of the coefficients $D_{\alpha\beta\gamma} \equiv \text{tr}[\{T_\alpha, T_\beta\}T_\gamma]$ are non-zero, this anomaly is cancelled by anomalies due to other 'spectator' massless fermions, which do not feel the strong forces that trap the constituents of the composite particles. Physical processes at energies much less than the characteristic energy scale Λ' of the trapping interactions will be described by an effective Lagrangian in which the trapped fermions do not appear. If the symmetries with generators T_α, T_β, etc. are not spontaneously broken, then there are no Goldstone bosons, so the only particles in this effective Lagrangian will be the weakly coupled gauge bosons and spectator fermions, plus any massless bound states of the trapped fermions and strongly interacting gauge bosons. The consistency of the effective field theory requires that it must be anomaly-free, but the spectator fermions by assumption have an anomaly constant equal to $-D_{\alpha\beta\gamma}$, so there must be massless bound states, to provide an anomaly constant equal and opposite to this, and hence equal to that of the original trapped fermions.** Note that this argument

** It remains to show that these particles must have spin $\frac{1}{2}$. We are assuming that there are no Goldstone bosons here, and that other spinless particles would not naturally be massless. The elementary gauge bosons of the theory are assumed neutral under the anomalous symmetry transformations, so they could not contribute to the anomaly. Composite particles of spin $j \geq 1$ are ruled out by a different argument.[16a] This is works no matter how weak are the gauge interactions with generators T_α, T_β, etc., so these gauge bosons and the untrapped fermions do not have to be real particles to reach the conclusion that *the massless spin $\frac{1}{2}$ bound states reproduce the anomalies of the trapped elementary spin $\frac{1}{2}$ fermions of which they are composed.*

As a simple example, suppose that the underlying theory contains n 'flavors' of massless fermions, each of which has both left- and right-handed parts in the defining representation N of an asymptotically free $SU(N)$ gauge group. We require N to be odd so that there can be untrapped $SU(N)$-neutral fermionic bound states. Just as in quantum chromodynamics, this theory has an automatic global $SU_L(n)\times SU_R(n)\times U_V(1)$ symmetry, with the left- and right-handed massless fermions respectively in its $(n,1)$ and $(1,n)$ representations, both having equal values for the $U_V(1)$ quantum number, which can be taken as unity. There are non-vanishing anomaly constants in the underlying theory for the $SU(n)_L$–$SU(n)_L$–$U(1)_Y$ and $SU(n)_R$–$SU(n)_R$–$U(1)_Y$ current triplets, which have the values

$$D_{aL,bL,0} = D_{aR,bR,0} = N\delta_{ab} \; ,$$

where a, b, etc. label the $SU(n)$ generators λ_a, normalized so that in the defining n-component representation $\mathrm{tr}\{\lambda_a\lambda_b\} = \frac{1}{2}\delta_{ab}$. For $n > 2$ there are also non-vanishing anomaly constants for the $SU(n)_L$–$SU(n)_L$–$SU(n)_L$ and $SU(n)_R$–$SU(n)_R$–$SU(n)_R$ currents, equal to

$$D_{aL,bL,cL} = D_{aR,bR,cR} = N\mathrm{tr}[\{\lambda_a, \lambda_b\}, \lambda_c] \; .$$

We suppose here that the $SU_L(n) \times SU_R(n) \times U(1)$ symmetry is not spontaneously broken. Because of trapping, the only fermionic bound states in the physical spectrum will be contained in representations of this symmetry that can be formed from m_L and m_R elementary fermions of helicity $+\frac{1}{2}$ and $-\frac{1}{2}$, respectively, and $\bar{m}_L$ and $\bar{m}_R$ of their antiparticles, with

$$m_L + m_R - \bar{m}_L - \bar{m}_R = kN \; , \tag{22.5.2}$$

where k is any positive or negative odd integer. In consequence, the only irreducible representations (r,s) of $SU(n)_L \times SU_R(n)$ encountered are those for which r is in the direct product of m_L of the defining representations of

a theory in which the anomalous currents can be constructed as Lorentz four-vector functions of the elementary spin $\frac{1}{2}$ fields, and in order to contribute to the anomaly these currents would have to have non-vanishing matrix elements between any massless composite particles of spin $j = 1$, which would violate Lorentz invariance. Massless composite particles of spin $j \geq 3/2$ are ruled out because in this theory it is also possible to construct a conserved energy-momentum tensor, which would have to have non-vanishing matrix elements between these massless composite particles, which for $j \geq 3/2$ would also violate Lorentz invariance.

$SU(n)$ and $\bar{m}_L$ of their complex conjugates, while s is in the direct product of m_R of the defining representations of $SU(n)$ and $\bar{m}_R$ of their complex conjugates, and the $U(1)_V$ quantum number is kN, with k, m_L, m_R, $\bar{m}_L$, and $\bar{m}_R$ subject to Eq. (22.5.2). Let $p(r,s,k)$ be the number of times an irreducible representation (r,s) of $SU_L(n) \times SU_R(n)$ with $U(1)_V$ quantum number kN appears among the helicity $+\frac{1}{2}$ bound states. Eq. (22.5.1) then reads

$$\sum_{r,s,k} p(r,s,k) d_s \operatorname{tr}^{(r)}[\{\mathcal{T}_a, \mathcal{T}_b\}\mathcal{T}_c] = N \operatorname{tr}[\{\lambda_a, \lambda_b\}\lambda_c] \,, \tag{22.5.3}$$

$$\sum_{r,s,k} p(r,s,k) d_s k \operatorname{tr}^{(r)}[\{\mathcal{T}_a, \mathcal{T}_b\}] = \operatorname{tr}[\{\lambda_a, \lambda_b\}] \,, \tag{22.5.4}$$

where $\operatorname{tr}^{(r)}$ denotes the trace in the irreducible representation r of $SU(n)$, and d_s is the dimensionality of representation s of $SU(N)$. The only other constraint on the $p(r,s,k)$ is that they must be positive integers.

The complex conjugate $(\bar{r}, \bar{s}, -k)$ of any representation (r,s,k) of $SU(2)_L \times SU(2)_R \times U(1)_V$ has values of the traces $\operatorname{tr}^{(r)}[\{\mathcal{T}_a, \mathcal{T}_b\}\mathcal{T}_c]$ and $k \operatorname{tr}^{(r)}[\{\mathcal{T}_a, \mathcal{T}_b\}]$ opposite to those of the representation (r,s,k), so Eqs. (22.5.3) and (22.5.4) only restrict the values of

$$\ell(r,s,k) \equiv p(r,s,k) - p(\bar{r},\bar{s},-k) \,.$$

Recall that in the notation we have been using here, these traces are over all massless bound states of helicity $+\frac{1}{2}$, including the antiparticles of the massless bound states of helicity $-\frac{1}{2}$, which transform according to the complex conjugate representations. The complex conjugate of any representation of $SU(2)_L \times SU(2)_R \times U(1)_V$ has values of $\operatorname{tr}^{(r)}[\{\mathcal{T}_a, \mathcal{T}_b\}\mathcal{T}_c]$ and $k \operatorname{tr}^{(r)}[\{\mathcal{T}_a, \mathcal{T}_b\}]$ opposite to those of that representation. Thus we can sum in (22.5.3) and (22.5.4) only over the representations with $U(1)_V$ quantum number $kN > 0$

$$\sum_{r,s,k>0} \ell(r,s,k)\, d_s \operatorname{tr}^{(r)}[\{\mathcal{T}_a, \mathcal{T}_b\}\mathcal{T}_c] = N \operatorname{tr}[\{\lambda_a, \lambda_b\}\lambda_c] \,, \tag{22.5.5}$$

$$\sum_{r,s,k>0} \ell(r,s,k)\, d_s k \operatorname{tr}^{(r)}[\{\mathcal{T}_a, \mathcal{T}_b\}] = \operatorname{tr}[\{\lambda_a, \lambda_b\}] \,, \tag{22.5.6}$$

with $\ell(r,s,k)$ equal to the number of times an irreducible representation (r,s) of $SU_L(n) \times SU_R(n)$ with $U(1)_V$ quantum number $kN > 0$ appears among the helicity $+\frac{1}{2}$ bound states, *minus* the number of times the same representation appears among the helicity $-\frac{1}{2}$ bound states. (If parity is unbroken then a representation r,s,k must occur among the helicity $-\frac{1}{2}$ bound states as often as the representation s,r,k occurs among the helicity $+\frac{1}{2}$ bound states, so in this case $\ell(r,s,k) = -\ell(s,r,k)$.)

First, consider the case of $n = 2$ flavors. There is no way of coupling three three-vectors together symmetrically to make an $SU(2)$-invariant, so both sides of (22.5.5) vanish automatically, leaving only the condition (22.5.6). The defining two-component representation of $SU(2)$ is contained in the product of any odd number of these two-component representations, so we can always find a solution of (22.5.6) with $\ell(r, s, k) = 0$ for all $SU(2)_L \times SU(2)_R \times U(1)_V$ representations with r non-trivial except the one in which r is the defining representation, s is the trivial representation, and $k = 1$, for which we take $\ell = 1$. Unfortunately, this solution is far from unique — there is an infinite number of ways of reproducing the anomalies of the underlying theory.

For general n and N it is usual to find many solutions of Eqs. (22.5.5) and (22.5.6), but there are some cases where no solutions can be found. In such theories we can reach the conclusion that the $SU(n)_L \times SU(n)_R \times U(1)_V$ symmetry *must* be partly or completely spontaneously broken. This has particularly interesting implications for quantum chromodynamics, so let us now specialize to the case of an $SU(3)$ gauge group.

To be concrete, we will focus on the representations, with $k = 1$ and $\bar{m}_L = \bar{m}_R = 0$, that can be formed from just three elementary fermions and no antifermions. These are:†

(a) r is the symmetric third-rank $SU(n)$ tensor; s is the trivial representation.

(b) r is the antisymmetric third-rank $SU(n)$ tensor; s is the trivial representation.

(c) r is the third-rank $SU(n)$ tensor of mixed symmetry; s is the trivial representation.

(d) r is the symmetric second-rank $SU(n)$ tensor; s is the $SU(n)$ vector.

(e) r is the antisymmetric second-rank $SU(n)$ tensor; s is the $SU(n)$ vector.

(f) r is the $SU(n)$ vector; s is the symmetric second-rank $SU(n)$ tensor.

(g) r is the $SU(n)$ vector; s is the antisymmetric second-rank $SU(n)$ tensor.

(h) r is the trivial representation; s is the symmetric third-rank $SU(n)$ tensor.

(i) r is the trivial representation; s is the antisymmetric third-rank $SU(n)$ tensor.

(j) r is the trivial representation; s is the third-rank $SU(n)$ tensor of mixed symmetry.

† All $SU(N)$ vectors and tensors here are understood to be contravariant.

For $n > 2$, Eqs. (22.5.5) and (22.5.6) here read [††]

$$\tfrac{1}{2}(n+3)(n+6)\ell_a + \tfrac{1}{2}(n-3)(n-6)\ell_b + (n^2-9)\ell_c + n(n+4)\ell_d$$
$$+n(n-4)\ell_e + \tfrac{1}{2}n^2(n+1)\ell_f + \tfrac{1}{2}n^2(n-1)\ell_g = 3 \tag{22.5.7}$$

and

$$\tfrac{1}{2}(n+2)(n+3)\ell_a + \tfrac{1}{2}(n-2)(n-3)\ell_b + (n^2-3)\ell_c + n(n+2)\ell_d$$
$$+n(n-2)\ell_e + \tfrac{1}{2}n(n+1)\ell_f + \tfrac{1}{2}n(n-1)\ell_g = 1\,. \tag{22.5.8}$$

There is no problem in satisfying Eq. (22.5.7), but note that if n is a multiple of three then for all values of the ℓs, each term on the left-hand side of Eq. (22.5.8) is also a multiple of three, which makes it impossible to satisfy this condition. We conclude in particular that *the $SU(3)_L \times SU(3)_R \times U(1)_V$ symmetry of quantum chromodynamics with three flavors of massless quarks must be spontaneously broken.* This result is not limited to the representations **(a)**–**(j)**, but applies to any representations of $SU(3)_L \times SU(3)_R \times U(1)_V$ that can be formed from color-neutral combinations of quarks and antiquarks.

* * *

Aside from special cases such as an $SU(3)$ gauge group with $n = 3$ elementary fermion $SU(3)$-triplets, the 't Hooft anomaly matching condition is not very restrictive; in general it allows a wide variety of massless bound states when chiral symmetry is unbroken. 't Hooft also proposed a *decoupling condition,* which requires that when one or more of the elementary fermion flavors become very heavy there should be no unbroken chiral symmetries that prevent composite particles that contain heavy elementary fermions from acquiring masses. For instance, in the case of an $SU(3)$ color gauge group, if we give one of the n quark types a large mass then those three-fermion bound states that contain a single massive quark will furnish representations (r', s') of the group $SU(n-1) \times SU(n-1)$ that are either

(v) r' is the symmetric second rank $SU(n-1)$ tensor; s' is the trivial representation,

(w) r' is the antisymmetric second rank $SU(n-1)$ tensor; s' is the trivial representation,

(x) r' and s' are $SU(n)$ vectors,

[††] In Ref. 16, 't Hooft assumed that parity is not spontaneously broken, so he gave these formulas for the case where $\ell_a = -\ell_h$, $\ell_b = -\ell_i$, $\ell_c = -\ell_j$, $\ell_d = -\ell_f$, and $\ell_e = -\ell_g$. As we see here, the main conclusion does not depend on parity conservation.

(y) r' is the trivial representation, s' is the symmetric second rank $SU(n-1)$ tensor,

(z) r' is the trivial representation, s' is the antisymmetric second rank $SU(n-1)$ tensor.

In order for the three-fermion mass state to acquire a large mass, it is necessary that $\ell'(r', s') = 0$, where $\ell'(r', s')$ is the number of times that an irreducible representation (r', s') occurs among the helicity $+\frac{1}{2}$ bound states minus the number of times the same representation appears among the helicity $-\frac{1}{2}$ bound states. By inspecting the list of three-fermion representations **(a)**–**(j)** of $SU(n) \times SU(n)$ to see what representations of $SU(n-1) \times SU(n-1)$ they yield, we see then that the 't Hooft decoupling condition requires that†

$$\begin{aligned} 0 = \ell'_v &= \ell_a + \ell_c + \ell_d\,, \\ 0 = \ell'_w &= \ell_b + \ell_c + \ell_e\,, \\ 0 = \ell'_x &= \ell_f + \ell_g + \ell_d + \ell_e\,, \\ 0 = \ell'_y &= \ell_f + \ell_h + \ell_j\,, \\ 0 = \ell'_z &= \ell_g + \ell_i + \ell_j\,. \end{aligned} \tag{22.5.9}$$

Unfortunately, in most cases there are still an infinite number of solutions, though there are no solutions in which the ℓs are n-independent integers.

The decoupling condition seems quite plausible, but its use by 't Hooft was questioned[17] on the ground that, as one or more fermion masses increase, one usually encounters phase transitions that make the mass spectrum different from what it would be for small fermion masses. There is a stronger condition, known as the *persistent mass condition*, which requires that when one or more of the elementary fermion flavors acquire *any* mass, there should be no unbroken chiral symmetries that prevent composite particles that contain these massive elementary fermions from acquiring some masses.[17] If valid, the persistent mass condition would lead to the same consequences, such as Eqs. (22.5.9), that were described by 't Hooft, with no chance of being invalidated by phase transitions.

It is easy to construct non-realistic models in which the persistent mass condition is violated, such as theories with spontaneously broken *non-chiral* symmetries, which lead to massless Goldstone bosons formed as composites of massive fermions.[17] (These models also exhibit phase transitions with increasing fermion masses, which invalidate 't Hooft's conclusions from the decoupling condition.) But the work of Vafa and

† In the parity conserving case studied by 't Hooft, the fourth and fifth equations are identical to the first and second, while the third is empty.

Witten discussed in Section 19.9 shows that, in a variety of more realistic QCD-like theories, non-chiral symmetries cannot be spontaneously broken, so this should not be taken as a serious objection to the persistent mass condition.

22.6 Consistency Conditions

The numerical coefficient appearing in the anomaly for any symmetry depends on the matter content of the theory. On the other hand, the *form* of the anomaly is largely independent of the details of the theory, because it is governed by consistency conditions, first presented in 1971 by Wess and Zumino.[18]

Even when we are interested in anomalies in global symmetry currents, in order to derive the consistency conditions it is convenient to imagine that all symmetry currents are coupled to gauge fields, which for non-Abelian symmetries also couple to each other, in such a way that these symmetries become *local* symmetries of the Lagrangian density. We can always return to the case of global symmetry by letting the corresponding gauge coupling constants become infinitesimal. Apart from anomalies, the effective action $\Gamma[A]$ in a background gauge field $A_{\alpha\mu}(x)$ will in this formalism be invariant under infinitesimal transformations $A_{\beta\mu}(y) \to A_{\beta\mu}(y) + i\int d^4x\, \epsilon_\alpha(x)\mathscr{T}_\alpha(x)\, A_{\beta\mu}(y)$ on the gauge field, where in order to reproduce the transformation (15.1.9) we must take

$$-i\mathscr{T}_\alpha(x) = -\frac{\partial}{\partial x^\mu}\frac{\delta}{\delta A_{\alpha\mu}(x)} - C_{\alpha\beta\gamma}A_{\beta\mu}(x)\frac{\delta}{\delta A_{\gamma\mu}(x)}\,. \tag{22.6.1}$$

Taking anomalies into account, $\mathscr{T}_\alpha(x)$ no longer annihilates $\Gamma[A]$, but rather

$$\mathscr{T}_\alpha(x)\Gamma[A] = G_\alpha[x;A]\,, \tag{22.6.2}$$

where $G_\alpha[x;A]$ represents the effect of the anomaly. Eq. (22.6.2) may also be written as a formula for the covariant divergence of the expectation value of the current:

$$D_\mu\langle J^\mu_\alpha(x)\rangle = -iG_\alpha[x;A]\,, \tag{22.6.3}$$

where

$$\langle J^\mu_\alpha(x)\rangle \equiv \frac{\delta}{\delta A_{\alpha\mu}(x)}\Gamma[A] \tag{22.6.4}$$

and D_μ is the gauge-covariant derivative (15.1.10), taken here in the adjoint representation with $(t_\beta)_{\gamma\alpha} = -iC_{\alpha\beta\gamma}$.

The Wess–Zumino consistency conditions follow from the commutation relations*

$$[\mathscr{T}_\alpha(x), \mathscr{T}_\beta(y)] = iC_{\alpha\beta\gamma}\delta^4(x-y)\mathscr{T}_\gamma(x)\,. \tag{22.6.5}$$

From Eqs. (22.6.2) and (22.6.5), we derive the general consistency condition

$$\mathscr{T}_\alpha(x)G_\beta[y;A] - \mathscr{T}_\beta(y)G_\alpha[x;A] = iC_{\alpha\beta\gamma}\delta^4(x-y)G_\gamma[y;A]\,. \tag{22.6.6}$$

These consistency conditions were originally derived by Wess and Zumino for the chiral $SU(3)\times SU(3)$ symmetry of the strong interactions, a special case of physical as well as historical importance. Here the generators $\mathscr{T}_\alpha(x)$ acting on the gauge fields consist of the even parity generators $\mathscr{Y}_a(x)$ defined by Eq. (22.3.31), and the odd parity generators $\mathscr{X}_a(x)$ defined by (22.3.32).** They satisfy the commutation relations

$$[\mathscr{Y}_a(x), \mathscr{Y}_b(y)] = i\delta^4(x-y)f_{abc}\mathscr{Y}_c(x)\,,$$

$$[\mathscr{Y}_a(x), \mathscr{X}_b(y)] = i\delta^4(x-y)f_{abc}\mathscr{X}_c(x)\,,$$

$$[\mathscr{X}_a(x), \mathscr{X}_b(y)] = i\delta^4(x-y)f_{abc}\mathscr{Y}_c(x)\,,$$

where f_{abc} are the $SU(3)$ structure constants. Since the $SU(3)$ subgroup generated by the $\mathscr{Y}_a$ is not spontaneously broken, it is convenient to treat the integration over fermion momenta in such a way as to preserve invariance under gauge transformations generated by $\mathscr{Y}_a$, so that

$$\mathscr{Y}_a(x)\Gamma = 0\,,$$

leaving us with the non-zero anomaly

$$\mathscr{X}_a(x)\Gamma = G_a(x)\,.$$

The non-trivial consistency conditions are then

$$\mathscr{Y}_a(x)G_b(y) = i\delta^4(x-y)f_{abc}G_c(x)$$

and

$$\mathscr{X}_a(x)G_b(y) - \mathscr{X}_b(x)G_a(y) = 0\,.$$

The first of these simply says that $G_a(x)$ transforms like an octet under ordinary $SU(3)$ transformations. The second condition imposes other strong constraints on $G_a(x)$. The reader may check that this condition is

---

* The factor $-i$ was inserted in Eq. (22.6.1) in order to provide the conventional factor $+i$ accompanying the structure constant $C_{\alpha\beta\gamma}$ in this commutation relation. A reminder: we are using a basis for the Lie algebra in which the structure constants are totally antisymmetric, and so we do not distinguish between upper and lower gauge indices.

** Another reminder: as mentioned in Section 22.3, the $\mathscr{X}_a$ and $\mathscr{Y}_a$ used here are what were called Y_a and X_a in Ref. 8.

satisfied by the Bardeen formula (22.3.34) for $G_a(x)$. We will not go into this here, but will instead use as an illustration a general gauge theory with all currents treated symmetrically. In Section 22.3 we quoted the formula (22.3.38) for the anomaly in this case, but did not derive the terms in this formula of higher than second order in the gauge fields. Here we shall show that these terms are dictated by the consistency conditions (22.6.6).

For this purpose, and also to allow for further generalizations, it is very convenient to reformulate the set of Wess–Zumino consistency conditions as a condition of invariance under the BRST transformations described in Section 15.7. Let us introduce a ghost field ω_α, and define the nilpotent BRST operator s for a general gauge theory by

$$sA_{\alpha\mu} = \partial_\mu \omega_\alpha + C_{\alpha\beta\gamma} A_{\beta\mu} \omega_\gamma \,, \tag{22.6.7}$$

$$s\omega_\alpha = -\, \tfrac{1}{2} C_{\alpha\beta\gamma} \omega_\beta \omega_\gamma \,, \tag{22.6.8}$$

it being understood that s satisfies the distributive rule $s(AB) = (sA)B \pm A(sB)$, the sign being negative where A is a fermionic quantity like ω_α, and otherwise positive. In place of the anomaly function $G_\alpha[x;A]$, we shall work with a functional

$$G[\omega, A] = \int \omega_\alpha(x)\, G_\alpha[x;A]\, d^4x \,. \tag{22.6.9}$$

Then (keeping in mind that ω_α is fermionic)

$$\begin{aligned} sG[\omega, A] =& -\, \tfrac{1}{2} C_{\alpha\beta\gamma} \int d^4x\; \omega_\beta(x)\, \omega_\gamma(x)\, G_\alpha[x;A] \\ & - \int d^4x\; \omega_\alpha(x) \int d^4y\; \left[\frac{\partial \omega_\beta(y)}{\partial y^\mu} + C_{\beta\gamma\delta} A_{\gamma\mu}(y) \omega_\delta(y)\right] \frac{\delta G_\alpha[x;A]}{\delta A_{\beta\mu}(y)} \\ =& \int d^4x \int d^4y\; \omega_\alpha(x)\, \omega_\beta(y) \Big[-\, \tfrac{1}{2} C_{\alpha\beta\gamma} \delta^4(x-y) G_\gamma[x;A] \\ & + i\, \mathscr{T}_\beta(y) G_\alpha[x;A] \Big] \,. \end{aligned}$$

Because the ghost fields anticommute with each other, this can be written

$$\begin{aligned} sG[\omega, A] =& -\, \tfrac{1}{2} i \int d^4x \int d^4y\; \omega_\alpha(x)\, \omega_\beta(y) \\ & \times \Big[i C_{\alpha\beta\gamma} \delta^4(x-y) G_\gamma[x;A] + \mathscr{T}_\beta(y) G_\alpha[x;A] - \mathscr{T}_\alpha(x) G_\beta[y;A] \Big] \,. \end{aligned}$$

We see that the consistency condition (22.6.6) will hold if and only if $G[\omega, A]$ is BRST-invariant

$$sG[\omega, A] = 0 \tag{22.6.10}$$

for all ghost fields $\omega_\alpha(x)$.

Now consider the possibility that the anomaly $G[\omega; A]$ could be written as the BRST operator s acting on a local functional $F[A]$:

$$G[\omega; A] = sF[A] . \tag{22.6.11}$$

(Note that the functional F is necessarily independent of the ghost field, because the operator s adds one ghost field factor, and the anomaly functional G is already linear in the ghost field.) The BRST operator satisfies $s^2 = 0$, so such an anomaly would satisfy the consistency condition $sG = 0$. If $F[A]$ is a *local* functional† of the gauge field, then it could be subtracted from the action, thus cancelling the anomaly. The same is true of any term in the anomaly that can be written as the BRST operator s acting on a local functional; such a term satisfies the consistency condition by itself, and can be cancelled by adding a local term to the action. The possible anomalies that interest us are thus the local functionals $G[\omega; A]$ of ghost number unity that satisfy the consistency condition (22.6.10), modulo terms that can be expressed as s acting on some local functional of ghost number zero. In accordance with the usual terminology for nilpotent operators, the equivalence classes of such functionals form what is called the *cohomology* of the s operator at ghost number unity.

We can also express this in terms of the local densities themselves. We can write the anomaly (or any term in the anomaly) as $G = \int d^4x\, \mathscr{G}(x)$, where $\mathscr{G}(x)$ is a power series in the gauge and ghost fields and their derivatives at the spacetime point x. The condition $sG = 0$ then is equivalent to the statement that

$$s\mathscr{G}(x) = \partial_\mu \mathscr{J}^\mu(x) \tag{22.6.12}$$

for some function $\mathscr{J}^\mu(x)$ of fields and field derivatives. Likewise, the terms in $\mathscr{G}$ that can be cancelled by adding local terms to the action are those that are of the form $s\mathscr{F}$ up to possible derivatives. Thus the anomalies that interest us are the local functions of ghost number unity that satisfy the consistency condition (22.6.12), modulo terms that can be expressed as s acting on some local functional of ghost number zero, modulo possible derivatives. This is known as the cohomology of s at ghost number unity in the space of local functions, modulo derivatives, and denoted $H^1(s|d)$.

Algebraic methods have been used to work out the cohomology of the BRST operator s, and thereby deduce the form of the anomaly in general gauge theories.[19] This approach leaves unknown only constant coefficients that depend on the matter content of the theory and need to be calculated by the methods of Sections 22.2 or 22.3. Since we have already calculated the terms in the anomaly of second order in the gauge fields for general gauge theories, including their constant coefficients, here we will use the

† By a local functional is meant the integral of a local function, that is, a function of fields and field derivatives at a given point.

consistency condition (22.6.12) to calculate the terms of higher order in the fields.

We saw in Section 22.3 that when all currents are treated symmetrically, the terms of second order in the gauge fields are one-third the expression (22.3.26). This is an operator of dimensionality four (in units of mass), while the Wess–Zumino consistency condition (22.6.6) relates only operators of the same dimensionality, so to satisfy this condition we should add only terms of higher order in the gauge field that have the same dimensionality. We therefore seek a solution of the consistency conditions in the (not necessarily unique) form

$$G_\alpha = i[\langle\partial_\mu J^\mu_\alpha\rangle]_{\rm anom} = -\frac{i}{24\pi^2}\epsilon^{\kappa\nu\lambda\rho}\,{\rm Tr}\Big\{T_\alpha\Big[\partial_\kappa A_\nu\,\partial_\lambda A_\rho + ic_1\partial_\kappa A_\nu\,A_\lambda A_\rho$$
$$+ic_2 A_\kappa\,\partial_\nu A_\lambda\,A_\rho + ic_3 A_\kappa A_\nu \partial_\lambda A_\rho - c_4 A_\kappa A_\nu A_\lambda A_\rho\Big]\Big\}\,, \tag{22.6.13}$$

where $A_\mu \equiv A_{\alpha\mu}T_\alpha$, and the c_i are constants to be determined.

In order to save a great deal of effort, it will be convenient to rewrite this in the language of differential forms. (See Section 8.8.) We introduce a set of c-number parameters dx^μ that are taken to anticommute with themselves and all fermionic fields, such as the ghost field ω_α. The dx^μ then also anticommute with the BRST operator s. Because $dx^\kappa dx^\nu dx^\lambda dx^\rho$ is totally antisymmetric, it may be written as

$$dx^\kappa dx^\nu dx^\lambda dx^\rho = \epsilon^{\kappa\nu\lambda\rho}d^4x\,, \qquad d^4x = dx^0dx^1dx^2dx^3\,. \tag{22.6.14}$$

We also introduce the exterior derivative

$$d \equiv dx^\mu\frac{\partial}{\partial x^\mu}\,,$$

which since derivatives commute is nilpotent as well as anticommuting with s:

$$d^2 = 0\,, \qquad ds + sd = 0\,. \tag{22.6.15}$$

Finally, we introduce the anticommuting quantities

$$A \equiv iA_\mu dx^\mu = iA_{\alpha\mu}T_\alpha dx^\mu\,, \qquad \omega \equiv i\omega_\alpha T_\alpha\,. \tag{22.6.16}$$

In this notation, Eq. (22.6.13) reads

$$G[\omega, A] = \frac{1}{24\pi^2}\int {\rm Tr}\Big\{\omega\,\Big[(dA)^2 + c_1(dA)A^2$$
$$+c_2A(dA)A + c_3A^2(dA) + c_4A^4\Big]\Big\}\,. \tag{22.6.17}$$

In order to implement the consistency condition (22.6.10), we note that the BRST transformation rules (22.6.7) and (22.6.8) may be written as

$$sA = -d\omega + \{A, \omega\}\,, \tag{22.6.18}$$

$$s\omega = \omega^2\,. \tag{22.6.19}$$

Now, the BRST transformation of the last term in Eq. (22.6.17) is given by

$$\begin{aligned} s\mathrm{Tr}\left[\omega A^4\right] &= \mathrm{Tr}\Big[\omega^2 A^4 - \omega\{A,\omega\}A^3 + \omega A\{A,\omega\}A^2 \\ &\quad -\omega A^2\{A,\omega\}A + \omega A^3\{A,\omega\}\Big] + \omega d\omega A^3 \text{ terms} \\ &= \mathrm{Tr}\left[\omega^2 A^4\right] + \omega d\omega A^3 \text{ terms}\,. \end{aligned}$$

There is no other contribution to sG proportional to $\mathrm{Tr}\left[\omega^2 A^4\right]$, so the consistency condition (22.6.10) can only be satisfied if $c_4 = 0$. With $c_4 = 0$, a straightforward calculation gives

$$\begin{aligned} sG = \frac{1}{24\pi^2}\int \mathrm{Tr}\Big\{ &-(dA)^2\omega^2 + \omega\, d\omega\, A\, dA - d\omega\,\omega\, dA\, A \\ &-A\omega\, dA\, d\omega - \omega A\, d\omega\, dA \\ &+c_1\Big[\omega\, dA\, d\omega\, A - \omega\, dA\, A\, d\omega\Big] \\ &+c_2\Big[\omega\, d\omega\, dA\, A - \omega A\, dA\, d\omega\Big] \\ &+c_3\Big[\omega\, d\omega\, A\, dA - \omega A\, d\omega\, dA\Big] \\ &-c_1\Big[-\omega A\, d\omega\, A^2 + \omega\, d\omega\, A^3 + \omega\, dA\, A^2\omega\Big] \\ &-c_2\Big[\omega A^2\, d\omega\, A - \omega A\, d\omega\, A^2 + \omega A\, dA\, d\omega\Big] \\ &-c_3\Big[-\omega A^3\, d\omega + \omega A^2\, d\omega\, A + \omega A^2\, dA\,\omega\Big]\Big\}\,. \end{aligned}$$

We do not need to assume that the integrand vanishes, but only that it is the derivative of some local function, so that its integral vanishes. This condition must be satisfied separately for the terms involving two derivatives and those involving one derivative, since no cancellation can occur between terms with different numbers of derivatives. It is not hard to see that the terms in the integrand involving just one derivative are of the form $d\mathscr{F}$ if we take $c_1 = -c_2 = +c_3 \equiv c$. The remaining terms are a total derivative if $c = -1/2$, thus justifying the previously quoted result

(22.3.38). This result is often expressed more compactly as

$$G[\omega, A] = \frac{1}{24\pi^2}\int \mathrm{Tr}\left\{\omega\, d\left[A\,dA - \frac{1}{2}A^3\right]\right\}$$
$$= \frac{1}{24\pi^2}\int \mathrm{Tr}\left\{\omega\, d\left[AF + \frac{1}{2}A^3\right]\right\}, \tag{22.6.20}$$

where F is the matrix-valued field-strength two-form

$$F \equiv \tfrac{1}{2} i t_\alpha F_{\alpha\mu\nu}\, dx^\mu\, dx^\nu = dA - A^2 . \tag{22.6.21}$$

The anomaly does not have to be put in the form (22.6.13), so Eq. (22.6.20) is not the unique result for $G[\omega, A]$. Results quoted in the next section show that, for any subgroup H of G for which $\mathrm{Tr}\left\{t_i\{t_j, t_k\}\right\} = 0$ for all generators of H, it is possible to add local terms to the action in such a way that the anomaly G_i vanishes when t_i is any generator of H.

There is an elegant algebraic tool, known as the *Stora–Zumino descent equations,*[18a] for constructing a solution of the consistency conditions. It is just as easy to describe this method in a spacetime of any even dimensionality as in four spacetime dimensions, so we will take the spacetime dimensionality to be $2n$. To start, one must imagine that at least two additional variables have been added to the $2n$ coordinates of space and time, in order to give meaning to the $(2n+2)$-form $\mathrm{Tr}\, F^{n+1}$. Note that

$$dF = -d(A^2) = -(dA)A + A(dA) = [A, F]\,, \tag{22.6.22}$$

so that $\mathrm{Tr}\, F^{n+1}$ is closed:

$$d\mathrm{Tr}\, F^{n+1} = (n+1)\mathrm{Tr}\left\{(dF)F^n\right\} = \mathrm{Tr}\left\{[A, F^{n+1}]\right\} = 0\,. \tag{22.6.23}$$

As long as the extended spacetime is simply connected, Poincaré's theorem then tells us that $\mathrm{Tr}\left\{F^{n+1}\right\}$ is exact, in the sense that there is a $(2n+1)$-form Ω_{2n+1} (known as the *Chern–Simons form*), for which

$$\mathrm{Tr}\left\{F^{n+1}\right\} = d\,\Omega_{2n+1}\,. \tag{22.6.24}$$

Further, $\mathrm{Tr}\left\{F^{n+1}\right\}$ is manifestly gauge-invariant and depends only on the gauge field, so it is BRST-invariant:

$$s\mathrm{Tr}\left\{F^{n+1}\right\} = 0\,. \tag{22.6.25}$$

The 'differentials' dx^μ are understood to anticommute with fermionic fields like the ghost field ω_α, so the operator d anticommutes with the operator s defined by Eqs. (22.6.7) and (22.6.8):

$$sd + ds = 0\,.$$

Because s is nilpotent, it follows then that $s\,\Omega_{2n+1}$ is also closed:

$$d(s\,\Omega_{2n+1}) = -s\mathrm{Tr}\left\{F^{n+1}\right\} = 0\,.$$

Again using Poincaré's theorem, this means that there must be a $2n$-form Ω^1_{2n}, of first order in ω_α, for which

$$s\,\Omega_{2n+1} = d\Omega^1_{2n}\,. \qquad (22.6.26)$$

Furthermore, $d(s\,\Omega^1_{2n}) = -s^2\Omega_{2n+1} = 0$, so there is also a $(2n-1)$-form Ω^2_{2n-1}, of second order in the ghost field, for which

$$s\,\Omega^1_{2n} = d\Omega^2_{2n-1}\,. \qquad (22.6.27)$$

It follows then that the integral of Ω^1_{2n} over the $2n$ dimensions of space and time is BRST-invariant:

$$s\int_{\text{spacetime}} \Omega^1_{2n} = 0\,, \qquad (22.6.28)$$

even though Ω^1_{2n} is not itself BRST-invariant. We can thus find a candidate $\int \Omega^1_{2n}$ for the anomaly functional $G[\omega, A]$ by integrating the two first-order differential equations $d\Omega_{2n+1} = \text{Tr}\{F^{n+1}\}$ and $d\Omega^1_{2n} = s\,\Omega_{2n+1}$. General (not unique) solutions of these equations are

$$\Omega_{2n+1} = (n+1)\int_0^1 dt\, \text{Tr}\left\{A\,F_t^n\right\}, \qquad (22.6.29)$$

$$\Omega^1_{2n} = -(n+1)\sum_{r=0}^{n-1}\int_0^1 dt\,(1-t)\,\text{Tr}\left\{\omega\, d(F_t^r\, A\, F_t^{n-1-r})\right\}, \qquad (22.6.30)$$

where $F_t \equiv tF + (t-t^2)A^2$. Evaluation of the integral (22.6.30) shows that Eq. (22.6.20) gives a result for $G[\omega, A]$ proportional to $\int \Omega^1_4$ in the case of four spacetime dimensions.

We can continue this descent, and derive other useful results. In particular, it follows from Eq. (22.6.27) and the nilpotence of s that $d(s\Omega^2_{2n-1}) = 0$, so Poincaré's theorem tells us that $s\Omega^2_{2n-1}$ is of the form $d\Omega^3_{2n-2}$, so the integral of Ω^2_{2n-1} over the $2n-1$ coordinates of *space* is BRST-invariant:

$$s\int_{\text{space}} \Omega^2_{2n-1} = 0\,. \qquad (22.6.31)$$

Such BRST-invariant functionals of second order in the ghost fields are candidates[18b] for so-called *Schwinger terms.*[18c] Schwinger terms of the sort that concern us here arise as anomalous terms $S_{\alpha,\beta}(\mathbf{x},\mathbf{y})$ in the equal-time commutation relations of the time components of two symmetry currents:

$$\left[J^0_\alpha(\mathbf{x},t),\, J^0_\beta(\mathbf{y},t)\right] = iC_{\alpha\beta\gamma}J^0_\gamma(\mathbf{x},t)\delta^{2n-1}(\mathbf{x}-\mathbf{y}) + S_{\alpha\beta}(\mathbf{x},\mathbf{y},t)\,. \qquad (22.6.32)$$

(All operators in this paragraph are taken at the same time t, which will henceforth not be shown explicitly.) From the antisymmetry of the commutator we have $S_{\alpha\beta}(\mathbf{x},\mathbf{y}) = -S_{\beta\alpha}(\mathbf{y},\mathbf{x})$, so all information about

$S_{\alpha\beta}(\mathbf{x},\mathbf{y})$ is contained in the functional

$$S[\omega] \equiv \int d^{2n-1}x\, d^{2n-1}y\; \omega_\alpha(\mathbf{x})\, \omega_\beta(\mathbf{y})\, S_{\alpha\beta}(\mathbf{x},\mathbf{y})\,. \qquad (22.6.33)$$

Note that $S_{\alpha\beta}(\mathbf{x},\mathbf{y})$ depends in general on various matter and gauge fields, so $S[\omega]$ generally depends on these fields as well as on the ghost field $\omega_\alpha(\mathbf{x})$. Taking the commutator of Eq. (22.6.32) with a third current $J^0_\gamma(\mathbf{z})$, contracting with $\omega_\alpha(\mathbf{x})\, \omega_\beta(\mathbf{y})\, \omega_\alpha(\mathbf{z})$, integrating over $\mathbf{x}$, $\mathbf{y}$, and $\mathbf{z}$, and using the Jacobi identities, we find

$$0 = \int d^{2n-1}x\, d^{2n-1}y\, d^{2n-1}z\; \omega_\alpha(\mathbf{x})\, \omega_\beta(\mathbf{y})\, \omega_\gamma(\mathbf{z})$$
$$\times \Big[iC_{\alpha\beta\delta}\, S_{\gamma\delta}(\mathbf{z},\mathbf{x})\delta^{2n-1}(\mathbf{x}-\mathbf{y}) + [J^0_\gamma(\mathbf{z}),\, S_{\alpha\beta}(\mathbf{x},\mathbf{y})] \Big]\,.$$

On functionals of gauge and matter fields like $S_{\alpha\beta}(\mathbf{x},\mathbf{y})$, the action of the BRST operator s is the same as that of a gauge transformation with transformation parameter ω_α, so

$$s\, S_{\alpha\beta}(\mathbf{x},\mathbf{y}) = i\Big[\int d^{2n-1}z\; \omega_\gamma(\mathbf{z})\, J^0_\gamma(\mathbf{z}),\, S_{\alpha\beta}(\mathbf{x},\mathbf{y}) \Big]\,.$$

Recalling Eq. (22.6.8), we find that the functional (22.6.33) is BRST-invariant

$$s\, S[\omega] = 0\,. \qquad (22.6.34)$$

Also, by adding terms to the currents we can change $S[\omega]$ by terms of the form $sT[\omega]$, so the set of possible Schwinger terms that could *not* be removed by adding terms to the currents is given by the cohomology of the BRST operator s at ghost number two — that is, by BRST-invariant functionals S of second order in the ghost field that are not themselves of the form sT. Eq. (22.6.31) shows that $\int \Omega^2_{2n-1}$ is a candidate for such a functional.

* * *

The analysis of anomalies given so far in this section is strictly applicable only to anomalies in one-loop order. It is true that a theory with one-loop anomalies in currents to which quantum gauge fields are coupled is inconsistent, and therefore does not need to be studied in higher orders. But the converse does not hold; if a theory with quantum gauge fields is anomaly-free in one-loop order, we still need to show that anomalies are absent in higher orders. Also, there is nothing inconsistent in theories with anomalies in global symmetries, such as the chiral symmetries of quantum chromodynamics, and for these we need to know whether higher-order corrections affect the anomalies.

Since BRST transformations act non-linearly on the fields, even in the absence of anomalies we would not necessarily expect the quantum effective action $\Gamma[\omega, A]$ to be BRST-invariant beyond the one-loop approximation. As we saw in Section 17.1, beyond this approximation we need to consider functionals not only of gauge and ghost fields but also of their antifields. (The introduction of antifields is sometimes important even in one-loop order for another reason: a local functional of fields alone that satisfies the Wess–Zumino consistency conditions, and that is not expressible as the BRST operator acting on a local functional of fields alone, will not be a candidate anomaly if it *can* be expressed as the antibracket of the action with a local functional of fields and antifields, because in this case the anomaly can be cancelled by subtracting this term from the action. This corresponds to a change in the action of the fields combined with a change in the gauge symmetry obeyed by this action.)

The study of anomalies with antifields included in the action turns out also to have a cohomological formulation.[20,21] The analysis of this problem is based on the Zinn-Justin version (17.1.10) of what Batalin and Vilkovisky called the master equation. In the absence of anomalies $(\Gamma, \Gamma) = 0$, so that in one-loop order $(S, \Gamma_1) = 0$, where S is the zeroth-order action and Γ_1 is the one-loop contribution to the quantum effective action. In the presence of anomalies we have instead

$$(S, \Gamma_1) = G_1 \, , \tag{22.6.35}$$

where G_1 is some functional of fields and antifields which, since S and Γ_1 have ghost number zero, must have ghost number unity. The action is assumed to satisfy the classical master equation $(S, S) = 0$, so the antibracket operation in Eq. (22.6.35) is nilpotent, and therefore $(S, G_1) = 0$. But if $G_1 = (S, F_1)$ for some local functional F_1 of ghost number zero then we can cancel the anomaly to one-loop order by subtracting the term F_1 (which is treated as a quantum correction of order $\hbar$) from the action and hence from Γ_1. (Of course, $G_1 = (S, \Gamma_1)$, but in the presence of massless particles Γ_1 is not a local functional.) Thus the candidate anomalies are those local functionals G_1 of ghost number unity that are closed, in the sense that they satisfy $(S, G_1) = 0$, but that are not exact, that is, that cannot be expressed as $G_1 = (S, F_1)$ for local functionals F_1 of ghost number unity. In other words, the candidate anomalies correspond to the cohomology of the antibracket operation $X \mapsto (S, X)$ at ghost number unity on the space of local functionals of the fields and antifields.

This is just like the result found earlier in this section, but with the antibracket $(S, \cdots)$ replacing the BRST operator s. If we set the antifields equal to zero in Eq. (22.6.35) and recall that $(\delta\Gamma_1/\delta\chi^n)_{\chi^\ddagger=0} = s\chi^n$, we find that the condition (22.6.35) in fact yields the condition $s\Gamma_1 = 0$, which as we have seen is equivalent to the requirement that Γ_1 satisfies the Wess–Zumino consistency conditions. But there is a sense in which the analysis based on Eq. (22.6.35) can be extended to higher orders.

To see this, suppose that it is found that the antibracket operation $X \mapsto (S, X)$ has an empty cohomology at ghost number unity in the space of local functionals, and that we redefine the action as described above to make $G_1 = (S, \Gamma_1) = 0$. An anomaly that violates the master equation $(\Gamma, \Gamma) = 0$ in two-loop order is represented by a function G_2 for which

$$\left(\Gamma_1, \Gamma_1\right) + 2\left(S, \Gamma_2\right) = G_2 \,.$$

But since $(S, \Gamma_1) = 0$ and $(S, S) = 0$, any such G_2 would satisfy $(S, G_2) = 0$. On the assumption of an empty cohomology, this means that it can be expressed as $G_2 = (S, F_2)$ with F_2 a local functional of ghost number zero, so in this order the anomaly can be cancelled by subtracting F_2 from the action.

This argument can be extended to all orders. Suppose we have cancelled the anomalies in the master equation up to order $N - 1$, so that

$$0 = G_M = \sum_{L=0}^{M} \left(\Gamma_L, \Gamma_{M-L}\right) = 2\left(S, \Gamma_M\right) + \sum_{L=1}^{M-1} \left(\Gamma_L, \Gamma_{M-L}\right) ,$$

for all $M < N$. The Nth-order term in the antibracket (Γ, Γ) is likewise

$$G_N = 2\left(S, \Gamma_N\right) + \sum_{M=1}^{N-1} \left(\Gamma_M, \Gamma_{N-M}\right) ,$$

so, using the Jacobi identity (15.9.21) (in which for three bosonic operators all signs are $-$) and the above formula for (S, Γ_M), we find:

$$\left(S, G_N\right) = -2 \sum_{M=1}^{N-1} \left(\left(S, \Gamma_M\right), \Gamma_{N-M}\right) = \sum_{M=1}^{N-1} \sum_{L=1}^{M-1} \left(\left(\Gamma_L, \Gamma_{M-L}\right), \Gamma_{N-M}\right) .$$

This can be written in a more symmetric way

$$\left(S, G_N\right) = -\sum_{M_1=1}^{N-2} \sum_{M_2=1}^{N-2} \sum_{M_3=1}^{N-2} \delta_{N, M_1+M_2+M_3} \left(\Gamma_{M_1}, \left(\Gamma_{M_2}, \Gamma_{M_3}\right)\right) .$$

Since the ranges of M_1, M_2, and M_3 are the same, we can write the double antibracket in this sum as a sum over the 3! permutations over these indices, which vanishes according to the Jacobi identity (15.9.21), leading to the conclusion that $(S, G_N) = 0$. If as assumed the cohomology is empty then this implies that there is a local functional F_N for which $G_N = (S, F_N)$, so by subtracting F_N from the action the anomaly can be cancelled in order N, as was to be shown.

Using purely algebraic methods Barnich, Brandt, and Henneaux[22] have succeeded in showing that for Yang–Mills theories (in four spacetime dimensions) based on semisimple gauge groups, the cohomology of the antibracket operation $X \mapsto (S, X)$ (at ghost number unity on the space of local functionals) consists entirely of a linear combination of terms of the form (22.6.20), one for each simple subgroup of the gauge group, with unknown coefficients.†† This shows without any reference to the matter content of the theory that in one-loop order, where the anomaly G_1 automatically satisfies $(S, G_1) = 0$, the anomaly for a semi-simple gauge group must be a linear combination of terms of the form (22.6.20), with only the constant coefficient for each simple subgroup left to be determined by detailed calculations that take into account the matter content of the theory. Further, we saw in Section 22.4 that there are gauge groups for which the trace in Eq. (22.6.20) vanishes automatically for any menu of fermion fields. (These are the semisimple gauge groups with no $SU(n)$ factors with $n \geq 3$.) In such cases the theorem of Ref. 22 shows that the cohomology of the antibracket operation $X \mapsto (S, X)$ at ghost number unity is zero. As we have seen, this means that in such theories there is no anomaly in any order of perturbation theory.

There is another sense in which the anomaly is related to the cohomology of the antibracket operation.[23] In deriving the Slavnov–Taylor identity (16.4.6), we assumed that the measure $\prod_{n,x} d\chi^n(x)$ is invariant under the symmetry transformation in question. In Section 15.9 the Zinn-Justin equation was derived from the Slavnov–Taylor identity for the symmetry transformation $\chi^n \to \chi^n + \theta\delta S/\delta\chi_n^\ddagger$, so the derivation of the Zinn-Justin equation given in Section 15.9 breaks down unless $\prod_{n,x} d\chi^n(x)$ is invariant under this transformation, or in other words unless $\Delta S = 0$, where Δ is the operator (15.9.34). Where $\Delta S \neq 0$, it still may be possible to save the Zinn-Justin equation by adding local functionals to S that violate the classical master equation $(S, S) = 0$ in such a way as to cancel the effect of the non-invariance of the measure. It turns out that the condition for this cancellation is nothing but the quantum master equation (15.9.35). To construct an action S that satisfies this equation, we start with a zeroth-order action S_0 that satisfies the classical master equation, $(S_0, S_0) = 0$, and add quantum corrections. In the case in which the cohomology of the operation $X \to (S_0, X)$ (at ghost number unity in the space of local functionals) is empty, the same proof that was used above to show the absence in this case of anomalies to all orders can be used to show that a local functional can be added to S_0 so that the quantum master equation is satisfied to all orders.

†† It is not necessary to specify the representation of the gauge algebra in which the trace in Eq. (22.6.20) is to be calculated, because this trace is the same up to a constant coefficient for all representations of a simple Lie algebra,[19a] and the constant coefficient is not determined anyway by this cohomology theorem.

22.7 Anomalies and Goldstone Bosons

In the same 1971 paper in which they introduced the consistency conditions, Wess and Zumino[18] also noted that the possibility of anomalies has important consequences for the interactions of Goldstone bosons. To understand their point, it is helpful to apply the 'anomaly matching' argument developed by 't Hooft[16] in 1979, which has already been used in Section 22.5.

Consider a broken global symmetry group G that is realized linearly in some underlying theory of trapped massless fermions, as for instance the global chiral $SU(3) \times SU(3)$ symmetry in quantum chromodynamics with three massless quarks. In this underlying theory introduce fictitious gauge fields, so that aside from possible anomalies the global symmetry G becomes local. In general this local symmetry *will* be broken by anomalies, since in the real world the symmetry is purely global, and there is no reason why a global symmetry should admit an extension to an anomaly-free local symmetry. However these anomalies can be cancelled by adding suitable massless spectator fermions. As long as the gauge couplings introduced in this way are sufficiently small, and the spectator fermions have only these very weak gauge interactions, the dynamics of the theory will not be substantially changed by these modifications.

Next consider the effective field theory that describes physics at low energy, where the trapped fermions are unobservable. The only degrees of freedom in this theory will be the massless particles: the fictitious gauge bosons and spectator fermions, and a set of Goldstone bosons with fields ξ_a, one for each independent broken symmetry. Since the underlying theory had been made gauge-invariant and anomaly-free, the same must be true of the effective field theory. But the spectator fermions produce an anomaly which had previously cancelled the anomaly due to the trapped massless fermions in the underlying theory, so in order for them to cancel the anomaly due to the Goldstone bosons, *the gauged effective field theory of the Goldstone bosons must have an anomaly for the fictitious local symmetries which is equal to that produced by the trapped fermions in the underlying theory.* That is, in place of Eq. (22.6.2), the effective action $\Gamma[\xi, A]$ of the fictitious gauge fields and Goldstone bosons is subject to the condition

$$\mathscr{T}_\beta(x)\Gamma[\xi, A] = G_\beta[x; A] , \tag{22.7.1}$$

where $G_\beta[x; A]$ is the anomaly function of the underlying theory, in which there are no Goldstone bosons, and $\mathscr{T}_\beta$ is a generator of the gauge group G, now acting on both gauge and Goldstone boson fields. (The index β on $\mathscr{T}_\beta$ runs over values i labelling a complete set of independent generators $\mathscr{Y}_i$ of the unbroken symmetry subgroup H, together with values

a labelling a set of independent broken symmetry generators $\mathscr{X}_a$; there is one Goldstone boson field ξ_a for each $\mathscr{X}_a$.)

Of course, Eq. (22.7.1) may also be used to study the interactions of Goldstone bosons with real weakly coupled gauge fields. For instance, where the underlying theory includes electroweak gauge bosons coupled to the quarks, we would identify some of what we have called 'fictitious' gauge fields with the electroweak gauge fields. In such cases some of the 'spectator' fermions must also be real, in order to cancel the anomalies produced by loops of trapped fermions in the gauge symmetries of the real weakly coupled gauge fields, as for instance the leptons cancel the electroweak anomalies produced by the quarks.

Now let us turn to the implications of Eq. (22.7.1). To calculate the gauge symmetry generator $\mathscr{T}_\beta(x)$ in this context, note that under a general group transformation $g = \exp(-i \int \zeta_\beta(x)\mathscr{T}_\beta(x)\, d^4x\,)$, the Goldstone boson fields $\xi_a(x)$ transform into fields $\xi'_a(x)$ given by Eq. (19.6.18), and the gauge fields $A^\mu_\beta(x)$ transform into gauge-transformed fields $A'^\mu_\beta(x)$, so that

$$\mathscr{T}_\beta(x) = \mathscr{T}^A_\beta(x) + \mathscr{T}^\xi_\beta(x)\,. \tag{22.7.2}$$

Here $\mathscr{T}^A_\beta(x)$ acts on the gauge fields and is given by Eq. (22.6.1):

$$-\,i\mathscr{T}^A_\beta(x) = -\frac{\partial}{\partial x^\mu}\frac{\delta}{\delta A_{\beta\mu}(x)} - C_{\beta\gamma\alpha}A_{\gamma\mu}(x)\frac{\delta}{\delta A_{\alpha\mu}(x)}\,, \tag{22.7.3}$$

where $C_{\alpha\beta\gamma}$ are the totally antisymmetric structure constants of the gauge group G, while $\mathscr{T}^\xi_\beta(x)$ acts on the Goldstone fields and is given by the infinitesimal limit of Eq. (19.6.17), which (for the exponential parameterization $\gamma(\xi) = \exp(i\xi_a X_a)$) reads:*

$$T_\beta \exp\left(i\xi_a(x)X_a\right) = -\mathscr{T}^\xi_\beta(x)\exp\left(i\xi_a(x)X_a\right) + \exp\left(i\xi_a(x)X_a\right)\theta_{\beta i}(x)Y_i\,. \tag{22.7.4}$$

Here T_β are matrices representing the generators of G in any representation; they are divided into sets X_a and Y_i, which represent the broken and unbroken symmetry generators, respectively. Also the $\theta_{\beta i}(x)$ are ξ-dependent functions whose form will not concern us here.

About the anomaly functions $G_\beta[x;A]$, we will assume only the consistency conditions (22.6.6)

$$\mathscr{T}_\alpha(x)G_\beta[y;A] - \mathscr{T}_\beta(y)G_\alpha[x;A] = iC_{\alpha\beta\gamma}\delta^4(x-y)G_\gamma[y;A] \tag{22.7.5}$$

* The minus sign in the first term on the right-hand side of Eq. (22.7.4) and the factor $-i$ on the left-hand side of Eq. (22.7.3) appear because it is $\exp[-i\int \Lambda_\beta(x)\mathscr{T}_\beta(x)]$ that induces a gauge transformation (15.1.17) with gauge parameter Λ_β, as can be seen by requiring the $\mathscr{T}_\beta(x)$ to satisfy the commutation relations (22.6.5).

and the absence of anomalies in the unbroken symmetries

$$G_i[x;A] = 0\,. \tag{22.7.6}$$

As mentioned in the previous section, as long as the trace $\mathrm{Tr}\,[T_i\{T_j,T_k\}]$ vanishes for the generators of unbroken symmetry subgroup (as it does for the non-chiral generators of $SU(3)\times SU(3)$), it is always possible to add a local functional to the action so that Eq. (22.7.6) is satisfied.

Under the assumptions (22.7.5) and (22.7.6), it is always possible to find a solution of the anomalous Slavnov–Taylor identities (22.7.1):

$$\Gamma[\xi,A] = -i\int_0^1 dt \int \xi_b(y)\, G_b[y;A_{-t\xi}]\, d^4y\,, \tag{22.7.7}$$

where $[A_{-t\xi}(x)]_\mu$ is the result of acting on $A_\mu \equiv T_\beta A_{\mu\beta}$ with a gauge transformation (15.1.17) having $\Lambda_a = -t\xi_a$ and $\Lambda_i = 0$:

$$\begin{aligned}[A_{-t\xi}(x)]_\mu = {}& \exp(-itX_a\xi_a(x))\, A_\mu(x) \exp(itX_a\xi_a(x)) \\ & -i\Big[\partial_\mu \exp(-itX_a\xi_a(x))\Big] \exp(itX_a\xi_a(x))\,.\end{aligned} \tag{22.7.8}$$

In contrast with the case in which unbroken symmetries have anomalies, Eqs. (22.7.7) and (22.7.8) define a *local* (though complicated) functional of gauge and Goldstone boson fields. Any other solution of Eq. (22.7.1) will differ from this one by an anomaly-free functional.

Here is an outline of the proof[24] that the action (22.7.7) satisfies Eq. (22.7.1). Instead of working with the local generator $\mathscr{T}_\beta(x)$, it is convenient to introduce an arbitrary function $\eta_\beta(x)$, and define

$$\mathscr{T}[\eta] = \int d^4x\, \eta_\beta(x)\mathscr{T}_\beta(x)\,. \tag{22.7.9}$$

To evaluate $\mathscr{T}[\eta]\xi_b(x)$, we introduce the matrix

$$\eta_{-t\xi}(x) \equiv \exp\Big(-iX_a\xi_a(x)t\Big)\Big[\eta(x)+\mathscr{T}[\eta]\Big]\exp\Big(iX_a\xi_a(x)t\Big) \equiv [\eta_{-t\xi}(x)]_\beta T_\beta\,, \tag{22.7.10}$$

where $\eta(x) \equiv \eta_\beta(x)T_\beta$. Then

$$\frac{\partial}{\partial t}\eta_{-t\xi}(x) = -i\Big[X_a\xi_a(x), \eta_{-t\xi}(x)\Big] + i\Big(\mathscr{T}[\eta]\xi_a(x)\Big)X_a\,,$$

so that

$$\mathscr{T}[\eta]\xi_b(x) = -i\frac{\partial}{\partial t}[\eta_{-t\xi}(x)]_b + iC_{a\gamma b}\,\xi_a(x)\,[\eta_{-t\xi}(x)]_\gamma\,. \tag{22.7.11}$$

To evaluate $\mathscr{T}[\eta]G_b[y,A]$, we apply $\mathscr{T}[\eta]$ to the gauge field, and after a straightforward calculation find that

$$\mathscr{T}[\eta][A_{-t\xi}(x)]_\mu = \Big(\mathscr{T}^A[\eta_{-t\xi}]A_\mu(x)\Big)_{A\to A_{-t\xi}}\,, \tag{22.7.12}$$

so, using the consistency condition (22.7.5),

$$\begin{aligned}\mathscr{T}[\eta]G_b[y;A_{-t\xi}] &= \int d^4x\,[\eta_{-t\xi}(x)]_\gamma\left(\mathscr{T}^A_\gamma(x)\,G_b[y;A]\right)_{A\to A_{-t\xi}}\\ &= \int d^4x\,[\eta_{-t\xi}(x)]_a\left(\mathscr{T}^A_b(y)\,G_a[x;A]\right)_{A\to A_{-t\xi}}\\ &\quad +iC_{\gamma ba}\,[\eta_{-t\xi}(y)]_\gamma G_a[y;A_{-t\xi}]\,. \qquad (22.7.13)\end{aligned}$$

The structure constant terms in Eqs. (22.7.11) and (22.7.13) cancel, yielding

$$\begin{aligned}\mathscr{T}[\eta]\Gamma[\xi,A] = \int_0^1 dt\int d^4y\,\Bigg\{&-\left[\frac{\partial}{\partial t}[\eta_{-t\xi}(y)]_b\right]G_b[y;A_{-t\xi}]\\ &-i[\eta_{-t\xi}(y)]_a\left(\mathscr{T}[\xi]\,G_a[y;A]\right)_{A\to A_{-t\xi}}\Bigg\}\,. \qquad (22.7.14)\end{aligned}$$

Another straightforward calculation shows that

$$\frac{\partial}{\partial t}[A_{-t\xi}(x)]_\mu = i\left(\mathscr{T}^A(\xi)A_\mu(x)\right)_{A\to A_{-t\xi}}\,, \qquad (22.7.15)$$

so the terms in the integrand in Eq. (22.7.14) add up to a t-derivative

$$\begin{aligned}\mathscr{T}[\eta]\Gamma[\xi,A] &= -\int_0^1 dt\int d^4y\,\frac{\partial}{\partial t}\Big\{[\eta_{-t\xi}(y)]_b\,G_b[y;A_{-t\xi}]\Big\}\\ &= -\int d^4y\Big[[\eta_{-t\xi}(y)]_b\,G_b[y;A_{-t\xi}]\Big]_{t=0}^{t=1}\,. \qquad (22.7.16)\end{aligned}$$

At $t = 1$ we have

$$\eta_{-\xi}(x) = \exp\Big(-iX_a\xi_a(x)\Big)\Big[\eta(x)+\mathscr{T}[\eta]\Big]\exp\Big(iX_a\xi_a(x)\Big)\,.$$

From Eq. (22.7.4) we see that this is a linear combination of the generators of the unbroken symmetry subgroup H, so the coefficient $[\eta_{-\xi}(y)]_b$ of any broken symmetry generator X_b vanishes. Also, Eqs. (22.7.10) and (22.7.8) show immediately that at $t = 0$, $\eta_{-t\xi}(y) = \eta(y)$ and $[A_{-t\xi}(y)]_\mu = A_\mu(y)$, so Eq. (22.7.16) yields

$$\mathscr{T}[\eta]\Gamma[\xi,A] = \int d^4y\,[\eta_0(y)]_b\,G_b[y;A_0] = \int d^4y\,\eta_b(y)\,G_b[y;A]\,, \quad (22.7.17)$$

which with Eq. (22.7.6) is equivalent to the desired result (22.7.1), as was to be proved.

The solution (22.7.7) of Eq. (22.7.1) is not unique, but it *is* the unique solution that vanishes for $\xi = 0$. To see this, note that

$$\exp\left[-i\int \eta_\beta(x)\,\mathscr{T}_\beta(x)\,d^4x\right]\Gamma[\xi, A] = \Gamma[\xi', A'] , \tag{22.7.18}$$

where the primes indicate a gauge transformation with transformation parameter η_β. It is convenient to represent the exponential here as

$$\exp(z) = 1 + \int_0^1 dt\,\exp(zt)\,z ,$$

so Eqs. (22.7.1) and (22.7.18) yield

$$\Gamma[\xi, A] - i\int_0^1 dt\,\exp\left[-it\int \eta_\beta(x)\mathscr{T}^A_\beta(x)\,d^4x\right]\int \eta_b(y)G_b[y;A]d^4y$$
$$= \Gamma[\xi', A'] . \tag{22.7.19}$$

In particular, if we take $\eta_a = -\xi_a$ and $\eta_i = 0$, then $\xi'_a = 0$, in which case by assumption Eq. (22.7.19) has a vanishing right-hand side, and therefore yields the formula:

$$\Gamma[\xi, A] = -i\int_0^1 dt\,\exp\left[it\int \xi_a(x)\mathscr{T}^A_a(x)\,d^4x\right]\int \xi_b(y)G_b[y;A]\,d^4y . \tag{22.7.20}$$

The functional operator $\exp[it\int \xi_a(x)\mathscr{T}^A_a(x)\,d^4x]$ in Eq. (22.7.20) simply produces a gauge transformation (15.1.17) with gauge parameter $\Lambda_\beta(x) = -t\xi_a(x)$, so that Eq. (22.7.20) may be written as in Eq. (22.7.7).

Eq. (22.7.7) may be applied to study the electroweak interactions of the octet of pseudoscalar Goldstone bosons, but it has important implications for the interactions of the Goldstone bosons themselves, in the absence of real gauge fields. In the case where $A = 0$, Eq. (22.7.8) becomes a 'pure gauge' field

$$[A_{-t\xi}(x)]_\mu = -i\Big[\partial_\mu \exp\left(-itX_a\xi_a(x)\right)\Big]\exp\left(itX_a\xi_a(x)\right)$$
$$= -i\Big[\partial_\mu V\Big(t\xi(x)\Big)\Big]\,V^{-1}\Big(t\xi(x)\Big) , \tag{22.7.21}$$

where

$$V\Big(t\xi(x)\Big) \equiv \exp\Big(-itX_a\xi_a(x)\Big) . \tag{22.7.22}$$

With $A^\mu_\alpha(x) = 0$, Eq. (22.7.1) gives

$$\mathscr{T}_\beta(x)\Gamma[\xi, 0] = 0 , \tag{22.7.23}$$

so the result of using Eq. (22.7.21) in Eq. (22.7.7) is a G-invariant local functional of the Goldstone boson field $\xi_a(x)$, though, as we shall see, it is not in general the integral over spacetime of a G-invariant function of $\xi_a(x)$ and its derivatives.

The simplest example is the case of a completely broken symmetry group. Here the condition (22.7.6) is empty, and we can use the symmetric form (22.3.38) for the anomaly:

$$G_a[x;A] = \frac{i}{24\pi^2}\epsilon^{\kappa\nu\lambda\rho}\,\mathrm{Tr}\Big\{T_a\Big[\partial_\kappa A_\nu(x)\,\partial_\lambda A_\rho(x) - \tfrac{1}{2}i\partial_\kappa A_\nu(x)\,A_\lambda(x)A_\rho(x)$$
$$+\tfrac{1}{2}iA_\kappa(x)\,\partial_\nu A_\lambda(x)\,A_\rho(x) - \tfrac{1}{2}iA_\kappa(x)A_\nu(x)\partial_\lambda A_\rho(x)\Big]\Big\}\,, \quad (22.7.24)$$

where now T_a is the specific representation of the group generator furnished by the left-handed fermions (including antifermions, where the distinction is relevant) of the theory. In this case when we use Eq. (22.7.21) in Eq. (22.7.24), we find that the only terms in the trace in (22.7.24) that survive when contracted with $\epsilon^{\kappa\nu\lambda\rho}$ are then all proportional to

$$\mathrm{Tr}\left\{T_a(\partial_\kappa V)V^{-1}(\partial_\nu V)V^{-1}(\partial_\lambda V)V^{-1}(\partial_\rho V)V^{-1}\right\}\,,$$

with coefficients -1, $+\frac{1}{2}$, $-\frac{1}{2}$ and $+\frac{1}{2}$, respectively, so Eq. (22.7.7) becomes here

$$\Gamma[\xi,0] = -\frac{1}{48\pi^2}\epsilon^{\kappa\nu\lambda\rho}\int d^4y\;\xi_a(y)\int_0^1 dt\;\mathrm{Tr}\Big\{T_a\left[\partial_\kappa V\big(t\xi(y)\big)\right]$$
$$\times V^{-1}\big(t\xi(y)\big)\left[\partial_\nu V\big(t\xi(y)\big)\right]V^{-1}\big(t\xi(y)\big)\left[\partial_\lambda V\big(t\xi(y)\big)\right]$$
$$\times V^{-1}\big(t\xi(y)\big)\left[\partial_\rho V\big(t\xi(y)\big)\right]V^{-1}\big(t\xi(y)\big)\Big\}\,. \quad (22.7.25)$$

As promised, this is not the integral of an invariant function of fields and field derivatives. For instance, for small Goldstone boson fields Eq. (22.7.25) becomes

$$\Gamma[\xi,0] = -\frac{1}{240\pi^2}\epsilon^{\kappa\nu\lambda\rho}\,\mathrm{Tr}\left\{T_aT_bT_cT_dT_e\right\}\int d^4y\;\xi_a\,\partial_\kappa\xi_b\,\partial_\nu\xi_c\,\partial_\lambda\xi_d\,\partial_\rho\xi_e$$
$$+O(\xi^6)\,. \quad (22.7.26)$$

Any function of the covariant derivatives of Goldstone boson fields would have a term of lowest order in the fields which would be simply a product of partial derivatives of the Goldstone fields, and therefore could not have a lowest-order term of the form (22.7.26).

For an example of greater practical importance, consider the $SU(3) \times SU(3)$ chiral symmetry of quantum chromodynamics with massless u, d, and s quarks, spontaneously broken to the diagonal $SU(3)$ subgroup of Gell-Mann and Ne'eman. In order to use the results of this section, we *must* label internal momenta in the fermion loop integral so that the vector currents of the diagonal $SU(3)$ subgroup are anomaly-free, in which case the anomaly takes the Bardeen form (22.3.34):

$$G_a[V,A] = \frac{in}{16\pi^2}\epsilon^{\mu\nu\rho\sigma}\mathrm{Tr}\left\{t_a\Big[V_{\mu\nu}V_{\rho\sigma} + \tfrac{1}{3}A_{\mu\nu}A_{\rho\sigma} - \tfrac{32}{3}A_\mu A_\nu A_\rho A_\sigma \right.$$
$$\left. + \tfrac{8}{3}i(A_\mu A_\nu V_{\rho\sigma} + A_\mu V_{\rho\sigma}A_\nu + V_{\rho\sigma}A_\mu A_\nu)\Big]\right\}, \tag{22.7.27}$$

where V_μ, A_μ, $V_{\mu\nu}$, and $A_{\mu\nu}$ are defined by Eqs. (22.3.35)–(22.3.37); t_a is here half the Gell-Mann matrix λ_a, given by Eq. (19.7.2); and n is the number of quark types ('colors') of each flavor. (We are now using a lower case t for the matrices representing the group generators, because in this trace we sum only over the left-handed quarks, not the left-handed antiquarks.) For a pure gauge field like (22.7.21) the field strengths $V_{\mu\nu}$ and $A_{\mu\nu}$ vanish, so that using Eq. (22.7.27) in Eq. (22.7.7) yields the anomalous effective Goldstone boson action

$$\Gamma[\xi,0] = -\frac{2n}{3\pi^2}\epsilon^{\mu\nu\rho\sigma}\int_0^1 dt \int d^4x\ \mathrm{Tr}\left\{\xi_a t_a[A_{-t\xi}]_\mu[A_{-t\xi}]_\nu[A_{-t\xi}]_\rho[A_{-t\xi}]_\sigma\right\}, \tag{22.7.28}$$

where 'A' now denotes the axial-vector as opposed to the vector gauge field. To find $[A_{-t\xi}]_\mu$ we use Eq. (22.7.21), which here reads

$$[V_{-t\xi}(x)]_\mu + \gamma_5[A_{-t\xi}(x)]_\mu = -i\Big[\partial_\mu \exp\Big(-it\gamma_5 t_a\xi_a(x)\Big)\Big]\exp\Big(it\gamma_5 t_a\xi_a(x)\Big). \tag{22.7.29}$$

Multiplying with $(1+\gamma_5)/2$ and $(1-\gamma_5)/2$ and taking the difference yields the axial-vector term

$$\begin{aligned}[A_{-t\xi}(x)]_\mu &= -\tfrac{1}{2}i\Big[\partial_\mu \exp\Big(-it\xi_a t_a\Big)\Big]\exp\Big(it\xi_a t_a\Big) \\ &\quad + \tfrac{1}{2}i\Big[\partial_\mu \exp\Big(it\xi_a t_a\Big)\Big]\exp\Big(-it\xi_a t_a\Big) \\ &= \tfrac{1}{2}i\exp\Big(it\xi_a t_a\Big)U^{-1}\Big(t\xi(x)\Big)\Big[\partial_\mu U\Big(t\xi(x)\Big)\Big]\exp\Big(-it\xi_a t_a\Big),\end{aligned} \tag{22.7.30}$$

where

$$U\Big(t\xi\Big) \equiv \exp\Big(2it_a\xi_a t\Big). \tag{22.7.31}$$

Using this in Eq. (22.7.28) yields the effective anomalous action

$$\begin{aligned}\Gamma[\xi,0] = -\frac{n}{24\pi^2}\epsilon^{\mu\nu\rho\sigma}\int_0^1 dt\int d^4x\ \mathrm{Tr}\Big\{&\xi_a t_a U^{-1}\Big(t\xi(x)\Big)\Big[\partial_\mu U\Big(t\xi(x)\Big)\Big] \\ &\times U^{-1}\Big(t\xi(x)\Big)\Big[\partial_\nu U\Big(t\xi(x)\Big)\Big]U^{-1}\Big(t\xi(x)\Big)\Big[\partial_\rho U\Big(t\xi(x)\Big)\Big]U^{-1}\Big(t\xi(x)\Big) \\ &\times\Big[\partial_\sigma U\Big(t\xi(x)\Big)\Big]\Big\}.\end{aligned} \tag{22.7.32}$$

As noted by Witten,[25] this may be expressed in a convenient five-dimensional form. We take t as our fifth coordinate, and define $\xi_a(x,t) \equiv t\xi_a(x)$. Then Eq. (22.7.32) becomes

$$\Gamma[\xi,0] = -\frac{in}{240\pi^2}\epsilon^{ijklm}\int d^5z\ \mathrm{Tr}\Big\{U^{-1}\big(\xi(z)\big)\big[\partial_i U\big(\xi(z)\big)\big]$$
$$\times U^{-1}\big(\xi(z)\big)\big[\partial_j U\big(\xi(z)\big)\big]U^{-1}\big(\xi(z)\big)\big[\partial_k U\big(\xi(z)\big)\big]U^{-1}\big(\xi(z)\big)$$
$$\times \big[\partial_l U\big(\xi(z)\big)\big]U^{-1}\big(\xi(z)\big)\big[\partial_m U\big(\xi(z)\big)\big]\Big\}\,, \tag{22.7.33}$$

where i, j, etc. run over the values 1, 2, 3, 0, 5, with $z^i = x^i$ for $i = 1$, 2, 3, 0 and $z^5 = t$, and the integral is over the region $0 \leq z^5 \leq 1$. (An extra factor $1/5$ appears in Eq. (22.7.33) to take account of the fact that any one of the five indices i, j, k, l, or m may take the value 5.) Since $\xi_a(z)$ takes a fixed value zero for $z^5 = 0$ and all values of the other components z^μ of z^i, we can identify these values of z^i as a single point, and consider the region of integration in Eq. (22.7.33) as a five-dimensional ball, with the four-dimensional boundary $z^5 = 1$ taken as ordinary spacetime. Eq. (22.7.33) is thus a special case of the 'Wess–Zumino–Witten' action given by Eqs. (19.8.1) and (19.8.3), which was also proportional to an integer n; the only difference is that n is now identified as the number of colors. We saw in Section 19.8 that the integral (19.8.3) depends only on the values of $\xi_a(z)$ on the spacetime boundary of the five-ball, so in deriving Eq. (22.7.33) we have shown that Eq. (19.8.2) applies for any continuation of $\xi_a(x)$ into the interior of the five-ball, not just for $\xi_a(x,t) = t\xi(x)$.

* * *

More generally, consider an arbitrary gauge group G that is spontaneously broken to a subgroup H that by itself is anomaly-free — that is, for which the D-symbol (22.3.12) vanishes for any three generators of H. Chu, Ho, and Zumino[26] have shown that we can add a local functional $B[A]$ to the action, in such a way that the currents of the subgroup H are anomaly free, even when the gauge fields of the broken symmetries are taken into account. This functional is

$$B[A] = \frac{1}{48\pi^2}\epsilon_{\mu\nu\rho\sigma}\int d^4x\ \mathrm{Tr}\Big\{[A_h^\mu, A^\nu](F^{\rho\sigma} + F_h^{\rho\sigma}) + A^\mu A_h^\nu A_h^\rho A_h^\sigma$$
$$-A_h^\mu A^\nu A^\rho A^\sigma + \frac{1}{2}A_h^\mu A^\nu A_h^\rho A^\sigma\Big\}\,, \tag{22.7.34}$$

where again $A^\mu \equiv T_\alpha A_\alpha^\mu$ and $F^{\mu\nu} \equiv T_\alpha F_\alpha^{\mu\nu}$, while $A_h^\mu \equiv T_i A_i^\mu$ and $F_h^{\mu\nu} \equiv T_i F_i^{\mu\nu}$ are the terms in A^μ and $F^{\mu\nu}$ in the algebra of the unbroken symmetry subgroup H. The total anomaly is then

$$G'_\beta[x;A] = G_\beta[x;A] + \mathcal{T}_\beta(x)B[A] \tag{22.7.35}$$

(where $G_\beta[x;A]$ is the symmetrized anomaly (22.3.38)) and satisfies the desired condition

$$G'_i[x;A] = 0\,. \tag{22.7.36}$$

The anomalous part of the effective action for Goldstone bosons and gauge fields is given by using Eq. (22.7.35) in place of G_β in Eq. (22.7.7). In this way, Chu *et al.*[26] found an anomalous effective action

$$\Gamma'[\xi, A] = \Gamma[\xi, A] - B[A_{-\xi}] + B[A]\,, \tag{22.7.37}$$

where $\Gamma[\xi, A]$ is the previously derived effective action (22.7.7), and $A_{-\xi}(x)$ is obtained by setting $t = 1$ in Eq. (22.7.8). In particular, where the gauge field vanishes $A_{-\xi}$ is the pure gauge field

$$[A_{-\xi}(x)]_\mu = -i\Big[\partial_\mu V\Big(\xi(x)\Big)\Big]V^{-1}\Big(\xi(x)\Big)\,,$$
$$V\Big(\xi(x)\Big) = \exp\Big(-iX_a\xi_a(x)\Big)\,,$$

so here the Goldstone boson action is

$$\Gamma'[\xi, 0] = \Gamma[\xi, 0] - \frac{1}{48\pi^2}\epsilon_{\mu\nu\rho\sigma}\int d^4x\, \mathrm{Tr}\Big\{A^\mu_{-\xi}\,A^\nu_{-\xi\,h}\,A^\rho_{-\xi\,h}\,A^\sigma_{-\xi\,h}$$
$$-A^\mu_{-\xi\,h}\,A^\nu_{-\xi}\,A^\rho_{-\xi}\,A^\sigma_{-\xi} + \frac{1}{2}A^\mu_{-\xi\,h}\,A^\nu_{-\xi}\,A^\rho_{-\xi\,h}\,A^\sigma_{-\xi}\Big\}\,. \tag{22.7.38}$$

This result is not unique; in particular, in parity-conserving theories like quantum chromodynamics we can add additional local terms to the effective action to cancel any parity-non-conserving terms in Eq. (22.7.38).

Problems

1. Calculate the rate of the decay process $\eta \to \gamma + \gamma$, to leading order in m_s, with $m_u = m_d = 0$.

2. Consider a chiral $SU(3)$ symmetry under which the left-handed parts of the spin $\frac{1}{2}$ fields of a fermion-number-conserving theory form N defining representations **3** of $SU(3)$, while the right-handed parts are all singlets. Evaluate the anomaly in the $SU(3)$ symmetry. What is the anomaly if we add M fermion fields whose left-handed parts are singlets, and whose right-handed parts transform as symmetric traceless second-rank $SU(3)$ tensors?

3. Find a solution of the 't Hooft anomaly matching conditions (22.5.5) and (22.5.6) for the case of $n = 4$ flavors. Find a solution for $n = 2$ flavors other than the one given in the text.

4. Derive the Zinn-Justin equation from the quantum master equation, without assuming that $\Delta S = 0$.

References

1. J. Steinberger, *Phys. Rev.* **76**, 1180 (1949); Also see R. J. Finkelstein, *Phys. Rev.* **72**, 415 (1949); H. Fukuda and Y. Miyamoto, *Prog. Theor. Phys.* **4**, 347 (1949); J. Schwinger, *Phys. Rev.* **82**, 664 (1951); L. Rosenberg, *Phys. Rev.* **129**, 2786 (1963). A rough estimate of this decay rate was made by S. Sakata and Y. Tanikawa, *Phys. Rev.* **57**, 548 (1940), before the experimental discovery of the π^0 meson.

2. D. G. Sutherland, *Nucl. Phys.* **B2, 433** (1967)

3. M. Veltman, *Proc. Roy. Soc.* **A301**, 107 (1967).

4. J. S. Bell and R. Jackiw, *Nuovo Cimento* **60A**, 47 (1969). Also see R. Jackiw, in *Lectures on Current Algebra and its Applications*, (Princeton Press, Princeton, 1972).

5. S. Adler, *Phys. Rev.* **177**, 2426 (1969). Also see S. Adler, in *Lectures on Elementary Particles and Quantum Field Theory – 1970 Brandeis University Summer Institute in Theoretical Physics*, eds. S. Deser, M. Grisaru, and H. Pendleton (MIT Press, Cambridge, MA, 1970): Volume I.

6. K. Fujikawa, *Phys. Rev. Lett.* **42**, 1195 (1979).

6*a*. M. F. Atiyah and I. M. Singer, *Proc. Nat. Acad. Sci.* **81**, 2597 (1984).

7. L. Alvarez-Gaumè and P. Ginsparg, *Ann. Phys.* **161**, 423 (1985).

7*a*. For a more detailed argument, see Y. Frishman, A. Schwimmer, T. Banks, and S. Yankielowicz, *Nucl. Phys.* **B177**, 157 (1981).

8. W. A. Bardeen, *Phys. Rev.* **184**, 1848 (1969).

8*a*. See, e.g., B. Zumino, Y–S. Wu, and A. Zee, *Nucl. Phys.* **B239**, 477 (1984).

9. S. L. Adler and W. A. Bardeen, *Phys. Rev.* **182**, 1517 (1969).

10. See, e.g., H. Georgi, *Lie Algebras in Particle Physics* (Benjamin–Cummings, Reading, MA, 1982): pp. 15, 198. The original reference is I. Schur, *Sitz. Preuss. Akad.*, p. 406 (1905).

10*a*. M. F. Atiyah and I. M. Singer, Ref. 6a; B. Zumino, in *Relativity, Groups, and Topology II*, eds. B. S. De Witt and R. Stora (North-Holland, Amsterdam, 1984); L. Alvarez-Gaumé and P. Ginsparg, Ref. 7; W. Bardeen and B. Zumino, *Nucl. Phys.* **B244**, 421 (1984); R. Stora, in *Progress in Gauge Field Theory*, eds. G. 't Hooft *et al.* (Plenum, New York, 1984): 543; O. Alvarez, I. Singer, and B. Zumino, *Comm. Math. Phys* **96**, 409 (1984); L. Alvarez-Gaumé and P. Ginsparg, *Nucl. Phys.* **B262**, 439 (1985); J. Mañes, R. Stora, and B. Zumino, *Commun. Math. Phys.* **102**, 157 (1985); B. Zumino, *Nucl. Phys.* **B253**, 477 (1985); J. M. Bismut and D. S. Freed, *Commun. Math. Phys.* **106**, 159 (1986); **107**, 103 (1986); D. S. Freed, *Commun. Math. Phys.* **107**, 483 (1986).

10*b*. E. Witten, *Phys. Lett.* **117 B**, 324 (1982).

11. D. J. Gross and R. Jackiw, *Phys. Rev.* **96**, 477 (1969).

12. H. Georgi and S. L. Glashow, *Phys. Rev.* **D6**, 429 (1972).

13. M. L. Mehta, *J. Math. Phys.* **7**, 1824 (1966); M. L. Mehta and P. K. Srivastava, *J. Math. Phys.* **7**, 1833 (1966).

13*a*. The cancellation of anomalies in the four-quark version of the standard electroweak theory was shown by C. Bouchiat, J. Iliopoulos, and Ph. Meyer, *Phys. Lett* **38B**, 519 (1972); S. Weinberg, in *Fundamental Interactions in Physics and Astrophysics*, eds. G. Iverson *et al.* (Plenum Press, New York, 1973): p. 157.

14. H. Georgi, in *Particles and Fields — 1974*, ed. C. Carlson (Amer. Inst. of Physics, New York, 1975).

15. R. Delbourgo and A. Salam, *Phys. Lett.* **40B**, 381 (1972); T. Eguchi and P. Freund, *Phys. Rev. Lett.* **37**, 1251 (1976). Also see N. K. Nielsen, M. T. Grisaru, R. Romer, and P. van Nieuwenhuizen, *Nucl. Phys.* **B140**, 477 (1978); M. J. Perry, *Nucl. Phys.* **B143**, 114 (1978); S. W. Hawking and C. Pope, *Nucl. Phys.* **B146**, 381 (1978); S. M. Christensen and M. J. Duff, *Phys. Lett.* **76B**, 571 (1978); R. Critchley, *Phys. Lett.* **78B**, 410 (1978); A. J. Hanson and H. Romer, *Phys. Lett.* **80B**, 58 (1978). For a general review, see T. Eguchi, P. B. Gilkey, and A. J. Hanson, *Phys. Rep.* **66**, 213 (1980). In $4n+2$ dimensions there is also an anomaly in the divergence of the energy momentum tensor; see L. Alvarez-Gaumè and E. Witten, *Nuc. Phys.* **B234**, 269 (1984).

16. G. 't Hooft, lecture given at the Cargèse Summer Institute, 1979, in *Recent Developments in Gauge Theories*, eds. G 't Hooft *et al.* (Plenum, New York, 1980), reprinted in *Dynamical Gauge Symmetry Breaking*, eds. E. Farhi and R. Jackiw (World Scientific, Singapore, 1982), and in G. 't Hooft, *Under the Spell of the Gauge Principle* (World Scientific, Singapore, 1994). Also see S. Dimopoulos, S. Raby, and L. Susskind, *Nucl. Phys.* **B173**, 208 (1980); S. Coleman and E. Witten, *Phys. Rev. Lett.* **45**, 1000 (1980); Y. Frishman, A. Schwimmer, T. Banks, and S. Yankielowicz, *Nucl. Phys.* **B177**, 157 (1981); A. Zee, Univ. of Pennsylvania report, 1980 (unpublished); R. Barbieri, L. Maiani, and R. Petronzio, *Phys. Lett.* **96B**, 63 (1980); G. Farrar, *Phys. Lett.* **96B**, 273 (1980); R. Chanda and P. Roy *Phys. Lett.* **99B**, 453 (1981).

16*a*. S. Weinberg and E. Witten, *Phys. Lett.* **96B**, 59 (1980).

17. J. Preskill and S. Weinberg, *Phys. Rev.* **D24**, 1059 (1981).

18. J. Wess and B. Zumino, *Phys. Lett.* **37B**, 95 (1971).

18*a*. R. Stora, in *Progress in Gauge Field Theory*, eds. G 't Hooft *et al.* (Plenum, New York, 1984):543 ; B. Zumino, in *Relativity, Groups and Topology II*, eds. B. S. De Witt and R. Stora (Elsevier, Amsterdam, 1984):1293; J. Mañes, R. Stora, and B. Zumino, *Commun. Math. Phys.* **102**, 157 (1985).

18*b*. L.D. Faddeev, *Phys. Lett.* **145B**, 81 (1984); B. Zumino, *Nucl. Phys.* **B253**, 477 (1985).

18*c*. J. Schwinger, *Phys. Rev. Lett.* **3**, 296 (1959).

19. J. A. Dixon, unpublished preprints (1976–9); *Commun. Math. Phys.* **139**. 495 (1991); F. Brandt, N. Dragon, and M. Kreuzer, *Nucl. Phys.* **B332**, 224 (1990); M. Dubois-Violette, M. Henneaux, M. Talon, and C. M. Viallet, *Phys. Lett.* **B289**, 361 (1992); M. Dubois-Violette, M. Talon, and C. M. Viallet, *Phys. Lett.* **B158**, 231 (1985); *Commun. Math. Phys.* **102**, 105 (1985); J. Mañes and B. Zumino, in *Supersymmetry and its Applications: Superstrings, Anomalies, and Supergravity*, eds. G. W. Gibbons, S. W. Hawking, and P. K. Townsend (Cambridge University Press, Cambridge, 1986).

19*a*. See, e.g., H. Georgi, Ref. 10: Eq. (XXV11.9).

20. R. Stora, in *New Directions in Quantum Field Theory and Statistical Mechanics – Lectures at the 1976 Cargese Summer School*, eds. M. Lévy and P. Mitter (Plenum, New York, 1977).

21. J. A. Dixon, Ref. 19.

22. G. Barnich and M. Henneaux, *Phys. Rev. Lett.* **72**, 1588 (1994); G. Barnich, F. Brandt, and M. Henneaux, *Commun. Math. Phys.* **174**, 57, 93 (1995).

23. W. Troost, P. van Nieuwenhuizen, and A. Van Proeyen, *Nucl. Phys.* **B333**, 727 (1990).

24. I learned about this proof from B. Zumino, private communication.

25. E. Witten, *Nucl. Phys.* **B223**, 422 (1983).

26. C.-S. Chu, P.-M. Ho, and B. Zumino, *Nucl. Phys.* **B475**, 484 (1996).

23
Extended Field Configurations

Most of this book has been devoted to applications of quantum field theory that can at least be described in perturbation theory, whether or not the perturbation series actually works well numerically. In using perturbation theory, we expand the action around the usual spacetime-independent vacuum values of the fields, keeping the leading quadratic term in the exponential $\exp(iI)$, and treating all terms of higher order in the fields as small corrections. Starting in the mid-1970s, there has been a growing interest in effects that arise because there are extended spacetime-dependent field configurations, such as those known as instantons,[1] that are also stationary 'points' of the action. In principle, we must include these configurations in path integrals and sum over fluctuations around them. (In Section 20.7 we have already seen an example of an instanton configuration, applied in a different context.) Although such non-perturbative contributions are often highly suppressed, they are large in quantum chromodynamics, and produce interesting exotic effects in the standard electroweak theory.

There are also extended field configurations that occur, not only as correction terms in path integrals for processes involving ordinary particles, but also as possible components of actual physical states. These configurations include some that are particle-like, such as magnetic monopoles[2] and skyrmions,[3] which are concentrated around a point in space or, equivalently, around a world line in spacetime. There are also string-like configurations,[4] similar to the vortex lines in superconductors discussed in Section 21.6, which are concentrated around a line in space or, equivalently, around a world sheet in spacetime. Then there are configurations that are sheet-like, like the domain walls[5] between spatial regions in which discrete symmetries are broken in different ways. In contrast, the instantons mentioned above are *event-like*, concentrated about a point in spacetime, and therefore never appear as components of actual physical states.

Some extended field configurations are stabilized because of boundary conditions that are imposed by the nature of the problem in which they appear. An example is the 'bounce' solution, which appears in the analysis of vacuum decay,[6] and will be discussed in Section 23.8. Other configurations are stable because they carry a quantum number whose conservation forbids any possible decay mode.[7]

In this chapter we shall mostly be concerned with extended field configurations that are stabilized by their topology. In analyzing all such configurations we use the same topological tools, chiefly homotopy theory, so we shall begin by considering all topologically stabilized configurations together, in a space or spacetime of arbitrary dimensionality d.

23.1 The Uses of Topology

It often happens that the space of all possible field configurations may be given a non-trivial topology by the condition that some functional S of the various fields is finite. In classical field theory S is the potential energy (or in some cases the potential energy per unit area or per unit length); no finite perturbation can produce a configuration where this is infinite. In classical statistical mechanics S is the Hamiltonian, and in quantum field theory formulated in Euclidean spacetime S is the Euclidean action or is proportional to it. (Euclidean path integrals and some of their applications are discussed in Appendix A of this chapter.) We construct a perturbation theory by starting with some equilibrium field configuration for which the Euclidean action or Hamiltonian is finite, and then integrating over fluctuations which leave it finite.

Two field configurations are said to be topologically equivalent if it is possible to deform one of them continuously into the other without passing through forbidden configurations with S infinite. This is evidently an equivalence relation (in the sense of being reflexive, symmetric, and transitive), and therefore divides the set of all field configurations into equivalence classes, each consisting of configurations of the same topology. For example, if S is the potential energy (the Hamiltonian for time-independent fields) in d space dimensions, then topologically different field configurations are forbidden by an infinite energy barrier from being transformed into one another. In particular, extended configurations with a different topology from the usual spatially uniform vacuum fields cannot spread out to become spatially uniform.

The topological classification is also useful when we are looking for a local minimum of S. If we can find a configuration that minimizes S for all configurations of a given topological type, then that field configuration must be at least a local minimum of S for *all* configurations of any type, since no small variation of the fields can change their topological type. Such a configuration is therefore a solution of the field equations, which are equivalent to the condition that S is stationary. This sort of problem comes up not only in stability problems, where S is the Hamiltonian, but also in finding field configurations around which we may expand the field variables in path integrals in Euclideanized d-dimensional spacetime. Here S is the negative of the Euclidean action I, and we must look for a local minimum so that the leading term in the expansion around this configuration will be a quadratic free-field action with second-order terms of the right sign.

Here are some examples:[8]

(a) Skyrmions, etc. Consider the real Goldstone boson fields π_a associated with the spontaneous breakdown of a continuous global symmetry group G to a subgroup H. As we saw in Chapter 19, the potential energy for these Goldstone bosons in a Euclidean space of dimensionality $d > 2$ will take the form

$$S[\pi] = \int d^d x \left[\tfrac{1}{2} \sum_{ab} g_{ab}(\pi) \partial_i \pi_a \partial_i \pi_b + \ldots \right] , \tag{23.1.1}$$

where g_{ab} is a positive-definite matrix, and '+...' denotes possible terms of higher order in the derivatives of π. Alternatively, Eq. (23.1.1) can be regarded as minus the action for a Goldstone boson field in a d-dimensional Euclidean *spacetime*.

Field configurations of finite S must have $\partial_i \pi_a(\mathbf{x})$ vanishing at infinity faster than $|\mathbf{x}|^{-d/2}$ (where $|\mathbf{x}| \equiv \sqrt{x_i x_i}$), so that $\pi_a(\mathbf{x})$ must approach a constant $\pi_{a\infty}$ as $\mathbf{x} \to \infty$ with a remainder vanishing faster than $|\mathbf{x}|^{(2-d)/2}$. The Goldstone boson fields π_a at any point form a homogeneous space, the coset space G/H, for which it is possible to transform any one field value to any other by a transformation of G, so by a global G transformation it is always possible to arrange that the asymptotic limit $\pi_{a\infty}$ takes any specific value, say $\pi_{a\infty} = 0$. The field $\pi_a(\mathbf{x})$ thus represents a mapping of the whole d-dimensional space, with the sphere $r = \infty$ taken as a single point, into the manifold G/H of all field values.

Now, a d-dimensional Euclidean space with the $(d-1)$-dimensional spherical surface at infinity identified as a single point is topologically the same as S_d, the d-dimensional sphere (that is, the surface of a $(d+1)$-dimensional ball), in the sense that either can be continuously mapped into the other. The fields $\pi(\mathbf{x})$ that approach zero as $\mathbf{x} \to \infty$ may therefore be classified according to the topologically distinct mappings of S_d into the manifold G/H of the field variables, for which the point at infinity is mapped into zero. The set of classes of such topologically distinct mappings $S_d \mapsto \mathcal{M}$ with one point of S_d mapped into a fixed point of $\mathcal{M}$ is known as $\pi_d(\mathcal{M})$, the dth *homotopy group* of the manifold $\mathcal{M}$. These homotopy groups will be discussed (and their group structure explained) in the next section, and a list of homotopy groups for various manifolds is given in Appendix B of this chapter. For the present it will be enough to mention that, although when the manifold $\mathcal{M}$ is a linear space the homotopy group $\pi_d(\mathcal{M})$ is trivial (in the sense that any field configuration $\pi(\mathbf{x})$ that approaches a constant value as $\mathbf{x} \to \infty$ can be continuously deformed into one in which the field takes that value everywhere), the manifold $\mathcal{M} = G/H$ of the Goldstone boson fields often has a non-trivial homotopy group. In the cases relevant to quantum chromodynamics, of $SU(2) \otimes SU(2)$ broken to $SU(2)$ or of $SU(3) \otimes SU(3)$ broken to $SU(3)$, the manifold G/H is the same as $SU(2)$ or $SU(3)$, respectively, for which according to Appendix B the homotopy groups $\pi_3(H)$ are non-trivial. The topologically non-trivial fields at local minima of the potential energy with $d = 3$ are known as *skyrmions*.[3] Baryons like the proton may in some respects be regarded as skyrmion solutions in a pure meson theory.

The functional (23.1.1) does not have skyrmion stationary points, unless terms involving higher powers of $\partial_i \pi_a$ are included in the integrand. In the absence of such terms, any topologically non-trivial field configurations will have a continuum of values for S, extending down to a lower bound $S = 0$ at which π becomes singular, so topology cannot stabilize such configurations. This is generally known as *Derrick's theorem*.[9] To prove this theorem, note that for any field configuration $\pi_a(\mathbf{x})$ we may introduce another configuration with the same topology,

$$\pi_a^R(\mathbf{x}) \equiv \pi_a(\mathbf{x}/R)$$

with R an arbitrary real positive scale factor. Then for the terms shown explicitly in Eq. (23.1.1)

$$S[\pi^R] = R^{d-2} S[\pi] \,.$$

For $d > 2$ this is a decreasing function of R as $R \to 0$, so there is a continuum of values of $S[\pi^R]$ extending down to a value $S = 0$. Furthermore $S[\pi] > 0$, because $S[\pi]$ can only vanish if $\pi(\mathbf{x})$ is constant, which is not possible because we assumed that π is topologically non-trivial. Hence this lower bound is attained only at $R = 0$, at which $\pi_a^R(\mathbf{x})$ becomes singular.

Goldstone boson field configurations can be stabilized by adding higher derivative terms to S. For instance, if we take $S[\pi] = T[\pi] + D[\pi]$ with

$$T[\pi] \equiv \int d^d x \; \tfrac{1}{2} \sum_{ab} g_{ab}(\pi) \partial_i \pi_a \partial_i \pi_b \geq 0 \,,$$

$$D[\pi] \equiv \int d^d x \, f_{abcd}(\pi) \, \nabla \pi_a \cdot \nabla \pi_b \, \nabla \pi_c \cdot \nabla \pi_d \geq 0 \,,$$

then $D[\pi^R] = R^{d-4} D[\pi]$, while as before $T[\pi^R] = R^{d-2} T[\pi]$, so $S[\pi^R]$ reaches a minimum at a finite R if $2 < d < 4$, and in particular for the physically interesting case $d = 3$.

The problem with the theory of skyrmions is not that we have to include higher-derivative terms like $D[\pi]$ in the action. As discussed in Section 19.5, we expect such terms in the action of any effective field theory of Goldstone bosons. The problem is that there is no rationale for excluding an infinite number of other higher-derivative terms, all of which are generically of the same order of magnitude for configurations that are stabilized by a balance between terms with different numbers of derivatives, which makes realistic calculations impossible.

(b) Domain boundaries. When a *discrete* symmetry is broken, we have the possibility that the symmetry is broken in different ways in different domains, separated by domain boundaries in which the vacuum fields make a transition from one minimum of the potential to another. For instance, consider a flat domain boundary in the y–z plane, and suppose that the energy per unit area is given by

$$S[\phi] = \int_{-\infty}^{\infty} dx \left[\frac{1}{2}\left(\frac{d\phi}{dx}\right)^2 + V(\phi)\right] , \tag{23.1.2}$$

where $\phi(x)$ is a real scalar field that is assumed to depend only on the distance x along the direction normal to the boundary, and $V(\phi)$ is a potential satisfying the reflection symmetry $\phi \to -\phi$, with minima only at field values $\pm\bar{\phi}$. For convenience, we will adjust an additive constant in $V(\phi)$ so that the minimum value of $V(\phi)$ is zero, in which case $V(\phi) \geq 0$, and $V(\phi) = 0$ only at $\phi = \pm\bar{\phi}$. To keep S finite, it is necessary for ϕ to approach either $+\bar{\phi}$ or $-\bar{\phi}$ as $x \to \infty$, and also to approach either $+\bar{\phi}$ or $-\bar{\phi}$ as $x \to -\infty$. We then have four topologically distinct configurations, in two of which ϕ approaches the same limits as $x \to \pm\infty$, so that the configuration can be smoothly deformed into the vacuum configurations with $\phi(x)$ constant everywhere, and in two of which ϕ approaches opposite limits as $x \to \pm\infty$, which are topologically stable. Here we are classifying field configurations according to $\pi_0(G)$, where $\pi_0(\mathscr{M})$ for any manifold $\mathscr{M}$ is conventionally defined as the set of connected components of $\mathscr{M}$, and G is the symmetry group, which in our case is the Z_2 group of reflections $\phi \to -\phi$.

This is a good place to introduce a trick due to Belavin *et al.*[1] that will prove useful in Sections 23.3 and 23.5 in dealing with the more complicated cases of monopoles and instantons. Rewrite Eq. (23.1.2) in the form

$$S[\phi] = \frac{1}{2}\int_{-\infty}^{\infty} dx \left(\frac{d\phi}{dx} \mp \sqrt{2V(\phi)}\right)^2 \pm \int_{\phi(-\infty)}^{\phi(\infty)} \sqrt{2V(f)}\, df\, . \tag{23.1.3}$$

The integral in the second term on the right-hand side of Eq. (23.1.3) can be regarded as a 'topological charge,' which depends only on the values taken by the field at $x \to \pm\infty$. For configurations that approach the same limit as $x \to +\infty$ and $x \to -\infty$, this integral vanishes, and the minimum value of S is zero, reached for constant fields. For a field $\phi(x)$ that takes different values at $x = \pm\infty$, we can choose the $\pm$ sign in Eq. (23.1.3) to yield a lower bound

$$S[\phi] \geq \int_{-\bar{\phi}}^{\bar{\phi}} \sqrt{2V(f)}\, df\,. \tag{23.1.4}$$

This bound is reached when the first term in Eq. (23.1.3) vanishes, or in other words, when

$$x = \pm \int_0^{\phi(x)} \frac{df}{\sqrt{2V(f)}} + x_0\,, \tag{23.1.5}$$

where x_0 is an integration constant, which evidently gives the position of the center of the domain boundary. Note that Derrick's theorem is no obstacle to a solution here, because for domain boundaries $d = 1$, so for a rescaled field $\phi(x/R)$ the integrals of the two terms in the integrand in Eq. (23.1.2) go as R^{-1} and R^{+1}, respectively.

Eq. (23.1.5) could have been obtained more directly, by deriving a second-order differential equation for $\phi(x)$ from the condition that Eq. (23.1.2) must be stationary under small variations in $\phi(x)$, and then using this differential equation to show that the quantity $\frac{1}{2}(d\phi/dx)^2 - V(\phi)$ is constant in x. The advantage of the derivation based on the formula (23.1.3) is that it shows immediately that the solution (23.1.5) is stable against small perturbations that maintain the flatness of the boundary, aside from the 'zero-mode' associated with changes in the boundary location x_0. By adding a term $\frac{1}{2}(d\phi/dy)^2 + \frac{1}{2}(d\phi/dz)^2$ in the integrand of Eq. (23.1.2), we can see that this solution is also stable against *any* perturbation $\delta\phi(x,y,z)$, provided $\delta\phi(x,y,z) \to 0$ for $x \to \pm\infty$ with fixed y and z.

If there are discrete spontaneously broken symmetries then domain boundaries would have formed when these symmetries became broken in the early universe. If the domain boundaries did not disappear they would produce gross distortions of the observed isotropy and homogeneity of the present universe.[5] We do not now know of any exact discrete symmetries except CPT, or of any spontaneously broken approximate or exact discrete symmetries, so for the present this is not a problem.

(c) Instantons, etc. Now consider a gauge theory, with

$$S[A] = \tfrac{1}{4} \int d^d x\, F_{\alpha ij} F_{\alpha ij}\,, \tag{23.1.6}$$

where $F_{\alpha ij}$ is the usual field strength tensor, and we take $d \geq 4$. This can either be regarded as the action for quantum gauge fields in a Euclidean

d-dimensional spacetime, or the potential energy for classical gauge fields in temporal gauge, with $A^0_\alpha = 0$, in $(d+1)$-dimensional spacetime.

In order for $S[A]$ to be finite, $F_{\alpha ij}(\mathbf{x})$ must vanish as $\mathbf{x} \to \infty$. This can be achieved by having $A_{\alpha i}(\mathbf{x})$ vanish sufficiently rapidly as $\mathbf{x} \to \infty$, but even for $d \geq 4$ it is also possible for $S[A]$ to be finite for a field $A_{\alpha i}(\mathbf{x})$ that vanishes as slowly as $1/|\mathbf{x}|$, as long as the field approaches a pure gauge as $|\mathbf{x}| \to \infty$

$$it_\alpha A_{\alpha i}(\mathbf{x}) \to g^{-1}(\hat{\mathbf{x}})\, \partial_i g(\hat{\mathbf{x}})\,, \tag{23.1.7}$$

where $g(\hat{\mathbf{x}})$ is a direction-dependent element of the gauge group G. Furthermore $A_{\alpha i}(\mathbf{x})$ is unaffected if we replace $g(\hat{\mathbf{x}})$ with $g_0 g(\hat{\mathbf{x}})$ for any fixed group element $g_0 \in G$, so by choosing $g_0 = g^{-1}(\hat{\mathbf{x}}_1)$ we can arrange that $g(\hat{\mathbf{x}}_1) = 1$ for any one direction $\hat{\mathbf{x}}_1$. Each gauge field of finite $S[A]$ therefore defines a mapping from the unit sphere $|\hat{\mathbf{x}}| = 1$ to the group manifold, with the point $\hat{\mathbf{x}}_1$ mapped into the unit element of G. (In the case in which the gauge field vanishes faster than $1/|\mathbf{x}|$ as $|\mathbf{x}| \to \infty$, this mapping takes all points on the unit sphere into the identity element of the gauge group.) The set of classes of such topologically distinct mappings $S_{d-1} \mapsto G$, with one point of S_{d-1} mapped into a fixed element of G, is known as $\pi_{d-1}(G)$, the $(d-1)$th homotopy group of the group manifold. As indicated in Appendix B of this chapter, $\pi_3(G)$ is non-trivial for any semisimple Lie group G. The topologically non-trivial stationary points of $S[A]$ for $d = 4$ are known as *instantons*.[1] Their importance in quantum chromodynamics is discussed in Sections 23.5 and 23.6.

In order for $S[A]$ to be stationary at a field $A(\mathbf{x})$, it is necessary that $A(\mathbf{x})$ should satisfy the field equations

$$\partial_i F_{\alpha ij} = 0\,. \tag{23.1.8}$$

A simple scaling argument again limits the values of the dimensionality d where we can hope to find a topologically non-trivial local minimum of $S[A]$. Define $A^R(\mathbf{x}) \equiv A(\mathbf{x}/R)/R$. Then

$$S[A^R] = R^{d-4} S[A]\,,$$

so for $d \neq 4$ there can be no topologically non-trivial stationary point of $S[A]$ unless $S[A] = 0$. But if $S[A] = 0$ then $F_{\alpha ij} = 0$ everywhere, so by a gauge transformation we can make $A_{\alpha i}$ also vanish everywhere.

As we shall see in Section 23.5, for $d = 4$ it *is* possible to find instanton solutions where $S[A]$ (here identified as $-I[A]$) is stationary, with $F_{\alpha ij}$ not equal to zero except at infinity. The scaling argument above shows that if $A(\mathbf{x})$ is such an instanton solution then so is $A(\mathbf{x}/R)/R$, but this degeneracy is removed by quantum corrections.

(d) Monopoles, vortex lines, etc. Now consider a theory of gauge fields together with scalars that furnish a linear representation of the gauge group, with

$$S[\phi, A] = \int d^dx \left[\tfrac{1}{2}\sum_{ab} g_{ab}(\phi) D_i\phi_a D_i\phi_b + \tfrac{1}{4}F_{\alpha ij}F_{\alpha ij} + U(\phi) \right] , \quad (23.1.9)$$

where $g_{ab}(\phi)$ is a positive-definite matrix (usually ϕ-independent), $U(\phi)$ is bounded below, and shifted by a constant term so that its minimum value is zero, and $F_{\alpha ij}$ and D_i are the usual field strength and gauge-covariant derivative. We require that $U(\phi)$ is a scalar and $g_{ab}(\phi)$ is a tensor under transformations in the gauge group G. Again, Eq. (23.1.9) gives either the action for a quantum field theory in Euclidean d-dimensional spacetime, or the potential energy for a classical field theory in temporal gauge in $(d+1)$-dimensional spacetime.

For $S[\phi, A]$ to be finite it is necessary for $U(\phi(\mathbf{x}))$ to vanish as $\mathbf{x} \to \infty$. The set of ϕs at which $U(\phi)$ vanishes is invariant under G and may be discrete or continuous. In case (b) above we have dealt with an example where this set is discrete. Let us now consider the broken symmetry case where the zeros of $U(\phi)$ form a continuous manifold $\mathcal{M}_0$ consisting of fields related by transformations $g \in G$. In this case each $\phi(\hat{\mathbf{x}})$ may be obtained by a transformation $\gamma(\hat{\mathbf{x}}) \in G$ acting on the value $\phi(\hat{\mathbf{x}}_1)$ of the field in any one direction $\hat{\mathbf{x}}_1$. We may therefore consider the field $\phi(\mathbf{x})$ to define a mapping $S_{d-1} \mapsto G/H$ into a coset space G/H; in other words, into the group G with elements g_1 and g_2 identified if they differ only by right multiplication with some element h of the subgroup $H \subset G$ that leaves $\phi(\hat{\mathbf{x}}_1)$ invariant, that is, if $g_1 = g_2 h$. In particular, the point $\hat{\mathbf{x}}_1$ is mapped into the subgroup H, in order that $\gamma(\hat{\mathbf{x}}_1)$ acting on $\phi(\hat{\mathbf{x}}_1)$ should yield $\phi(\hat{\mathbf{x}}_1)$ itself. The fields that approach values on the manifold $\mathcal{M}_0$ as $\mathbf{x} \to \infty$ may therefore be classified according to the topologically distinct mappings of S_{d-1} into G/H that map the point $\hat{\mathbf{x}}_1$ into the fixed 'unit' element H of G/H. The set of classes of such topologically distinct mappings $S_{d-1} \mapsto G/H$ with one point of S_{d-1} mapped into a fixed element of G/H is known as $\pi_{d-1}(G/H)$, the $(d-1)$th homotopy group of the manifold G/H.

In this case $\partial_i\phi(\mathbf{x})$ goes as $1/|\mathbf{x}|$ for $\mathbf{x} \to \infty$. In order for $S[\phi]$ to be finite $D_i\phi$ must vanish faster than $|\mathbf{x}|^{-d/2}$ for $\mathbf{x} \to \infty$, so it is necessary for $it_\alpha A_{\alpha i}(\mathbf{x})$ to approach $\gamma^{-1}(\hat{\mathbf{x}})\partial_i\gamma(\hat{\mathbf{x}})$ faster than $|\mathbf{x}|^{-d/2}$ for $\mathbf{x} \to \infty$. This is a pure gauge field, so the field strength tensor $F_{\alpha ij}(\mathbf{x})$ vanishes faster than $|\mathbf{x}|^{-d/2-1}$, which is fast enough to make $\int d^dx\, F_{\alpha ij}F_{\alpha ij}$ converge.

Derrick's theorem does not apply for the gauge theory defined by Eq. (23.1.9), but it is interesting to see where the same reasoning takes us. For any given fields $\phi(\mathbf{x})$ and $A(\mathbf{x})$, again define $\phi^R(\mathbf{x}) \equiv \phi(\mathbf{x}/R)$, and now also $A^R(\mathbf{x}) \equiv A(\mathbf{x}/R)/R$. The three terms in the Hamiltonian (23.1.9) now have the scaling properties

$$T[\phi^R, A^R] = R^{d-2}T[\phi, A]\,, \quad K[A^R] = R^{d-4}K[A]\,, \quad V[\phi^R] = R^d V[\phi]\,,$$

where $T[\phi, A] \equiv \frac{1}{2}\int d^dx \sum_{ab} g_{ab}(\phi)D_i\phi_a D_i\phi_b$, $K[A] \equiv \frac{1}{4}\int d^dx F_{\alpha ij}F_{\alpha ij}$, and $V[\phi] \equiv \int d^dx\, U(\phi)$. Now, for $d > 4$, $S[\phi^R, A^R]$ has no minimum at any finite value of R, so there is no stable configuration with non-trivial topology. For $0 < d < 4$ there is no difficulty in finding a finite value of R at which $S[\phi^R, A^R]$ is a minimum.

In the physically interesting case $d = 3$, topologically non-trivial field configurations are classified according to the homotopy group $\pi_2(G/H)$, which is non-trivial for a simply connected group G (such as $SU(2)$) broken to an H containing the $U(1)$ of electromagnetism. The topologically non-trivial classical field configurations with $d = 3$ are known as *magnetic monopoles.*[2] As we shall see in Section 23.3, their magnetic pole strength is quantized, the different values corresponding to different elements of $\pi_2(G/H)$.

For $d = 2$, topologically non-trivial configurations correspond to elements of $\pi_1(G/H)$, which is non-trivial when G is a non-simply connected group like $U(1)$ or $SO(3)$, broken either completely or to a discrete subgroup. The topologically non-trivial classical field configurations with $d = 2$ are the cross sections of *vortex lines.* One example is provided by superconductivity, where $G = U(1)$ is spontaneously broken to $H = Z_2$. We have seen in Section 21.6 that vortex lines occur in type II superconductors for magnetic field strengths in a certain range, and that the magnetic flux carried by a vortex line is quantized, the different values of the flux corresponding to different elements of $\pi_1(U(1)/Z_2)$. Vortex lines can also occur in relativistic quantum field theories,[4] and may be produced in symmetry-breaking transitions in the early universe, in which case they are known as *cosmic strings.*[11]

Monopoles and vortex lines share a remarkable feature that can be deduced on purely topological grounds. In both cases the forms of the Goldstone boson fields $\pi_a(\mathbf{x})$ on large spheres (S_1 for vortex lines, S_2 for monopoles) surrounding the configurations are twisted, in such a way that they cannot smoothly be deformed into constants. In particular, it is impossible smoothly to reduce the radii of these spheres to zero without encountering some sort of singularity, because a non-singular field $\pi_a(\mathbf{x})$ on a sphere would have to become a constant as the radius of the sphere shrinks to zero. The singularity in both cases occurs in a core (a line or perhaps a tube for vortex lines, a point or perhaps a ball for monopoles) within which the group G is no longer broken, so that the system is no longer described by Goldstone boson fields, but by an order parameter that transforms linearly under G.

For $d = 4$, the function $S[\phi^R, A^R]$ of R can only have a minimum at some finite R if $T[\phi, A] = V[\phi] = 0$, which would require that $\phi(\mathbf{x})$ everywhere takes a value at which $U(\phi) = 0$. Assuming that these values form a continuum related by transformations in the gauge group G, by a gauge transformation they may be made constants, $\phi(\mathbf{x}) = \phi_0$. Then in this gauge the condition $T[\phi, A] = 0$ implies that $A_a(\mathbf{x}) = 0$ for all broken symmetries, for which $t_a\phi_0 \neq 0$. Both $T[\phi, A]$ and $V[\phi]$ are stationary at such a field configuration, so in order for $S[\phi, A]$ to be stationary $K[\phi, A]$ must also be stationary, which means that the non-zero gauge fields $A_{i\mu}$ (which belong to the subgroup $H \subset G$ that is unbroken by ϕ_0) must satisfy the Euclidean Yang–Mills field equations

$$\partial_\mu F_{i\mu\nu} = 0 . \tag{23.1.10}$$

This case therefore reduces to case (c), but with the gauge group G replaced with its unbroken subgroup H.

23.2 Homotopy Groups

We learned in the previous section to classify field configurations, at which the Hamiltonian or other functionals are finite, in correspondence with the elements of appropriate homotopy groups. But we have not yet explained in what sense the homotopy groups are *groups*, nor have we given any physical significance to the group structure. As we shall see, there is a natural definition of the multiplication rule for elements of homotopy groups, according to which two extended configurations of fields forming a manifold $\mathcal{M}$ in d dimensions, that belong to different elements c_1 and c_2 of $\pi_d(\mathcal{M})$, can only fuse continuously to form a configuration belonging to the element $c_1 \times c_2$ of $\pi_d(\mathcal{M})$.

We will begin by defining the first homotopy group $\pi_1(\mathcal{M})$ of an arbitrary manifold $\mathcal{M}$, also known as the *fundamental group of the manifold.* As we have seen, the existence of a non-trivial $\pi_1(G/H)$ for some coset space G/H is the condition for the topological stability of a vortex line in three dimensions (or a monopole in two dimensions). After considering $\pi_1(\mathcal{M})$, we will then move on to more general homotopy groups.

A connected manifold $\mathcal{M}$ is said to be multiply connected if there is some closed curve of points $p(z)$ on the manifold, parameterized by a single variable z with $0 \leq z \leq 1$ and $p(0) = p(1)$, which cannot be contracted to a point by a continuous deformation. Since on a connected manifold we can always continuously deform any such closed curve so that any one point on the curve is anywhere we like on the manifold, we may restrict our attention only to curves for which $p(0) = p(1) = p_0$, where p_0 is any fixed point of the manifold, known as the *base point*. Two such closed curves $p_1(z)$ and $p_2(z)$ are said to be homotopically equivalent if they *can* be deformed into each other, that is, if there is a *continuous* function $p(z,t)$ for $0 \leq t \leq 1$, with

$$p(z,0) = p_1(z)\,, \qquad p(z,1) = p_2(z)\,,$$

$$p(0,t) = p(1,t) = p_0\,.$$

The relation of homotopic equivalence is an equivalence relation in the sense of being symmetric, reflexive, and transitive, so it divides the space of closed curves on the manifold into equivalence classes: two closed curves are in the same class if and only if they are homotopically equivalent. The set of these equivalence classes is known as the first homotopy group of the manifold, $\pi_1(\mathcal{M})$.

To define the multiplication rule for $\pi_1(\mathcal{M})$, choose a standard curve $p[z;c]$ that starts and ends at the base point p_0 for each equivalence class c in $\pi_1(\mathcal{M})$. For any two equivalence classes c_1 and c_2, define the 'product' $c_1 \times c_2$ as the equivalence class containing the curve $p[z,c_1,c_2]$ that starts at p_0, follows $p[z,c_1]$ back to p_0, and then follows $p[z,c_2]$ back again to p_0. Formally, we take

$$p[z,c_1,c_2] \equiv \begin{cases} p[2z,c_1] & 0 \leq z \leq \frac{1}{2} \\ p[2z-1,c_2] & \frac{1}{2} \leq z \leq 1 \end{cases}\,.$$

We must now show that multiplication defined in this way satisfies the conditions for a group. First, let us check that this multiplication is *associative*. For this purpose, note that $(c_1 \times c_2) \times c_3$ is the equivalence class containing a curve $p[z,c_1 \times c_2,c_3]$ that goes along the standard curve $p[z,c_1 \times c_2]$ from the base point and back again, and then along the standard curve $p[z,c_3]$ from the base point and back again, while $c_1 \times (c_2 \times c_3)$ is the equivalence class containing a curve $p[z,c_1,c_2 \times c_3]$ that goes along the standard curve $p[z,c_1]$ from the base point and back again, and then along the standard curve $p[z,c_2 \times c_3]$ from the base point and back again. By definition, the curve $p[z,c_1 \times c_2]$ may be deformed into a curve that goes along $p[z,c_1]$ from the base point and back again and then along $p[z,c_2]$ from the base point and back again, while $p[z,c_2 \times c_3]$ may be deformed into a curve that goes along $p[z,c_2]$ from the base point and back again and then along $p[z,c_3]$ from the base point and back again. Hence both curves $p[z,c_1 \times c_2,c_3]$ and $p[z,c_1,c_2 \times c_3]$ may be deformed into the curve that goes along $p[z,c_1]$ from the base point and back again, then along $p[z,c_2]$ from the base point and back again, and finally along $p[z,c_3]$ from the base point and back again, and hence they may be deformed into each other, showing that

$$(c_1 \times c_2) \times c_3 = c_1 \times (c_2 \times c_3).$$

The unit element e of $\pi_1(\mathcal{M})$ is defined as the equivalence class containing the curve $p[z,e] = p_0$ that stays at the base point. To check that $e \times c = c$, note that

$$p[z,e,c] = \begin{cases} p_0 & 0 \leq z \leq \frac{1}{2} \\ p[2z-1,c] & \frac{1}{2} \leq z \leq 1 \end{cases} .$$

But this curve can be continuously deformed into $p[z,c]$ by taking

$$p(z,t) = \begin{cases} p_0 & 0 \leq z \leq t/2 \\ p[(2z-t)/(2-t),c] & t/2 \leq z \leq 1 \end{cases} ,$$

which is the same as $p[z,e,c]$ for $t = 1$ and the same as $p[z,c]$ for $t = 0$. The product $e \times c$ is the equivalence class containing $p[z,e,c]$, which we now see is the same as the equivalence class containing $p[z,c]$, which is just c. The proof that $c \times e = c$ is similar.

The 'inverse' c^{-1} of the equivalence class c is the equivalence class that contains a curve $p^{-1}[z,c]$ which goes around the same path as the standard curve $p[z,c]$ but in the opposite direction; that is,

$$p^{-1}[z,c] \equiv p[1-z,c].$$

This is not necessarily the same as the 'standard' curve $p[z,c^{-1}]$, but by the definition of the equivalence class c^{-1}, the two curves may be deformed into each other. To check that $c^{-1} \times c = e$, note that by deforming $p[z,c^{-1}]$ into $p^{-1}[z,c]$, the curve $p[z,c^{-1},c]$ may be deformed into

$$p[z,c^{-1},c] \to \begin{cases} p[1-2z,c] & 0 \leq z \leq \frac{1}{2} \\ p[2z-1,c] & \frac{1}{2} \leq z \leq 1 \end{cases} .$$

But this curve can be continuously deformed into $p[z,e] = p_0$ by taking

$$p(z,t) = \begin{cases} p[1-2tz,c] & 0 \leq z \leq \frac{1}{2} \\ p[2tz+1-2t,c] & \frac{1}{2} \leq z \leq 1 \end{cases} ,$$

which is the same as $p[z,c^{-1},c]$ for $t = 1$ and the same as $p[z,e]$ for $t = 0$. The product $c^{-1} \times c$ is the equivalence class containing $p[z,c^{-1},c]$, which we now see is the same as the equivalence class containing $p[z,e]$, which is just e. The proof that $c \times c^{-1} = e$ is similar. The existence of a unit element and inverses shows that these equivalence classes form a *group*.

The classic example of a manifold $\mathcal{M}$ with a non-trivial first homotopy group is the circle itself, $\mathcal{M} = S_1$. This may be parameterized by an angle θ with $\theta = 0$ and $\theta = 2n\pi$ (with n any positive or negative integer) identified as the same point. The homotopy groups consist of classes of functions $\theta(z)$ for $0 \leq z \leq 1$ that begin at some base point, $\theta(0) = \theta_0$, and end at the same base point, $\theta(1) = \theta_0 + 2n\pi$. Two such functions may be continuously deformed into each other if and only if they have the same value of n, so $\pi_1(S_1)$ consists of a denumerably infinite number of classes c_n labelled by the positive or negative integer n. Furthermore, the 'product' of two classes c_n and c_m consists of curves that go n times around the circle from the base point and back again, and then m times around the circle from the base point and back again, so here multiplication is just addition:

$$c_n \times c_m = c_{n+m} \tag{23.2.1}$$

and hence

$$\pi_1(S_1) = Z \,, \tag{23.2.2}$$

where Z is the group of addition of positive and negative integers. As an immediate physical application, we note that when a gauge group $SO(2)$ is completely spontaneously broken, the coset space is just $SO(2)$ itself, which has the topology of a circle, so in this case there is an infinite number of types of topologically stable vortex line, characterized by a positive or negative integer n. For instance, this is the case in a type II superconductor, where as we saw in Section 21.6 the $U(1)$ of electromagnetic gauge invariance is spontaneously broken to a discrete subgroup Z_2.

More generally, all spheres S_d with $d > 1$ are simply connected, which means that they have a trivial first homotopy group, a statement conventionally expressed as

$$\pi_1(S_d) = 0 \qquad \text{for } d > 1\,. \tag{23.2.3}$$

Only a few of the more familiar Lie groups are multiply connected:

$$\pi_1(G) = \begin{cases} Z & G = U(k) & k \geq 1 \\ Z_2 & G = SO(k) & k \geq 3 \\ 0 & G = Spin(k) & k \geq 3 \\ 0 & G = SU(k) & k \geq 2 \\ 0 & G = USp(2k) & k \geq 1 \\ 0 & G = G_2, F_4, E_6, E_7, E_8 & \end{cases} . \tag{23.2.4}$$

Here Z_2 is the group with two elements 1 and -1, with group multiplication defined as ordinary multiplication, and $Spin(n)$ is the simply connected covering group of $SO(n)$. (As we saw in Section 2.7, $Spin(3)$ is the same as $SU(2)$.) Also, for a direct product of two manifolds $\mathcal{M}$ and $\mathcal{M}'$, the fundamental group is

$$\pi_1(\mathcal{M} \times \mathcal{M}') = \pi_1(\mathcal{M}) \times \pi_1(\mathcal{M}')\,. \tag{23.2.5}$$

We can appreciate the physical significance of the group structure of $\pi_1(\mathcal{M})$ by asking what happens when two distant parallel vortex lines in three dimensions are brought together. When the vortex lines are sufficiently far apart their fields do not interact, so the configuration can be described by specifying the classes c' and c'' in $\pi_1(G/H)$ to which each belongs. The class c to which the whole configuration belongs is determined by the behavior of the fields on a very large circle surrounding both vortex lines. By a continuous deformation we can distort this very large circle into two large circles, one surrounding each vortex line, which intersect at a point midway between them. As we go around this closed curve in two-dimensional space we first trace out a curve in G/H which consists of one of the closed curves in the class c', and then one of the closed curves in the class c'', just as in the definition of the product of the classes. We conclude then that the whole configuration is in a class $c = c' \times c''$, and so the two vortex lines can only fuse together to form a vortex line of this class. In particular, they can only annihilate if $c'' = c'^{-1}$.

For instance, when $\pi_1(G/H) = Z$ (as in superconductivity, where $G/H = SO(2)/Z_2$), in three dimensions two vortex lines of classes c_n and c_m can fuse together to form a vortex line of class c_{n+m}, so they can only annihilate if $n = -m$. On the other hand, when $\pi_1(G/H) = Z_2$ (as when $G = SO(N)$ with $N \geq 3$ and H is trivial) there is only one kind of vortex line in three dimensions, corresponding to the element -1 of Z_2, and since $(-1)^2 = 1$, any two can annihilate.

Now let us consider the general homotopy group $\pi_k(\mathcal{M})$. This is much like $\pi_1(\mathcal{M})$, except that instead of considering mappings of the circle S_1 into a manifold $\mathcal{M}$, we consider mappings of the k-sphere S_k (the surface of a $(k+1)$-dimensional ball) into $\mathcal{M}$, again with one point of S_k always mapped into the same 'base point' p_0 of $\mathcal{M}$. Two such mappings are equivalent if one can be continuously deformed into the other, keeping the same point of S_k always mapped into the base point. The kth homotopy group $\pi_k(\mathcal{M})$ has elements consisting of equivalence classes of these mappings.

It is often convenient to picture the d-sphere S_d as the interior of a d-dimensional hypercube, with all points on the boundary identified as a single point. For instance, we have already seen that the circle S_1 can be treated as the interval $0 \leq \theta \leq 2\pi$, with points $\theta = 0$ and $\theta = 2\pi$ identified. Similarly we can make a map of an S_2 like the earth's surface by cutting out the south pole and spreading out the resulting sheet on the unit square $0 \leq z_1 \leq 1$, $0 \leq z_2 \leq 1$. Continuous mappings of this square into $\mathcal{M}$ must take all points on the boundary into the same point of $\mathcal{M}$, because all points on the boundary are the same point, the south pole. In general, two mappings $p(z_1, \cdots z_d)$ and $p'(z_1, \cdots z_d)$ of S_d into $\mathcal{M}$ are homotopically equivalent if one can be continuously deformed into the other, while keeping p on the boundary of the hypercube equal to the base point p_0.

As before, for each equivalence class c we choose a standard mapping $p(z_1, \cdots z_d; c)$. The product of c_1 and c_2 is defined as the equivalence class that contains the mapping

$$p(z_1, z_2, \cdots z_d; c_1, c_2) = \begin{cases} p(2z_1, z_2, \cdots z_d; c_1) & 0 \le z_1 \le \frac{1}{2} \\ p(2z_1 - 1, z_2, \cdots z_d; c_2) & \frac{1}{2} \le z_1 \le 1 \end{cases} . \tag{23.2.6}$$

The unit element e is defined as the equivalence class that contains the mapping with $p = p_0$ for all z, and the inverse c^{-1} of c is defined as the equivalence class that contains the mapping with

$$p^{-1}[z_1, z_2, \cdots z_d; c] = p[1 - z_1, z_2, \cdots z_d; c] . \tag{23.2.7}$$

In the same way as for π_1, it can be shown that this multiplication is associative, and that $e \times c = c \times e = c$ and $c^{-1} \times c = c \times c^{-1} = e$. All $\pi_n(\mathscr{M})$ for $n \ge 2$ are Abelian. (There are manifolds $\mathscr{M}$ for which $\pi_1(\mathscr{M})$ is non-Abelian, such as the plane with two or more points removed.)

In any case in which $\pi_k(\mathscr{M}) = Z$, there must be a one-to-one mapping of the k-sphere S_k into a k-sphere S'_k in $\mathscr{M}$, which corresponds to the element 'one' of Z (not the unit element, which is zero). The element v of Z with $v = 2, 3, \ldots$ corresponds to the mapping of S_k into the same k-sphere S'_k in $\mathscr{M}$, which covers S'_k v times, with the Jacobian of the transformation $S_k \to S'_k$ positive. The element v of Z with $v = -1, -2, \ldots$ corresponds to the mapping $S_k \to S'_k$, which covers S'_k $|v|$ times, with the Jacobian of the transformation $S_k \to S'_k$ negative.

For instance, we saw in the previous section that magnetic monopoles arise when a simply connected group G is broken to the $U(1)$ of electromagnetism. In this case the appendix shows that

$$\pi_2(G/U(1)) = \pi_1(U(1)) = Z , \tag{23.2.8}$$

so a magnetic monopole carries an integer-valued quantum number v, which as shown in Section 23.3 is proportional to the magnetic charge. This quantum number gives the number of times that a two-sphere of large radius surrounding the monopole is mapped into a two-sphere in the manifold $G/U(1)$ of Goldstone boson fields (with the relative orientation of the two two-spheres being the same or opposite for v positive or negative, respectively) and is therefore known as the *winding number*. The structure of the group Z shows that this quantum number is conserved, in the sense that a monopole of quantum number v can fuse with a monopole of quantum number v' only to form a monopole with quantum number $v + v'$.

If the unbroken subgroup is $SO(n)$ with $n \ge 3$, then according to Appendix B of this chapter

$$\pi_2(G/SO(n)) = \pi_1(SO(n)) = Z_2 . \tag{23.2.9}$$

In this case there is just one sort of 'monopole,' corresponding to the element -1 of Z_2, which can annihilate only in pairs. It is important to distinguish between this case and that in which $SO(n)$ is replaced with its simply connected covering group $Spin(n)$, for which there are no monoples. We will come back to this point at the end of the next section.

Another example: we saw in the previous section that the skyrmions in quantum chromodynamics with n light quarks correspond to elements of $\pi_3(SU(n))$, which according to the appendix is Z. Thus these skyrmions carry a conserved integer-valued quantum number ν, which perhaps may be identified with baryon number. Similarly, recall from the previous section that the instantons in a gauge theory based on a simple gauge group G correspond to the elements of $\pi_3(G)$, which according to the appendix is Z, so these instantons like skyrmions carry an integer-valued quantum number ν, also known as the winding number. In Section 23.5 we shall see how to express this quantum number as a local functional of the gauge field.

23.3 Monopoles

As a detailed example of a topologically non-trivial field configuration, we shall now consider the monopole of 't Hooft and Polyakov,[2] and its generalizations. We saw in Section 23.1 that when a simply connected gauge group G is spontaneously broken to the $U(1)$ of electromagnetism, the configurations of finite energy are classified according to the elements of the group $\pi_2(G/U(1)) = \pi_1(U(1)) = Z$. (The case of non-simply connected Lie groups is considered at the end of this section.) According to the physical interpretation of homotopy groups discussed in Section 23.2, this means that these configurations have a conserved additive quantum number. But we still need to show that any of these stationary configurations actually exist, and to give a physical interpretation of their topological quantum numbers.

As an illustrative example, consider a theory (like the Georgi–Glashow electroweak model[12]) in which an $SU(2)$ gauge group is spontaneously broken by the vacuum expectation value of an $SU(2)$ triplet of scalar fields ϕ_n. (It is explained at the end of this section why in this case we say that the gauge group is $SU(2)$ rather than $SO(3)$.) The Lagrangian density for the scalars and gauge fields in Minkowskian spacetime is taken as

$$\mathscr{L} = -\tfrac{1}{4}F_{n\,\mu\nu}F_n^{\mu\nu} - \tfrac{1}{2}D_\mu\phi_n D^\mu\phi_n - V(\phi_n\phi_n)\,, \tag{23.3.1}$$

where

$$F_{n\,\mu\nu} \equiv \partial_\mu A_{n\,\nu} - \partial_\nu A_{n\,\mu} + e\,\epsilon_{nml}A_{m\,\mu}A_{l\,\nu}\,, \tag{23.3.2}$$

$$D_\mu\phi_n \equiv \partial_\mu\phi_n + e\,\epsilon_{nml}A_{m\,\mu}\phi_l\,, \tag{23.3.3}$$

and the function $V(\phi_n\phi_n)$ is assumed to be positive, with the value zero at a non-vanishing value $\langle\phi\rangle$ (taken positive) of $\sqrt{\phi_n\phi_n}$. (In much work on monopoles V is taken to be the quartic polynomial $\lambda(\phi_n\phi_n - \langle\phi\rangle^2)^2$, with $\lambda > 0$, but we shall not make this assumption here.) Eq. (23.3.1) describes a theory with a spacetime-independent vacuum solution with $A_{n\mu} = 0$, in which a vacuum expectation value ϕ_n with $\phi_n\phi_n = \langle\phi\rangle^2$ breaks $SU(2)$ to its $U(1)$ subgroup, which can be identified with the gauge group of electrodynamics. Instead we here shall seek topologically non-trivial inhomogeneous but time-independent classical solutions in temporal gauge, with $A^0_n = 0$ but $A^i_n \neq 0$. The Lagrangian density in this case is the negative of the potential energy density $\mathscr{H}$, given by

$$\mathscr{H} = \tfrac{1}{4}F^2_{nij} + \tfrac{1}{2}(D_i\phi_n)^2 + V(\phi)\,, \tag{23.3.4}$$

with squares including obvious index contractions. Because each term in (23.3.4) is positive, the integral of each term must separately converge in a configuration of finite energy.

In particular, for the integral of $V(\phi_n\phi_n)$ to converge, the vector ϕ_n must have the fixed length $\langle\phi\rangle$ at infinity, so each configuration of finite energy defines a smooth mapping of a large two-sphere $\mathscr{S}$ surrounding the monopole configuration into the two-sphere of the ϕ_n with $\phi_n\phi_n = \langle\phi\rangle^2$. As $\hat{\mathbf{x}}$ runs over $\mathscr{S}$, ϕ_n may run over the sphere $\phi_n\phi_n = \langle\phi\rangle^2$ any integer number N of times, either with Jacobian $\text{Det}\,(\partial x/\partial\phi)$ positive, in which case we say that the winding number is N, or with Jacobian negative, in which case the winding number is $-N$.

To see what the winding number has to do with the magnetic monopole moment, we must first consider what in this theory is observed as 'the' magnetic field. Whatever the field configuration, we can introduce a gauge in which the scalar field ϕ_n points in some definite direction, say the three-direction, in any given finite region, so that in this region the gauge field associated with the unbroken $U(1)$ subgroup of $SU(2)$ is A_{3i} . 't Hooft[2] found a gauge-invariant tensor $\mathscr{F}_{\mu\nu}$ which reduces to the usual electromagnetic field strength tensor $\partial_\mu A_{3\nu} - \partial_\nu A_{3\mu}$ in this gauge:

$$\mathscr{F}_{\mu\nu} \equiv F_{n\mu\nu}\hat{\phi}_n - \frac{1}{e}\epsilon_{nml}\,\hat{\phi}_n\,D_\mu\hat{\phi}_m\,D_\nu\hat{\phi}_l\,, \tag{23.3.5}$$

where $\hat{\phi}_n \equiv \phi_n/\sqrt{\phi_m\phi_m}$. To check that $\mathscr{F}_{\mu\nu}$ is the ordinary electromagnetic field strength tensor in a gauge with constant $\hat{\phi}_n$ (and for later purposes) we use Eqs. (23.3.2) and (23.3.3) and the identity $\epsilon_{abc}\epsilon_{ade} = \delta_{bd}\delta_{ce} - \delta_{be}\delta_{cd}$ to write $\mathscr{F}_{\mu\nu}$ in the form[13]

$$\mathscr{F}_{\mu\nu} = \partial_\mu(\hat{\phi}_nA_{n\nu}) - \partial_\nu(\hat{\phi}_nA_{n\mu}) - \frac{1}{e}\epsilon_{nml}\,\hat{\phi}_n\,\partial_\mu\hat{\phi}_m\,\partial_\nu\hat{\phi}_l\,. \tag{23.3.6}$$

Thus in a gauge in which $\hat{\phi}_n$ is a fixed unit vector in the three-direction, we have as promised

$$\mathscr{F}_{\mu\nu} = \partial_\mu A_{3\nu} - \partial_\nu A_{3\mu} \,.$$

The magnetic monopole moment g of any localized field configuration is defined as $1/4\pi$ times the magnetic flux through a large closed surface $\mathscr{S}$ around the configuration:

$$4\pi g \equiv \frac{1}{2}\epsilon_{ijk}\int_{\mathscr{S}} \mathscr{F}_{ij}\, d^2S_k \,. \tag{23.3.7}$$

The first two terms in Eq. (23.3.6) for $\mathscr{F}_{ij}$ are derivatives, and therefore do not contribute to the integral (23.3.7), so that

$$g = -\frac{1}{8\pi e}\epsilon_{ijk}\epsilon_{nml}\int_{\mathscr{S}} \hat{\phi}_n\,\partial_i\hat{\phi}_m\,\partial_j\hat{\phi}_l\, d^2S_k \,. \tag{23.3.8}$$

This has the important property of being a topological invariant: integrating by parts where necessary, we can see that the change in g under an infinitesimal variation $\delta\hat{\phi}_n$ in $\hat{\phi}_n$ is

$$\delta g = -\frac{3}{8\pi e}\epsilon_{ijk}\,\epsilon_{nml}\int_{\mathscr{S}} \delta\hat{\phi}_n\,\partial_i\hat{\phi}_m\,\partial_j\hat{\phi}_l\, d^2S_k \,.$$

But because $\hat{\phi}$ is a unit vector, $\delta\hat{\phi}$ as well as $\partial_i\hat{\phi}$ and $\partial_j\hat{\phi}$ are all in the plane perpendicular to $\hat{\phi}$, so

$$\epsilon_{nml}\,\delta\hat{\phi}_n\,\partial_i\hat{\phi}_m\,\partial_j\hat{\phi}_l = 0\,,$$

and therefore $\delta g = 0$. The quantity (23.3.8) is related to the topological invariant known as the *Kronecker index*.

Because g is a surface integral, it is additive: for any two distant localized configurations, the surface $\mathscr{S}$ used to calculate g may be taken to be a pair of spheres, one surrounding each configuration, connected by a thin neck between them, so the value of g for the whole system will be the sum of the values of g_1, g_2 for the individual localized configurations. Furthermore, since g is a topological invariant, we will have $g = g_1 + g_2$ for any field configuration that is formed by a smooth fusion of two configurations with magnetic monopole moments g_1 and g_2. It follows that g must be proportional to the winding number. Arafune, Freund, and Goebel[13] verified this and calculated the coefficient of proportionality by using the formula (23.3.8) for general winding number. Here we will simply calculate the coefficient by studying the 't Hooft–Polyakov monopole,[2] in which the fields have unit winding number.

As we saw in the previous section, the 'identity' (as opposed to the unit) element $\mathscr{I}$ of $\pi_2(SU(2)/U(1))$, which corresponds to the element 'one' of Z, consists of configurations in which the two-sphere $\mathscr{S}$ at infinity is mapped once (with positive Jacobian) into the sphere described by $\hat{\phi}_n$. As a representative of this homotopy class, we may take a configuration in which at infinity ϕ_n is in the same direction as **x**. To construct this configuration let us impose a symmetry under joint rotations on $\boldsymbol{\phi} = \{\phi_1, \phi_2, \phi_3\}$ and **x**, as well as parity conservation, and make the ansatz:

$$\phi_n = \hat{x}_n \langle\phi\rangle F(r)\,, \tag{23.3.9}$$

$$A_{ni} = \frac{\epsilon_{nil}\hat{x}_l}{er} G(r)\,. \tag{23.3.10}$$

There is an important similarity between this field configuration and that for a vortex line in a superconductor. The solution $\phi = \pm\ell\varphi/2e$ for the Goldstone boson field in a vortex line found in Section 21.6 shows that, although gauge invariance and rotational invariance are spontaneously broken, the vortex line solution is invariant under the combination of a global gauge transformation for which $\phi \to \phi + \Lambda$ and a rigid rotation $\varphi \to \varphi \pm 2e\Lambda/\ell$. Similarly, a monopole solution of the form (23.3.9)–(23.3.10) is not invariant under rotations or gauge transformations, but it is invariant under the combination of a rigid three-dimensional spatial rotation and an equal global $SO(3)$ gauge transformation.

As already mentioned, in order for the integral of $V(\phi_n\phi_n)$ to converge it is necessary for $\phi_n\phi_n$ to approach $\langle\phi\rangle^2$ as $r \to \infty$, so in this limit $F(r) \to 1$. To derive the limiting behavior of $G(r)$, note that the covariant derivative of the scalar field is

$$D_i\phi_n = \pm\langle\phi\rangle\left[\Big(1 - G(r)\Big)\Big(\delta_{ni} - \hat{x}_n\hat{x}_i\Big)\frac{F(r)}{r} + \hat{x}_n\hat{x}_i F'(r)\right]\,, \tag{23.3.11}$$

so the scalar term in the Hamiltonian density is

$$\tfrac{1}{2}(D_i\phi_n)^2 = \langle\phi\rangle^2\left[\frac{F^2(1-G)^2}{r^2} + \frac{F'^2}{2}\right]\,. \tag{23.3.12}$$

For this to have a finite integral it is necessary also that $G(r) \to 1$ and $F'(r) \to 0$ as $r \to \infty$. Finally, the field strength (23.3.2) is

$$F_{nij} = \frac{\epsilon_{ijk}}{e}\left[-\frac{1}{r}G'(r)\Big(\delta_{nk} - \hat{x}_n\hat{x}_k\Big) - \frac{1}{r^2}\Big(2G(r) - G^2(r)\Big)\hat{x}_n\hat{x}_k\right]\,, \tag{23.3.13}$$

so the Yang–Mills term in the Hamiltonian density is

$$\tfrac{1}{4}(F_{nij})^2 = \frac{1}{e^2}\left[\frac{G'^2}{r^2} + \frac{(2G - G^2)^2}{2r^4}\right]\,. \tag{23.3.14}$$

This has an integral that converges at large distances as long as $G'(r)$ vanishes sufficiently rapidly as $r \to \infty$.

We can use these results to calculate the magnetic monopole moment of this configuration. Eq. (23.3.13), together with the limits $G(\infty) = 1$ and

$G'(\infty) = 0$, shows that for $r \to \infty$, we have

$$F_{nij} \to -\frac{\epsilon_{ijk}\hat{x}_k\hat{x}_n}{er^2} . \tag{23.3.15}$$

Since $D_i\phi_n$ vanishes rapidly as $r \to \infty$, the magnetic part of the field strength tensor in this gauge is given for $r \to \infty$ by the first term in Eq. (23.3.5), so at large distances the magnetic field becomes

$$B_i \equiv \tfrac{1}{2}\epsilon_{ijk}\mathscr{F}_{jk} \to \tfrac{1}{2}\epsilon_{ijk}\hat{\phi}_n F_{njk} \to -\frac{\hat{x}_i}{er^2} . \tag{23.3.16}$$

Thus *this configuration has magnetic monopole moment* $g = -1/e$. According to the general argument above, the magnetic moment of a configuration with winding number ν, corresponding to the element ν of Z, is then

$$g_\nu = -\nu/e . \tag{23.3.17}$$

It requires a detailed numerical calculation to find the stable configuration that minimizes the integral of the Hamiltonian density (23.3.4). However there is a limiting case where an analytic solution is available. To see this, it is useful first to derive a general lower bound due to Bogomol'nyi[10] on the monopole energy for a given magnetic monopole moment g. Note that (23.3.4) may be written

$$\mathscr{H} = \tfrac{1}{4}\Big(F_{nij} \mp \epsilon_{ijk}D_k\phi_n\Big)^2 \pm \tfrac{1}{2}\epsilon_{ijk}F_{nij}D_k\phi_n + V(\phi_n\phi_n) . \tag{23.3.18}$$

Using the Bianchi identity (15.3.9), the second term may be written

$$\pm \tfrac{1}{2}\epsilon_{ijk}F_{nij}D_k\phi_n = \pm \tfrac{1}{2}\epsilon_{ijk}D_k\Big(F_{nij}\phi_n\Big) = \pm \tfrac{1}{2}\epsilon_{ijk}\partial_k\Big(F_{nij}\phi_n\Big) ,$$

so its integral is given by the magnetic monopole moment g

$$\pm \tfrac{1}{2}\epsilon_{ijk}\int d^3x\, F_{nij}D_k\phi_n = \pm\langle\phi\rangle\int \mathbf{B}\cdot d\mathbf{A} = \pm 4\pi\langle\phi\rangle g .$$

Since every other term in $\mathscr{H}$ is positive, we have a general lower bound on the energy of a configuration with magnetic monopole moment g

$$E = \int d^3x\, \mathscr{H} \geq 4\pi\langle\phi\rangle|g| . \tag{23.3.19}$$

For $g = \pm 1/e$, this gives an energy $E \geq 4\pi\langle\phi\rangle/e$, which for small coupling constant e is much greater than the corrections due to quantum fluctuations, which are at most of order $\langle\phi\rangle$. This is why we can take such a classical configuration seriously as the leading term in a perturbation expansion.

Now, it is tempting to try to minimize the energy for a given magnetic monopole moment by setting the first term in Eq. (23.3.18) equal to zero, so that

$$F_{nij} = \pm\epsilon_{ijk}D_k\phi_n , \tag{23.3.20}$$

but in general this does not lead to a configuration at which the energy is stationary. The condition that the energy should be stationary with respect to variations in the scalar field is the field equation

$$D_k D_k \phi_n = 2\phi_n V'(\phi_n \phi_n)\,, \tag{23.3.21}$$

while Eq. (23.3.20) together with the Bianchi identity (15.3.9) would imply that $D_k D_k \phi_n = 0$. This argument suggests that in the special case where $V(\phi_n \phi_n)$ is very small, it *will* be possible nearly to reach the lower bound (23.3.19), and hence minimize the energy for a given magnetic monopole moment, by imposing the condition (23.3.20). (Where V is the quartic polynomial $\lambda(\phi_n\phi_n - \langle\phi\rangle^2)^2$, the assumption that V is small means that $\lambda \ll e^2$, as in a type I superconductor.) Such stable configurations were studied in this way by Bogomol'nyi[10]. They had been found earlier by Prasad and Sommerfeld[14] without direct use of Eq. (23.3.20), and are usually called *BPS monopoles.*

The Bogomol'nyi condition (23.3.20) provides first-order differential equations for $F(r)$ and $G(r)$, which are much easier to solve than the second-order field equations derived directly from the condition that the energy should be stationary. Using Eqs. (23.3.11) and (23.3.13), the terms in Eq. (23.3.20) proportional to $\epsilon_{ijk}[\delta_{kn} - \hat{x}_k\hat{x}_n]$ and $\epsilon_{ijk}\hat{x}_k\hat{x}_n$ respectively yield the differential equations

$$e\langle\phi\rangle F(1-G) = G'\,, \tag{23.3.22}$$

$$e\langle\phi\rangle r^2 F' = G(2-G)\,. \tag{23.3.23}$$

With the boundary condition that $F(r) \to 1$ and $G(r) \to 1$ as $r \to \infty$, these equations have the solution

$$F = \coth\rho - \frac{1}{\rho}\,, \qquad G = 1 - \frac{\rho}{\sinh\rho}\,, \tag{23.3.24}$$

where $\rho \equiv e\langle\phi\rangle r$. Note in particular that the field ϕ_n given by Eqs. (23.3.9) and (23.3.24) vanishes for $r \to 0$, so as remarked in Section 23.1, the $SU(2)$ symmetry is restored at the center of the monopole.

Let's now return to the case of a potential V of arbitrary strength. The 't Hooft–Polyakov monopole is stable, because there are no configurations of smaller topological quantum number into which it could decay. Configurations of higher magnetic monopole moment are generally unstable.[10] There are also interesting configurations with both magnetic monopole moments and electric charge, known as *dyons.*[15]

There is another way of understanding the value $1/e$ of the 't Hooft–Polyakov magnetic monopole moment, which goes back to the original work of Dirac on magnetic monopoles.[16] As mentioned earlier, instead of the gauge we have been using, we can make a gauge transformation $\phi_n \to R_{nm}(\mathbf{x})\phi_m$ that rotates ϕ_n to point in a fixed direction $\hat{\mathbf{v}}$, for instance the three-direction. Then the field strength $B_{nk} \equiv \frac{1}{2}\epsilon_{ijk}F_{nij}$ is transformed into $R_{nm}B_{mk}$, which approaches $\hat{v}_n\hat{x}_k/er^2$ for $r \to \infty$, so here we do not have to project this on a local unbroken symmetry direction. The price we need to pay for these conveniences is that the gauge transformation is singular; the rotation that takes a vector in the $\hat{\mathbf{x}}$ direction into some fixed direction $\hat{\mathbf{v}}$ is

$$R(\hat{\mathbf{x}};\hat{\mathbf{v}}) = 1 - \left(1 - \hat{\mathbf{v}}\cdot\hat{\mathbf{x}}\right)\hat{\mathbf{v}}\hat{\mathbf{v}}^T + \hat{\mathbf{v}}\left(\hat{\mathbf{x}} - (\hat{\mathbf{x}}\cdot\hat{\mathbf{v}})\hat{\mathbf{v}}\right)^T + \left(\hat{\mathbf{x}} - (\hat{\mathbf{x}}\cdot\hat{\mathbf{v}})\hat{\mathbf{v}}\right)\hat{\mathbf{v}}^T$$
$$- \frac{\left(\hat{\mathbf{x}} - (\hat{\mathbf{x}}\cdot\hat{\mathbf{v}})\hat{\mathbf{v}}\right)\left(\hat{\mathbf{x}} - (\hat{\mathbf{x}}\cdot\hat{\mathbf{v}})\hat{\mathbf{v}}\right)^T}{1 + \hat{\mathbf{x}}\cdot\hat{\mathbf{v}}}, \tag{23.3.25}$$

which is singular at $\hat{\mathbf{x}} = -\hat{\mathbf{v}}$. This R is not unique; for instance, we could perform the rotation $R(\hat{\mathbf{x}};-\hat{\mathbf{v}})$ that takes $\hat{\mathbf{x}}$ into the $-\hat{\mathbf{v}}$ direction, followed by a fixed rotation of 180° around some axis perpendicular to $\hat{\mathbf{v}}$, but this would become singular at $\hat{\mathbf{x}} = +\hat{\mathbf{v}}$. To avoid the singularities, we have to adopt different gauges in different regions; for instance, with $\hat{\mathbf{v}}$ in the three-direction, we can use a gauge for $0 < \theta < \theta_0$ that is singular at $\theta = \pi$ and a gauge for $\theta_0 < \theta < \pi$ that is singular at $\theta = 0$, where θ_0 is an arbitrary angle with $0 < \theta_0 < \pi$, often taken as $\pi/2$. Everywhere except at $\theta = 0$ and $\theta = \pi$ the magnetic field will be given at large distances by $B \to g\hat{\mathbf{x}}/r^2$, where g is the magnetic monopole strength. This can be written as the curl $\nabla \times \mathbf{A}$ of a vector potential whose only non-vanishing component is in the azimuthal φ direction. For $0 < \theta < \theta_0$ we must take $A_\varphi = g(1 - \cos\theta)/r\sin\theta$, which is only singular at $\theta = \pi$, while for $\theta_0 < \theta < \pi$ we must take $A_\varphi = -g(1 + \cos\theta)/r\sin\theta$, which is only singular at $\theta = 0$. The difference $\Delta\mathbf{A}$ between these two vector potentials is a gradient $\nabla\Lambda$ with $\Lambda = 2g\varphi$, which of course does not affect the magnetic field for $0 < \theta < \pi$, but could affect the dynamics of charged fields. A gauge transformation with $\Lambda = 2g\varphi$ will change a field of charge q by a factor $\exp(2iqg\varphi)$, which is not single-valued unless $2qg$ is an integer. This is the Dirac quantization condition; the existence of any magnetic monopole with monopole moment g would require all electric charges to be integer multiples of $(2g)^{-1}$. For the 't Hooft–Polyakov monopole this condition is automatically satisfied, because here $g = 1/e$, and all charges in the Georgi–Glashow model are integer multiples of $e/2$.

The Georgi–Glashow model was ruled out as a theory of weak and electromagnetic interactions by the discovery of neutral currents, but magnetic monopoles are expected to occur in other theories, where a simply connected group G is spontaneously broken not to $U(1)$, but to some subgroup $H' \times U(1)$, where H' is simply connected. (According to Appendix B of this chapter, for simply connected groups G we have $\pi_2(G/H) = \pi_1(H)$, which for $H = H' \times U(1)$ equals $\pi_1(H') \times \pi_1(U(1)) = \pi_1(U(1)) = Z$.) There are no monopoles produced in the spontaneous breaking of the gauge group

$SU(2) \times U(1)$ of the standard electroweak theory, which is not simply connected. (About this, more below.) But we do find monopoles when the simply connected gauge group G of theories of unified strong and electroweak interactions, such as $SU(4) \times SU(4)$ or $SU(5)$ or $Spin(10)$, is spontaneously broken to the gauge group $SU(3) \times SU(2) \times U(1)$ of the standard model. (See Section 21.5.) The monopoles in this case are expected to have a mass larger by an inverse square gauge coupling constant than the vector boson masses $M \approx 10^{15}$–10^{16} GeV produced by this symmetry breaking. Such monopoles would have been produced when the universe underwent a phase transition in which G was spontaneously broken to $SU(3) \times SU(2) \times U(1)$, at a temperature T of order M.

This poses a problem for some cosmological models.[17] The scalar fields before this phase transition would have necessarily been uncorrelated at distances larger than the horizon distance, the furthest distance that light could have travelled since the initial singularity. At an early time t in standard cosmological theories[18] the horizon distance is of order $t \approx (G_N T^4)^{-1/2}$ (where $G_N \simeq (10^{19}\text{ GeV})^{-2}$ is Newton's constant), so the number density of monopoles produced at this time would have been of order $t^{-3} \approx (G_N M^4)^{3/2}$, which is smaller than the photon density M^3 at $T \approx M$ by a factor of order $(G_N M^2)^{3/2}$. For $M \approx 10^{15}$ GeV this factor is of order 10^{-12}. If monopoles did not find each other to annihilate, then this ratio would remain roughly constant to the present, but with at least 10^9 microwave background photons per nucleon today, this would give at least 10^{-3} monopoles per nucleon, in gross disagreement with what is observed. This potential paradox was one of the factors leading to inflationary cosmological models,[19] in which there was a period of exponential expansion, which if it occurred before the monopoles were produced would have greatly extended the horizon, and if it occurred after the production of monopoles (but before a period of reheating) would have greatly diluted the monopole density.

The discovery of monopoles of any sort would create opportunities for the observation of remarkable phenomena, including the existence of fermion–monopole configurations of fractional fermion number,[20] and the violation of baryon conservation in fermion–monopole scattering.[21]

* * *

In the above discussion we have considered only monopoles associated with the spontaneous breakdown of a simply-connected gauge group G. This raises a question. For every Lie group G, whether simply connected or not, there is a simply-connected group $\overline{G}$ with the same Lie algebra, known as its *covering group*. (For examples, see Section 2.7.) Any non-simply connected group has fewer representations than its covering group (for instance, the doubly-connected groups $SO(n)$ have only scalar, vector, and tensor representations, while their covering groups $Spin(n)$ also have spinor representations). If a theory does not happen to involve fields that belong to the extra representations of the covering group, then we are free to consider the gauge group of the theory to be either the non-simply connected group G or its covering group $\overline{G}$? In particular, does the menu of possible monopoles depend on whether the theory contains only fields transforming as representations of a non-simply connected group G, or additional fields that would only furnish representations of its covering group $\overline{G}$? For instance, in the original Georgi–Glashow model[12] the only scalar fields belonged to a representation of $SO(3)$, the three-vector, but there were fermions that belonged to spinor representations of the covering group $Spin(3) = SU(2)$. Would the allowed values of the magnetic monopole moment change if we added scalar fields belonging to the spinor representations of $SU(2)$? Would it change if we removed the fermions?

The answer is that the menu of possible monopoles does not depend on whether we say that the gauge group is a non-simply connected group G or its covering group $\overline{G}$, and therefore is unaffected when we add or remove fields belonging to representations of $\overline{G}$ that are not representations of G. As we saw in Section 23.1, in general the topologically stable monopole-like configurations are classified according to the elements of $\pi_2(G/H)$. According to a result quoted in Appendix B of this chapter, this homotopy group consists of those elements of $\pi_1(H)$ that correspond to the trivial element of $\pi_1(G)$ when H is embedded in G. But if we replace G with its covering group $\overline{G}$, we also replace H with a different subgroup H', because some of the loops in H do not return to the base point when H is embedded in $\overline{G}$. These are just the loops that do not become trivial when H is embedded in G, so $\pi_2(G/H) = \pi_1(H')$, and thus as far as monopoles are concerned we could just as well say that the gauge group is $\overline{G}$ rather than G.

For instance, as far as the scalar fields are concerned, the gauge group of the Georgi–Glashow model might be considered to be the doubly-connected $SO(3)$ rather than its simply-connected covering group $SU(2)$. The unbroken subgroup is then $SO(2)$, in which we identify transformations that differ by a 360° rotation. Thus $\pi_1(SO(2))$ includes loops that extend from the unit element to a 360° rotation, which would not be loops when $SO(3)$ is embedded in $SU(2)$. But $\pi_2(SO(3)/SO(2))$ is not the same as $\pi_1(SO(2))$, but rather excludes the loops that are homotopically non-trivial when $SO(3)$ is embedded in $SU(2)$, which are just the loops that extend from the unit element to a 360° rotation, so $\pi_2(SO(3)/SO(2))$ is the $U(1)$ subgroup of $SU(2)$, just as if the gauge group were taken to be $SU(2)$ from the beginning.

It is convenient always to consider the gauge group G associated with any semi-simple gauge algebra to be the simply-connected covering group, so that we can use the simple result that $\pi_2(G/H) = \pi_1(H)$. As we have just seen, the connectivity properties of H are then fixed by its embedding in G, or more precisely, by the embedding of the Lie algebra of H in the Lie algebra of G. For instance, an $SU(3)$ gauge algebra might be spontaneously broken either into an $SU(2)$ subalgebra, under which the defining representation of $SU(3)$ transforms as a doublet plus a singlet, or into an $SO(3)$ subalgebra, under which the defining representation of $SU(3)$ transforms as a three-vector. In the first case we do not have the option of considering the unbroken subgroup to be $SO(3)$; since $\pi_1(SU(2)) = 0$, there are no monopoles here. In the second case the unbroken gauge group must be regarded as $SO(3)$, not $SU(2)$, so the theory does have monopole-like configurations, classified according to the elements of $\pi_1(SO(3)) = Z_2$. The nature of the unbroken subalgebra of H and its embedding in the gauge algebra G may be dynamically affected by the variety of field types that we introduce in the Lagrangian, but once the algebra of H and its embedding in the algebra of G are fixed, the menu of monopoles is otherwise entirely unaffected by the variety of fields in the theory.

In particular, the argument that led to the Dirac quantization condition shows that, in any theory in which a Lie algebra G is spontaneously broken to a subalgebra including the electric charge operator, the allowed magnetic monopole moments are integer multiples of the reciprocal of the smallest electric charge that appears in the representations of the covering group of G, whether or not there is actually any particle that carries that charge in the theory. If the algebra of G itself contains a $U(1)$ generator, then we must consider the covering group of this $U(1)$, which is the non-compact group of translations along the real line. If this $U(1)$ generator appears as a term in the electric charge operator, as in the standard electroweak theory, then there is no minimum electric charge in the representations of the covering group, and hence no monopoles.

23.4 The Cartan–Maurer Integral Invariant

In understanding the topology of various compact manifolds, it is a great help that there is often a topologically invariant quantity that can be written as an integral over the manifold. This will be important in our discussion of instantons in the next section, and it has already been used in studying Wess–Zumino–Witten terms in Sections 19.8 and 22.7.

Consider a mapping of an arbitrary compact manifold $\mathscr{S}$ of odd dimensionality d with coordinates $\theta^1, \theta^2, \cdots \theta^d$ into a manifold $\mathscr{M}$ of matrices $g(\theta^1, \theta^2, \cdots \theta^d)$ with $\operatorname{Det} g \neq 0$. (For the applications that concern us here,

$\mathscr{S}$ is usually a sphere S_d, and the g are the elements of a Lie group G in some representation.) We define a functional of $g(\theta)$, known as the *Cartan–Maurer form*:

$$\mathscr{I}[g] = \int d\theta^1\, d\theta^2 \cdots d\theta^d\, \epsilon^{i_1 i_2 \cdots i_d}$$
$$\times \operatorname{Tr}\left\{ g^{-1}(\theta)\frac{\partial g(\theta)}{\partial \theta^{i_1}}\, g^{-1}(\theta)\frac{\partial g(\theta)}{\partial \theta^{i_2}} \cdots g^{-1}(\theta)\frac{\partial g(\theta)}{\partial \theta^{i_d}} \right\}, \quad (23.4.1)$$

where $\epsilon^{i_1 i_2 \cdots i_d}$ is the totally antisymmetric quantity with $\epsilon^{12\cdots d} = 1$. From the fact that $\epsilon^{i_1 i_2 \cdots i_d} = -(-1)^d \epsilon^{i_2 \cdots i_d i_1}$, we see that $\mathscr{I}[g]$ vanishes where $\mathscr{S}$ is even-dimensional, so we will restrict ourselves here to the case where d is odd. The usefulness of this quantity arises from its several remarkable properties.

First, this integral is independent of the coordinate system used to parameterize the manifold $\mathscr{S}$. This follows rather obviously from the fact that $\epsilon^{i_1 i_2 \cdots i_d}$ is a contravariant tensor density, in the sense that

$$\epsilon^{i_1 i_2 \cdots i_d} \frac{\partial \theta'^{j_1}}{\partial \theta^{i_1}} \frac{\partial \theta'^{j_2}}{\partial \theta^{i_2}} \cdots \frac{\partial \theta'^{j_d}}{\partial \theta^{i_d}} = \operatorname{Det}\left(\frac{\partial \theta'}{\partial \theta}\right) \epsilon^{j_1 j_2 \cdots j_d}\,.$$

Second, the integral (23.4.1) is also invariant under small deformations of the mapping $\mathscr{S} \mapsto \mathscr{M}$. Using the properties of the trace, we see that under an infinitesimal change $g \to g + \delta g$ of the function $g(\theta)$, the change in each factor $g^{-1}\partial g/\partial\theta^i$ in (23.4.1) makes the same contribution to the change in $\mathscr{I}[g]$:

$$\delta\mathscr{I}[g] = d \int d\theta^1\, d\theta^2 \cdots d\theta^d\, \epsilon^{i_1 i_2 \cdots i_d}$$
$$\times \operatorname{Tr}\left\{ g^{-1}(\theta)\frac{\partial g(\theta)}{\partial \theta^{i_1}}\, g^{-1}(\theta)\frac{\partial g(\theta)}{\partial \theta^{i_2}} \cdots \delta\left(g^{-1}(\theta)\frac{\partial g(\theta)}{\partial \theta^{i_d}} \right) \right\}.$$

Now, the last factor in the trace is

$$\delta\left(g^{-1}(\theta)\frac{\partial g(\theta)}{\partial \theta^{i_d}} \right) = -g^{-1}(\theta)\, \delta g(\theta)\, g^{-1}(\theta)\frac{\partial g(\theta)}{\partial \theta^{i_d}} + g^{-1}(\theta)\frac{\partial \delta g(\theta)}{\partial \theta^{i_d}}$$
$$= g^{-1}(\theta)\frac{\partial}{\partial \theta^{i_d}}\left(\delta g(\theta) g^{-1}(\theta)\right) g(\theta)\,.$$

When we integrate by parts, the derivative $\partial/\partial\theta^{i_d}$ gives no contribution when acting on the partial derivatives $\partial g(\theta)/\partial\theta^{i_n}$ because $\epsilon^{i_1 i_2 \cdots i_d}$ is antisymmetric. The remaining $d-1$ terms where $\partial/\partial\theta^{i_d}$ acts on $g^{-1}(\theta)$ are all equal except for an alternating sign, so since there are an even number of them they add up to zero.

Finally, let us specialize to the case where $\mathscr{S}$ is the sphere S_d. Because $\mathscr{I}[g]$ is invariant under small variations of $g(\theta)$, it can be regarded as a function $\mathscr{I}(c)$ only of the homotopy class c to which $g(\theta)$ belongs. The integrals $\mathscr{I}(c)$ (or strictly speaking, $\exp\{\mathscr{I}(c)\}$) furnish a representation of the homotopy group $\pi_d(\mathscr{M})$, in the sense that

$$\mathscr{I}(c_a \times c_b) = \mathscr{I}(c_a) + \mathscr{I}(c_b) . \tag{23.4.2}$$

(If $g_a(\theta)$ and $g_b(\theta)$ are elements of the homotopy classes c_a and c_b respectively, then the homotopy class $c_a \times c_b$ consists of mappings homotopically equivalent to

$$g_{ab}(\theta) = \begin{cases} g_a(2\theta_1, \theta_2 \cdots \theta_d) & 0 \le \theta_1 \le \frac{1}{2} \\ g_b(2\theta_1 - 1, \theta_2 \cdots \theta_d) & \frac{1}{2} \le \theta_1 \le 1 \end{cases} .$$

The part of the integral $\mathscr{I}[g_{ab}]$ over the hemispheres with $0 \le \theta_1 \le 1/2$ and $1/2 \le \theta_1 \le 1$ can be done by changing variables to $\theta_1' = 2\theta_1$ and $\theta_1' = 2\theta_1 - 1$, respectively, yielding the terms $\mathscr{I}(c_a)$ and $\mathscr{I}(c_b)$ in (23.4.2).)

In particular, this tells us that the homotopy classes e, c, $c \times c$, etc., and c^{-1}, $c^{-1} \times c^{-1}$, etc., have

$$\mathscr{I}(c^n) = n\,\mathscr{I}(c) . \tag{23.4.3}$$

If $\mathscr{I}(c) \neq 0$ for some c, then these invariants are all different, so the classes c^n are all different, and therefore form a subgroup Z of $\pi_d(\mathscr{M})$. This goes far to explain the great difference between the sizes of the homotopy groups for odd and even dimensions shown in Appendix B of this chapter. For instance, $\pi_1(U(1)) = Z$, $\pi_3(G) = Z$ for all simple Lie groups G, and $\pi_5(SU(n)) = Z$ for all $n \ge 3$, while for all Lie groups G, $\pi_2(G) = 0$ and $\pi_4(G)$ is finite.

As a simple example where $\mathscr{I}(c) \neq 0$, consider the homotopy group $\pi_1(U(1))$, which is the same as the group $\pi_1(S_1)$ used as an example at the beginning of the previous section. Any mapping $S_1 \mapsto U(1)$ may be characterized by ν, the number of times that the phase of the $U(1)$ element goes counterclockwise around S_1 minus the number of times it goes around S_1 in the opposite direction, as the coordinate θ goes around S_1, two mappings being homotopically equivalent if and only if they have the same ν. The νth class contains the mapping $g_\nu(\theta) = \exp(2i\nu\pi\theta)$ with $0 \le \theta \le 1$, for which

$$\mathscr{I}[g_\nu] = \int_0^1 d\theta \, \exp(-2i\nu\pi\theta)\frac{d}{d\theta}\exp(2i\nu\pi\theta) = 2i\nu\pi ,$$

thus verifying that $\pi_1(U(1)) = Z$.

As an aid to calculating $\mathscr{I}(g)$ in less simple cases, suppose that we can continuously deform the manifold $\mathscr{M}$ into a Lie group H of dimensionality d. The result of performing the H transformation with parameters θ followed by the H transformation with parameters φ is an H transformation, say with parameters $\theta'(\theta, \varphi)$. In terms of a matrix representation

$g(\theta)$, this reads

$$g(\varphi)g(\theta) = g(\theta'(\theta, \varphi)) .$$

Differentiating with respect to θ' with φ fixed and multiplying on the left with the inverse of this equation gives

$$\frac{\partial \theta^i}{\partial \theta'^j} g^{-1}(\theta)\frac{\partial g(\theta)}{\partial \theta^i} = g^{-1}(\theta')\frac{\partial g(\theta')}{\partial \theta'^j} .$$

The integrand of $\mathscr{I}[g]$ at a point θ' is therefore

$$\epsilon^{j_1 j_2 \cdots j_d}\, \mathrm{Tr}\left\{g^{-1}(\theta')\frac{\partial g(\theta')}{\partial \theta'^{j_1}} g^{-1}(\theta')\frac{\partial g(\theta')}{\partial \theta'^{j_2}} \cdots g^{-1}(\theta')\frac{\partial g(\theta')}{\partial \theta'^{j_d}}\right\}$$
$$= \mathrm{Det}\left(\frac{\partial \theta}{\partial \theta'}\right) \epsilon^{i_1 i_2 \cdots i_d}\mathrm{Tr}\left\{g^{-1}(\theta)\frac{\partial g(\theta)}{\partial \theta^{i_1}} g^{-1}(\theta)\frac{\partial g(\theta)}{\partial \theta^{i_2}} \cdots g^{-1}(\theta)\frac{\partial g(\theta)}{\partial \theta^{i_d}}\right\} .$$

Now, every Lie group H has a metric $\gamma_{ij}(\theta)$ (not necessarily unique) which is form-invariant in the sense that

$$\gamma_{ij}(\theta') \equiv \frac{\partial \theta^k}{\partial \theta'^i}\frac{\partial \theta^\ell}{\partial \theta'^j}\gamma_{k\ell}(\theta) . \tag{23.4.4}$$

For instance, we can take

$$\gamma_{ij}(\theta) = -\frac{1}{2}\mathrm{Tr}\left\{g^{-1}(\theta)\frac{\partial g(\theta)}{\partial \theta^i} g^{-1}(\theta)\frac{\partial g(\theta)}{\partial \theta^j}\right\} . \tag{23.4.5}$$

For any choice of $\gamma_{ij}(\theta)$, the determinant of Eq. (23.4.4) gives

$$\mathrm{Det}\left(\frac{\partial \theta}{\partial \theta'}\right) = \sqrt{\frac{\mathrm{Det}\,\gamma(\theta')}{\mathrm{Det}\,\gamma(\theta)}} .$$

By replacing the coordinates θ in Eq. (23.4.1) with θ', we find

$$\mathscr{I}[g] = \epsilon^{i_1 i_2 \cdots i_d}\, \mathrm{Tr}\left\{g^{-1}(\theta)\frac{\partial g(\theta)}{\partial \theta^{i_1}} g^{-1}(\theta)\frac{\partial g(\theta)}{\partial \theta^{i_2}} \cdots g^{-1}(\theta)\frac{\partial g(\theta)}{\partial \theta^{i_d}}\right\}$$
$$\times \frac{1}{\sqrt{\mathrm{Det}\,\gamma(\theta)}} \int d^d\theta'\, \sqrt{\mathrm{Det}\,\gamma(\theta')} . \tag{23.4.6}$$

Since the parameters φ of the second H transformation are arbitrary, we may regard θ and θ' as independent variables, and evaluate the right-hand side of (23.4.6) at any value of θ, say $\theta^i = 0$. It will be convenient to normalize the generators t_i and coordinates θ^i so that for $\theta \to 0$,

$$g(\theta) \to 1 + 2i\theta^i t_i . \tag{23.4.7}$$

In this case, Eq. (23.4.6) reads

$$\mathscr{I}[g] = (2i)^d \epsilon^{i_1 i_2 \cdots i_d}\, \mathrm{Tr}\,\{t_{i_1} t_{i_2} \cdots t_{i_d}\} \frac{1}{\sqrt{\mathrm{Det}\,\gamma(0)}} \int d^d\theta'\, \sqrt{\mathrm{Det}\,\gamma(\theta')} . \tag{23.4.8}$$

We will be especially interested in the case $d = 3$. Bott[22] has shown that for any simple Lie group G, all continuous mappings $S_3 \mapsto G$ may be continuously deformed into mappings of S_3 into a 'standard' $SU(2)$ subgroup of G. (Where $G = SU(n)$, this standard $SU(2)$ subgroup is the one that acts only on the first two components of the defining representation of $SU(n)$. Not all $SU(2)$ subgroups of $SU(n)$ are equivalent to this one.) As remarked in Section 2.7, in a 2×2 representation the general element of $SU(2)$ may be written

$$g(\theta) = \begin{pmatrix} \theta_4 + i\theta_3 & \theta_2 + i\theta_1 \\ -\theta_2 + i\theta_1 & \theta_4 - i\theta_3 \end{pmatrix} = \theta_4 + 2i\boldsymbol{\theta} \cdot \mathbf{t} \,, \qquad (23.4.9)$$

where as usual

$$t_1 = \frac{1}{2}\begin{pmatrix} 0 & 1 \\ 1 & 0 \end{pmatrix}, \qquad t_2 = \frac{1}{2}\begin{pmatrix} 0 & -i \\ i & 0 \end{pmatrix}, \qquad t_3 = \frac{1}{2}\begin{pmatrix} 1 & 0 \\ 0 & -1 \end{pmatrix},$$

and θ_4 and $\boldsymbol{\theta}$ are real, with $(\theta_4)^2 = 1 - \boldsymbol{\theta}^2$. (Note that Eq. (23.4.9) is consistent with the normalization convention (23.4.7).) A straightforward calculation gives the metric (23.4.5) as

$$\gamma_{ij}(\theta) = \delta_{ij} + \frac{\theta_i \theta_j}{1 - \boldsymbol{\theta}^2} \,, \qquad (23.4.10)$$

so that

$$\text{Det}\; \gamma(\theta) = \frac{1}{1 - \boldsymbol{\theta}^2} \,. \qquad (23.4.11)$$

Hence Eq. (23.4.8) here reads

$$\mathscr{I}[g] = -8i\epsilon^{ijk}\; \text{Tr}\; \{t_i t_j t_k\} \int \frac{d^3\theta}{\sqrt{1 - \boldsymbol{\theta}^2}} \,.$$

Using $4t_i t_j = \delta_{ij} + 2i\epsilon^{ij\ell} t_\ell$ and $\text{Tr}\,\{t_\ell t_k\} = \frac{1}{2}\delta_{\ell k}$, we see that

$$8\epsilon^{ijk}\; \text{Tr}\; \{t_i\, t_j t_k\} = 2i\epsilon^{ijk}\epsilon^{ijk} = 12i \,.$$

Also, for the 'identity' mapping g_1, the integral here runs twice over the interior of the unit ball (because θ_4 can be positive or negative), and gives

$$\int \frac{d^3\theta}{\sqrt{1 - \boldsymbol{\theta}^2}} = 2\int_0^1 \frac{4\pi\, r^2\, dr}{\sqrt{1 - r^2}} = 2\pi^2 \,.$$

For the class c of mappings homotopic to g_1, we have then

$$\mathscr{I}(c) = 24\pi^2 \qquad (23.4.12)$$

and so

$$\mathscr{I}(c^\nu) = 24\pi^2 \nu \,. \qquad (23.4.13)$$

The integer ν is known as the *winding number*. This result is for a representation and normalization conventions for which the standard

$SU(2)$ subalgebra has generators t_i with structure constants ϵ_{ijk} and $\text{Tr}\,(t_i t_j) = \frac{1}{2}\delta_{ij}$. More generally, if $[t_i, t_j] = ig\epsilon_{ijk}t_k$ and $\text{Tr}\,(t_i t_j) = \frac{1}{2}Ng^2\delta_{ij}$, then

$$\mathscr{I}(c^\nu) = 24\pi^2 N\nu \,. \tag{23.4.14}$$

The results (23.4.13) or (23.4.14) show incidentally that for every simple Lie group, $\pi_3(G)$ contains Z. As listed in Appendix B to this chapter, $\pi_3(G) = Z$ for all simple Lie groups. Thus the homotopy class of $g(\theta)$ for any simple Lie group is entirely determined by its homotopy class when the group is deformed into its standard $SU(2)$ subgroup.

23.5 Instantons

As we saw in Section 23.1, the topologically non-trivial solutions of a pure gauge theory with a simple gauge group G in $d = 4$ Euclidean spacetime dimensions correspond to the elements of the homotopy group $\pi_3(G) = Z$. These are four-dimensional field configurations, known as instantons, that (for reasons discussed in the next section) must be included along with their fluctuations in path integrals. After Belavin, Polyakov, Schwarz, and Tyupkin[1] demonstrated the existence of these solutions, 't Hooft[23] showed that the inclusion of these configurations in path integrals solved the $U(1)$ problem outlined in Section 19.10. Here we shall first discuss the instantons themselves, and then consider their role in path integrals.

According to Eq. (23.1.7), in order for a topologically non-trivial gauge field to have finite action, the gauge fields must approach a pure gauge at $r \to \infty$ (where here $r \equiv \sqrt{x_i x_i}$, with i summed over the values 1, 2, 3, 4):

$$iA_i(x) \to g^{-1}(\hat{x})\,\partial_i g(\hat{x}) \,, \tag{23.5.1}$$

where $A_i \equiv t_\alpha A_{\alpha i}$ and $g(\hat{x})$ is a direction-dependent element of the gauge group G. The topological invariant discussed in the previous section may therefore be written in terms of the asymptotic behavior of the gauge field

$$\begin{aligned}\mathscr{I}[g] \equiv \int d\theta^1\, d\theta^2\, d\theta^3\; \epsilon^{abc} \\ \times\, \text{Tr}\left\{ g^{-1}(\theta)\frac{\partial g(\theta)}{\partial\theta^a} g^{-1}(\theta)\frac{\partial g(\theta)}{\partial\theta^b} g^{-1}(\theta)\frac{\partial g(\theta)}{\partial\theta^c} \right\} \\ = -i \lim_{r\to\infty} r^3 \int d\theta^1\, d\theta^2\, d\theta^3\; \epsilon^{abc} \frac{\partial \hat{x}_i}{\partial\theta^a}\frac{\partial \hat{x}_j}{\partial\theta^b}\frac{\partial \hat{x}_k}{\partial\theta^c}\, \text{Tr}\,\{A_i A_j A_k\} \,, \end{aligned} \tag{23.5.2}$$

where θ^a with $a = 1, 2, 3$ are any three parameters used to specify the direction of the unit four-vector $\hat{x}$. This surface integral can be evaluated using Gauss's theorem. In analogy with the current (22.2.29) in Minkowski spacetime, we can define a current in Euclidean spacetime

$$G_l \equiv \epsilon^E_{lijk}\left[A_{\gamma i}F_{\gamma jk} - \tfrac{1}{3}C_{\alpha\beta\gamma}\,A_{\alpha i}\,A_{\beta j}\,A_{\gamma k}\right], \tag{23.5.3}$$

whose divergence is

$$\partial_l G_l = \tfrac{1}{2}\epsilon^E_{ijkl}F_{\alpha ij}F_{\alpha kl}\,. \tag{23.5.4}$$

(Here ϵ^E_{ijkl} is the totally antisymmetric tensor with $\epsilon^E_{1234} \equiv 1$.) We are using a representation of the gauge group with totally antisymmetric structure constants, so that

$$\mathrm{Tr}\,\{t_\alpha t_\beta\} = \tfrac{1}{2}N\delta_{\alpha\beta}\,, \tag{23.5.5}$$

with N a constant that depends on the representation in which we calculate the trace in Eq. (23.5.2). Hence Eq. (23.5.3) may be written

$$G_l \equiv (2/N)\,\epsilon^E_{lijk}\;\mathrm{Tr}\left[A_iF_{jk} + (2i/3)A_i\,A_j\,A_k\right]. \tag{23.5.6}$$

For $r \to \infty$, the field strength F_{kl} vanishes, so

$$G_l \to (4i/3N)\,\epsilon^E_{lijk}\;\mathrm{Tr}\left[A_i\,A_j\,A_k\right]. \tag{23.5.7}$$

Hence Eq. (23.5.2) gives

$$\mathscr{I}[g] = -(3N/4)\int (d^4x)_E\;\partial_l\,G_l = -(3N/8)\,\epsilon^E_{ijkl}\int (d^4x)_E\;F_{\alpha ij}\,F_{\alpha kl}\,. \tag{23.5.8}$$

Thus in order to demonstrate the existence of topologically non-trivial field configurations, we have to show that there are configurations for which the Chern–Pontryagin density $\epsilon^E_{ijkl}\,F_{\alpha ij}\,F_{\alpha kl}$ has a non-vanishing integral.

For this purpose, it is very useful to take advantage of what is known as the *Bogomol'nyi inequality*.[10] From the fact that

$$0 \le \int \left(F_{\alpha ij} \mp \tfrac{1}{2}\epsilon^E_{ijkl}F_{\alpha kl}\right)^2 (d^4x)_E$$

(with squares indicating obvious index contractions), we have

$$S[A] \ge \tfrac{1}{8}\left|\epsilon^E_{ijkl}\int F_{\alpha ij}\,F_{\alpha kl}\,(d^4x)_E\right| = |\mathscr{I}[g]|/3N\,, \tag{23.5.9}$$

where $S[A]$ is (up to a factor) the Euclidean action

$$S[A] \equiv \tfrac{1}{4}\int F_{\alpha ij}\,F_{\alpha ij}\,(d^4x)_E\,. \tag{23.5.10}$$

The lower bound (23.5.9) is evidently reached if and only if the gauge field is self-dual or anti-self-dual, in the sense that

$$F_{\alpha ij} = \pm\,\tfrac{1}{2}\,\epsilon^E_{ijkl}\,F_{\alpha kl}\,. \tag{23.5.11}$$

Hence any solution of the first-order equation (23.5.11) is a minimum of $S[A]$ for gauge fields of winding number unity, and hence also a solution of the second-order Yang–Mills field equation.

Belavin *et al.*[1] found a solution of Eq. (23.5.11) of the form

$$iA_i(x) = \left(\frac{r^2}{r^2+R^2}\right) g_1^{-1}(\hat{x})\partial_i g_1(\hat{x})\,, \tag{23.5.12}$$

where R is an arbitrary scale factor, and $g_1(\hat{x})$ is an element of an $SU(2)$ subgroup of the gauge group, with

$$g_1(\hat{x}) = \left(\frac{x_4 + 2i\mathbf{x}\cdot\mathbf{t}}{r}\right)\,, \tag{23.5.13}$$

and

$$t_1 = \frac{1}{2}\begin{pmatrix} 0 & 1 \\ 1 & 0 \end{pmatrix}\,, \quad t_2 = \frac{1}{2}\begin{pmatrix} 0 & -i \\ i & 0 \end{pmatrix}\,, \quad t_3 = \frac{1}{2}\begin{pmatrix} 1 & 0 \\ 0 & -1 \end{pmatrix}\,. \tag{23.5.14}$$

It is clear that this solution has the asymptotic behavior (23.5.1), with $g(\hat{\mathbf{x}})$ the same as the 'identity' mapping (23.4.9), so this solution belongs to the homotopy class of the identity map, and therefore as we saw in the previous section has winding number $\nu = 1$. From Eqs. (23.4.14) and (23.5.9) (which is here an equality) we have then

$$S[A] = 8\pi^2\,. \tag{23.5.15}$$

From Eqs. (23.5.10) and (23.5.11) (with positive sign), we also have

$$\epsilon^E_{ijkl}\int F_{\alpha ij}\,F_{\alpha kl}\,(d^4x)_E = 64\pi^2\,. \tag{23.5.16}$$

This solution is not unique, because it can be translated or subjected to a gauge transformation, but aside from these degrees of freedom, there are no other solutions of the field equations with winding number unity.[24]

Because we have found field configurations with $\nu = 1$, we know that there are also field configurations with any integer ν. For instance, solutions with ν a positive integer $\mathcal{N}$ can be constructed by superimposing $\mathcal{N}$ solutions with $\nu = 1$ with centers so far apart that at these distances the non-linearities of the field equations become unimportant. A solution with $\nu = -1$ negative can be found by replacing g_1 in Eq. (23.5.12) with g_1^{-1}, and solutions with ν a negative integer $-\mathcal{N}$ can be found by superimposing $\mathcal{N}$ of these at large separations. For general winding number, Eqs. (23.5.15) and (23.5.16) become

$$S[A] = 8\pi^2|\nu|\,, \tag{23.5.17}$$

$$\epsilon^E_{ijkl}\int F_{\alpha ij}\,F_{\alpha kl}\,(d^4x)_E = 64\pi^2\nu\,. \tag{23.5.18}$$

These results are for a gauge field normalized as in Eqs. (23.5.11)–(23.5.13). With this normalization, the action $I[A]$ is not $-S[A]$, but

$$I[A] = -S[A]/g^2 = -8\pi^2|\nu|/g^2\,, \tag{23.5.19}$$

where g is the conventional coupling constant. If we had used our usual convention of including a factor g in the generators and structure constant, then the action $I[A]$ would be the same as $-S[A]$, but with $A_{\alpha\mu}$ and $F_{\alpha\mu\nu}$ carrying a factor $1/g$ we would have $S[A] = 8\pi^2/g^2$ instead of Eq. (23.5.15), so in this case again the action would be $-8\pi^2/g^2$, but now in place of Eq. (23.5.18) we would have

$$\epsilon^E_{ijkl} \int F_{\alpha ij}\, F_{\alpha kl}\, (d^4x)_E = 64\pi^2 \nu / g^2 \,. \tag{23.5.20}$$

We shall see in the next section that in path integrals we must sum the effects of instantons of all winding numbers. The contribution of configurations of winding number $\nu \neq 0$ to Euclidean path integrals is suppressed by a factor $\exp(I[A]) = \exp(-8|\nu|\pi^2/g^2)$. In Section 23.7 we will see that the coefficient of this exponential is a negative power $-n$ of g: in quantum chromodynamics, $n = 12$. The function $g^{-n}\exp(-8|\nu|\pi^2/g^2)$ and all its derivatives with respect to g vanish at $g = 0$, so such contributions are non-perturbative — they will never be encountered in any order of perturbation theory.

This does not necessarily mean that these contributions are small. As we saw in Chapter 18, in quantum chromodynamics the coupling g is not a fixed dimensionless parameter, but a function of a sliding energy scale, and becomes large at low energies. The effective energy scale to use in the coupling in Eq. (23.5.19) is determined by quantum fluctuations, the subject of Section 23.7, but on dimensional grounds it cannot be very different from $1/R$, where R is the instanton size in Eq. (23.5.12). The instanton size is not fixed, but must be integrated over, with some weight function that depends on the process under consideration. In quantum chromodynamics the running coupling constant g_μ is given for large μ by Eq. (18.7.7) as

$$g_\mu^2 = \frac{8\pi^2}{\beta_0 \ln(\mu/\Lambda)} \,,$$

where $\beta_0 = 11 - 2n_f/3$ and $\Lambda \approx 250$ MeV is the quantum chromodynamic scale factor. The factor $\exp(-8\pi^2/g^2_{1/R})$ is therefore, for small instantons,

$$\exp\Big(-8\pi^2/g^2_{1/R}\Big) = \Big(R\Lambda\Big)^{\beta_0} .$$

We cannot calculate this factor for large instantons, with $R\Lambda \gg 1$, but it is clear that in this case there is no suppression of instanton effects.

Despite initial hopes, the discovery of instantons has not led to much improvement in our ability to do quantitative calculations in quantum chromodynamics. On the other hand, as we shall now see, it has produced spectacular *qualitative* changes in our understanding of quantum chromodynamics and other gauge theories.

The mere fact that there are solutions of the field equations for which the integral (23.5.15) does not vanish is enough to provide a solution of the $U(1)$ problem discussed in Section 19.10. Under *global* $U(1)$ transformations $\psi \to \exp(i\gamma_5\alpha)\psi$, the measure for integration over quark fields undergoes a change given by Eq. (22.2.10):

$$[d\psi][d\bar{\psi}] \to \exp\left\{i\alpha\int \mathscr{A}(x)\,(d^4x)_E\right\}[d\psi][d\bar{\psi}]\,, \tag{23.5.21}$$

where (with the matrix t here equal to unity) the anomaly function $\mathscr{A}(x)$ is given by Eq. (22.2.45):

$$\mathscr{A}(x) = \frac{1}{16\pi^2}\,\epsilon^E_{ijkl}\,F_{ij\alpha}\,F_{kl\beta}\;\mathrm{tr}\,t_\alpha t_\beta\;. \tag{23.5.22}$$

The existence of this anomaly would not in itself solve the $U(1)$ problem, because the quantity (23.5.22) is a total derivative, and so would have a zero integral for a non-singular gauge field that vanishes sufficiently rapidly at infinity. The instanton solution vanishes only as $1/r$, and yields a non-vanishing value for this integral, given by Eqs. (23.5.22), (23.5.18), and (23.5.5) as

$$\int (d^4x)_E\,\mathscr{A}(x) = 2N\nu\,, \tag{23.5.23}$$

showing that the anomaly *does* violate the $U(1)$ chiral symmetry.

As we saw in Section 22.4, the currents of baryon and lepton number also contain anomalies due to the interaction of quarks and leptons with the $SU(2)\times U(1)$ gauge fields of the standard model. Instanton configurations of the $SU(2)$ gauge field therefore produce violations of baryon and lepton conservation.[25] As noted in Section 22.4, there are various currents whose conservation is not violated by anomalies or anything else, such as baryon number minus lepton number, and the differences of electron, muon, and tau lepton numbers, so these will be conserved in any baryon- and lepton-non-conserving process. For instance, the decay of a proton or deuteron is forbidden, but the decay $He^3 \to e^+ + \mu^+ + \bar{\nu}_\tau$ is allowed. The amplitudes for these effects are suppressed by the same factor $\exp(-8|\nu|\pi^2/g^2)$ as before, but now with g the $SU(2)$ coupling constant $e/\sin\theta$, evaluated not at a sliding scale but at the natural scale for electroweak processes, roughly of order m_Z. Taking $e^2/4\pi = 1/129$ (see Section 18.2) and $\sin^2\theta = 0.23$, the suppression factor for $|\nu| = 1$ is $\exp(-373)$. He^3 decay into three antileptons is not likely to be observed.

* * *

There is another approach[26] that illuminates some aspects of instantons. In temporal gauge, with $A_{\alpha 4}(\mathbf{x}, x_4) = 0$, the gauge field for $x_4 \to \pm\infty$ is expected to approach time-independent pure gauges for $x_4 \to \pm\infty$:

$$iA(\mathbf{x}, x_4) \equiv it_\alpha A_\alpha(\mathbf{x}, x_4) \to g_\pm^{-1}(\mathbf{x})\nabla g_\pm(\mathbf{x}) \,, \tag{23.5.24}$$

where $g_\pm(\mathbf{x})$ are group elements in the representation generated by t_α. Assuming that $\mathbf{A}_\alpha(\mathbf{x}, x_4)$ vanishes for $\mathbf{x} \to \infty$, the group elements $g_\pm(\mathbf{x})$ must approach constants $g_\pm$ for $\mathbf{x} \to \infty$, so the three-spaces at $x_4 \to \infty$ may be regarded as three-spheres, with the point at infinity regarded as an ordinary point. Following the same argument that led to Eq. (23.4.13), we then have

$$\epsilon_{ijk} \int d^3x \, \mathrm{Tr}\left\{ g_\pm^{-1}(\mathbf{x})\, \partial_i g_\pm(\mathbf{x})\, g_\pm^{-1}(\mathbf{x})\, \partial_j g_\pm(\mathbf{x})\, g_\pm^{-1}(\mathbf{x})\, \partial_k g_\pm(\mathbf{x}) \right\} = 24\pi^2 n_\pm \,, \tag{23.5.25}$$

where $n_\pm$ are integers. The integral over the boundary of four-space in Eq. (23.5.2) may be regarded as the integral over the 'plane' $x_4 = +\infty$ minus the integral over the 'plane' $x_4 = -\infty$, so from (23.5.8) and (23.5.18) (with $N = 1$) we have

$$\nu = n_+ - n_- \,. \tag{23.5.26}$$

The exponential factor $\exp(-8|\nu|\pi^2/g^2)$ from Eq. (23.5.19) may therefore be regarded as the amplitude for a transition from a configuration with spatial winding number n_- at $x_4 \to -\infty$ to one of spatial winding number n_+ at $x_4 \to +\infty$. The exponential form of this factor for $\nu \neq 0$ reflects the fact that this is a *tunneling* process; no continuous sequence of pure gauge fields can take us from a configuration of one spatial winding number to a configuration with a different spatial winding number.

The interpretation of the factor $\exp(-8|\nu|\pi^2/g^2)$ as a tunneling amplitude suggests that baryon- and lepton-non-conserving processes may proceed rapidly at temperatures above about 1 TeV, where instead of having to tunnel through the barrier, thermal fluctuations can take the vacuum *over* the barrier.[27] This process may be of cosmological importance, but it is still subject to the selection rules mentioned above: thermal fluctuations will not change the density of baryon number minus lepton number, or the differences of the densities of the three varieties of lepton number.

23.6 The Theta Angle

We have seen that there are configurations of arbitrary integer winding number, but how do we know that these configurations must be included in the path integral? To keep an open mind, suppose we add up configurations with arbitrary weight factors $f(\nu)$ for each winding number, leaving open the possibility that some or most of these weight factors might vanish. The expectation value of a local observable $\mathcal{O}$ located within a large Euclidean spacetime volume Ω is then

$$\langle\mathcal{O}\rangle_\Omega = \frac{\sum_\nu f(\nu)\, \int_\nu [d\phi] \exp\left(I_\Omega[\phi]\right)\mathcal{O}[\phi]}{\sum_\nu f(\nu)\, \int_\nu [d\phi] \exp\left(I_\Omega[\phi]\right)}\,, \tag{23.6.1}$$

where ϕ stands for all the fields of the theory; the subscript ν on the integrals over ϕ indicates that we are to include only configurations of winding number ν; and $I_\Omega[\phi]$ is the integral of the Lagrangian density over the spacetime volume Ω. Now suppose that Ω is divided into two very large volumes Ω_1 and Ω_2, with $\mathcal{O}$ in the volume Ω_1. The integral over all fields with winding number ν may be written as an integral over all fields with winding number ν_1 in volume Ω_1 and winding number ν_2 in volume Ω_2, with ν_1 and ν_2 summed over all values with $\nu_1 + \nu_2 = \nu$, and so to a good approximation Eq. (23.6.1) becomes

$$\langle\mathcal{O}\rangle_\Omega = \frac{\sum_{\nu_1,\nu_2} f(\nu_1+\nu_2)\, \int_{\nu_1} [d\phi] \exp\left(I_{\Omega_1}[\phi]\right)\mathcal{O}[\phi]\, \int_{\nu_2} [d\phi] \exp\left(I_{\Omega_2}[\phi]\right)}{\sum_{\nu_1,\nu_2} f(\nu_1+\nu_2)\, \int_{\nu_1} [d\phi] \exp\left(I_{\Omega_1}[\phi]\right)\, \int_{\nu_2} [d\phi] \exp\left(I_{\Omega_2}[\phi]\right)}\,. \tag{23.6.2}$$

But then for general weight factors the average is not the same as if we omitted the volume Ω_2, in contradiction with our general ideas about cluster decomposition. (See Chapter 4.) In order for the factors involving the volume Ω_2 to cancel in this ratio, we must have

$$f(\nu_1 + \nu_2) = f(\nu_1) f(\nu_2)\,.$$

This will be the case if and only if $f(\nu)$ is of the form

$$f(\nu) = \exp(i\theta\nu)\,, \tag{23.6.3}$$

where θ is a free parameter. Thus in particular we cannot arbitrarily discard all configurations with non-zero winding number, because then an instanton of winding number ν in one region would have to be balanced with an instanton of winding number $-\nu$ in some other region, making it impossible to calculate expectation values without considering what is happening far from the location of the operators being measured.

The factor $f(\nu)$ may be put in a more familiar form. According to Eq. (23.5.18), with gauge fields normalized so that in the standard $SU(2)$ subgroup the structure constants are ϵ_{ijk}, the winding number may be written as an integral:

$$\nu = \frac{1}{64\pi^2}\int (d^4x)_E\, \epsilon^E_{ijkl}\, F_{\alpha ij}\, F_{\alpha kl}\,. \tag{23.6.4}$$

This can be expressed in terms of a Minkowskian path integral; since $(d^4x)_E = i\,d^4x$; $F_{\alpha 34} = -iF_{\alpha 30}$; and $\epsilon^{1230} = -1$, Eq. (23.6.4) may be written

$$\nu = -\frac{1}{64\pi^2}\int d^4x\, \epsilon^{\kappa\lambda\rho\sigma}F_{\alpha\kappa\lambda}F_{\alpha\rho\sigma}\,. \tag{23.6.5}$$

Thus inclusion of the weighting factor (23.6.3) is therefore equivalent to adding a term

$$\mathscr{L}_\theta = -\frac{\theta}{64\pi^2}\epsilon^{\kappa\lambda\rho\sigma}F_{\alpha\kappa\lambda}F_{\alpha\rho\sigma} \tag{23.6.6}$$

to the Lagrangian density. But as mentioned at the beginning of Section 15.2, we might have included such a term in the Lagrangian of any non-Abelian gauge theory anyway, with arbitrary real θ.

The inclusion of a term (23.6.6) in the Lagrangian density would violate P and CP conservation. We could of course simply set $\theta = 0$, but this would invalidate what in Section 18.7 was scored as one of the successes of quantum chromodynamics: it made it *automatic* for P and CP to be conserved by the strong interactions even though they are evidently violated by the weak interactions.

To assess the physical consequences of including the term (23.6.6) in the Lagrangian, consider the effect of redefinition of all the fermion fields

$$\psi_f \to \exp(i\gamma_5\alpha_f)\psi_f\,, \tag{23.6.7}$$

where f is a flavor index, and α_f are a set of real phases. According to Eqs. (22.2.10) and (22.2.24), the effect on the measure for path integrals over fermion fields is

$$[d\psi][d\bar\psi] \to \exp\left\{\frac{-i}{32\pi^2}\int d^4x\, \epsilon_{\mu\nu\rho\sigma}\, F_\alpha^{\mu\nu}\, F_\alpha^{\rho\sigma}\sum_f \alpha_f\right\}[d\psi][d\bar\psi]\,. \tag{23.6.8}$$

(We are using generators normalized so that $\mathrm{Tr}\,(t_\alpha t_\beta) = \delta_{\alpha\beta}/2$.) Comparing this with Eq. (23.6.6) shows that this is equivalent to shifting θ by

$$\theta \to \theta + 2\sum_f \alpha_f\,. \tag{23.6.9}$$

The redefinition of the fermion fields will also change the mass terms in the Lagrangian density. In order to take account of masses involving γ_5, let us write the fermion mass term in the Lagrangian as

$$\mathscr{L}_m = -\tfrac{1}{2}\sum_f \mathscr{M}_f\bar\psi_f(1+\gamma_5)\psi_f - \tfrac{1}{2}\sum_f \mathscr{M}_f^*\bar\psi_f(1-\gamma_5)\psi_f\,, \tag{23.6.10}$$

with mass parameters $\mathscr{M}_f$ which, if complex, would violate P and CP conservation. Then the redefinition (23.6.7) changes these parameters by

$$\mathscr{M}_f \to \exp(2i\alpha_f)\mathscr{M}_f\,. \tag{23.6.11}$$

A mere change of path integration variables cannot have any physical effect, so observable quantities cannot depend separately on θ or the phases of the mass parameters $\mathscr{M}_f$, but only on the combination

$$\exp(-i\theta) \prod_f \mathscr{M}_f \,. \tag{23.6.12}$$

In particular, we can always define the fermion fields so that $\theta = 0$, but at the price of perhaps introducing P and CP violating phases in the mass parameters.

This discussion shows that if any of the quark masses were to vanish, then the theta angle would have no effect and there would be no P or CP non-conservation in quantum chromodynamics. The analysis of quark mass ratios in Section 19.7 indicates that the quarks all have non-zero masses, though it is sometimes suggested that the inclusion of terms of second order in m_s in this analysis might allow m_u to vanish.[28] But there is no question that the u and d quarks are quite light, which leads to some suppression of the effects of a non-zero theta angle. We saw in Section 19.4 that m_u and m_d are roughly of order m_π^2/m_N (where m_N is used here as a typical quantum chromodynamic mass scale), so we might expect the effects of the theta angle to be suppressed by four factors of m_π, but this is not quite correct. With P and CP not conserved there would be a non-zero amplitude for the π^0 to disappear into the vacuum, proportional to m_π^4, but the 'tadpole' graphs with pion lines ending in such vacuum transition vertices would be enhanced by a factor m_π^{-2} from the pion propagator, giving a net effect proportional to m_π^2, not m_π^4. In particular, if we define fermion fields so that all $\mathscr{M}_f$ are real, then a non-zero theta angle would produce a P- and T-non-conserving neutron electric dipole moment d_n proportional to $|\theta|$ and m_π^2, and hence on dimensional grounds of order[29]

$$d_n \approx |\theta| e m_\pi^2/m_N^3 \approx 10^{-16}|\theta|\, e\ \mathrm{cm}\,. \tag{23.6.13}$$

The neutron electric dipole moment is known to be less than about 10^{-25} e cm, so $|\theta| < 10^{-9}$.

In order to explain in a natural way why θ is so small, Peccei and Quinn[30] proposed a theory in which θ became in effect a dynamical variable, which could relax to a minimum of the effective potential, at which P and CP would be conserved. Their idea was taken up by Wilczek and myself,[31] who noted that it would require the existence of a light spinless particle, the *axion*. The axion that appeared in the original Peccei–Quinn model has been ruled out by experiment, but there are more general possibilities,[32] in which the axion couples too weakly to ordinary matter to have been observed.

The common feature of all versions of the axion theory is that there is some $U(1)$ symmetry that is spontaneously broken at energies much higher than those associated with quantum chromodynamics, and is also broken by an anomaly involving the gluon fields. According to the general formalism of Chapters 19 and 22, the low energy effective field theory will contain a Goldstone boson field ϕ, so that, under the symmetry transformation

$$\phi \to \phi + F_\phi \epsilon\,, \tag{23.6.14}$$

the effective Lagrangian undergoes the transformation

$$\delta\mathscr{L}_{\rm eff} = -\frac{\epsilon A}{64\pi^2}\epsilon_{\mu\nu\rho\sigma}\, F_\alpha^{\mu\nu}\, F_\alpha^{\rho\sigma}\,, \tag{23.6.15}$$

where A is a dimensionless constant of order unity, characterizing the anomaly, and F_ϕ is a constant of the order of the energy scale at which the symmetry is spontaneously broken (with ϕ defined to be canonically normalized.) Then the terms in the effective Lagrangian involving ϕ are

$$\mathscr{L}_\phi = -\frac{1}{2}\partial_\mu\phi\partial^\mu\phi - \frac{1}{64\pi^2}\frac{\phi}{M}\epsilon_{\mu\nu\rho\sigma}\, F_\alpha^{\mu\nu}\, F_\alpha^{\rho\sigma} + \cdots\,, \tag{23.6.16}$$

where $M \equiv F_\phi/A$ and '$\cdots$' denotes possible interactions involving derivatives of ϕ. Comparing Eqs. (23.6.16) and (23.6.6), we see that, for a constant ϕ, all observables will be functions not of ϕ and θ separately, but only of $\theta + \phi/M$ (This is with fermion fields defined to make the mass parameters $\mathscr{M}_f$ all real; otherwise observables will depend on $\theta - \sum_f \mathrm{Arg}\mathscr{M}_f + \phi/M$.) If everything in the theory but the theta term (23.6.6) and the ϕ interaction in Eq. (23.6.16) conserves P or CP, then the effective potential will be even in $\theta + \phi/M$, so it will have a stationary point at $\theta + \phi/M = 0$, preserving the conservation of both P and CP. In the real world P and CP are not exact, but the only observed violations are in the weak interactions, and would shift the expectation value of ϕ only a small way[33] from $-M\theta$.

Even without specifying the underlying theory, by using effective field theory techniques it is possible to say a fair amount about the general properties of the axion. The most general Lagrangian for the axion field ϕ, now including the theta term and all interactions with u and d quarks up to order $1/M$, is of the form

$$\mathscr{L}_\phi = -\frac{1}{2}\partial_\mu\phi\partial^\mu\phi + \frac{1}{64\pi^2}\left[\frac{\phi}{M} + \theta\right]\epsilon_{\mu\nu\rho\sigma}\, F_\alpha^{\mu\nu}\, F_\alpha^{\rho\sigma}$$
$$-\frac{if_u}{M}\,\partial_\mu\phi\,\bar{u}\gamma_5\gamma^\mu\, u - \frac{if_d}{M}\,\partial_\mu\phi\,\bar{d}\gamma_5\gamma^\mu\, d\,, \tag{23.6.17}$$

where f_u and f_d are dimensionless coupling constants, expected to be of order unity. As shown by Eq. (23.6.8), the redefinition (23.6.7) of the quark fields has the effect of subjecting the quantity $\phi(x)/M + \theta$ in the second term of the Lagrangian (23.6.17) to the replacement

$$\frac{\phi(x)}{M} + \theta \rightarrow \frac{\phi(x)}{M} + \theta + 2\Big[\alpha_u(x) + \alpha_d(x)\Big] . \qquad (23.6.18)$$

By choosing $\alpha_f = -(\theta + \phi/M)c_f/2$ with constant coefficients c_f satisfying $c_u + c_d = 1$, we eliminate the term in the Lagrangian involving $\epsilon_{\mu\nu\rho\sigma} F_\alpha^{\mu\nu} F_\alpha^{\rho\sigma}$, and change the mass term in the low energy quantum chromodynamic Lagrangian density to

$$\mathscr{L}_m = -m_u \bar{u} \exp\Big[-ic_u(\theta + \phi/M)\gamma_5\Big]u - m_d \bar{d} \exp\Big[-ic_d(\theta + \phi/M)\gamma_5\Big]d . \qquad (23.6.19)$$

In addition, from the kinematic part of the quark Lagrangian we pick up a derivative interaction term

$$\tfrac{1}{2}ic_u\,(\bar{u}\gamma^\mu\gamma_5)u\,\partial_\mu\phi/M + \tfrac{1}{2}ic_d\,(\bar{d}\gamma^\mu\gamma_5)d\,\partial_\mu\phi/M , \qquad (23.6.20)$$

so that f_u and f_d in Eq. (23.6.17) are replaced with $f'_u = f_u - c_u/2$ and $f'_d = f_d - c_d/2$. To derive an effective Lagrangian for low-energy pions and axions, we follow the procedure of Section 19.5 (but now with u and d quarks in place of protons and neutrons), and make the replacements:

$$\begin{aligned} &\bar{u}u \rightarrow v\cos(\pi^0/F_\pi) , && \bar{d}d \rightarrow v\cos(\pi^0/F_\pi) , \\ &\bar{u}\gamma_5 u \rightarrow iv\sin(\pi^0/F_\pi) , && \bar{d}\gamma_5 d \rightarrow -iv\sin(\pi^0/F_\pi) , \\ &i\bar{u}\gamma^\mu\gamma_5 \rightarrow \tfrac{1}{2}F_\pi\partial^\mu\pi^0 + \cdots , && i\bar{d}\gamma^\mu\gamma_5 \rightarrow \tfrac{1}{2}F_\pi\partial^\mu\pi^0 + \cdots , \end{aligned} \qquad (23.6.21)$$

where v is the constant $v = \langle\bar{u}u\rangle = \langle\bar{d}d\rangle$, and '$\cdots$' denotes terms that do not have the one-pion pole. From Eqs. (23.6.19) and (23.6.20) we find the effective pion–axion Lagrangian

$$\begin{aligned} \mathscr{L}_{\text{eff}} = &-\frac{1}{2}\partial_\mu\pi^0\partial^\mu\pi^0 - \frac{1}{2}\partial_\mu\phi'\partial^\mu\phi' - \left(\frac{f'_u - f'_d}{2M}\right) F_\pi\partial_\mu\phi'\partial^\mu\pi^0 \\ &-m_u v\cos\left(\frac{\pi^0}{F_\pi} - \frac{c_d\phi'}{M}\right) - m_d v\cos\left(\frac{\pi^0}{F_\pi} + \frac{c_d\phi'}{M}\right) , \end{aligned} \qquad (23.6.22)$$

where ϕ' is the difference between the axion field and its expectation value

$$\phi' \equiv \phi - \langle\phi\rangle = \phi + M\theta . \qquad (23.6.23)$$

Since c_u and c_d are arbitrary except for the condition that $c_u + c_d = 1$, we are free to eliminate the $\partial_\mu\phi\partial^\mu\pi^0$ cross term by taking $c_u = \frac{1}{2} + f_u - f_d$ and $c_d = \frac{1}{2} + f_d - f_u$, so that $f'_u = f'_d$. The quadratic part of the Lagrangian (23.6.22) is

$$\mathscr{L}_{\text{quad}} = -\frac{1}{2}\partial_\mu\pi^0\partial^\mu\pi^0 - \frac{1}{2}\partial_\mu\phi'\partial^\mu\phi' - \frac{1}{2}\begin{pmatrix}\pi^0 \\ \phi'\end{pmatrix}^T M_0^2 \begin{pmatrix}\pi^0 \\ \phi'\end{pmatrix} , \qquad (23.6.24)$$

where

$$M_0^2 = \begin{pmatrix} (m_u + m_d)v/F_\pi^2 & (-m_u c_u + m_d c_d)v/F_\pi M \\ (-m_u c_u + m_d c_d)v/F_\pi M & (m_u c_u^2 + m_d c_d^2)v/M^2 \end{pmatrix} . \tag{23.6.25}$$

For $M \gg F_\pi$ one eigenvalue of M_0^2 is $(m_u + m_d)v/F_\pi^2$, which according to Eq. (19.7.20) is the π^0 squared mass. The other eigenvalue is then the axion mass

$$m_a^2 = \frac{v}{M^2}\frac{m_d m_u}{m_d + m_u} = \frac{F_\pi^2}{M^2}\frac{m_d m_u}{(m_d + m_u)^2}m_\pi^2 . \tag{23.6.26}$$

For the quark mass ratio derived in Section 19.7, this gives $m_a = 13\ \text{MeV}/M(\text{GeV})$.

We can also use this formalism to say something about the interactions of axions with hadrons. The eigenvector of M_0^2 with eigenvalue m_a^2 has a component along the original π^0 direction equal to $(m_u c_u - m_d c_d)F_\pi/(m_u + m_d)M$. As mentioned earlier, because of the one-pion pole this is the dominant axion–hadron coupling. We see that the ratio of the axion and pion production interaction amplitudes will typically be of order F_π/M. The fact that axions are not observed in such collisions indicates that $M > 3$ TeV, in contradiction with the original expectation[30,31] that the anomalous $U(1)$ symmetry is spontaneously broken by the same scalar vacuum expectation values of order 0.3 TeV that break the electroweak $SU(2) \times U(1)$ symmetry. It is possible to explain why axions are not found in reactor or accelerator experiments by taking M as an independent parameter,[32] much larger than the electroweak breaking scale, but there are still astrophysical limitations. Limits on the rate of cooling of red giant stars give[34] $M > 10^7$ GeV, while observations of the supernova SN1987A indicate[35] that $M > 10^{10}$ GeV.* Cosmological arguments suggest[36] an *upper* bound $M < 10^{12}$ GeV, leaving an open but narrow window of allowed axion parameters.

---

* For $M > 10^7$ GeV the axion mass would be less than about 1 eV, so that stars are hot enough to produce axions. The ratio of the axion and π^0 decay rates into two photons is expected to be of order $(F_\pi/M)^2$ times a phase space ratio of order $(m_a/m_\pi)^3$, or

$$\frac{\Gamma(a \to \gamma + \gamma)}{\Gamma(\pi^0 \to \gamma + \gamma)} = \left(\frac{F_\pi}{M}\right)^2 \left(\frac{m_a}{m_\pi}\right)^3 \approx \left(\frac{F_\pi}{M}\right)^5 .$$

Hence for $M > 10^7$ GeV the axion lifetime is expected to be longer than about 10^{24} s, which is ample time for the axion to travel even cosmological distances before decaying.

23.7 Quantum Fluctuations around Extended Field Configurations

Topologically non-trivial four-dimensional field configurations such as instantons provide only zeroth-order contributions to path integrals. Now we must consider the effect of quantum fluctuations around these configurations.

To start in a very general way, consider a set of fields, collectively called $\phi(x)$, whose dynamics is described by the Euclidean action $I[\phi]$, and suppose we have a set of configurations $\phi_{v,u}$ at which $I[\phi]$ is stationary, where v labels the topological type of the configuration, and u represents a set of continuous collective parameters on which the configuration depends. For instance, for instantons v is the winding number, an integer, and u includes the position and scale as well as its 'direction' in the gauge group. Euclidean path integrals may then be written

$$\int [d\phi]\, \exp\Big(I[\phi]\Big)\mathscr{O} = \sum_v \int du \int_u [d\phi'] \exp\Big(I[\phi_{v,u} + \phi']\Big)\mathscr{O}\,, \qquad (23.7.1)$$

where the subscript on the integral over the fluctuation ϕ' indicates that we are to integrate only over fluctuations that do not entail changes in the collective parameters, and $\mathscr{O}$ stands for any product of local functions of field operators. Because $I[\phi]$ is stationary at $\phi = \phi_{v,u}$, its expansion to second order in the fluctuations takes the form

$$I[\phi_{v,u} + \phi'] \simeq I_v - \tfrac{1}{2} \int d^4x\, d^4y\; K_{xl,ym}(v,u)\, \phi'_l(x)\, \phi'_m(y)\,, \qquad (23.7.2)$$

where l and m here include spin and species indices, and $I_v \equiv I[\phi_{v,u}]$ is a function of v alone because the action is supposed to be stationary at all field configurations $\phi_{v,u}$. The integral over the fluctuation ϕ' in Eq. (23.7.1) will then yield a sum of terms from contractions of the fields in $\mathscr{O}$, times (for real bosonic fields) an over-all factor $[\text{Det}\, K(v,u)]^{-1/2}$, it being understood that K now acts only in the subspace of fluctuations that do not entail changes in the collective parameters u. This factor can be written as a product over the eigenvalues $\lambda_n(v,u)$ of the 'matrix' $K(v,u)$:

$$\Big(\text{Det}\, K(v,u)\Big)^{-1/2} = \prod_n{}' \lambda_n^{-1/2}(v,u)\,, \qquad (23.7.3)$$

the prime here indicating that we are to exclude the 'zero modes', the zero eigenvalues of K, for which the eigenfunctions correspond to changes in the collective parameters.

These remarks allow us to refine the results given in Section 23.5 for the coupling-constant dependence of instanton contributions with different winding numbers. Suppose we define the fields of our theory in such a way that the action takes the form $I[\phi, g] = g^{-2} I_1[\phi]$ where $I_1[\phi]$ is independent of coupling constants. (For instance, as discussed at the end of Section 15.2, in Yang–Mills theories we would use as a gauge field the canonically normalized gauge field times a factor g.) Then all eigenvalues of K are proportional to g^{-2}, and the factor (23.7.3) is proportional to g to a power equal to the number of non-zero eigenvalues of K. This power is of course infinite, but it can be written as the number of *all* eigenvalues of K, which is also infinite but is independent of ν, minus the number $\mathcal{N}(\nu)$ of zero modes, equal to the number of collective parameters, which is finite and dependent on ν. We conclude then that aside from factors that do not depend on ν, the contribution of fluctuations around configurations of topological type ν is a factor with a coupling-constant dependence that is given in terms of the number $\mathcal{N}(\nu)$ of collective parameters by

$$g^{-\mathcal{N}(\nu)} . \tag{23.7.4}$$

This estimate is based on the approximation (23.7.2), which corresponds to a one-loop approximation, so when higher-order terms are taken into account the factor (23.7.4) will be multiplied with a power series in g, whose details depend on the operators $\mathcal{O}$ appearing in the path integral.

Let's see how this applies to instantons. The configuration with $\nu = 0$ of course has no collective parameters, and so its contribution to path integrals is just a power series in g. The configuration with $\nu = 1$ has four collective parameters giving the spacetime location of the instanton, one collective parameter giving the scale of the instanton, and a number N_1 of collective parameters corresponding to the rotations and/or global gauge group transformations that do not leave the instanton invariant, so (now including the factor $\exp(I_\nu)$ given by Eq. (23.5.19)), the coupling-constant dependence for $\nu = 1$ instantons is $g^{-5-N_1} \exp(-8\pi^2/g^2)$. For the $\nu = 1$ instanton (23.5.12) in an $SU(2)$ Yang–Mills theory, there are three independent rotations and three independent $SU(2)$ transformations, but since the instanton is invariant under three combined rotations and $SU(2)$ transformations, we have $N_1 = 3$, so fluctuations around the $\nu = 1$ instanton yield a coupling-constant dependence $g^{-8} \exp(-8\pi^2/g^2)$. In an $SU(3)$ Yang–Mills theory we have three independent rotations and eight independent $SU(3)$ transformations, but again the instanton is invariant under three independent combined rotations and $SU(3)$ transformations in the standard $SU(2)$ subgroup (say, acting on the first two components of the defining representation of $SU(3)$), and now also invariant under one additional $SU(3)$ transformation that (like hypercharge) commutes with all generators of the standard $SU(2)$ subgroup. Hence $N_1 = 3+8-3-1 = 7$, and the $\nu = 1$ instanton yields a factor proportional to $g^{-12} \exp(-8\pi^2/g^2)$.

Now suppose that the action I also contains a term

$$-\int d^4x \int d^4y \; \bar{\psi}_l(x)\, \mathscr{K}_{lx,my}(v,u)\, \psi_m(y)\,,$$

involving independent fermion fields ψ and $\bar{\psi}$. If none of these fields appear in $\mathcal{O}$ then the integral over these fields yields a factor $\text{Det}\,\mathscr{K}(v,u)$, which vanishes if $\mathscr{K}(v,u)$ has any zero modes. This result can be simply understood in terms of the rules for integration over fermionic parameters. By expanding $\psi(x)$ and $\bar{\psi}(x)$ in the eigenmodes of $\mathscr{K}$, we can write the integral over $\psi(x)$ and $\bar{\psi}(x)$ as an integral over the coefficients in these expansions. The coefficients of the zero modes do not appear in the quadratic approximation to the action, so for each such fermionic zero mode we have an integral over a fermionic parameter that does not appear in the integrand, which vanishes according to the general rules of Section 9.5. The only terms in an integral over fermionic variables that do not vanish are those for which the integrand contains a single factor of each integrated variable. Hence the integral over fermionic fields in Eq. (23.7.1) will not vanish only if there is a single fermionic field in $\mathcal{O}$ for each zero mode of $\mathscr{K}$. For instantons there are fermionic zero modes whose numbers and chiralities are governed by index theorems, like the Atiyah–Singer index theorem derived in Section 22.2, so only certain processes are allowed for a given winding number. It was on this basis that 't Hooft[25] showed that the baryon- and lepton-nonconserving effective interaction produced by $\nu = 1$ instantons in the electroweak standard model must involve just one of each lepton flavor.

23.8 Vacuum Decay

A vacuum state is stable if its scalar field expectation values are at a true minimum of the effective potential. But if the scalar field expectation values are at a local minimum that is higher than the true minimum then this vacuum will be metastable. A metastable 'false' vacuum state corresponding to a local minimum will decay into the stable 'true' vacuum corresponding to the true minimum by a process of barrier penetration, analogous to nuclear alpha decay or spontaneous fission. This is not a process that can be observed in our laboratories, but it has presumably occurred several times in the history of the universe as various symmetries have become spontaneously broken, so it is important to be able to calculate the rate of such false vacuum decay. As we shall now see, this calculation involves consideration of yet another extended field configuration.[6]

Let us concentrate on the component ϕ of the scalar field multiplet that acquires an expectation value $\langle\phi\rangle$ in the true vacuum. For instance, in the theory of broken chiral symmetry discussed in Section 19.5, ϕ would be the fourth component of a chiral four-vector. It will turn out that, where barrier penetration is strongly suppressed, the other scalar fields (including those of any Goldstone bosons) do not affect the dominant suppression factor in the decay rate. For definiteness, we will take the Lagrangian density in the form

$$\mathscr{L} = -\tfrac{1}{2}\partial_\mu\phi\,\partial^\mu\phi - V(\phi)\,. \tag{23.8.1}$$

We assume that the lowest-order effective potential $V(\phi)$ has a true minimum at $\phi = \langle\phi\rangle$ and a local minimum at $\phi = 0$, and we adjust an additive constant in the Lagrangian density so that $V(0) = 0$, in which case $V(\langle\phi\rangle) < 0$. We want to calculate the rate at which the false vacuum state with scalar field expectation value zero will decay into the true vacuum state with scalar field expectation value $\langle\phi\rangle$.

The results of Appendix A of this chapter (Eqs. (23.A.6), (23.A.21), and (23.A.23)) show that the energy E_0 of the false vacuum state in which the scalar field vacuum expectation value vanishes is given by

$$E_0 = -\lim_{T\to\infty}\frac{1}{T}\ln\left[\int \exp\Big(-S[\phi;T]\Big)\prod_{\mathbf{x},t} d\phi(\mathbf{x},t)\right]\,, \tag{23.8.2}$$

where $S[\phi;T]$ is the Euclidean action derived from Eq. (23.8.1)

$$S[\phi;T] = \int d^3x \int_{-T/2}^{+T/2} dt\,\left[\frac{1}{2}\left(\frac{\partial\phi}{\partial t}\right)^2 + \frac{1}{2}(\nabla\phi)^2 + V(\phi)\right]\,, \tag{23.8.3}$$

and the integral in Eq. (23.8.2) is over all fields $\phi(\mathbf{x},t)$ satisfying the conditions

$$\phi(\mathbf{x},T/2) = \phi(\mathbf{x},-T/2) = 0\,. \tag{23.8.4}$$

The energy (23.8.2) is complex; its imaginary part will give us the decay rate.

To calculate the functional integral in Eq. (23.8.2), we look for a stationary 'point' of the Euclidean action $S[\phi,T]$. The fields at which (23.8.3) is stationary satisfy the field equations

$$0 = \frac{\delta S}{\delta\phi} = -\frac{\partial^2\phi}{\partial t^2} - \nabla^2\phi + \frac{dV(\phi)}{d\phi}\,, \tag{23.8.5}$$

subject to the boundary conditions (23.8.4). Because of these boundary conditions, such a solution is known as a *bounce*.

We shall look for these bounce solutions by making the ansatz that $\phi(\mathbf{x},t)$ is invariant under rotations around a point $\mathbf{x}_0$, t_0 in four dimensions:

$$\phi(\mathbf{x},t) = \phi(\rho) \quad\text{where}\quad \rho \equiv \sqrt{(\mathbf{x}-\mathbf{x}_0)^2 + (t-t_0)^2}\,. \tag{23.8.6}$$

There are solutions that are not rotationally invariant in four dimensions, but these other solutions have higher values[37] of S, and so become negligible for large T. Using Eq. (23.8.6) in Eq. (23.8.5) yields the ordinary differential equation

$$\frac{d^2\phi}{d\rho^2} + \frac{3}{\rho}\frac{d\phi}{d\rho} = V'(\phi)\,. \tag{23.8.7}$$

Strictly speaking, a solution of this form is consistent with the boundary conditions (23.8.4) only when T is very large compared with the characteristic time associated with $V(\rho)$, in which case we can take T to be infinite in Eq. (23.8.4), which then becomes the condition that $\phi(\rho)$ must vanish when $\rho \to \infty$. Also $\phi(\mathbf{x}, t)$ must be an analytic function of $\mathbf{x}$ near $\mathbf{x} = 0$ at all t including at $t = t_0$, so $\phi(\rho)$ is a power series in ρ^2 for $\rho \to 0$, and in particular $d\phi/d\rho = 0$ at $\rho = 0$. With these conditions, Eq. (23.8.7) is the equation of motion of a particle of unit mass with 'position' ϕ at 'time' ρ, moving under the influence of a potential $-V(\phi)$ and a viscous force $-(3/\rho)d\phi/d\rho$, that travels from rest at some finite initial value ϕ_0 of ϕ at $\rho = 0$, and just reaches $\phi = 0$ at $\rho \to \infty$, losing its initial 'energy' $-V(\phi_0) > 0$ to viscosity along the way. The Euclidean action (23.8.2) for such a solution is

$$B \equiv \int_0^\infty 2\pi^2\rho^3 d\rho \left[\frac{1}{2}\left(\frac{d\phi}{d\rho}\right)^2 + V(\phi)\right]\,. \tag{23.8.8}$$

It is crucial to determine the sign of B. For this purpose,[38] we use the same device that was used to prove Derrick's theorem in Section 23.1, and consider the action (23.8.8) for a modified field $\phi_R(\rho) \equiv \phi(\rho/R)$. By rescaling the variable of integration, we find

$$S[\phi_R] = \int_0^\infty 2\pi^2\rho^3 d\rho \left[\frac{R^2}{2}\left(\frac{d\phi}{d\rho}\right)^2 + R^4 V(\phi)\right]\,. \tag{23.8.9}$$

If $\phi(\rho)$ is the solution of Eq. (23.8.7) then the action must be stationary with respect to any variation in ϕ, so that $dS[\phi_R]/dR$ must vanish at $R = 1$, and therefore

$$\int_0^\infty \rho^3 d\rho \left(\frac{d\phi}{d\rho}\right)^2 = -4\int_0^\infty \rho^3 d\rho\, V(\phi)\,. \tag{23.8.10}$$

It follows then that

$$B = \frac{\pi^2}{2}\int_0^\infty \rho^3 d\rho \left(\frac{d\phi}{d\rho}\right)^2 > 0\,. \tag{23.8.11}$$

We will come back later to an explicit approximate solution for $\phi(\rho)$, but first let's consider how such solutions are to be used.

We have here not just a single one-bounce configuration at which the Euclidean action is stationary, but a continuum, characterized by the collective coordinates $\mathbf{x}_0$ and t_0. According to the results of Section 23.7, we must integrate over these parameters, which, since B is independent of $\mathbf{x}_0$ and t_0, yields factors of $\mathscr{V}$ and T in a box of spatial volume $\mathscr{V}$. The contribution of all one-bounce configurations to the functional integral in Eq. (23.8.2) is given in one-loop order by

$$\mathscr{V} T A \exp(-B)\,. \tag{23.8.12}$$

The coefficient A is proportional* to a product $\prod_n' \lambda_n^{-1/2}$, where the λ_n are the eigenvalues of the kernel $\delta^2 S[\phi]/\delta\phi(\mathbf{x},t)\delta\phi(\mathbf{x}',t')$, and the prime indicates that we are to omit the zero eigenvalues, which correspond to changes in the collective coordinates $\mathbf{x}_0$ and t_0. Using Eq. (23.8.10), we can see that the second derivative of Eq. (23.8.9) with respect to R is negative at $R = 1$, so there is at least one (and in fact just one[39]) negative eigenvalue, and hence to this order A is imaginary. We will not attempt to calculate A here, but will just note that A unlike $\exp(-B)$ does not have a dramatic dependence on the parameters of the theory, so that we can estimate it on dimensional grounds to be roughly of order iM^{-4}, where M is some characteristic mass scale of the theory.

For large T and $\mathscr{V}$ we can find additional stationary configurations by superimposing any number N of these bounce configurations, yielding a contribution that is the Nth power of the quantity (23.8.12), divided by $N!$ to take account of the fact that in integrating over N of the $\mathbf{x}_0$ and t_0 we are summing over configurations that only differ by permutations of the N identical bounces. Summing over N then gives the exponential of the quantity (23.8.12), so the energy (23.8.2) is just the quantity (23.8.12) divided by $-T$

$$E_0 = -\mathscr{V} A \exp(-B)\,. \tag{23.8.13}$$

Since A is of order iM^{-4}, the decay rate per volume of the false vacuum is thus of order

$$\Gamma/\mathscr{V} \approx M^{-4} \exp(-B)\,. \tag{23.8.14}$$

Note that this is a decay rate per volume, because the decay does not occur by a change in the scalar field simultaneously everywhere in space, but by the occurrence of bubbles of true vacuum in a false vacuum background.

* Where a continuous symmetry G is spontaneously broken to a subgroup H, there are additional collective coordinates giving the orientation of H within G. The integration over these parameters yields an additional factor in A equal to the volume of the coset space G/H.

The result (23.8.14) is chiefly useful in the case where B is large, so that barrier penetration is strongly suppressed, and we can estimate the suppression factor as simply $\exp(-B)$. Fortunately, the most natural circumstance in which B is large is one in which it is possible to calculate B in closed form. This is the case in which the energy $V(\langle\phi\rangle) \equiv -\epsilon$ of the true vacuum is only slightly below the zero energy of the false vacuum, but $V(\phi)$ is positive and not small between $\phi = 0$ and $\phi = \langle\phi\rangle$. To minimize the Euclidean action (23.8.3) in this case, we must take ϕ to be near $\langle\phi\rangle$ within a four-dimensional ball with a large radius R, at which ϕ drops to zero within a shell of thickness given by some length $L \approx M^{-1}$ characteristic of the potential in the limit $\epsilon \to 0$. (This is sometimes called the 'thin wall approximation,' but perhaps a better term would be the big bubble approximation.) The action (23.8.3) in this approximation is

$$S(R) \simeq -\tfrac{1}{2}\pi^2 R^4 \epsilon + 2\pi^2 R^3 \mathscr{S} , \tag{23.8.15}$$

where $\mathscr{S}$ is a surface tension, equal to the shell contribution to the action per area. The second term on the left-hand side of Eq. (23.8.7) becomes negligible for $\rho \approx R$, so this is now essentially a one-dimensional problem. We can therefore take the surface tension from Eq. (23.1.4), which for a solution of the field equations is an equality, and in present notation reads

$$\mathscr{S} = \int_0^{\langle\phi\rangle} \sqrt{2V(f)}\, df \, . \tag{23.8.16}$$

The action (23.8.15) is stationary at a radius

$$R \simeq 3\mathscr{S}/\epsilon \, , \tag{23.8.17}$$

so the action at its stationary point has the value

$$B \simeq \frac{27\pi^2 \mathscr{S}^4}{2\epsilon^3} \, . \tag{23.8.18}$$

Note that B is large for small ϵ, so in this case the decay rate of the false vacuum is strongly suppressed. After the barrier has been penetrated the bubble of true vacuum will grow at the speed of light, colliding eventually with other bubbles, until all space is in the state of lowest energy.

Appendix A Euclidean Path Integrals

This appendix will outline the use of Euclidean path integrals in quantum field theory. As mentioned in Section 9.1, it is possible to formulate quantum field theory in a four-dimensional Euclidean spacetime. Instead of going into the non-trivial analytic continuation needed to calculate S-matrix elements in this approach, here we shall illustrate the use of

Euclidean path integrals by addressing a problem for which they are naturally suited.

We consider a set of Hermitian canonical variables Q_a and P_a, with commutation relations

$$[Q_a, P_b] = i\delta_{ab} , \tag{23.A.1}$$

$$[Q_a, Q_b] = [P_a, P_b] = 0 . \tag{23.A.2}$$

In quantum field theory the index a is understood, as in Section 9.1, to consist of a spatial position $\mathbf{x}$ and any discrete Lorentz and species indices m, and the Kronecker delta in Eq. (23.A.1) is understood as $\delta_{\mathbf{x}m,\,\mathbf{y}n} = \delta^3(\mathbf{x}-\mathbf{y})\delta_{mn}$. We define eigenstates of the Q_a

$$Q_a|q\rangle = q_a|q\rangle , \tag{23.A.3}$$

normalized so that

$$\langle q'|q\rangle = \delta(q'-q) \equiv \prod_a \delta(q'_a - q_a) , \tag{23.A.4}$$

and likewise for eigenstates $|p\rangle$ of the P_a.

The problem we consider is the calculation of the matrix element

$$F(q',q;T) \equiv \langle q'|\exp\Big(-H[Q,P]T\Big)|q\rangle , \tag{23.A.5}$$

where H is the Hamiltonian, and T is an arbitrary positive constant. One application is to the study of ground-state energies. If the smallest eigenvalue of H is E_0, with eigenvector $|0\rangle$, then for $T \to \infty$

$$F(q',q;T) \to \langle q'|0\rangle\langle 0|q\rangle \exp(-E_0 T) ,$$

so

$$E_0 = -\mathrm{lim}_{T\to\infty}\left(\frac{\ln F(q',q;T)}{T}\right) . \tag{23.A.6}$$

Also, we can calculate the partition function of statistical mechanics from the trace:

$$Z(\beta) \equiv \mathrm{Tr}\,\exp(-\beta H) = \int \left[\prod_a dq_a\right] F(q,q;\beta) , \tag{23.A.7}$$

where $1/\beta$ is the temperature.

To derive a path integral formula for $F(q',q;T)$ we define Euclidean time dependent operators

$$Q_a(t) \equiv e^{Ht}Q_a e^{-Ht} , \qquad P_a(t) \equiv e^{Ht}P_a e^{-Ht} , \tag{23.A.8}$$

and corresponding left- and right-eigenstates

$$|q,t\rangle \equiv \exp(Ht)|q\rangle , \qquad \langle q,t| \equiv \langle q|\exp(-Ht) , \tag{23.A.9}$$

$$|p,t\rangle \equiv \exp(Ht)|p\rangle\,, \qquad \langle p,t| \equiv \langle p|\exp(-Ht)\,, \tag{23.A.10}$$

such that

$$Q_a(t)|q,t\rangle = q_a|q,t\rangle\,, \qquad \langle q,t|Q_a(t) = q_a\langle q,t|\,, \tag{23.A.11}$$

and

$$P_a(t)|p,t\rangle = p_a|p,t\rangle\,, \qquad \langle p,t|P_a(t) = p_a\langle p,t|\,. \tag{23.A.12}$$

One difference between this and the usual Minkowskian formalism is that the 'time' evolution of the operators is governed by a non-unitary similarity transformation (23.A.8), so there is no simple relation between the right-eigenstates $\langle q,t|$ of $Q(t)$ and the Hermitian adjoint of the left-eigenstates $|q,t\rangle$, except at $t=0$.

In this language, the definition (23.A.5) of $F(q',q;T)$ may be rewritten

$$F(q',q;T) = \langle q',T/2|q,-T/2\rangle\,. \tag{23.A.13}$$

Let us first calculate $\langle q',t+dt|q,t\rangle$ for infinitesimal dt. Adopting the convention that $H(Q,P)$ is written with all Qs to the left of all Ps, we have

$$\langle q',t+dt|q,t\rangle = \langle q',t|\exp\Big(-H(q',P)dt\Big)|q,t\rangle\,.$$

Let us expand $|q,t\rangle$ in a complete set of eigenstates of the $P_a(t)$ operators. From Eq. (23.A.1) we have as usual

$$\langle q,t|p,t\rangle = \prod_a \frac{\exp(ip_a q_a)}{\sqrt{2\pi}}\,, \qquad \langle p,t|q,t\rangle = \prod_a \frac{\exp(-ip_a q_a)}{\sqrt{2\pi}}\,,$$

so

$$\langle q',t+dt|q,t\rangle = \int\left(\prod_a \frac{dp_a}{2\pi}\right)\exp\left(i\sum_a p_a(q'_a-q_a) - H(q',p)dt\right).$$

(The summation convention is suspended here.)

As in Section 9.1, we divide the time interval from $-T/2$ to $T/2$ into a large number of very small intervals and insert a sum over Q eigenstates for each interval. Defining functions $q(t)$ and $p(t)$ that interpolate between the values of the Q and P eigenvalues at each interval, we obtain the path integral expression for F in its most general form

$$F(q',q;T) = \int_{q(-T/2)=q,\,q(T/2)=q'}\Big(\prod_{a,t} dq_a(t)\Big)\int\Big(\prod_{a,t}\frac{dp_a(t)}{2\pi}\Big)$$
$$\times\exp\left(\int_{-T/2}^{T/2} dt\,\Big[i\sum_a \dot q_a(t)p_a(t) - H\Big(q(t),\,p(t)\Big)\Big]\right). \tag{23.A.14}$$

To calculate the partition function (23.A.7) we would integrate over the ps and qs subject only to the condition that $q(t)$ is periodic with period equal to the inverse temperature β:

$$Z(\beta) = \int_{q(\beta/2)=q(-\beta/2)} \Big(\prod_{a,t} dq_a(t)\Big) \int \Big(\prod_{a,t} \frac{dp_a(t)}{2\pi}\Big)$$
$$\times \exp\left(\int_{-\beta/2}^{\beta/2} dt \Big[i\sum_a \dot{q}_a(t)p_a(t) - H\big(q(t), p(t)\big)\Big]\right) . \quad (23.A.15)$$

Eqs. (23.A.14) and (23.A.15) look a little odd, with one term in the exponent real, and the other imaginary. These formulas begin to look more familiar after we do the path integral over the $p_a(t)$s. This integral is trivial in the important class of theories where $H(q,p)$ is quadratic in the Ps:

$$H(q,p) = \frac{1}{2}\sum_{a,b} A_{ab}(q)p_a p_b + \sum_a B_a(q)p_a + C(q) . \quad (23.A.16)$$

As shown in the appendix to Chapter 9, the integral over the ps in Eq. (23.A.14) yields

$$F(q',q;T) = \int_{q(-T/2)=q,\, q(T/2)=q'} \Big(\prod_{a,t} dq_a(t)\Big) \Big(\text{Det}\big[2i\pi\mathscr{A}(q)\big]\Big)^{-1/2}$$
$$\times \exp\left(\int_{-T/2}^{T/2} dt \Big[i\sum_a \dot{q}_a(t)\bar{p}_a(t) - H\big(q(t), \bar{p}(t)\big)\Big]\right), \quad (23.A.17)$$

where $\mathscr{A}(q)$ is the 'matrix'

$$\mathscr{A}(q)_{at',bt} = \delta(t'-t)A_{ab}(q(t)) , \quad (23.A.18)$$

and $\bar{p}(t)$ is the stationary 'point' of the argument of the exponential in Eq. (23.A.17) — that is, the solution of the equation

$$i\dot{q}_a(t) = \left.\frac{\delta H\big(q(t), p\big)}{\delta p_a}\right|_{p=\bar{p}(t)} . \quad (23.A.19)$$

The factor i here should not be surprising, because Eq. (23.A.19) is the same equation that is satisfied by the non-Hermitian operators (23.A.8):

$$i\dot{Q}_a(t) = i[H, Q_a(t)] = \frac{\delta H\big(Q(t), P(t)\big)}{\delta P_a(t)} .$$

For a Hamiltonian in the general quadratic form (23.A.16), the solution of Eq. (23.A.19) is

$$\bar{p}_a = \sum_b \big[A^{-1}(q)\big]_{ab}\big(i\dot{q}_b - B_b(q)\big) , \quad (23.A.20)$$

and so Eq. (23.A.17) takes the form

$$F(q',q;T) = \int_{q(-T/2)=q,\,q(T/2)=q'} \left(\prod_{a,t} dq_a(t)\right) \times \left(\mathrm{Det}\left[2i\pi\mathscr{A}(q)\right]\right)^{-1/2} \exp\left(-S[q]\right), \quad (23.A.21)$$

where $S[q]$ is the action

$$S[q] \equiv \int_{-T/2}^{T/2} dt \left[\frac{1}{2}\sum_{ab} A_{ab}^{-1}(q)\dot{q}_a\dot{q}_b + i\sum_{ab} A_{ab}^{-1}(q)B_a(q)\dot{q}_b - \frac{1}{2}\sum_{ab} A_{ab}^{-1}(q)B_a(q)B_b(q) + C(q)\right]. \quad (23.A.22)$$

In the special (but common) case where $B_a(q) = 0$, this simplifies to

$$S[q] \equiv \int_{-T/2}^{T/2} dt \left[\frac{1}{2}\sum_{ab} A_{ab}^{-1}(q)\dot{q}_a\dot{q}_b + C(q)\right]. \quad (23.A.23)$$

Thus in this case the 'Lagrangian' appearing in the path integral is equal to what the Hamiltonian would be in Minkowskian spacetime when the *p*s are expressed in terms of *q*s and $\dot{q}$s.

Appendix B A List of Homotopy Groups

This appendix presents a list of homotopy groups[40] for various manifolds. Here Z denotes the group of the integers, with group composition defined by addition, so that zero is the unit element. Also, Z_n is the group of the integers modulo n. The trivial group, consisting of the element 0, is denoted 0. Homotopy groups for direct products of manifolds may be obtained from the homotopy groups of the manifolds themselves by the product rule:

$$\pi_n(\mathscr{M}_1 \times \mathscr{M}_2) = \pi_n(\mathscr{M}_1) \times \pi_n(\mathscr{M}_2)\,.$$

Spheres

$$\pi_n(S_m) = 0 \text{ for } n < m$$

$$\pi_n(S_n) = Z$$

$$\pi_{n+1}(S_n) = Z_2 \text{ except } \pi_2(S_1) = 0;\ \pi_3(S_2) = Z$$

$$\pi_{n+2}(S_n) = Z_2 \text{ except } \pi_3(S_1) = 0$$

$$\pi_{n+3}(S_n) = Z_{24} \text{ except } \pi_4(S_1) = 0;\ \pi_5(S_2) = Z_2;\ \pi_6(S_3) = Z_{12};\ \pi_7(S_4) = Z \times Z_{12}$$

$$\pi_n(S_1) = 0 \text{ except } \pi_1(S_1) = Z$$

Lie Group Manifolds

$$\pi_1(G) = \begin{cases} Z & G = U(1) \\ Z_2 & G = SO(n)\ (n \geq 3) \\ 0 & \text{other simple compact connected Lie groups} \end{cases}$$

$$\pi_2(G) = 0 \qquad G \text{ any compact connected Lie group}$$

$$\pi_3(G) = Z \qquad G \text{ any compact connected simple Lie group}$$

$$\pi_4(G) = \begin{cases} Z_2 \times Z_2 & G = SO(4),\ Spin(4) \\ Z_2 & G = USp(2n),\ SU(2),\ SO(3),\ Spin(5),\ SO(5) \\ 0 & G = SU(n)\ (n \geq 3),\ SO(n)\ (n \geq 6),\ G_2,\ F_4,\ E_n \end{cases}$$

$$\pi_{4n+2}(USp(2n)) = \begin{cases} Z_{(2n+2)!} & n \text{ even} \\ Z_{2(2n+2)!} & n \text{ odd} \end{cases} \qquad \pi_{2n}(SU(n)) = Z_{n!}$$

Bott Periodicity Theorems
For $n \geq (k-1)/4$, $k \geq 2$,

$$\pi_k(USp(2n)) = \begin{cases} Z & k = 3, 7\ (mod\ 8) \\ Z_2 & k = 4, 5\ (mod\ 8) \\ 0 & k = 0, 1, 2, 6\ (mod\ 8) \end{cases}$$

For $n \geq k+2$, $k \geq 2$,

$$\pi_k(SO(n)) = \begin{cases} Z & k = 3, 7\ (mod\ 8) \\ Z_2 & k = 0, 1\ (mod\ 8) \\ 0 & k = 2, 4, 5, 6\ (mod\ 8) \end{cases}$$

For $n \geq (k+1)/2$, $k \geq 2$,

$$\pi_k(SU(n)) = \begin{cases} Z & k \text{ odd} \\ 0 & k \text{ even} \end{cases}$$

Coset Spaces
For any Lie group G and any Lie subgroup $H \subset G$,

$$\pi_2(G/H) = \ker\left\{\pi_1(H) \mapsto \pi_1(G)\right\}.$$

That is, $\pi_2(G/H)$ is the subgroup of $\pi_1(H)$ that maps into the trivial element of $\pi_1(G)$ when H is embedded in G. As a special case,

$$\pi_2(G/H) = \pi_1(H) \qquad \text{for } \pi_1(G) = 0\,.$$

Problems

1. What sort of term would need to be added to the action for Goldstone

boson fields in four-dimensional Euclidean spacetime to make it possible to have topologically non-trivial field configurations at which the action is stationary?

2. Consider a theory of scalar fields in six space dimensions in which the chiral symmetry $SU(2) \times SU(2)$ of the Lagrangian is spontaneously broken to the $SU(2)$ of isospin. Suppose that enough higher derivative terms are added to the Lagrangian so that skyrmions are stabilized. What sort of conservation law do these skyrmions obey? (Hint: Note that, as shown in Section 2.7, $SU(2)$ is topologically the same as S_3.)

3. Show that all $\pi_n(\mathscr{M})$ for $n > 1$ and arbitrary manifolds $\mathscr{M}$ are Abelian.

4. What is the coupling-constant dependence of the contribution of instantons of unit winding number in an $SU(4)$ gauge theory?

5. Derive a formula for the axion mass in the case in which m_u, m_d, and m_s are all small.

6. Consider the theory of a real scalar field ϕ with Lagrangian density

$$\mathscr{L} = -\tfrac{1}{2}\partial_\mu\phi\partial^\mu\phi + \tfrac{1}{2}m^2\phi^2 - \tfrac{1}{4}\lambda\phi^4 - g\phi\,,$$

where m^2 and λ are positive, and g is very small. Calculate the exponential suppression factor $\exp(-B)$ in the rate of decay of the metastable vacuum state in terms of m^2, λ, and g.

References

1. A. A. Belavin, A. M. Polyakov, A. S. Schwarz and Yu. S. Tyupkin, *Phys. Lett.* **59B**, 85 (1975).

2. G. 't Hooft, *Nucl. Phys.* **B79**, 276 (1974); A. M. Polyakov, *JETP. Letters* **20**, 194 (1974).

3. T. H. R. Skyrme, *Proc. Roy. Soc. London* **A260**, 127 (1961).

4. H. Nielsen and P. Oleson, *Nucl. Phys.* **B61**, 45 (1973).

5. Ya. B. Zel'dovich, I. Yu. Kobzarev, and L. B. Okun', *Sov. Phys. JETP* **40**, 2 (1975).

6. S. Coleman, *Phys. Rev.* **D15**, 2929 (1977); C. G. Callan and S. Coleman, *Phys. Rev.* **D16**, 1762 (1977). The discussion in Section 23.8 is based on these references. For earlier work on the formation

of regions of broken symmetry, see T. D. Lee and G. C. Wick, *Phys. Rev.* **D9**, 2291 (1974); M. B. Voloshin, I. Yu. Kobzarev, and L. B. Okun', *Sov. J. Nucl. Phys.* **20**, 644 (1975).

7. S. Coleman, *Nucl. Phys.* **B262**, 263 (1985); T. D. Lee and Y. Pang, *Phys. Rep.* **221**, 251 (1992), and other references by Lee and collaborators quoted therein.

8. For an extremely lucid review of these topics and references to much of the original literature, see S. Coleman, *Aspects of Symmetry* (Cambridge University Press, Cambridge, 1985): Chapters 6 and 7. A good, more recent review is given by E. J. Weinberg, *Ann. Rev. Nucl. Part. Sci.* **42**, 177 (1992).

9. G. H. Derrick, *J. Math. Phys.* **5**, 1252 (1964).

10. E. B. Bogomol'nyi, *Sov. J. Nucl. Phys.* **24**, 449 (1976).

11. Homotopy theory has been used to study symmetry breaking in the early universe by T. W. B. Kibble, *J. Phys. A* **9**, 1387 (1976); A. Vilenkin and E. P. S. Shellard, *Cosmic Strings and Other Topological Defects* (Cambridge University Press, Cambridge, 1994), and references quoted therein.

12. H. Georgi and S. L. Glashow, *Phys. Rev. Lett.* **28**, 1494 (1972).

13. J. Arafune, P. G. O. Freund, and C. J. Goebel, *J. Math. Phys.* **16**, 433 (1975).

14. M. K. Prasad and C. M. Sommerfeld, *Phys. Rev. Lett.* **35**, 760 (1975).

15. B. Julia and A. Zee, *Phys. Rev.* **D11**, 276 (1974); **D13**, 819 (1976).

16. P. A. M. Dirac, *Proc. Roy. Soc. London* **A133**, 60 (1934); *Phys. Rev.* **74**, 817 (1948).

17. Ya. B. Zel'dovich and M. Yu. Khlopov, *Phys. Lett.* **79B**, 239 (1978); J. Preskill, *Phys. Rev. Lett.* **43**, 1365 (1979). For a review, see J. Preskill, *Annual Rev. Nucl. Part. Science* **34**, 461 (1984).

18. See, e. g., S. Weinberg, *Gravitation and Cosmology* (Wiley, New York, 1972): Chapter 15.

19. A. Guth, *Phys. Rev.* **D23**, 347 (1981); A. D. Linde, *Phys. Lett.* **108B**, 389 (1982); A. Albrecht and P. Steinhardt, *Phys. Rev. Lett.* **48**, 1220 (1982).

20. R. Jackiw and C. Rebbi, *Phys. Rev.* **D13**, 3398 (1976).

21. V. A. Rubakov, *JETP Lett.* **33**, 644 (1981); *Nucl. Phys.* **B212**, 391 (1982); C. G. Callan, *Phys. Rev.* **D25**, 2141 (1982); *Phys. Rev.* **D26**, 2058 (1982); *Nucl. Phys.* **B212**, 391 (1982).

22. R. Bott, *Bull. Soc. Math. France* **84**, 251 (1956).

23. G. 't Hooft, *Phys. Rev.* **D14**, 3432 (1976); *Phys. Rep.* **142**, 357 (1986).

24. M. Atiyah and R. Ward, *Commun. Math. Phys.* **55**, 117 (1977).

25. G. 't Hooft, *Phys. Rev. Lett.* **37**, 8 (1976). Instead of the identification of anomaly-free conserved currents used here, 't Hooft's derivation of the selection rules was based on an analysis of the zero modes of the Dirac operator (see Section 23.7), which gives the change in each quantum number for an instanton of a given winding number.

26. C. G. Callan, R. F. Dashen, and D. J. Gross, *Phys. Lett.* **63B**, 334 (1976); R. Jackiw and C. Rebbi, *Phys. Rev. Lett.* **37**, 172 (1976).

27. V. A. Kuzmin, V. A. Rubakov, and M. E. Shaposhnikov, *Phys. Lett.* **155B**, 36 (1985). The transition is dominated by field configurations known as sphalerons; see N. S. Manton, *Phys. Rev.* **D28**, 2019 (1983); F. R. Klinkhammer and N. S. Manton, *Phys. Rev.* **D30**, 2212 (1984); R. F. Dashen, B. Hasslacher, and A. Neveu, *Phys. Rev.* **D10**, 4138 (1974).

28. H. Georgi and I. McArthur, unpublished (1981); D. B. Kaplan and A. V. Manohar, *Phys. Rev. Lett.* **56**, 2004 (1986); K. Choi, *Nucl. Phys.* **B383**, 58 (1992).

29. For a more detailed calculation, see V. Baluni, *Phys. Rev.* **D19**, 2227 (1978); R. J. Crewther. P. Di Vecchia, G. Veneziano, and E. Witten, *Phys. Lett.* **88B**, 123 (1979).

30. R. D. Peccei and H. Quinn, *Phys. Rev. Lett.* **38**, 1440 (1977); *Phys. Rev.* **D16**, 1791 (1977).

31. S. Weinberg, *Phys. Rev. Lett.* **40**, 223 (1978); F. Wilczek, *Phys. Rev. Lett.* **40**, 279 (1978).

32. J. E. Kim, *Phys. Rev. Lett.* **43**, 103 (1979); M. A. Shifman, A. I. Vainshtein, and V. I. Zakharov, *Nucl. Phys.* **B166**, 493 (1980); A. R. Zhitnitsky, *Sov. J. Nucl. Phys.* **31**, 260 (1980); M. Dine, W. Fischler, and M. Srednicki, *Phys. Lett.* **104B**, 199 (1981).

33. S. Weinberg, in *Neutrinos '78*, ed. E. C. Fowler (Purdue, Lafayette, 1978).

34. D. A. Dicus, E. W. Kolb, V. I. Teplitz, and R. V. Wagoner, *Phys. Rev.* **D18**, 1829 (1978); **D22**, 839 (1980).

35. For reviews, see M. S. Turner, *Phys. Rept.* **197**, 67 (1990); G. G. Raffelt, *Phys. Rept.* **198**, 1 (1990).

36. J. Preskill, M. B. Wise, and F. Wilczek, *Phys. Lett.* **120B**, 127 (1983); L. F. Abbott and P. Sikivie, *Phys. Lett.* **120B**, 127 (1983); M. Dine and W. Fischler, *Phys. Lett.* **120B**, 137 (1983). For reviews, see J. E. Kim, *Phys. Rep.* **150**, 1 (1987); M. S. Turner, *Phys. Rep.* **197**, 68 (1990).

37. S. Coleman, V. Glaser, and A. Martin, *Commun. Math. Phys.* **58**, 211 (1978).

38. S. Coleman, Ref. 8, Section 6.2.

39. S. Coleman, Ref. 8, Section 6.5.

40. A. Actor, *Rev. Mod. Phys.* **51**, 461 (1979); *Encyclopedic Dictionary of Mathematics* (MIT Press, Cambridge, 1980): Appendix A.

Author Index

Where page numbers are given in *italics*, they refer to publications cited in lists of references.

Subject Index

Printed in the United States
By Bookmasters